AF389327

eHealth in Chronic Diseases

eHealth in Chronic Diseases

Editors

Irene Torres-Sanchez
Marie Carmen Valenza

MDPI • Basel • Beijing • Wuhan • Barcelona • Belgrade • Manchester • Tokyo • Cluj • Tianjin

Editors
Irene Torres-Sanchez   Marie Carmen Valenza
University of Granada   University of Granada
Spain   Spain

Editorial Office
MDPI
St. Alban-Anlage 66
4052 Basel, Switzerland

This is a reprint of articles from the Special Issue published online in the open access journal *International Journal of Environmental Research and Public Health* (ISSN 1660-4601) (available at: https: //www.mdpi.com/journal/ijerph/special_issues/ehealth_diseases).

For citation purposes, cite each article independently as indicated on the article page online and as indicated below:

LastName, A.A.; LastName, B.B.; LastName, C.C. Article Title. *Journal Name* **Year**, *Volume Number*, Page Range.

ISBN 978-3-0365-2902-8 (Hbk)
ISBN 978-3-0365-2903-5 (PDF)

Contents

International Journal of
**Environmental Research
and Public Health**

Article

Digital Health Transition in Rheumatology: A Qualitative Study

Felix Mühlensiepen [1,2,*], Sandra Kurkowski [3], Martin Krusche [4], Johanna Mucke [5], Robert Prill [2,6], Martin Heinze [1,7], Martin Welcker [8], Hendrik Schulze-Koops [9], Nicolas Vuillerme [10,11,12], Georg Schett [13,14] and Johannes Knitza [10,13,14]

[1] Center for Health Services Research, Brandenburg Medical School Theodor Fontane, 15562 Rüdersdorf, Germany; martin.heinze@immanuelalbertinen.de
[2] Faculty of Health Sciences Brandenburg, Brandenburg Medical School Theodor Fontane, 14476 Potsdam, Germany; Robert.Prill@mhb-fontane.de
[3] Department of Palliative Medicine, CCC Erlangen-EMN, Universitätsklinikum Erlangen, Friedrich-Alexander-Universität Erlangen-Nürnberg (FAU), 91054 Erlangen, Germany; Sandra.Kurkowski@uk-erlangen.de
[4] Rheumatology and Clinical Immunology, Charité Universitätsmedizin Berlin, 10117 Berlin, Germany; martin.krusche@charite.de
[5] Policlinic and Hiller Research Unit for Rheumatology, Heinrich-Heine-University, 40225 Duesseldorf, Germany; Johanna.Mucke@med.uni-duesseldorf.de
[6] Department of Orthopaedics and Trauma Surgery, Brandenburg Medical School Theodor Fontane, Municipal Clinic Brandenburg, 14770 Brandenburg, Germany
[7] Department of Psychiatry and Psychotherapy, Brandenburg Medical School Theodor Fontane, Immanuel Klinik Rüdersdorf, 15562 Rüdersdorf, Germany
[8] Medizinisches Versorgungszentrum für Rheumatologie Dr. M. Welcker GmbH, 82152 Planegg, Germany; Martin.Welcker@rheumatologie-welcker.de
[9] Division of Rheumatology and Clinical Immunology, Department of Medicine IV, Ludwig-Maximilians-Universität München, 80336 Munich, Germany; Hendrik.Schulze-Koops@med.uni-muenchen.de
[10] AGEIS, Faculty of Medicine, Université Grenoble Alpes, 38706 Grenoble, France; Nicolas.Vuillerme@univ-grenoble-alpes.fr (N.V.); johannes.knitza@uk-erlangen.de (J.K.)
[11] Institut Universitaire de France, 75006 Paris, France
[12] LabCom Telecom4Health, Université Grenoble Alpes & Orange Labs, 38400 Grenoble, France
[13] Department of Internal Medicine 3—Rheumatology and Immunology, Friedrich-Alexander University Erlangen-Nürnberg and Universitätsklinikum Erlangen, 91054 Erlangen, Germany; georg.schett@uk-erlangen.de
[14] Deutsches Zentrum für Immuntherapie, Friedrich-Alexander University Erlangen-Nürnberg and Universitätsklinikum Erlangen, 91054 Erlangen, Germany
[*] Correspondence: felix.muehlensiepen@mhb-fontane.de

Citation: Mühlensiepen, F.; Kurkowski, S.; Krusche, M.; Mucke, J.; Prill, R.; Heinze, M.; Welcker, M.; Schulze-Koops, H.; Vuillerme, N.; Schett, G.; et al. Digital Health Transition in Rheumatology: A Qualitative Study. *Int. J. Environ. Res. Public Health* **2021**, *18*, 2636. https://doi.org/10.3390/ijerph18052636

Academic Editor: Irene Torres-Sanchez

Received: 20 January 2021
Accepted: 2 March 2021
Published: 5 March 2021

Publisher's Note: MDPI stays neutral with regard to jurisdictional claims in published maps and institutional affiliations.

Abstract: The global COVID-19 pandemic has led to drastic changes in the management of patients with rheumatic diseases. Due to the imminent risk of infection, monitoring intervals of rheumatic patients have prolonged. The aim of this study is to present insights from patients, rheumatologists, and digital product developers on the ongoing digital health transition in rheumatology. A qualitative and participatory semi-structured fishbowl approach was conducted to gain detailed insights from a total of 476 participants. The main findings show that digital health and remote care are generally welcomed by the participants. Five key themes emerged from the qualitative content analysis: (1) digital rheumatology use cases, (2) user descriptions, (3) adaptation to different environments of rheumatology care, and (4) potentials of and (5) barriers to digital rheumatology implementation. Codes were scaled by positive and negative ratings as well as on micro, meso, and macro levels. A main recommendation resulting from the insights is that both patients and rheumatologists need more information and education to successfully implement digital health tools into clinical routine.

Keywords: rheumatology; chronic disease; digital health; eHealth; telemedicine; remote care; patient perspective; qualitative research; fishbowl discussion; content analysis

1. Introduction

COVID-19 has led to drastic changes in the management of patients with rheumatic diseases. Due to the imminent risk of infection, monitoring intervals of rheumatic patients have prolonged [1,2]. Some were cancelled or postponed, whereas others were replaced by video consultations [1,3], thereby catalyzing digital disruption in healthcare. Besides the COVID-19-induced uptake of digital health [4], the Digital Health Act, passed in December 2019, reshapes the German healthcare system [5]. Measures include the introduction of digital health applications into the statutory health insurance scheme, the implementation of digital patient records, and the promotion of video consultation hours [3,5]. Due to aggravating challenges, such as the declining number of rheumatologists [6], the aging population, and the need for early diagnosis [7] and continuous monitoring [8], rheumatology is considered to have a great potential to benefit from a digital health transition [2,9].

In a response to the current challenges and uptake of digitization, a virtual fishbowl discussion was organized at the annual scientific conference of the German Society of Rheumatology (DGRh Congress), which took place in September 2020. Patients, rheumatologists, and industry stakeholders had the opportunity to exchange information and experiences on digital health in rheumatology. The aim of this study is to present insights from patients, rheumatologists, and digital product developers on the ongoing digital health transition and innovation potentials in rheumatology in Germany.

2. Materials and Methods

The authors conducted a virtual fishbowl discussion [10] on the question "How does the internet affect the doctor–patient relationship?" at the first virtual annual conference of the German Society for Rheumatology 2020 (9–12 September 2020). The event was scheduled for 90 min. It consisted of two introductory key notes on digital health services in rheumatology and the actual fishbowl discussion. The discussion was recorded, transcribed verbatim, and later examined using qualitative content analysis [11].

Recently, the fishbowl technique has been successfully implemented in national rheumatology conferences [12]. It is a validated method, fostering open group discussions and engagement of all audience members [13]. By including an expert panel (inner circle) and one empty chair for an alternating audience member, the technique promotes a dynamic and direct exchange with the audience. The authors intended to hold a face-to-face fishbowl discussion at DGRh Congress 2020. Due to the COVID-19 pandemic, the entire conference was held virtually, using Zoom™-based software (Zoom Video Communications Inc, San Jose, CA, USA) [14]. Hence, the authors successfully tested the fishbowl discussion as an online format (Figure 1).

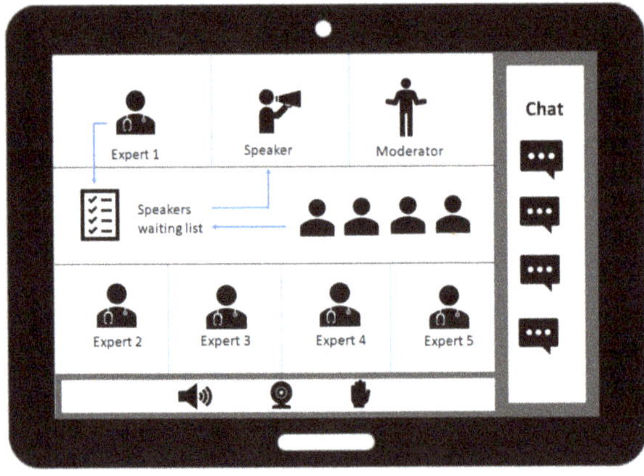

Figure 1. Virtual fishbowl discussion (in accordance with Muehlensiepen et al. [10]).

The inner circle of the fishbowl discussion consisted of one patient representative, two rheumatologists, and two digital product developers. To contribute to the discussion, they used the "hand raise" function to communicate with the moderator (F.M.). Audience members (the outer circle) who wanted to take part in the discussion and take the empty seat in the inner circle had to report this interest in the chat and were registered in the speaker list. When it was their turn, the participants' cameras and microphones were activated by tech support. After their statement, the participants left the inner circle of the discussion again.

As the fishbowl was part of the official congress program, delegates could also participate in the discussion without taking part in the study: either by not contributing verbally or by indicating that their contribution would be deleted from the recording. However, delegates did not use this option.

Analysis was based on qualitative content analysis by Philip Mayring [11]. The fishbowl discussion was audio-recorded and subsequently anonymized and transcribed verbatim. The transcript was imported into MAXQDA 2020 Software (VERBI GmbH, Berlin, Germany) [15]. The data analysis involved four steps (Figure 2).

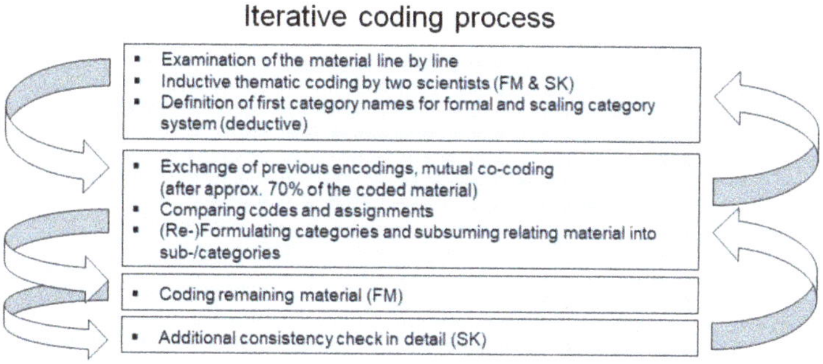

Iterative coding process

- Examination of the material line by line
- Inductive thematic coding by two scientists (FM & SK)
- Definition of first category names for formal and scaling category system (deductive)

- Exchange of previous encodings, mutual co-coding (after approx. 70% of the coded material)
- Comparing codes and assignments
- (Re-)Formulating categories and subsuming relating material into sub-/categories

- Coding remaining material (FM)

- Additional consistency check in detail (SK)

Figure 2. Qualitative content analysis: iterative coding process.

Step 1: Two scientists with a background in healthcare science (F.M., male, PhD) and social science (S.K., female, M.Sc.) independently examined the material line by line, coded inductively with regard to the content (thematic coding) and formulated as closely to the notes as possible. Additionally, codes were scaled in accordance with Kurkowski et al. [16]. Formal and scaled codes were assigned simultaneously and deductively to matching text sections. Positive and negative connotations were differentiated (scaling). Codes were scaled to the category "positive rating" if they contained clearly positive expressions (e.g., "ideal," "great opportunity"), whereas the codes were scaled as "negative" if they expressed a clearly negative connotation (e.g., "we do not wish," "a negative example"). If codes could not be assigned to a positive or negative scaling, they were not considered in the further analysis. Step 2, after approximately 70% of the material was coded, the two scientists (F.M. and S.K.) exchanged their codes and compared their codes and assignments. After this, they defined the main categories and subsumed relating codes. The main categories represent the central themes and aspects of the discussion: (1) digital rheumatology use cases, (2) user descriptions, (3) adaptation to different environments of rheumatology care, and (4) potentials of and (5) barriers to digital rheumatology implementation (Figure 3). The categories, potentials (4), and barriers (5) of digital rheumatology implementation were grouped into micro, meso, and macro levels. Codes were assigned to the macro level if they related to the German health system. The meso level referred to rheumatology care as a whole. The micro level concerned the individual patient–practitioner level. Step 3: After reworking the code system, one scientist (F.M.) coded the remaining notes. Step 4: S.K. performed an additional consistency check, and inconsistencies were resolved. To ensure validity of the data analysis, the results of the analysis were reviewed and confirmed by participants of the fishbowl discussion (M.W., H.S.K., and J.K.) before drafting the manuscript. For the presentation of the results, representative quotes of the discussion transcript were selected, translated, and included in the text. The manuscript has been compiled in accordance with the Consolidated Criteria for Reporting Qualitative Research (COREQ) [17] (please refer to Supplementary Materials 1).

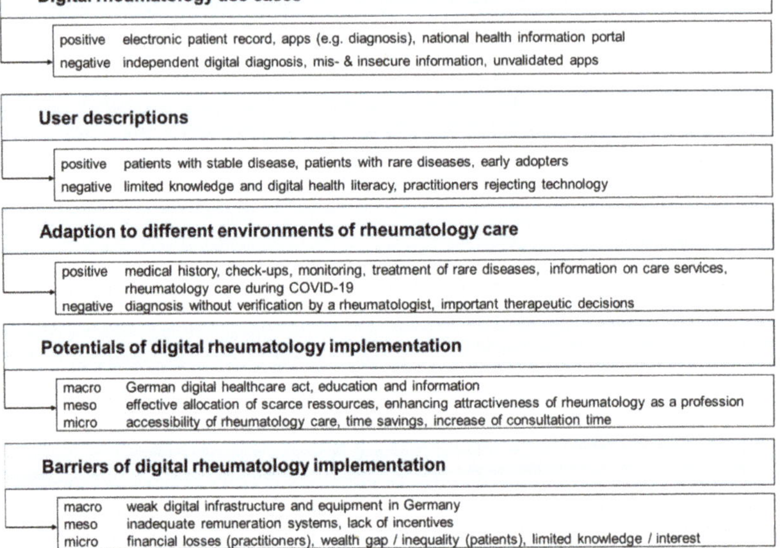

Figure 3. Qualitative content analysis: main categories (according to Kurkowski et al. 2020 [16]).

Ethical approval was not sought for the present study as this manuscript reports on the results of a discussion held during a virtual medical conference. Despite this, all authors declared to adhere to the Declaration of Helsinki in its current form in all steps of this project [18]. Discussion participants were informed of the audio recording, transcription, and subsequent analysis of the discussion. In the course of transcription, all data and personal references were removed. There was no risk involved in participating in the discussion. The manuscript does not allow identification of individuals.

3. Results

A total of 476 delegates attended the 90-min session. The actual fishbowl discussion lasted 56 min. In total, there were 19 content contributions by 12 delegates: 9 rheumatologists, 2 digital product developers, and 1 patient representative. Seven of the 19 contributions came from audience members from the outer circle (37%).

3.1. Digital Rheumatology Use Cases

Delegates discussed various examples of how digital health could complement rheumatology care. Implementation of electronic health records in Germany was considered to have great potential for boosting doctor–patient relationships as well as tracking previous medical care, resulting in more efficient use of scarce time:

> *"Electronic patient records—the next big thing. I see this as an extremely great opportunity to strengthen the doctor–patient relationship and this precious therapy time of five minutes and so on." (Patient representative I)*

Participants pointed out the importance of implementing a digital national health portal where patients could access validated and evidence-based information about their medical conditions:

"I consider the concept as a chance to provide sensible content to patients, to people, in a centralized way." (Rheumatologist I)

The introduction of healthcare apps into German Social Health Insurance system is expected to have great potential and a positive impact on rheumatology care as well:

"I think this extension and supplementation of care through digital health apps, DiGAs, will be a nice, good, viable addition." (Digital product developer I)

"On the one hand sensitivity and specificity [of diagnosis apps] is very poor at 50 percent. But on the other hand it's still 20 percent higher than the referral of the non-experienced general practitioner, with a specificity of about 30 percent. That means there's a clear potential for improvement without any physical examination, as you said, with the methods of [name of a certain digital health application] or [name of another application] or other diagnostic apps. But the power has already been demonstrated." (Rheumatologist II)

Delegates also discussed negative examples of digital healthcare in chronic diseases. Approaches were seen as negative when they are disconnected from the "regular" health system and digital healthcare is not controlled by medical practitioners:

"But that's what you read now in magazines again and again in a striking way: A patient has foam in her urine, feels somehow bad, weary, and so on, goes to the convenience store, somehow hands in her sample and then receives her results digitally, oh, she knows it's lupus? That is, I think, not exactly the good start to a chronic patient's career and good care that we actually want." (Patient representative I)

Additionally, delegates identified insecure sources and misinformation on diseases as a danger of digital health. Furthermore, they expressed their opposition to invalidated digital health applications without proof of benefit. Interestingly, video consultations were not a main discussion topic.

3.2. User Descriptions

Discussants identified patients with a stable course of their chronic rheumatic disease as potential users of digital health tools:

"And, for a well-cared for or well-adjusted patient like me, a doctor's contact is really necessary only once a year for therapy monitoring reasons, and if there is something beyond that, like maybe checking whether my blood pressure needs to be adjusted, talking to the doctor once in a while, that would of course be ideal to cover the perhaps necessary two to three doctor's contacts a year once briefly via a secure telemedicine platform." (Patient representative I)

Digital health can also significantly help considering rare diseases:

"Particularly in the area of rare diseases or unusual symptoms, patients also do research on the internet if they do their research well, find good sources or do research at organizations or patient self-help organizations. Patients who use swarm intelligence can also arrive more quickly at more or less target-oriented diagnostic suggestions. That is a knowledge gain overall." (Patient representative I)

Among physicians, early adopters were mentioned as potential users of digital health. Less technophile rheumatologists are not likely to use digital health very soon.

"My rheumatologist is like that; she tells me that she won't invest in any digital infrastructure here and I know five, six, seven rheumatologists who are of the same age." (Patient representative I)

According to the patient representative, monetary incentives or even sanctions for practitioners who decline to connect to digital infrastructure would be strategies to spread digitization in rheumatology care. According to the delegates, knowledge and digital health literacy are low among both practitioners and patients.

3.3. Adaptation to Different Environments of Rheumatology Care

Throughout the fishbowl discussion, participants agreed that digital health can complement current rheumatology care, but not replace face-to-face appointments. Digital health tools should be directed by medical practitioners. Diagnosis as well as important therapy decisions should be carried out by rheumatologists after a thorough clinical examination. Diagnosis, exclusively based on digital health, was strongly rejected by the discussants:

> *"Initial diagnosis is only possible by the rheumatologist in direct physical contact and after detailed clinical examination. And then later on, you can talk about how the patient can be followed up by the internet or video consultation or anything else. That's my opinion at least." (Rheumatologist III)*

> *"Hey, a qualified initial diagnosis should please, please, please be made by a qualified rheumatologist. I think that is the consensus here in the panel." (Patient representative I)*

However, on the other side, participants mentioned different environments of rheumatology care, where digital healthcare could contribute. These are medical history, follow-up appointments, and disease monitoring:

> *"Many patients have no idea what kind of medication they are taking, what kind of therapy they actually want, and I often spend most of my time gathering all this information. Electronic patient records could be a huge relief and I think that it is the key to speed up digitization." (Rheumatologist II)*

> *"Yes, I am well-adjusted and my CRP is fine. But with new digital tools, for example the app that [name of a digital health application] is going to launch in the area of disease monitoring or something like [name of another application] . . . new solutions for disease monitoring are entering the market, which will contribute to maintaining the ability to work." (Patient representative I)*

Participants of the fishbowl discussion also highlighted the importance of digital approaches to ensure adequate and safe rheumatology care during the COVID-19 pandemic. They believe that secure and validated information on the internet can help people access adequate rheumatology care. Most importantly, this applies to the therapy of rare diseases.

3.4. Potentials of Digital Rheumatology Implementation

In regard to the healthcare system (macro level), the discussants highlighted the German Digital Health Act, which is anticipated to contribute to a digital transition of healthcare in the coming years:

> *"It has been passed and eight months later the first products are about to be approved. I think that is the speed of light for the health system and shows again how the legislator wants to support this transition." (Digital product developer I)*

To increase the knowledge of medical practitioners and patients with regard to digital health services, the discussants call for education and information campaigns:

> *"It can only be done with education and information campaigns. How should patients and practitioners receive the information, that there is the possibility of prescribing apps . . . and which ones are available . . . and how good they are?" (Digital product developer II)*

Despite this momentum, the benefits of digital health applications still have to be proven:

> *"Digitization is not positive in itself, but it might be positive in its effect and this effect still has to be proven, in other words, evaluated. And in the digitization law and specifically also for the introduction of apps, there is still a necessity to prove the benefit." (Rheumatologist I)*

According to the participants, a broader spectrum of rheumatology care services (meso level) will be available through digital transition. This provides the opportunity to effectively allocate the scarce human resources available in rheumatology care. The effective use of digital health in rheumatology can be supported by the innovative medical professional associations and active patient organizations:

> *"How do the individual specialist areas position themselves in regard to digitalization and apps? I think that the rheumatologists—to praise them once again—are pretty much at the forefront here, as is the group of Young Rheumatologists and the Digital Commission. Ultimately, I consider it very state of the art to deal with the topic in this way." (Digital product developer II)*

In addition, digital health was also seen as a way to enhance the attractiveness of rheumatology as a profession in general:

> *"I think this is precisely where digital aids could offer a very exciting aspect and expansion: To the extent that the job description of a rheumatologist becomes much more interesting and attractive when one knows that one can deal with digital, innovative tools in the treatment of patients." (Digital product developer I)*

On the micro level, respectively, individual level, patients and physicians could benefit from an improved accessibility to rheumatology care and the reduction of "unnecessary" appointments:

> *"Because I also think it might be quite pleasant for patients to do without one or two unneeded doctor's appointments." (Rheumatologist III)*

> *"As long as things are fine, I don't need to travel through the city to a non-accessible doctor's office, anyways." (Patient representative I)*

Since much of the relevant information can be provided digitally in advance, the digital transition could increase consultation time:

> *"Digitization allows us to have more time to talk. That is the quality of care, which is also increasing, because if I have already answered all the questions asked beforehand, then I have maybe five more minutes in the conversation with the patient, and that is an essential five minutes of conversation to improve the perceived care." (Rheumatologist I)*

3.5. Barriers to Digital Rheumatology Implementation

At the macro level, the discussants identified limited digital infrastructure and equipment as barriers to digital health transformation in Germany:

> *"In many parts of society, that not only include special circumstances, but also students and the general population, from patients to doctors, who simply do not have the technical equipment. And I don't even want to mention the 5G network, which is also not available in Germany." (Rheumatologist IV)*

On the meso level, the discussants pointed out that the current remuneration systems in rheumatology care are not designed for digital approaches and does not incentivize them:

> *" ... also the remuneration of digital services. And I think this is the challenge right now, especially in rheumatology ... set it up digitally and get away from 'well, I only want to see the patients who also bring me the money, but I try to care for the patients according to their needs with digital support'." (Digital product developer I)*

This results in a barrier on the micro level at the same time. Practitioners could suffer financial losses through the use of digital health. Furthermore, digital health could lead to a wealth gap on the part of patients, which results in health opportunities being distributed unequally, a digital divide in healthcare:

> *"We also have to consider that digitalization creates a wealth gap: people who cannot afford large contracts, good mobile phones, good tablets, do not have good access. And this is also evident in telemedicine and applications, where these systems are not being used. Thus, I believe that we also have to consider the social aspect." (Rheumatologist I)*

Finally, the confusing numbers of available digital tools, limited individual knowledge and interest concerning digital health represent a barrier to digital transition in daily rheumatology practice:

"These apps that we are supposed to recommend—to pick up on this—I personally have a very, very hard time with them in my day-to-day practice. Surprisingly, not a single patient has asked me for an app so far. So, all this information is not yet there and then I once again ask you: Why has no patient asked for it yet?" (Rheumatologist V)

4. Discussion

We performed a qualitative study using a virtual fishbowl discussion on digital transition in rheumatology with 476 participants at the first virtual conference of German Society for Rheumatology 2020. The qualitative approach provides a unique insight on the status quo of digital rheumatology from the perspectives of patient representatives, rheumatologists, and digital product developers.

The participants revealed a positive view toward digital transition. They identified digital health as a valuable addition to current rheumatology care, mentioning various use cases that could support existing services: electronic patient records, digital health applications (diagnosis assistance/monitoring), and implementation of a health information portal. The participants of the fishbowl discussion rejected digital approaches if these were decoupled from the traditional healthcare system. Final diagnosis and therapeutic decisions should exclusively be made by rheumatologists. Digital health should be integrated into rheumatology care routines but must not entirely eradicate personal patient–doctor interaction [2]. Digital rheumatology is considered as supportive regarding anamnesis, follow-up appointments, monitoring, and (validated) information on services and diseases. The discussants highlighted patients with stable disease courses as potential telemedicine users. This is in line with the research by Kulcsar et al. [19]. The importance of digitally provided information for patients with rare diseases was stressed. The accelerating and connecting effect of digitization—"digital crowdsourcing"—to inform patients is consistent with previous results by Krusche et al. [20] and Ruffer et al. [21]. On the other hand, with regard to the medical practitioners, less technophile doctors would not be likely to adopt digital health in their clinical routines, according to the discussants. Similarly, low eHealth literacy and skepticism was observed among older rheumatic patients [22]. Overall, knowledge on digital health is still very limited among practitioners and patients according to the discussants. Previous survey data [23,24] confirm this lack of knowledge as a main barrier. It also appears that in particular highly affected patient groups are significantly less interested in eHealth [25]. On the macro level, the participants in the fishbowl discussion pointed out that German Digital Health Care Act as well as education and information campaigns could foster digital transition. The weak digital infrastructure and equipment in Germany is a barrier from the discussants' perspectives. Regarding rheumatology care as a whole (meso level), digital transition could support effective allocation of scarce rheumatology workforce and enhance the attractiveness of rheumatology as a profession. It has been demonstrated that telemedicine-based monitoring is not clinically inferior to standard follow-up [26]. Furthermore, it has been demonstrated that digitally complemented lectures increase students' interest in rheumatology [27]. The participants identified inadequate remuneration systems and a lack of incentives as a potential barrier to digital health on the meso level. This underlines the high relevance of innovative and sustainable funding systems for large-scale implementation of eHealth [28]. On the individual patient–practitioner level (micro level), digital health could improve accessibility of rheumatology care services, increase individual consultation time, and at the same time enable the care of more patients overall by time savings in clinical practice. This is in line with recent qualitative findings that point to the potential of digital health contributing to patient-centered care in rheumatology [29]. The opportunities of digital health in rheumatology care are contrasted by possible financial losses among practitioners and limited knowledge about digital health as barriers. Finally, the influence of the digital transition on social disparities and the allocation of health opportunities is controversially discussed and needs to be intensively observed [30–32].

Study findings could have direct implications for informing guidelines on remote care in rheumatology and can help to successfully implement digital health tools in the management of rheumatic diseases. A main strength of this study is the inclusion of potential users of digital health and in particular patient representatives. To the best of our knowledge, this is the first qualitative study in rheumatology that included the three stakeholders: patients, rheumatologists, and digital product developers. We also highly recommend the virtual fishbowl format [10], as it is an outstanding option for multi-perspective, interactive discussions and, during COVID-19, a solid alternative to physical face-to-face debates. Compared to other discussion formats, i.e., panel discussions or large debates, fishbowl discussions encourage active participation of non-experts and enable the inclusion of multiple perspectives [12].

The study design has several limitations. The study reports on the content of one 54-min discussion session. Because conference delegates chose to participate in the discussion based on their interest in the topic, a self-selection bias and a rather positive attitude toward digital health is likely to exist. The two introductory key notes on digital services in rheumatology might have positively biased the subsequent fishbowl and our results. Although fishbowl is an inclusive, participatory discussion technique, not all attendees participated equally in the discussion. The discussion was held at a virtual medical conference including mainly rheumatologists. The patient perspective is underrepresented, as mostly physicians expressed their views and only one patient representative actively took part in the discussion.

5. Conclusions

German patients, rheumatologists, and digital health developers expressed a generally positive attitude toward the digital health transition in rheumatology. It could contribute to effective allocation of scarce resources, improve accessibility, and enhance the attractiveness of rheumatology as a profession. Digital health may certainly complement rheumatology care, but (final) diagnosis and key decisions are to remain in the hands of rheumatologists, according to the participants of the fishbowl discussion. Patients and rheumatologists need more information and education to successfully implement digital health tools into clinical routine.

Supplementary Materials: The following are available online at https://www.mdpi.com/1660-4601/18/5/2636/s1: Table S1: COREQ Checklist.

Author Contributions: Study conception and design, F.M., J.M., M.W., H.S.-K., and J.K.; acquisition of data, F.M., M.K., J.M., M.W., H.S.-K., and J.K.; analysis and interpretation of data, F.M., S.K., M.K., R.P., M.H., M.W., N.V., H.S.-K., G.S., and J.K. All authors have read and agreed to the published version of the manuscript.

Funding: This research received no external funding.

Institutional Review Board Statement: Ethical approval was not sought for the present study because the manuscript describes findings of a discussion at a virtual medical conferences. There was no risk involved in participating in the discussion. The study was conducted according to the guidelines of the Declaration of Helsinki.

Informed Consent Statement: Discussion participants were informed of the audio recording, transcription, and subsequent analysis of the discussion. In the course of transcription, all data and personal references were removed. The manuscript does not allow identification of individuals.

Data Availability Statement: All data relevant to the study are included in the article. For further questions regarding the reuse of data, please contact the corresponding author (felix.muehlensiepen@mhb-fontane.de).

Acknowledgments: The authors would like to thank all participants for their valuable contributions, as well as the organizers of the DGRh Congress 2020 and the technical support. The present work is part of the PhD thesis of the last author, J.K. (AGEIS, Université Grenoble Alpes, Grenoble, France).

Conflicts of Interest: The authors declare no conflict of interest.

References

1. Dejaco, C.; Alunno, A.; Bijlsma, J.W.; Boonen, A.; Combe, B.; Finckh, A.; Machado, P.M.; Padjen, I.; Sivera, F.; Stamm, T.A.; et al. Influence of COVID-19 pandemic on decisions for the management of people with inflammatory rheumatic and musculoskeletal diseases: A survey among EULAR countries. *Ann. Rheum. Dis.* **2020**. [CrossRef]

2. Schulze-Koops, H.; Specker, C.; Krueger, K. Telemedicine holds many promises but needs to be developed to be accepted by patients as an alternative to a visit to the doctor. Response to: 'Patient acceptance of using telemedicine for follow-up of lupus nephritis in the COVID-19 outbreak' by So et al. *Ann. Rheum. Dis.* **2020**. [CrossRef]

3. Aries, P.; Welcker, M.; Callhoff, J.; Chehab, G.; Krusche, M.; Schneider, M.; Specker, C.; Richter, J.G. Stellungnahme der Deutschen Gesellschaft für Rheumatologie e. V. (DGRh) zur Anwendung der Videosprechstunde in der Rheumatologie [Statement of the German Society for Rheumatology (DGRh) on the use of video consultations in rheumatology]. *Z. Rheumatol.* **2020**, *79*, 1078–1085. [CrossRef] [PubMed]

4. Kernder, A.; Morf, H.; Klemm, P.; Vossen, D.; Haase, I.; Mucke, J.; Meyer, M.; Kleyer, A.; Sewerin, P.; Bendzuck, G.; et al. Digital rheumatology in the era of COVID-19: Results of a national patient and physician survey. *RMD Open* **2021**, *7*, e001548. [CrossRef] [PubMed]

5. Gesetz für Eine Bessere Versorgung durch Digitalisierung und Innovation [Digital Health Act]. 2019. Available online: https://dip21.bundestag.de/dip21/btd/19/134/1913438.pdf (accessed on 11 January 2021).

6. Krusche, M.; Sewerin, P.; Kleyer, A.; Mucke, J.; Vossen, D.; Morf, H. Facharztweiterbildung quo vadis? [Specialist training quo vadis?]. *Z. Rheumatol.* **2019**, *78*, 692–697. [CrossRef]

7. Combe, B.; Landewe, R.; Daien, C.I.; Hua, C.; Aletaha, D.; Álvaro-Gracia, J.M.; Bakkers, M.; Brodin, N.; Burmester, G.R.; Codreanu, C.; et al. 2016 update of the EULAR recommendations for the management of early arthritis. *Ann. Rheum. Dis.* **2016**, *76*, 948–959. [CrossRef] [PubMed]

8. Smolen, J.S.; Aletaha, D.; Bijlsma, J.W.; Breedveld, F.C.; Boumpas, D.; Burmester, G.; Combe, B.; Cutolo, M.; de Wit, M.; Dougados, M.; et al. Treating rheumatoid arthritis to target: Recommendations of an international task force. *Ann. Rheum. Dis.* **2010**, *69*, 631–637. [CrossRef] [PubMed]

9. Knitza, J.; Knevel, R.; Raza, K.; Bruce, T.; Eimer, E.; Gehring, I.; Mathsson-Alm, L.; Poorafshar, M.; Hueber, A.J.; Schett, G.; et al. Toward Earlier Diagnosis Using Combined eHealth Tools in Rheumatology: The Joint Pain Assessment Scoring Tool (JPAST) Project. *JMIR Mhealth Uhealth* **2020**, *8*, e17507. [CrossRef] [PubMed]

10. Muehlensiepen, F.; Mucke, J.; Krusche, M.; Kurkowski, S.; Bendzuck, G.; Koetter, I.; Lemarié, V.; Grahammer, M.; Heinze, M.; Schulze-Koops, H.; et al. The virtual fishbowl: Bringing back dynamic debates to medical conferences. *Ann. Rheum. Dis.* **2020**. [CrossRef]

11. Mayring, P. Qualitative Inhaltsanalyse. *Forum Qual. Soc. Res.* **2000**, *1*, 20. Available online: https://www.qualitative-research.net/index.php/fqs/article/view/1089/2383 (accessed on 11 January 2021).

12. Mucke, J.; Anders, H.J.; Aringer, M.; Chehab, G.; Fischer-Betz, R.; Heipe, F.; Lorenz, H.-M.; Schwarting, A.; Specker, C.; Voll, R.E.; et al. Swimming against the stream: The fishbowl discussion method as an interactive tool for medical conferences: Experiences from the 11th European lupus meeting. *Ann. Rheum. Dis.* **2019**, *78*, 713–714. [CrossRef]

13. Dutt, K.M. The Fishbowl Motivates students to participate. *Coll. Teach.* **1997**, *45*, 143. [CrossRef]

14. Zoom Video Communications, Inc. ZOOM Cloud Meetings. Available online: https://zoom.us/ (accessed on 11 January 2021).

15. MAXQDA. Software für qualitative Datenanalyse. Available online: https://www.maxqda.de/ (accessed on 11 January 2021).

16. Kurkowski, S.; Radon, J.; Vogt, A.R.; Weber, M.; Stiel, S.; Ostgathe, C.; Heckel, M. Hospital end-of-life care: Families' free-text notes. *BMJ Support Palliat Care* **2020**. [CrossRef] [PubMed]

17. Tong, A.; Sainsbury, P.; Craig, J. Consolidated criteria for reporting qualitative research (COREQ): A 32-item checklist for interviews and focus groups. *Int. J. Qual. Health Care* **2007**, *19*, 349–357. [CrossRef]

18. World Medical Association. World Medical Association Declaration of Helsinki: Ethical principles for medical research involving human subjects. *JAMA* **2013**, *310*, 2191–2194. [CrossRef]

19. Kulcsar, Z.; Albert, D.; Ercolano, E.; Mecchella, J.N. Telerheumatology: A technology appropriate for virtually all. *Semin. Arthritis Rheum.* **2016**, *46*, 380–385. [CrossRef] [PubMed]

20. Krusche, M.; Burmester, G.R.; Knitza, J. Digital crowdsourcing: Unleashing its power in rheumatology. *Ann. Rheum. Dis.* **2020**, *79*, 1139–1140. [CrossRef]

21. Ruffer, N.; Knitza, J.; Krusche, M. #Covid4Rheum: An analytical twitter study in the time of the COVID-19 pandemic. *Rheumatol. Int.* **2020**, *40*, 2031–2037. [CrossRef] [PubMed]

22. Knitza, J.; Simon, D.; Lambrecht, A.; Raab, C.; Tascilar, K.; Hagen, M.; Kleyer, A.; Bayat, S.; Derungs, A.; Amft, O.; et al. Mobile Health Usage, Preferences, Barriers, and eHealth Literacy in Rheumatology: Patient Survey Study. *JMIR Mhealth Uhealth* **2020**, *8*, e19661. [CrossRef] [PubMed]

23. Krusche, M.; Klemm, P.; Grahammer, M.; Mucke, J.; Vossen, D.; Kleyer, A.; Sewerin, P.; Knitza, J. Acceptance, Usage, and Barriers of Electronic Patient-Reported Outcomes Among German Rheumatologists: Survey Study. *JMIR Mhealth Uhealth* **2020**, *8*, e18117. [CrossRef] [PubMed]

24. Muehlensiepen, F.; Knitza, J.; Marquardt, W.; Engler, J.; Hueber, A.; Welcker, M. Acceptance of Telerheumatology: Results of a nationwide Survey of Rheumatologists and General Practitioners. *J. Intern. Med.* under review. [CrossRef]

25. Mangin, D.; Parascandalo, J.; Khudoyarova, O.; Agarwal, G.; Bismah, V.; Orr, S. Multimorbidity, eHealth and implications for equity: A cross-sectional survey of patient perspectives on eHealth. *BMJ Open* **2019**, *9*, e023731. [CrossRef] [PubMed]

26. de Thurah, A.; Stengaard-Pedersen, K.; Axelsen, M.; Fredberg, U.; Schougaard, L.M.V.; Hjollund, N.H.I.; Pfeiffer-Jensen, M.; Laurberg, T.B.; Tarp, U.; Lomborg, K.; et al. Tele-Health Followup Strategy for Tight Control of Disease Activity in Rheumatoid Arthritis: Results of a Randomized Controlled Trial. *Arthritis Care Res.* **2018**, *70*, 353–360. [CrossRef] [PubMed]

27. Knitza, J.; Kleyer, A.; Klüppel, M.; Krasuer, M.; Wacker, J.; Schett, G.; Simon, D. Online-Ultraschalllernmodule in der Rheumatologie [Online ultrasound learning modules in rheumatology]. *Z. Rheumatol.* **2020**, *79*, 276–279. [CrossRef]

28. Melchiorre, M.G.; Lamura, G.; Barbabella, F. eHealth for people with multimorbidity: Results from the ICARE4EU project and insights from the "10 e's" by Gunther Eysenbach. *PLoS ONE* **2018**, *13*, e0207292. [CrossRef] [PubMed]

29. Granström, E.; Wannheden, C.; Brommels, M.; Hvitfeldt, H.; Nyström, M.E. Digital tools as promoters for person-centered care practices in chronic care? Healthcare professionals' experiences from rheumatology care. *BMC Health Serv Res.* **2020**, *20*, 1108. [CrossRef]

30. Mehta, B.; Jannat-Khah, D.; Fontana, M.A.; Moezinia, C.J.; Mancuso, C.A.; Bass, A.R.; Antao, V.C.; Gibofsky, A.; Goodman, S.M.; Ibrahim, S. Impact of COVID-19 on vulnerable patients with rheumatic disease: Results of a worldwide survey. *RMD Open* **2020**, *6*, e001378. [CrossRef] [PubMed]

31. Kataria, S.; Ravindran, V. Digital health: A new dimension in rheumatology patient care. *Rheumatol. Int.* **2018**, *38*, 1949–1957. [CrossRef] [PubMed]

32. Kavadichanda, C.; Shah, S.; Daber, A.; Bairwa, D.; Mathew, A.; Dunga, S.; Das, A.C.; Gopal, A.; Ravi, K.; Kar, S.S.; et al. Tele-rheumatology for overcoming socioeconomic barriers to healthcare in resource constrained settings: Lessons from COVID-19 pandemic. *Rheumatology (Oxford)* **2020**. [CrossRef] [PubMed]

International Journal of
Environmental Research and Public Health

Assessment of Glycosylated Hemoglobin Outcomes Following an Enhanced Medication Therapy Management Service via Telehealth

Jennifer M. Bingham [1,2,*], Jennifer Stanislaw [2], Terri Warholak [2], Nicole Scovis [1], David R. Axon [2], Jacques Turgeon [3,4] and Srujitha Marupuru [2]

[1] Tabula Rasa HealthCare, Tucson, AZ 85701, USA; nscovis@trhc.com
[2] College of Pharmacy, University of Arizona, Tucson, AZ 85721, USA; jstanislaw@pharmacy.arizona.edu (J.S.); warholak@pharmacy.arizona.edu (T.W.); axon@pharmacy.arizona.edu (D.R.A.); marupuru@pharmacy.arizona.edu (S.M.)
[3] Tabula Rasa HealthCare, Precision Pharmacotherapy Research and Development Institute, Orlando, FL 32827, USA; jturgeon@trhc.com
[4] Faculty of Pharmacy, Université de Montréal, Montreal, QC H3C 3J7, Canada
* Correspondence: jbingham@trhc.com; Tel.: +1-520-955-8587

Abstract: (1) Background: Regular contact with a medication therapy management (MTM) pharmacist is shown to improve patients' understanding of their condition; however, continued demonstration of the value of a pharmacist delivered comprehensive medication review (CMR) using enhanced MTM services via telehealth is needed. The study aimed to describe a pilot program designed to improve type 2 diabetes mellitus (T2DM) management through enhanced condition specific MTM services. (2) Methods: This retrospective study included patients with T2DM aged 40–75 years who received a pharmacist-delivered CMR between January and December 2018. An evaluation of glycosylated hemoglobin (HbA1c) values 3 months pre- and post-CMR was performed. Wilcoxon signed-rank and chi-square tests were used. (3) Results: Of 444 eligible patients, a majority were female (58%) with a median age of 70 years. Median HbA1c values post-CMR were lower than pre-CMR (median 7.1% range 4.5–13.6; median 7.4% range 4.5–13.9, respectively; $p = 0.009$). There were fewer participants with HbA1c >9% post-CMR ($n = 66$) than pre-CMR ($n = 80$; $p < 0.001$) and more with HbA1C <6.5% post-CMR ($n = 151$) than pre-CMR ($n = 130$; $p < 0.001$). (4) Conclusion: This program evaluation highlighted the value of an enhanced condition specific MTM service via telehealth. Patients had improved HbA1c values three months after receiving a single pharmacist delivered CMR.

Keywords: type 2 diabetes; pharmacist; glycosylated hemoglobin; medication therapy management; comprehensive medication review; telehealth; T2DM

Citation: Bingham, J.M.; Stanislaw, J.; Warholak, T.; Scovis, N.; Axon, D.R.; Turgeon, J.; Marupuru, S. Assessment of Glycosylated Hemoglobin Outcomes Following an Enhanced Medication Therapy Management Service via Telehealth. *Int. J. Environ. Res. Public Health* **2021**, *18*, 6560. https://doi.org/10.3390/ijerph18126560

Academic Editors: Irene Torres-Sanchez and Marie Carmen Valenza

Received: 19 May 2021
Accepted: 17 June 2021
Published: 18 June 2021

Publisher's Note: MDPI stays neutral with regard to jurisdictional claims in published maps and institutional affiliations.

1. Introduction

Medication therapy management (MTM) services aim to optimize medication use, reduce medication-related problems, and reduce overall healthcare costs. One component of MTM services is an annual comprehensive medication review (CMR), which may be conducted by pharmacists or other qualified providers in a variety of ways, including in-person at provider offices and community pharmacies, or via the telephone [1,2]. It is known that MTM pharmacists play an integral role in assessing the patient's understanding of his/her conditions and therapeutic regimens through regular contact and accessibility [3]. Yet, patients do not typically receive regular follow-up contact from MTM pharmacists beyond the annual CMR encounter.

Evidence linking pharmacist involvement to improved clinical outcomes and patient empowerment has led to the expansion of the pharmacist's role in chronic disease management for several chronic conditions [4–9]. An example of this is in the management

13

of type 2 diabetes [10]. For example, one study found that, for patients with type 2 diabetes, pharmacist delivered care significantly improved their ability to self-manage their condition, as seen by decreased glycosylated hemoglobin (HbA1c) clinical values 6 months post intervention [11]. Research also shows that through MTM counseling on lifestyle modifications (e.g., physical activity, nutrition), disease state management, and medication adherence, pharmacists make an important contribution to the diabetes care team [12]. Hence, MTM services are a crucial component of ensuring optimal outcomes in diabetes care given their ability to improve medication safety and effectiveness.

Although previous work has demonstrated the economic, clinical, and humanistic benefits of pharmacist delivered MTM services [8–11], an enhanced, condition specific MTM program for patients with type 2 diabetes mellitus (T2DM) has not yet been evaluated in the literature. A pilot, condition specific, pharmacist delivered MTM program was therefore implemented by a national MTM provider to empower self-management of tT2DM, optimize medication regimens, address gaps in care continuity, improve glycemic control, reduce risks of diabetes-related complications, and improve chronic condition management for patients with T2DM. Patients were identified by their respective health plan based on eligibility criteria set forth by the Centers for Medicare and Medicaid Services and then contacted by the national MTM provider. The purpose of this paper is to describe, within the context of MTM, a pilot program designed to improve T2DM management through enhanced condition specific MTM services integrated into an annual pharmacist delivered CMR service.

2. Materials and Methods

2.1. Description of Program

The CMR included an interactive, systematic assessment of patient-specific health information and medications to identify and resolve drug therapy problems and to improve health outcomes. The CMR was conducted using an audio-only telehealth (i.e., telephone call) application. Quality checks were performed on CMRs conducted by the pharmacists to ensure consistency and standardization.

First, the medication list was reconciled within the software program (RxCompanion™, Tucson, AZ, USA) and medication names, strengths, doses, routes, frequencies, and indications were recorded and assessed for drug-drug interactions, medication safety concerns, and appropriateness using telehealth. Next, patient allergies and reactions were recorded and assessed for contraindications. Pharmacists at the call centers were assisted by proprietary software that raised alerts for medication non-adherence, therapeutic duplications, and missing guideline-directed therapy.

Next, the pharmacist conducted a condition specific review via telephone. If the patient was diagnosed with T2DM, pharmacists were expected to address each of the following through teach-back education: (1) fasting blood sugars; (2) patient understanding of how to identify and manage hypoglycemia and hyperglycemia; (3) patient knowledge of how their most recent HbA1c compared to the previous value; (4) whether their HbA1C was uncontrolled based on provider recommendations; and (5) a review of results with self-monitoring of blood glucose. The pharmacist used these findings to have meaningful conversations about lifestyle factors to improve blood glucose management and address specific times of the day that were problematic for the patient.

Based on the patient response, the pharmacist provided teach-back education and an individualized written summary to empower self-management of T2DM. The patient was mailed a personalized medication list in the Centers for Medicare & Medicaid Services standardized CMR format upon completion of the CMR. The pharmacist directly contacted the provider via telephone or facsimile if they identified patient issues related to access to medication, therapeutic recommendations warranting a dosage change. The pharmacist referred the patient to their provider for further diabetes management and/or suggested a diabetes educator or dietician as needed.

2.2. Description of Sites

The MTM service provider offered an annual standardized CMR to qualified patients based on benefits eligibility. The MTM service provider started providing services via telehealth in 2006. Five national MTM call centers employed by the MTM service provider were included in the pilot program and presented a suite of MTM services to meet the performance needs of health plans and patients, mainly through pharmacist-delivered telehealth medication reviews. The centers consisted of a team dedicated to improving health, wellness, and chronic disease management through MTM services adopting an interprofessional team model that included: pharmacists, pharmacy technicians, student pharmacists, pharmacy residents, nursing students, and registered nurses.

2.3. Study Population

Patients were included in this program evaluation if they received an annual CMR in 2018 from a pharmacist employed by the MTM service provider. Additional patient criteria included: age 40 to 75 years; diagnosis of T2DM denoted by international classification of disease (ICD)-10 coding (e.g., E11.0–E11.9); and, had a presence of one or more oral antihyperglycemic prescriptions based on claims data. The Centers for Medicare & Medicaid Services (CMS) assess certain oral antihyperglycemics in performance metrics to analyze medication adherence for diabetes medications, which include: biguanides, sulfonylureas, thiazolidinediones, dipeptidyl peptidase-4 (DPP) inhibitors, incretin mimetics, meglitinides, or sodium-glucose co-transporter-2 (SGLT2) inhibitors. Medicare beneficiaries who only use insulin are not included in this CMS metric, and were subsequently excluded from the study. Additionally, patients who were 40 years of age or younger and older than 75 years were excluded from the study.

2.4. Data Collection and Analysis

This program evaluation used retrospective patient data from one MTM service provider between 1 January 2018 and 31 December 2018. The variables of interest included: (1) age; (2) gender; (3) race/ethnicity; (4) number of medications; (5) HbA1c three months pre-CMR; and (6) HbA1c three months post-CMR. Health marker data was provided by health plan to the national MTM provider. A pre-post study design was used. For nominal data, a chi-square test was used. For internal level data with a non-normal distribution, a Wilcoxon signed-rank test was used. Descriptive data were also calculated. All tests used an a-priori α level of 0.05. Data analysis was conducted using IBM SPSS Statistics v2015 (IBM Corp, Armonk, NY, USA). This program evaluation was approved by the institutional review board (No. 1911128095).

3. Results

A total of 444 patients were included in the program evaluation, of which 264 were female and 180 male. The median age of patients was 70 years of age (range 40–75). The median number of medications per patient was 21 (range 9–60). Additional baseline patient demographic information is outlined in Table 1.

Median HbA1c values post-CMR were lower than pre-CMR scores (median 7.1% range 4.5–13.6 and median 7.4% range 4.5–13.9, respectively; $Z = -2.60, p = 0.009$). An HbA1c threshold of greater than 9% was used to identify patients at highest risk for microvascular and macrovascular complications, as well as mortality. There were fewer participants with HbA1c >9% post-CMR ($n = 66$) compared to pre-CMR ($n = 80$; $p < 0.001$), and there were more participants with HbA1C <6.5% post-CMR ($n = 151$) compared to pre-CMR ($n = 130$; $p < 0.001$). A subgroup analysis was conducted to assess the impact in adults and the elderly. Of 444 patients, 104 (23%) were adults and 340 (75%) were older adults. Additional data concerning the number and percent of patients in each HbA1c strata pre- and post-CMR are in featured in Table 2.

Table 1. Baseline characteristics for participants (*n* = 444).

Characteristic	Median (Range)
	N (%)
Gender	
Male	180 (40)
Female	264 (58)
Race/Ethnicity	
Asian	8 (2)
Black	32 (7)
White	37 (8)
Unknown/Other	337 (76)
Hispanic	30 (7)

Note: Percentages may not equal 100% due to rounding and missing values.

Table 2. Comparison of HbA1c values pre-comprehensive medication review (CMR) intervention and post-CMR intervention.

	Pre-CMR	Post-CMR	*p*-Value
Total population N = 444			
HbA1c	N (%)	N (%)	
<6.5%	130 (29)	151 (34)	<0.001
6.5–9.0%	234 (52)	227 (51)	<0.001
>9.0%	80 (18)	66 (15)	<0.001
HbA1c *	Median (range)	Median (range)	*p*-value
	7.4 (4.5–13.9)	7.1 (4.5–13.6)	0.009
Adults (40–64 years of age) N = 104			
HbA1c	N (%)	N (%)	
<6.5%	24 (23)	40 (38)	<0.001
6.5–9.0%	51 (49)	46 (44)	<0.001
>9.0%	29 (28)	18 (17)	<0.001
HbA1c *	Median (range)	Median (range)	*p*-value
	7.8 (4.8–13.2)	7.1 (4.5–12.1)	0.002
Elderly (65–75 years of age) N = 340			
HbA1c	N (%)	N (%)	
<6.5%	106 (31)	111 (33)	<0.001
6.5–9.0%	183 (54)	181 (53)	<0.001
>9.0%	51 (15)	48 (14)	<0.001
HbA1c *	Median (range)	Median (range)	*p*-value
	7.2 (4.5–13.9)	7.4 (5.0–13.6)	0.256

HbA1c: glycosylated hemoglobin; * HbA1c data had a skewed distribution, hence median and range were reported. Differences between the pre- and post-groups were calculated using the Wilcoxon signed-rank test given that the data were dependent (i.e., the same people were included in the pre- and post-groups).

4. Discussion

Our study highlights the effect of an enhanced condition specific MTM service designed to improve HbA1C control in patients with T2DM via telehealth. The most important finding was the significant difference between median HbA1c values pre- and post-CMR. This provides preliminary evidence to support the value of an enhanced diabetes specific MTM service in preventing the risk of microvascular and macrovascular complications associated with an HbA1c >9% [13]. The evidence also supports the value in assisting patients to achieve a desired HbA1c <6.5%, especially those who have a longer remaining life expectancy, fewer comorbidities, and a lower risk of hypoglycemia. Hence, we demonstrated in this study that the telehealth approach, by itself, had a significant impact on HbA1C control and represents another complimentary option to the use of medical devices, biosensors, or smart phone applications to support patients in their own homes.

MTM services are a crucial component of ensuring optimal outcomes in diabetes care [14]. The MTM pharmacist plays an integral role in assessing the patient's understanding of his/her condition and therapeutic regimens through regular contact and accessibility [3]. One study found that, for patients with T2DM, pharmacist-delivered information significantly decreased HbA1c values and improved patient knowledge [11]. Another study found improved indicators following pharmacist provided MTM services [15]. Anderson et al. also found improved T2DM health markers following telehealth MTM pharmacist services [8]. In addition to managing T2DM, MTM service provider pharmacists also help reduce cardiovascular related complications and morbidity. Research shows that through MTM counseling on lifestyle modifications, disease state management, and medication adherence, pharmacists make an important contribution to the diabetes care team [12].

It is well known that preventative care is one of the most efficient ways to maintain patient health and to reduce healthcare costs [16]. Moczygemba et al. found that telephonic MTM programs reduced both medication- and health-related problems [10]. Thus, CMRs via telehealth can be thought of as preventative healthcare. The results of this study suggest that pharmacists can provide preventative care that may help to address unmet patient needs stemming from primary care provider shortages in the United States.

Telephonic CMRs are an effective method to enhance patient care and to reduce healthcare expenditures because they can optimize medication therapy and empower patients [16]. Another benefit of telephonic CMRs is that they can provide access to pharmacists who speak the patient's preferred language, thus increasing culturally responsive patient care [17]. This is applicable because the MTM service provider in our study provides CMRs via telehealth to patients in over 30 languages. An additional advantage of telephonic CMRs is improved ease for patients to reconcile medication lists while remaining in their residences, rather than transporting pill bottles to another site and unintentionally forgetting one at home [1]. Future research should evaluate the relative association of each of these components on patient outcomes.

The results of our pilot program further suggest a role for MTM pharmacists in providing preventative care that may help to address unmet patient needs stemming from primary care provider shortages in the United States. One novel aspect of this program was the use of a telephonic condition specific review designed to assist patients with poorly controlled T2DM. The condition review allowed the pharmacist to discover patient specific challenges in glucose management and offered customized solutions based on their HbA1c and fasting blood glucose goals. In addition, the condition reviews were an effective method to enhance patient care by optimizing medication therapy and empowering patients. These results support the need for pharmacists to fill gaps in care subsequent to national provider shortages in the United States of America [16].

The successes of the pilot program provides lessons learned to other MTM providers aiming to improve quality measures for patients with T2DM. These findings parallel other studies that have shown pharmacist-delivered medication counseling and healthcare education to be valuable in improving clinical values [17].

Limitations

This pilot evaluation study design prohibited the establishment of a causal relationship between the CMR and diabetes outcomes, as it did not control for length of the CMR, class of antihyperglycemic medication, specific changes to the medication list, patient-specific responses to the condition review, other comorbidities, or whether the patient sought services from their provider during the three-month time post-CMR. Hence, the study was not able to capture the clinical significance, nor the cost-effectiveness of the outcomes observed as there were limitations in data available to the research team. The study also did not capture longitudinal effects on clinical outcomes beyond three months post-CMR. In addition, the large proportion of patients with unknown race and ethnicity limits generalizability. Thus, these pilot program findings are not generalizable to all Medicare beneficiaries. In future studies, it would be prudent to collate data from the telehealth application regarding the length of the consultation. It is also suggested that researchers collect specifics from the software program on the type of medication and compare differences in the list pre- and post-intervention. One final suggestion is to integrate a complimentary follow-up consultation with the patient to assess for diabetes management from other healthcare providers.

5. Conclusions

This program evaluation highlights the value of pharmacist delivered CMR telehealth services aimed to decrease HbA1C values in patients with T2DM. Future research should evaluate the impact in a more diverse population in other countries over a longer time to determine the cost effectiveness and clinical significance of the intervention.

Author Contributions: Methodology, J.S., J.M.B., T.W. and N.S.; validation, J.S., D.R.A. and S.M.; formal analysis, J.S. and S.M.; investigation, J.S.; resources, J.S.; writing—original draft preparation, J.S., J.M.B. and S.M.; writing—review and editing, J.S., J.M.B., D.R.A., T.W., N.S., S.M. and J.T.; supervision, J.M.B.; project administration, J.S. All authors have read and agreed to the published version of the manuscript.

Funding: This research received no external funding.

Institutional Review Board Statement: Ethical review and approval were waived for this study, due to human research exemption.

Informed Consent Statement: Not applicable.

Data Availability Statement: The data are not publicly available due to proprietary concerns.

Acknowledgments: The authors would like to acknowledge Orsula V. Knowlton, Dana M. Filippoli, and Ann M. Taylor for their contributions to the project.

Conflicts of Interest: Terri Warholak received funding from Arizona Department of Health Services, Merck and Co., Pharmacy Quality Alliance and Tabula Rasa HealthCare Group. Jennifer M. Bingham and Nicole Scovis have disclosed an outside interest in Tabula Rasa HealthCare Group. Conflicts of interest resulting from this interest are being managed by The University of Arizona in accordance with its policies. David R. Axon received funding from the American Association of Colleges of Pharmacy, Arizona Department of Health Services, Merck and Co., Pharmacy Quality Alliance, and Tabula Rasa HealthCare Group. Srujitha Marupuru received funding from Tabula Rasa HealthCare Group and Merck and Co. The other authors did not receive any specific grant from funding agencies in the public, commercial, or not-for-profit sectors for this study.

References

1. Centers for Medicare & Medicaid Services. Available online: https://www.cms.gov/Medicare/Prescription-DrugCoverage/PrescriptionDrugCovContra/MTM (accessed on 24 June 2020).
2. Ai, A.L.; Carretta, H.; Beitsch, L.M.; Watson, L.; Munn, J.; Mehriary, S. Medication therapy management programs: Promises and pitfalls. *J. Manag. Care Spec. Pharm.* **2014**, *20*, 1162–1182. [CrossRef] [PubMed]
3. Inamdar, S.; Kulkarni, R.; Karajgi, S.; Manvi, F.; Ganachari, M.; Mahendra, B. Medication Adherence in Diabetes Mellitus: An Overview on Pharmacist Role. *Am. J. Adv. Drug Deliv.* **2013**, *1*, 238–250.
4. Taylor, A.M.; Bingham, J.; Schussel, K.; Axon, D.R.; Dickman, D.J.; Boesen, K.; Warholak, T.L. Integrating innovative telehealth solutions into an interprofessional team-delivered chronic care management pilot program. *J. Manag. Care Spec. Pharm.* **2018**, *24*, 813–818. [CrossRef] [PubMed]
5. Bingham, J.M.; Axon, D.R.; Scovis, N.; Taylor, A.M. Evaluating the effectiveness of clinical pharmacy consultations on nutrition, physical activity, and sleep in improving patient-reported psychiatric outcomes for individuals with mental illnesses. *Pharmacy* **2019**, *7*, 2. [CrossRef] [PubMed]

6.	Axon, D.R.; Taylor, A.M.; Vo, D.; Bingham, J. Initial assessment of an interprofessional team-delivered telehealth program for patients with epilepsy. *Epilepsy Res.* **2019**, *158*, 106235. [CrossRef] [PubMed]
7.	Tetuan, C.; Axon, D.R.; Bingham, J.; Boesen, K.; Lipsy, R.; Scovis, N.; Leal, S. Assessing the effect of a telepharmacists' recommendations during an integrated, interprofessional telehealth appointment and their alignment with quality measures. *J. Manag. Care Spec. Pharm.* **2019**, *25*, 1334–1339. [PubMed]
8.	Anderson, E.J.; Axon, D.R.; Taylor, A.M.; Towers, V.; Warholak, T.; Johnson, M.; Manygoats, T. Impact evaluation of a four-year academic-community partnership in provision of medication management and tertiary prevention services for rural patients with diabetes and/or hypertension. *Prev. Med. Rep.* **2020**, *17*, 101038. [CrossRef]
9.	Bingham, J.M.; Taylor, A.M.; Boesen, K.P.; Axon, D.R. Preliminary investigation of pharmacist-delivered, direct-to-provider interventions to reduce co-prescribing of opioids and benzodiazepines among a Medicare population. *Pharmacy* **2020**, *8*, 25. [CrossRef]
10.	Moczygemba, L.R.; Barner, J.C.; Lawson, K.A.; Brown, C.M.; Gabrillo, E.R.; Godley, P.; Johnsrud, M. Impact of telephone medication therapy management on medication and health-related problems, medication adherence, and Medicare Part D drug costs: A 6-month follow up. *Am. J. Geriatr. Pharmacother.* **2011**, *9*, 328–338. [CrossRef]
11.	Michiels, Y.; Bugnon, O.; Chicoye, A.; Dejager, S.; Moisan, C.; Allaert, F.A.; Vergès, B. Impact of a community pharmacist-delivered information program on the follow-up of type-2 diabetic patients: A cluster randomized controlled study. *Adv. Ther.* **2019**, *36*, 1291–1303. [CrossRef]
12.	Isetts, B.J.; Schondelmeyer, S.W.; Artz, M.B.; Lenarz, L.A.; Heaton, A.H.; Wadd, W.B.; Cipolle, R.J. Clinical and economic outcomes of medication therapy management services: The Minnesota experience. *J. Am. Pharm. Assoc.* **2008**, *48*, 203–211. [CrossRef] [PubMed]
13.	Nicholas, J.; Charlton, J.; Dregan, A.; Gulliford, M.C. Recent HbA1c values and mortality risk in type 2 diabetes. population-based case-control study. *PLoS ONE* **2013**, *8*, e68008. [CrossRef] [PubMed]
14.	Manganelli, J. The Role of the Clinical Pharmacist in Achieving Clinical and Quality Outcomes in Diabetes Management. *Am. J. Manag. Care* **2016**, *22*, SP128. [PubMed]
15.	Rodis, J.L.; Sevin, A.; Awad, M.H.; Porter, B.; Glasgow, K.; Hornbeck Fox, C.; Pryor, B. Improving chronic disease outcomes through medication therapy management in federally qualified health centers. *J. Prim. Care Commun. Health* **2017**, *8*, 324–331. [CrossRef]
16.	Manolakis, P.G.; Skelton, J.B. Pharmacists' contributions to primary care in the United States collaborating to address unmet patient care needs: The emerging role for pharmacists to address the shortage of primary care providers. *Am. J. Pharm. Educ.* **2010**, *74*, S7. [CrossRef] [PubMed]
17.	DeZeew, E.; Coleman, A.; Nahata, M. Impact of telephonic comprehensive medication reviews on patient outcomes. *Am. J. Manag. Care* **2018**, *24*, e54–e58.

International Journal of
Environmental Research and Public Health

MDPI

Article

Monitoring Patients Reported Outcomes after Valve Replacement Using Wearable Devices: Insights on Feasibility and Capability Study: Feasibility Results

Honoria Ocagli [1], Giulia Lorenzoni [1], Corrado Lanera [1], Alessandro Schiavo [2], Livio D'Angelo [2], Alessandro Di Liberti [2], Laura Besola [3], Giorgia Cibin [4], Matteo Martinato [1], Danila Azzolina [1,5], Augusto D'Onofrio [4], Giuseppe Tarantini [2], Gino Gerosa [4], Ester Cabianca [6] and Dario Gregori [1,*]

[1] Unit of Biostatistics, Epidemiology and Public Health, Department of Cardiac, Thoracic, Vascular Sciences and Public Health, University of Padova, 35121 Padova, Italy; honoria.ocagli@unipd.it (H.O.); giulia.lorenzoni@unipd.it (G.L.); corrado.lanera@unipd.it (C.L.); matteo.martinato@unipd.it (M.M.); danila.azzolina@uniupo.it (D.A.)

[2] Department of Cardiac, Thoracic and Vascular Sciences, University of Padua Medical School, 35121 Padua, Italy; alessandro.schiavo.2@gmail.com (A.S.); livio.doc@gmail.com (L.D.); alessandro.diliberti@gmail.com (A.D.L.); giuseppe.tarantini@unipd.it (G.T.)

[3] Saint Paul's Hospital, University of British Columbia, Vancouver, BC V6Z 1Y6 VBC, Canada; laura.besola@hotmail.it

[4] Cardiac Surgery Unit, Department of Cardiac, Thoracic, Vascular Sciences and Public Health, University of Padova, 35121 Padova, Italy; cibin.gio@gmail.com (G.C.); augusto.donofrio@unipd.it (A.D.); gino.gerosa@unipd.it (G.G.)

[5] Department of Translational Medicine, University of Piemonte Orientale, 28100 Novara, Italy

[6] Cardiology Unit, Dipartimento Strutturale Cardio-vascolare, Azienda ULSS 8 Berica, 36100 Vicenza, Italy; ester.cabianca@aulss8.veneto.it

* Correspondence: dario.gregori@unipd.it; Tel.: +39-049-8275384

Citation: Ocagli, H.; Lorenzoni, G.; Lanera, C.; Schiavo, A.; D'Angelo, L.; Liberti, A.D.; Besola, L.; Cibin, G.; Martinato, M.; Azzolina, D.; et al. Monitoring Patients Reported Outcomes after Valve Replacement Using Wearable Devices: Insights on Feasibility and Capability Study: Feasibility Results. *Int. J. Environ. Res. Public Health* **2021**, *18*, 7171. https://doi.org/10.3390/ijerph18137171

Academic Editors: Irene Torres-Sanchez and Marie Carmen Valenza

Received: 25 April 2021
Accepted: 29 June 2021
Published: 4 July 2021

Publisher's Note: MDPI stays neutral with regard to jurisdictional claims in published maps and institutional affiliations.

Abstract: Wearable devices (WDs) can objectively assess patient-reported outcomes (PROMs) in clinical trials. In this study, the feasibility and acceptability of using commercial WDs in elderly patients undergoing transcatheter aortic valve replacement (TAVR) or surgical aortic valve replacement (SAVR) will be explored. This is a prospective observational study. Participants were trained to use a WD and a smartphone to collect data on their physical activity, rest heart rate and number of hours of sleep. Validated questionnaires were also used to evaluate these outcomes. A technology acceptance questionnaire was used at the end of the follow up. In our participants an overall good compliance in wearing the device (75.1% vs. 79.8%, SAVR vs. TAVR) was assessed. Half of the patients were willing to continue using the device. Perceived ease of use is one of the domains that scored higher in the technology acceptance questionnaire. In this study we observed that the use of a WD is accepted in our frail population for an extended period. Even though commercial WDs are not tailored for clinical research, they can produce useful information on patient behavior, especially when coordinated with intervention tailored to the single patient.

Keywords: surgical aortic valve replacement; transcatheter aortic valve replacement; physical function; wearable devices; feasibility

1. Introduction

Use of consumer wearable devices (WDs) is increasing both in daily usage and in clinical trials [1]. Fuller et al. [1], in a recent systematic review, showed that various brands of WDs were used in clinical studies, offering accurate measures for steps and heart rate.

The International Data Corporation (IDC) reported a Year-Over-Year Growth of 35.1% in 2020 in the global market of WDs [2]. In clinical setting, the introduction of WDs has brought new challenges to face. At first, the European Medicines Agency did not release any specific guidance addressing the use of WDs. While still recognizing the importance of their use for drugs development, the appropriateness for a specific population, the validity of collected data and the management of large amounts of data collected through the device, issues concerning the choice of the suitable device depending on safety have emerged [3].

A recent US national survey on the use of wearable healthcare devices showed that 82.38% of the people involved in the study are willing to share their health data recorded by their WDs with their healthcare professionals [4]. The use of these technologies tends to decline with advancing age, although the elderly, especially those with a chronic disease, are one of the populations that could benefit most from continuous monitoring in their daily setting. The elderly, however, have poor knowledge on the use of WDs [5]. Various studies have investigated the impact of commercial WDs on the elderly. For example, WDs appear to be accurate in measuring step counts and activity duration in community-dwelling adults [6] and have a positive impact on health as their use had increased physical activity in obese patients [7]. Since WDs are often not tailored for the elderly [8], it is important to evaluate their acceptability in their daily life. A study evaluating the acceptance of WDs among the elderly showed that they seem to accept them and understand the importance of their use in healthcare setting [9].

Patients-reported outcome measures (PROMs) have been suggested in literature to be integrated in clinical trials [10], since clinical outcomes do not measure the patients' perception of their health status or functional being. The assessment of PROMs (i.e., physical function and quality of life), should be encouraged in elderly patients [11]. Physical function assessment can be useful for evaluating outcomes not directly related to the disease, but still relevant to maintaining personal dependence [12]. WDs can objectively assess physical function in daily clinical practice [13,14].

High levels of physical function are essential for the success of cardiac procedures, so much so that ad hoc cardiac rehabilitation programs have been established to improve patients' functional recovery through exercise therapy [15]. In the available literature some trials have shown that, for example, after both transcatheter aortic valve replacement (TAVR) and surgical aortic valve replacement (SAVR), physical function is improved. SAVR and TAVR are both highly valuable options for patients with heart failure (HF), a condition that affect mostly elderly patients as showed in a study on time trends in first hospitalization for HF in a community-based population [16]. Patients who had a higher level of physical function before the procedure showed more favorable trajectories [12]. Various studies have compared the two procedures in both high [17,18] and low surgical risk patients [19]. In these studies, New York Heart Association (NYHA) score or, more often self-reporting questionnaire [12,20,21] were used to evaluate physical functions of these patients. However, these instruments have some limitations related to self-reporting, for example in recall and desirability biases.

The present work aims to describe the feasibility of using WDs to monitor PROMs in patients who undergoing TAVR/SAVR enrolled in the run-in phase of the Capability study [22]. In this study, the feasibility and acceptability of using a commercial wearable device in elderly patients undergoing TAVR or SAVR will be explored.

2. Materials and Methods

2.1. Study Design and Inclusion Criteria

The study design characteristics and inclusion criteria are described elsewhere [22]. This is a multicenter prospective observational study, that enrolled patients undergoing SAVR and TAVR according to the evaluation of the local heart team since March 2018 at the University hospital of Padova and the hospital of Vicenza.

2.2. Data Collection and Procedures

Patients were enrolled at least one week prior to the procedure. Questions related to socio-demographic characteristics, risk factors, physical activity and clinical characteristics were the information collected at baseline assessment. WDs along with the smartphone were delivered at baseline assessment. Participants completed the same assessments, except for demographic characteristics, at one month, three months, six months and twelve months after the procedure. From March 2020 to the end of the follow up period, assessments were carried out only by telephone given the restrictions related to the COVID-19 pandemic.

2.3. Ethical Considerations

The study is registered in Clinicaltrials.gov (NCT03843320) and approved by the hospital ethics committee with the protocol No 943 (4 January 2019). Written informed consent was obtained from all patients for study participation and for data collection through their Garmin© device account after proper explanation of the study outcomes by a physician.

2.4. Device

A Garmin© Vívoactive® 3 smartwatch activity tracker device and a smartphone with the Garmin Connect© application (GARMIN, Olathe, KS, U.S.) installed in the smartphone for data transfer were provided to each patient at baseline assessment. The device used in the study was commercially available and was chosen for its ability to estimate steps count and sleep duration in free living environments both in the general and elderly population [23]. Both the patient and his/her caregiver were provided with information and trained on the use of the WD and the smartphone.

2.5. Device Setup and Usage

A personal account of the Garmin Connect© application has been created for each patient. The app contains only patient demographics (i.e., gender, age, weight, height, wrist of usage of the WD). Both the participants and their caregivers were not granted access to the account. All notifications were disabled in the device and the smartphone was cleaned by all the unrelated applications, to avoid affecting behaviors of the participants. Participants were asked to wear the device on their wrist 24 h a day, including while showering and sleeping, except while charging. They were asked to charge the device daily and sync data weekly. A member of the study assisted the patients at each of follow up for connectivity issues and for collecting data, while also remaining available by phone and possibly in person to troubleshoot the device at any time.

Patients were also informed that they would be asked to return the device at the end of the study. However, at the last follow up, participants were asked to choose whether to continue using the device with the smartphone for private use. After the end of the study, we collected no further data from the device.

2.6. Measurements

Physical function was assessed through a series of standard and validated tests: Duke Activity Status Index (DASI) [24], Activity of Daily Living (Barthel Index) and Instrumental Activity of Daily Living (IADL)). The DASI is a measure of functional capacity that can be obtained by self-administered questionnaire and already used in patients that underwent TAVR [25]. The Barthel Index (BI) evaluates activity of daily living [26] and has been used in aortic valve replacement [27]. The DASI score ranges from 0 to 58.2, Barthel Index ranges from 0 to 100, for both the instrument, the higher the score, the higher the functional status. IADLs were evaluated with the scale of Lawton and Brody [28]. The higher the score, the greater the person's abilities. The Epworth sleepiness scale (ESS) was used to evaluate the "subject's general level of daytime sleepiness" [29], a higher score means higher sleepiness and could be interpreted as follows: 0–5 lower normal daytime sleepiness, 6–10 higher normal daytime sleepiness, 11–12 mild excessive daytime sleepiness, 13–15 moderate excessive daytime sleepiness, 16–24 severe excessive daytime sleepiness [30].

2.7. Acceptance of the Technology

Compliance with the use of the WD (i.e., the time the device was worn) was determined by calculating the proportion of days the device was active and the total number of days before and after the procedure. At the end of the follow up, patients were given a questionnaire on technology acceptance, based on the work of Puri et al. [9]. The instrument investigated six key dimensions for WD acceptance: perceived usefulness, perceived ease of use, privacy concerns, perceived risks, facilitating conditions and equipment characteristics. Device acceptance was assessed with the question L33 as suggested in the validation study [9], "Would you use the device you used during the last year to continue to monitor or track your physical activity or health?"

2.8. Statistical Analysis

Continuous variables were reported as I, II (median) and III quartiles, categorical variables were presented as absolute numbers and percentages. Wilcoxon's test and Chi-squared test were used to evaluate differences between TAVR and SAVR, respectively, for continuous and categorical variables. The compliance on the use of the WDs was summarized computing the number of days before the procedure divided by the number of days that data was synchronized in that period, the same was carried out for the follow up period. Significance was evaluated for p-value lower than 0.05. Data analysis was performed with R software (version 4.0.3) [31].

3. Results

3.1. Baseline Characteristics

The patients considered eligible for the study were 17, 12 TAVR and 5 SAVR, 4 from the center of Vicenza and the remaining from the University hospital of Padova. Eight patients completed the entire follow up period (Flowchart Figure 1) with an enrolment rate of 47%. Four patients were found to be ineligible after completion of the baseline assessment.

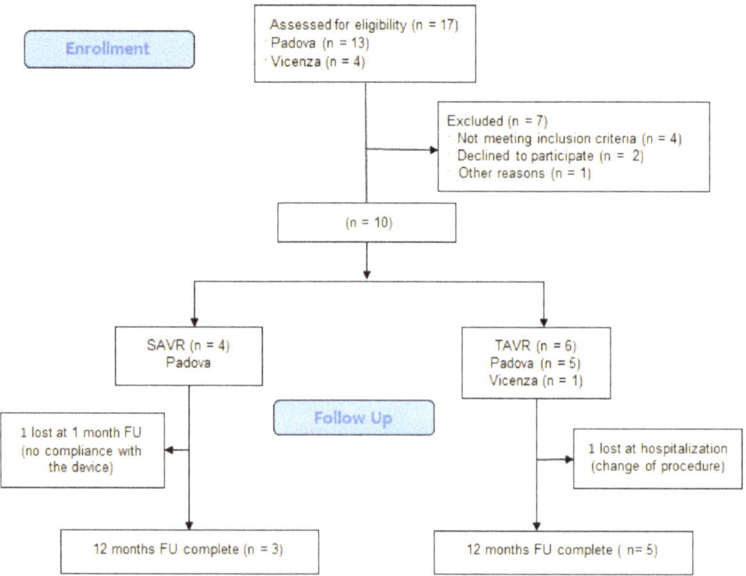

Figure 1. Flowchart of eligible patients in the two centers, Padova and Vicenza.

Table 1 reports patients baseline demographic characteristics by type of procedure, TAVR or SAVR. The overall sample had a median age of 79 years, 78 in the SAVR group and 82 in the TAVR group (p-value = 0.046).

Table 1. Baseline characteristics of the sample according to the type of intervention, surgical aortic valve replacement (SAVR) or transcatheter aortic valve replacement (TAVR).

		N	SAVR (N = 5)	TAVR (N = 12)	Overall (N = 17)	*p* Value
Center	Padova	17	100% (5)	67% (8)	76% (13)	0.14
	Vicenza		0% (0)	33% (4)	24% (4)	
Drop out	Yes	17	40% (2)	50% (6)	47% (8)	0.71
Gender	Female	17	40% (2)	67% (8)	59% (10)	0.31
	Male		60% (3)	33% (4)	41% (7)	
Age		17	76/78/79	79/82/85	78/79/83	0.046
Marital status	Married cohabitant	17	80% (4)	67% (8)	71% (12)	0.58
	Widowed unmarried		20% (1)	33% (4)	29% (5)	
Educational level	Primary	16	40% (2)	64% (7)	56% (9)	0.38
	Secondary		60% (3)	36% (4)	44% (7)	
Risk factors	Diabetes	17	0% (0)	8% (1)	6% (1)	0.47
	Hypertension		100% (5)	75% (9)	82% (14)	
	Smoker		0% (0)	17% (2)	12% (2)	
Status	Elective	16	100% (5)	82% (9)	88% (14)	0.31
	Urgent		0% (0)	18% (2)	12% (2)	
Clinical frailty scale	Well	17	40% (2)	8% (1)	18% (3)	0.25
	Managing well		40% (2)	17% (2)	24% (4)	
	Vulnerable		0% (0)	33% (4)	24% (4)	
	Mildly Frail		20% (1)	25% (3)	24% (4)	
	Moderate Frail		0% (0)	17% (2)	12% (2)	
NYHA class	1	16	20% (1)	9% (1)	12% (2)	0.82
	2		60% (3)	64% (7)	62% (10)	
	3		20% (1)	27% (3)	25% (4)	
COPD		17	0% (0)	8% (1)	6% (1)	0.78
Ejection fraction		15	57/60/62	50/58/61	54/58/62	0.49

Abbreviations: NYHA = New York heart association; COPD = Chronic obstructive pulmonary disease.

Except for age, there were no statistically differences between the SAVR and the TAVR groups. Patients were mainly female (10, 59%), married or cohabiting (12, 71%). Hypertension was the main risk factors in both groups, (14, 82%). Only two patients underwent TAVR procedure in urgent status. Patients, according to clinical frailty scale were more than vulnerable in the TAVR group (4, 33%; vulnerable, 3, 25% mildly frail and, 2, 17% moderate frail). Ejection fraction was similar in the two groups, median of 60 and 58, respectively, in SAVR and TAVR (*p*-value 0.49).

3.2. Score Trend

Table 2 reports the trend of the Barthel Index, DASI score, IADL index and ESS score at each follow up according to the type of procedure. SAVR patients reported higher physical function levels than TAVR especially at 12-month follow up (BI 100 vs. 85, DASI 19.9 vs. 12.8, respectively, for SAVR and TAVR). TAVR recorded lower levels of physical function at baseline according to the DASI score (30.4 vs. 14.4). Physical function level was similar at 12 months of follow up compared to the baseline assessment in both groups according to BI. The DASI score instead showed a decrease for both groups: 30.4 vs. 19.9 and 14.4 vs. 12.8, respectively, for SAVR and TAVR. As for the IADL, there was a decrease at 12 months from baseline both for TAVR and SAVR (6 vs. 4 for SAVR, 6 vs. 5 for TAVR). The ESS score was highest at 12 months follow up for both SAVR and TAVR, both groups had lower normal daytime sleepiness at each follow up.

Table 2. Baseline characteristics according to the type of intervention. Data are reported for that underwent SAVR (3 patients) and TAVR (5 patients).

		Baseline	**1 Month**	**3 Months**	**6 Months**	**12 Months**
Barthel Index	SAVR	88/95/98	92/95/98	95/100/100	72/95/98	80/100/100
	TAVR	90/90/100	70/70/85	80/85/90	90/95/95	85/85/100
DASI score	SAVR	21.6/30.4/31.9	8.2/16.4/20.8	14.5/16.2/19.8	12.6/23.4/25.2	12.2/19.9/22.6
	TAVR	10.7/14.4/20.4	7.2/7.2/14.4	12.7/16.4/32.5	10.7/15.4/26.9	7.2/12.8/24.4
IADL score	SAVR	4.5/6.0/6.5	2.5/5.0/6.0	2.5/3.0/4.0	1.5/3.0/3.5	3.5/5.0/5.5
	TAVR	5/6/6	2/4/4	2/4/6	2/3/6	3/5/8
ESS score	SAVR	1.0/2.0/6.0	2.5/4.0/5.0	2.5/3.0/4.5	3.0/4.0/6.0	3.0/3.0/5.5
	TAVR	3/3/3	3/3/4	3/5/5	4/4/5	5/5/5

Abbreviations: SAVR = Surgical aortic valve replacement; TAVR = transcatheter aortic valve replacement.

3.3. Device's Data

Figures 2–4 show the daily trend for each patient of rest heart rate, the number of steps and the number of hours of sleep, respectively, recorded by the device divided by baseline and follow up period and according to the procedure. Patients that underwent TAVR had a median rest heart rate higher than SAVR in the follow up period (median 64 vs. 57) (Figure 2).

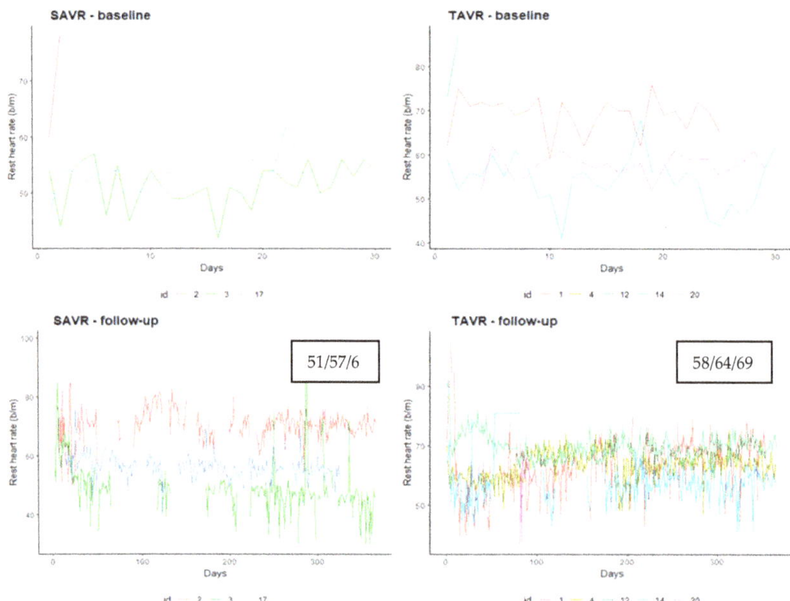

Figure 2. Rest heart rate trend recorded by the device according to procedure and baseline vs. follow-up period. Data in the box are the I, median and III quartile of rest heart rate.

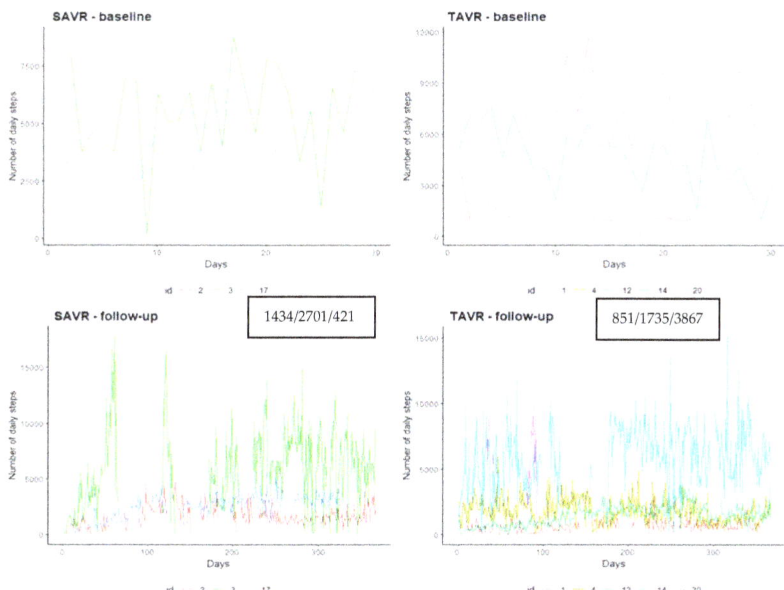

Figure 3. Number of steps trend recorded by the device according to procedure and baseline vs. follow-up period. Data in the box are the I, median and III quartile of number of steps.

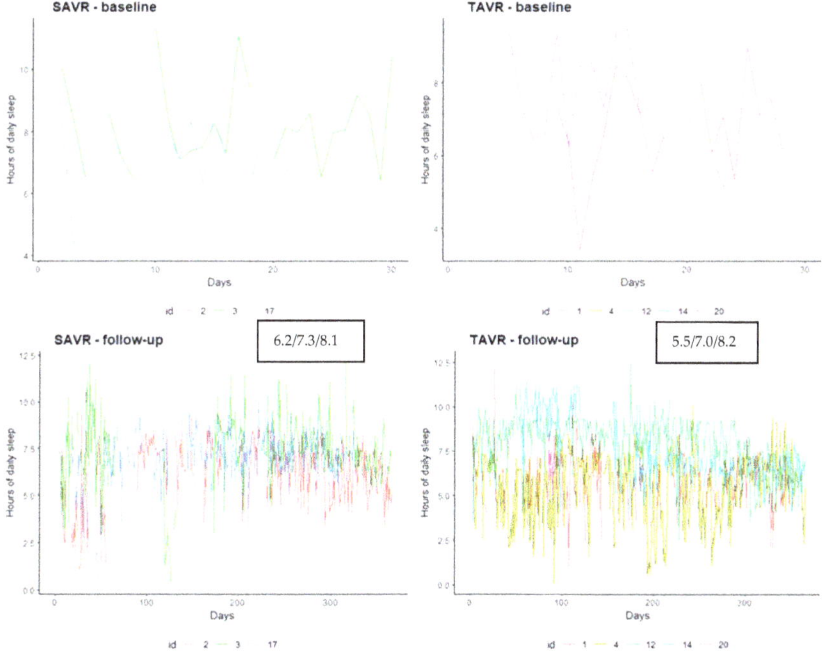

Figure 4. Number of hours slept trend recorded by the device according to procedure and baseline vs. follow-up period. Data in the box are the I, median and III quartile of hours of sleep.

As for the number of steps, the SAVR group seems more active, median of 2701 steps/day vs. 1735 of TAVR (Figure 3).

Regarding the number of hours of sleep, both groups after the procedure had a similar median number of hours of sleep (7.3 SAVR vs. 7.0 TAVR) (Figure 4).

3.4. Compliance and Acceptance of the Technology

Data were mainly synchronized by the caregivers in both groups, only one TAVR patient synchronized by himself the device. The WDs and the smartphones were charged mainly by the caregivers, two TAVR patients charged their devices themselves. None of the patients or their caregivers had used a wearable device prior to the study. The compliance, evaluated as the percentage of days device use divided by the total number of days, was similar in both groups and reached a median of 75.1% in SAVR and 79.8% in TAVR (Table 3).

Table 3. Compliance on the use of the device according to the procedure, SAVR vs. TAVR. Data are reported as I quartile, median and III quartile, median and standard deviations.

Period.	SAVR (N = 3)	TAVR (N = 5)	Combined (N = 8)	*p* Value
Overall	76/82/96 84+/13	39/83/100 71+/34	73/82/100 76+/28	0.92
Pre	89/100/100 93+/12	25/86/100 67+/39	65/93/100 77+/33	0.47
Post	71.1/75.1/80.1 75.8+/0.091	79.5/79.8/99.5 75.9+/32.4	73.1/79.6/88.8 75.9+/25.0	0.5

Abbreviations: SAVR = Surgical aortic valve replacement; TAVR = transcatheter aortic valve replacement.

Figure 5 shows an overview of WDs wear time for the entire study period. Valid and missing data days are showed for each participant. The percentages of missing data vary from 0, no data, up to 70.15%. Noticeably, the patients with only 0.55% and 0% of missing data were those who could rely on the daily assistance of the caregiver. Only one patient collected less than 30% of data: in this case the caregiver was not often in contact with the patient and the patient was not able to use the smartphone autonomously.

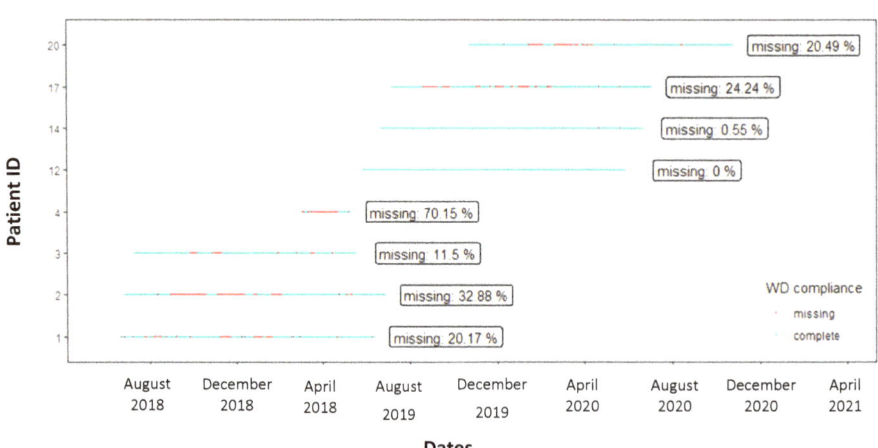

Figure 5. Wearable device wear time for the entire follow-up period for each participant. A line that represents the period of enrolment is reported for each participant, in blue are reported the days without missing data, in red the days with missing data.

One patient, not reported in the analysis, stopped using the activity tracker before the surgery due to the tightness of the strap and the caregiver unwillingness to synchronize data. Missing data varied between 11.5% and 32.88%: in these cases, caregivers were available on a weekly basis. The patient who collected data by himself eventually reduced his compliance due to the long waiting-time for the surgery. In the preoperative period, SAVR patients synchronized data more often than TAVR ones, however this result could be affected by the difference in the length of the pre-operative period, which was longer for SAVR patients. In the follow up period, the compliance was similar on both groups, 75.1% vs. 75.9% for SAVR and TAVR, respectively.

3.5. Technology Acceptance Questionnaire

Table 4 reports the technology acceptance questionnaire scores for each of the seven dimensions based on the type of procedure. Except for perceived risk dimension (median 10 vs. 5.5 SAVR vs. TAVR, *p*-value = 0.016) there were no differences in all the dimensions among the two groups. The perceived risk dimensions score was higher in patients that underwent SAVR. Four patients reported that they would wish to continue to use the device (3 TAVR, 1 SAVR). The device characteristics satisfied the patients, the equipment characteristics reached for both group the maximum score, the same for perceived ease of use, also perceived usefulness showed good levels. Conversely, privacy concern (median 8 vs. 9, SAVR vs. TAVR), perceived risk (10 vs. 5 SAVR vs. TAVR) and subjective norm (10 vs. 9, SAVR vs. TAVR) showed lower level of satisfaction.

Table 4. Differences in the technology acceptance questionnaire scores per dimension according to the procedure for the overall sample. For each dimension is reported the maximum score. Data are reported as I, II and III quartiles. In parenthesis is reported the highest score for each dimension.

	N	SAVR (N = 3)	TAVR (N = 5)	Combined (N = 8)	*p* Value
Perceived usefulness (30)	7	16/16/16	17/17/17	17/17/17	0.16
Perceived ease of use (35)	7	29/30/31	32/32/32	30/32/32	0.57
Equipment characteristics (10)	7	30/31/32	30/31/32	30/31/32	0.73
Privacy concern (15)	7	7.0/8.0/9.0	9.0/9.0/9.0	8.0/9.0/9.5	0.72
Perceived risk (15)	7	10/10/10	5/5/5	5/5/8	0.009
Facilitating conditions (10)	7	5.8/6.5/7.2	7.0/7.0/7.0	6.5/7.0/7.5	0.86
Subjective norm (15)	7	9.5/10.0/10.5	8.0/9.0/10.0	8.5/9.0/10.5	0.48

Abbreviations: SAVR = Surgical aortic valve replacement; TAVR = transcatheter aortic valve replacement.

4. Discussion

This study explores the feasibility and acceptability of using a commercial wearable device in elderly patients undergoing TAVR or SAVR.

The main findings of our study showed that there was an overall good compliance in wearing the device (75.1% vs. 79.8%, SAVR vs. TAVR) with half of the patients willing to continue using the device. Physical function decreased after both SAVR and TAVR, more in TAVR patients. The ESS score increased after the procedure but remained as "normal daytime sleepiness" according to the score, the WDs showed an increase in the number of hours slept up to 7 h per night for both groups.

Our results showed a good overall acceptance of wearing the device during the follow up period. These results are in line with what was reported in a recent literature review which showed a high-level adherence in long term daily use [32]. The acceptance rate in our study was high compared to other studies using the same device [33]. Our results considered a longer follow up period. Studies in the literature, in order to evaluate the compliance and acceptance of a device, considered shorter periods ranging from a few days [34] to weeks [5]. The number of valid days in the follow up period was lower than in the study of Henriksen et al. [35] (265 vs. 292). This result is promising in our population as data synchronization was usually performed by a caregiver. However, this high compliance can be explained by the desire of the patients to use of the WD to contribute to the study.

The main discomfort reported by our patients were related to the need to use another smartphone to synchronize data with the website and connectivity issues, as also reported in a review of activity trackers for senior citizens [36]. This could easily be avoided by allowing the patients or the caregiver to download the app directly to their own smartphone.

Other technical problems were related to changing accidentally the setup of the WD or of the smartphone or forgetting to charge them. In one case, the main problem was related to the unavailability of a WI-FI connection at home. To solve these problems the presence of a technically skilled staff member available on request has proven to be a key point. Other studies showed that technical problems could reduce compliance in the use of WDs [37].

There is a growing interest in evaluating the acceptability of wearable devices both in the general and elderly population. A greater acceptance of these devices can improve the quality of real time data collection [38]. Various studies have shown that older adults accept the use of wearable devices, especially after facing acceptance barriers [9,35,39]. Perceived ease of use is one of the domains that has higher level in our results. This is in line with the findings of a recent study [40] and related to the fact that, while technology use has been increasing also in elderly people, they still need additional information and support to adopt it [41]. A recent study showed that commercial wearable devices are reliable for measuring physical activity level in elderly patients in real-life setting [42]. Despite this increase and the fact that elderly population is the one that can benefit the most from the use of these devices, especially when chronic illnesses are present [34], very few elderlies currently use daily WDs [5] or consider them in health monitoring [43]. However, the perceived ease of use recorded with the questionnaire was high for both groups, respectively, median of 30 for SAVR and 32 points for TAVR (maximum 35).

Even though the compliance was high, the perceived usefulness was not as high in both groups, median of 16 and 17, respectively, for SAVR and TAVR (64% and 68% of the overall score). This is likely sue to the fact that most of the device-related procedures were performed by caregivers and not by patients. During the follow up encounters, the researcher reinforced the importance of collecting data from the activity tracker. Caregivers, on the other hand, reported that having the possibility to see heart rate, number of steps and number of hours of sleep was useful for obtaining information on the health of the patient.

Participants when asked How much would you be willing to pay for the device you wore during the last year? (question L35) answered mainly 0 euro (4, 50%), 2 from 1 to 50 euros and 1 from 51 to 100 euros. This contrasts with what Kekade [5] reported. This may be related to the fact that in our study the population was extremely elderly (median 78 and 82, respectively, for SAVR and TAVR), while the Kekade study considered elderly over 65 years of age.

Cardiac rehabilitation after both TAVR and SAVR have shown improved functional capacity in a recent review [27]. A reduction in physical function level both according to BI and DASI score from baseline to one year follow up was showed in our sample. This result is in contradiction of what reported in other metanalysis for TAVR [21] and both TAVR and SAVR [12]. Cardiac rehabilitation was required only for two participants, both underwent SAVR. So, the sole adoption of the wearable devices is not enough to improve physical activity in these patients. In the future it would be useful to help patients in recognizing the long-term benefits of the device, along with social support as suggested by Kononova [44] in his study on tracker perceptions among older adults. Moreover, it would be helpful to use the commercial wearable activity tracker in a broader physical activity intervention as shown by the review of Brickwood [45].

Our data showed that patients that underwent SAVR were more active than TAVR patients. A functional decline or lack of improvement after the procedure was found, as already found by Kim in his study evaluating changes in functional status in the year after aortic valve replacement [12]. These differences between SAVR and TAVR and no change in functional status for TAVR may be related to the fact that TAVR group had a higher median age and none of them had rehabilitation after the procedure.

Limitations

The main limitations of this study derived from the difficulties to recruit patients with our strictly inclusion criteria. Moreover, the study design was adapted to the limitations derived from the spread of the pandemic of COVID-19.

5. Conclusions

Given the importance of developing a proper observational study to evaluate the use of wearable devices in elderly patients that underwent TAVR or SAVR, our feasibility study provided useful insights on how to implement further our project. In this study we have observed that the use of a WD is accepted in our frail population for an extended period. The use of a WD for collecting data allows the collection of data on daily basis, directly at home, with improved quality since data does not have to be manually entered and checked. However, the use of WDs in clinical trials requires an additional effort on behalf of the research team. A researcher must be available to set up the device at the beginning of the study and to solve problems related to the device. The collection of data through a WD requires always to have an application for collecting data: this could cause problems related to connectivity and device communication. Moreover, the transmission of data requires a minimum ability in the use of technologies in participants. The help of a caregiver is required especially when participants are elderly and sometimes this is not possible.

However, the presence of the device alone is not enough to encourage healthy behaviors. Therefore, it would be useful to create a coordinated intervention with a physiatrist to implement a physical activity program tailored to the single patient. As also reported in the literature, elderly people need more training session than young people, this is particularly true in population such as ours, who are not used to any type of electronic device. Considering that older people, especially frail ones, often rely on the help of a caregiver, they also need to be trained in the potential of these tools. Another suggestion could be the creation of custom reports to allow patients to view their progress in terms of physical activity. In our experience it would be faster to download the app directly on the smartphone of patients or assistants to avoid connectivity problems.

Even though commercial wearable devices were not tailored for clinical research, they could produce useful information on the behavior of the patient. Furthermore, the implementation of the use of these devices, especially in elderly, will minimize the need to attend the clinical study center, with the potential of reducing the time and costs related to person-by-person visits.

Author Contributions: Conceptualization, D.G., G.T., D.A. and G.L.; methodology, D.G. and G.L.; software, C.L.; formal analysis, D.A. and H.O.; investigation, H.O., A.S., L.D., A.D.L., L.B. and G.C.; resources, A.D., G.T., G.G. and E.C.; data curation, H.O.; writing—original draft preparation, H.O.; writing—review and editing, G.L., D.A. and C.L.; visualization, C.L.; supervision, D.G.; project administration, M.M.; funding acquisition, G.T. All authors have read and agreed to the published version of the manuscript.

Funding: This research was funded by an unrestricted grant provided by Edwards Lifesciences. The funding body was not involved in the design of the study and collection, analysis and interpretation of data and in writing the manuscript.

Institutional Review Board Statement: The study was conducted according to the guidelines of the Declaration of Helsinki, and approved by the Ethics Committee of the hospital of Vicenza (protocol code 943, 4 January 2019).

Informed Consent Statement: Informed consent was obtained from all subjects involved in the study.

Data Availability Statement: The data presented in this study are available on request from the corresponding author. The data are not publicly available due to privacy restrictions.

Conflicts of Interest: The authors declare no conflict of interest.

References

1. Fuller, D.; Colwell, E.; Low, J.; Orychock, K.; Tobin, M.A.; Simango, B.; Buote, R.; Van Heerden, D.; Luan, H.; Cullen, K.; et al. Reliability and Validity of Commercially Available Wearable Devices for Measuring Steps, Energy Expenditure, and Heart Rate: Systematic Review. *JMIR mHealth uHealth* **2020**, *8*, e18694. [CrossRef] [PubMed]

2.   Shipments of Wearable Devices Leap to 125 Million Units, Up 35.1% in the Third Quarter, According to IDC [Internet. IDC Prem Glob Mark Intell Co. Available online: https://www.idc.com/getdoc.jsp?containerId=prUS47067820 (accessed on 13 December 2020).

3.   Byrom, B.; Watson, C.; Doll, H.; Coons, S.J.; Eremenco, S.; Ballinger, R.; Mc Carthy, M.; Crescioni, M.; O'Donohoe, P.; Howry, C. Selection of and Evidentiary Considerations for Wearable Devices and Their Measurements for Use in Regulatory Decision Making: Recommendations from the ePRO Consortium. *Value Health* **2018**, *21*, 631–639. [CrossRef] [PubMed]

4.   Chandrasekaran, R.; Katthula, V.; Moustakas, E. Patterns of Use and Key Predictors for the Use of Wearable Health Care Devices by US Adults: Insights from a National Survey. *J. Med. Internet Res.* **2020**, *22*, e22443. [CrossRef] [PubMed]

5.   Kekade, S.; Hseieh, C.-H.; Islam, M.d.M.; Atique, S.; Mohammed Khalfan, A.; Li, Y.-C.; Abdul, S.S. The usefulness and actual use of wearable devices among the elderly population. *Comput. Methods Programs Biomed.* **2018**, *153*, 137–159. [CrossRef] [PubMed]

6.   Straiton, N.; Alharbi, M.; Bauman, A.; Neubeck, L.; Gullick, J.; Bhindi, R.; Gallagher, R. The validity and reliability of consumer-grade activity trackers in older, community-dwelling adults: A systematic review. *Maturitas* **2018**, *112*, 85–93. [CrossRef]

7.   Cadmus-Bertram, L.A.; Marcus, B.H.; Patterson, R.E.; Parker, B.A.; Morey, B.L. Randomized Trial of a Fitbit-Based Physical Activity Intervention for Women. *Am. J. Prev. Med.* **2015**, *49*, 414–418. [CrossRef]

8.   Barton, J.; O'Flynn, B.; Tedesco, S. A review of physical activity monitoring and activity trackers for older adults. *Stud. Health Technol. Inform.* **2017**, *242*, 748–754.

9.   Puri, A.; Kim, B.; Nguyen, O.; Stolee, P.; Tung, J.; Lee, J. User Acceptance of Wrist-Worn Activity Trackers Among Community-Dwelling Older Adults: Mixed Method Study. *JMIR mHealth uHealth* **2017**, *5*, e173. [CrossRef]

10.   Anker, S.D.; Agewall, S.; Borggrefe, M.; Calvert, M.; Jaime Caro, J.; Cowie, M.R.; Ford, I.; Paty, J.A.; Riley, J.P.; Swedberg, K.; et al. The importance of patient-reported outcomes: A call for their comprehensive integration in cardiovascular clinical trials. *Eur. Heart J.* **2014**, *35*, 2001–2009. [CrossRef]

11.   Deutsch, M.-A.; Bleiziffer, S.; Elhmidi, Y.; Piazza, N.; Voss, B.; Lange, R.; Krane, M. Beyond adding years to life: Health-related quality-of-life and functional outcomes in patients with severe aortic valve stenosis at high surgical risk undergoing transcatheter aortic valve replacement. *Curr. Cardiol. Rev.* **2013**, *9*, 281–294. [CrossRef] [PubMed]

12.   Kim, D.H.; Afilalo, J.; Shi, S.M.; Popma, J.J.; Khabbaz, K.R.; Laham, R.J.; Grodstein, F.; Guibone, K.; Lux, E.; Lipsitz, L.A. Evaluation of Changes in Functional Status in the Year After Aortic Valve Replacement. *JAMA Intern. Med.* **2019**, *179*, 383–391. [CrossRef]

13.   Baig, M.M.; GholamHosseini, H.; Moqeem, A.A.; Mirza, F.; Lindén, M. A Systematic Review of Wearable Patient Monitoring Systems—Current Challenges and Opportunities for Clinical Adoption. *J. Med. Syst.* **2017**, *41*, 115. [CrossRef] [PubMed]

14.   Izmailova, E.S.; Wagner, J.A.; Perakslis, E.D. Wearable Devices in Clinical Trials: Hype and Hypothesis. *Clin. Pharm.* **2018**, *104*, 42–52. [CrossRef] [PubMed]

15.   Russo, N.; Compostella, L.; Tarantini, G.; Setzu, T.; Napodano, M.; Bottio, T.; D'Onofrio, A.; Isabella, G.; Gerosa, G.; Iliceto, S. Cardiac rehabilitation after transcatheter versus surgical prosthetic valve implantation for aortic stenosis in the elderly. *Eur. J. Prev. Cardiol.* **2014**, *21*, 1341–1348. [CrossRef] [PubMed]

16.   Lorenzoni, G.; Azzolina, D.; Lanera, C.; Brianti, G.; Gregori, D.; Vanuzzo, D.; Baldi, I. Time trends in first hospitalization for heart failure in a community-based population. *Int. J. Cardiol.* **2018**, *271*, 195–199. [CrossRef]

17.   Liu, Z.; Kidney, E.; Bem, D.; Bramley, G.; Bayliss, S.; de Belder, M.A.; Cummins, C.; Duarte, R. Transcatheter aortic valve implantation for aortic stenosis in high surgical risk patients: A systematic review and meta-analysis. *PLoS ONE* **2018**, *13*, e0196877. [CrossRef]

18.   Schymik, G.; Heimeshoff, M.; Bramlage, P.; Herbinger, T.; Würth, A.; Pilz, L.; Schymik, J.S.; Wondraschek, R.; Süselbeck, T.; Gerhardus, J.; et al. A comparison of transcatheter aortic valve implantation and surgical aortic valve replacement in 1,141 patients with severe symptomatic aortic stenosis and less than high risk. *Catheter. Cardiovasc. Interv.* **2015**, *86*, 738–744. [CrossRef] [PubMed]

19.   Levett, J.Y.; Windle, S.B.; Filion, K.B.; Brunetti, V.C.; Eisenberg, M.J. Meta-Analysis of Transcatheter Versus Surgical Aortic Valve Replacement in Low Surgical Risk Patients. *Am. J. Cardiol.* **2020**, *125*, 1230–1238. [CrossRef] [PubMed]

20.   Olsson, K.; Nilsson, J.; Hörnsten, Å.; Näslund, U. Patients' self-reported function, symptoms and health-related quality of life before and 6 months after transcatheter aortic valve implantation and surgical aortic valve replacement. *Eur. J. Cardiovasc. Nurs.* **2017**, *16*, 213–221. [CrossRef] [PubMed]

21.   Straiton, N.; Jin, K.; Bhindi, R.; Gallagher, R. Functional capacity and health-related quality of life outcomes post transcatheter aortic valve replacement: A systematic review and meta-analysis. *Age Ageing* **2018**, *47*, 478–482. [CrossRef]

22.   Lorenzoni, G.; Azzolina, D.; Fraccaro, C.; Di Liberti, A.; D'Onofrio, A.; Cavalli, C.; Fabris, T.; D'Amico, G.; Cibin, G.; Nai Fovino, L.; et al. Using Wearable Devices to Monitor Physical Activity in Patients Undergoing Aortic Valve Replacement: Protocol for a Prospective Observational Study. *JMIR Res. Protoc.* **2020**, *9*, e20072. [CrossRef]

23.   Tedesco, S.; Sica, M.; Ancillao, A.; Timmons, S.; Barton, J.; O'Flynn, B. Validity Evaluation of the Fitbit Charge2 and the Garmin vivosmart HR+ in Free-Living Environments in an Older Adult Cohort. *JMIR mHealth uHealth* **2019**, *7*, e13084. [CrossRef]

24.   Hlatky, M.A.; Boineau, R.E.; Higginbotham, M.B.; Lee, K.L.; Mark, D.B.; Califf, R.M.; Cobb, F.R.; Pryor, D.B. A brief self-administered questionnaire to determine functional capacity (the Duke Activity Status Index). *Am. J. Cardiol.* **1989**, *64*, 651–654. [CrossRef]

25. Mok, M.; Nombela-Franco, L.; Dumont, E.; Urena, M.; DeLarochellière, R.; Doyle, D.; Villeneuve, J.; Côté, M.; Ribeiro, H.B.; Allende, R.; et al. Chronic obstructive pulmonary disease in patients undergoing transcatheter aortic valve implantation: Insights on clinical outcomes, prognostic markers, and functional status changes. *JACC Cardiovasc. Interv.* **2013**, *6*, 1072–1084. [CrossRef]
26. Mahoney, F.I.; Barthel, D.W. Functional evaluation: The Barthel Index. *Md. State Med. J.* **1965**, *14*, 61–65. [PubMed]
27. Ribeiro, G.S.; Melo, R.D.; Deresz, L.F.; Dal Lago, P.; Pontes, M.R.; Karsten, M. Cardiac rehabilitation programme after transcatheter aortic valve implantation versus surgical aortic valve replacement: Systematic review and meta-analysis. *Eur. J. Prev. Cardiol.* **2017**, *24*, 688–697. [CrossRef] [PubMed]
28. Lawton, M.P.; Brody, E.M. Assessment of older people: Self-maintaining and instrumental activities of daily living. *Gerontologist* **1969**, *9*, 179–186. [CrossRef] [PubMed]
29. Johns, M.W. A new method for measuring daytime sleepiness: The Epworth sleepiness scale. *Sleep* **1991**, *14*, 540–545. [CrossRef] [PubMed]
30. About the ESS—Epworth Sleepiness Scale. Available online: https://epworthsleepinessscale.com/about-the-ess/ (accessed on 1 December 2020).
31. Team, RC. R: A Language and Environment for Statistical Computing. Vienna, Austria. 2020. Available online: https://www.R-project.org/ (accessed on 1 December 2020).
32. Alharbi, M.; Straiton, N.; Smith, S.; Neubeck, L.; Gallagher, R. Data management and wearables in older adults: A systematic review. *Maturitas* **2019**, *124*, 100–110. [CrossRef] [PubMed]
33. Finley, D.J.; Fay, K.A.; Batsis, J.A.; Stevens, C.J.; Sacks, O.A.; Darabos, C.; Cook, S.B.; Lyons, K.D. A feasibility study of an unsupervised, pre-operative exercise program for adults with lung cancer. *Eur. J. Cancer Care (Engl.)* **2020**, *29*, e13254. [CrossRef]
34. Mercer, K.; Giangregorio, L.; Schneider, E.; Chilana, P.; Li, M.; Grindrod, K. Acceptance of Commercially Available Wearable Activity Trackers Among Adults Aged Over 50 and With Chronic Illness: A Mixed-Methods Evaluation. *JMIR mHealth uHealth* **2016**, *4*, e7. [CrossRef] [PubMed]
35. Henriksen, A.; Sand, A.-S.; Deraas, T.; Grimsgaard, S.; Hartvigsen, G.; Hopstock, L. Succeeding with prolonged usage of consumer-based activity trackers in clinical studies: A mixed methods approach. *BMC Public Health* **2020**, *20*, 1300. [CrossRef] [PubMed]
36. Tedesco, S.; Barton, J.; O'Flynn, B. A Review of Activity Trackers for Senior Citizens: Research Perspectives, Commercial Landscape and the Role of the Insurance Industry. *Sensors* **2017**, *17*, 1277. [CrossRef] [PubMed]
37. Hermsen, S.; Moons, J.; Kerkhof, P.; Wiekens, C.; De Groot, M. Determinants for Sustained Use of an Activity Tracker: Observational Study. *JMIR mHealth uHealth* **2017**, *5*, e164. [CrossRef]
38. Baig, M.M.; Afifi, S.; GholamHosseini, H.; Mirza, F. A Systematic Review of Wearable Sensors and IoT-Based Monitoring Applications for Older Adults—A Focus on Ageing Population and Independent Living. *J. Med. Syst.* **2019**, *43*, 233. [CrossRef]
39. Preusse, K.C.; Mitzner, T.L.; Fausset, C.B.; Rogers, W.A. Older Adults' Acceptance of Activity Trackers. *J. Appl. Gerontol.* **2017**, *36*, 127. [CrossRef]
40. Kim, S.; Choudhury, A. Comparison of Older and Younger Adults' Attitudes Toward the Adoption and Use of Activity Trackers. *JMIR mHealth uHealth* **2020**, *8*, e18312. [CrossRef]
41. Batsis, J.A.; Naslund, J.A.; Zagaria, A.B.; Kotz, D.; Dokko, R.; Bartels, S.J.; Carpenter-Song, E. Technology for Behavioral Change in Rural Older Adults with Obesity. *J. Nutr. Gerontol. Geriatr.* **2019**, *38*, 130–148. [CrossRef] [PubMed]
42. Martinato, M.; Lorenzoni, G.; Zanchi, T.; Bergamin, A.; Buratin, A.; Azzolina, D.; Gregori, D. Assessment of Physical Activity in the Elderly Population: Usability and Accuracy of a Smartwatch. *JMIR mHealth uHealth* **2021**, *9*, e20966. [CrossRef]
43. Kristoffersson, A.; Lindén, M. A Systematic Review on the Use of Wearable Body Sensors for Health Monitoring: A Qualitative Synthesis. *Sensors* **2020**, *20*, 1502. [CrossRef]
44. Kononova, A.; Li, L.; Kamp, K.; Bowen, M.; Rikard, R.V.; Cotten, S.; Peng, W. The Use of Wearable Activity Trackers Among Older Adults: Focus Group Study of Tracker Perceptions, Motivators, and Barriers in the Maintenance Stage of Behavior Change. *JMIR mHealth uHealth* **2019**, *7*, e9832. [CrossRef] [PubMed]
45. Brickwood, K.-J.; Watson, G.; O'Brien, J.; Williams, A.D. Consumer-Based Wearable Activity Trackers Increase Physical Activity Participation: Systematic Review and Meta-Analysis. *JMIR mHealth uHealth* **2019**, *7*, e11819. [CrossRef] [PubMed]

International Journal of
Environmental Research and Public Health

MDPI

Article

Pattern of Use of Electronic Health Record (EHR) among the Chronically Ill: A Health Information National Trend Survey (HINTS) Analysis

Rose Calixte [1,*], Sumaiya Islam [2], Zainab Toteh Osakwe [3], Argelis Rivera [4] and Marlene Camacho-Rivera [5]

1   Department of Epidemiology and Biostatistics, SUNY Downstate Health Sciences University, Brooklyn, NY 11203, USA
2   CUNY School of Medicine, City College of New York, New York, NY 10031, USA; sislam011@csom.cuny.edu
3   College of Nursing and Public Health, Adelphi University, Garden City, NY 11530, USA; zosakwe@adelphi.edu
4   Department of Medicine, Mount Sinai Icahn School of Medicine, New York, NY 10011, USA; argelis.rivera@mountsinai.org
5   Department of Community Health Sciences, SUNY Downstate Health Sciences University, Brooklyn, NY 11203, USA; marlene.camacho-rivera@downstate.edu
*   Correspondence: rose.calixte@downstate.edu

Citation: Calixte, R.; Islam, S.; Osakwe, Z.T.; Rivera, A.; Camacho-Rivera, M. Pattern of Use of Electronic Health Record (EHR) among the Chronically Ill: A Health Information National Trend Survey (HINTS) Analysis. *Int. J. Environ. Res. Public Health* **2021**, *18*, 7254. https://doi.org/10.3390/ijerph18147254

Academic Editors: Irene Torres-Sanchez and Marie Carmen Valenza

Received: 4 June 2021
Accepted: 4 July 2021
Published: 7 July 2021

Publisher's Note: MDPI stays neutral with regard to jurisdictional claims in published maps and institutional affiliations.

Abstract: Effective patient–provider communication is a cornerstone of patient-centered care. Patient portals provide an effective method for secure communication between patients or their proxies and their health care providers. With greater acceptability of patient portals in private practices, patients have a unique opportunity to manage their health care needs. However, studies have shown that less than 50% of patients reported accessing the electronic health record (EHR) in a 12-month period. We used HINTS 5 cycle 1 and cycle 2 to assess disparities among US residents 18 and older with any chronic condition regarding the use of EHR for secure direct messaging with providers, to request refills, to make clinical decisions, or to share medical records with another provider. The results indicate that respondents with multimorbidity are more likely to share their medical records with other providers. However, respondents who are 75 and older are less likely to share their medical records with another provider. Additionally, respondents who are 65 and older are less likely to use the EHR for secure direct messaging with their provider. Additional health care strategies and provider communication should be developed to encourage older patients with chronic conditions to leverage the use of patient portals for effective disease management.

Keywords: chronic conditions; patient portal; health communication

1. Introduction

Patient–physician communication, particularly patients' satisfaction with physicians' communication approaches, is important for better outcomes in patient-centered health care organizations [1,2]. Finding effective ways of maintaining communication between health care providers and patients outside of the health care organization is important for disease management and care coordination [1,2]. With legislation mandating the meaningful use of electronic health records (EHR), by 2015, 98% of hospitals and 78% of private practices in the United States offered patient portals [3,4]. Patient portals also provide effective communication tools with the potential to increase patient engagement in self-care management. Patients can also use patient portals for prescription renewal, appointment management, checking lab results, and messaging their providers [5]. Therefore, the use of patient portals is an increasingly common approach in patient-centered care practices [1,2].

With the aging of the US population, certain diseases are becoming more prevalent [6]. Additionally, with people living longer, the prevalence of multimorbidity is also increasing [7–12]. Patients with multiple chronic conditions usually require a team of specialists

35

for managing their conditions [13,14]. Maintenance of their treatment plans requires effective communication with patients and coordination of care among their team of providers to achieve better outcomes [15]. Prior research has shown that access to patient portals resulted in improved diabetes-related outcomes and adherence to hypertensive medication [16,17]. Additionally, a recent study of patients with chronic kidney disease (CKD) found that the use of EHR improves patient-centered outcomes such as CKD-specific knowledge, while reducing CKD-related stress [5]. The ongoing COVID-19 pandemic has accelerated the use of technologies such as patient portals for health care delivery [18,19]. Despite the increasing reliance on the use of technology for both preventive and follow-up care, some patients lack resources to engage effectively in telehealth [18].

Despite the proliferation of health-related internet use (HRIU) and the widespread use of the internet by the general population, disparities still exist in terms of access and use of internet for disease management [20–25]. Recent studies suggest that despite the increasing trend in the use of patient portals, the acceptability of patient portals in the general population remains unusually low [18]. Additionally, sociodemographic disparities exist in assessing patient portals as a tool for disease management and communication with providers [26–28]. A recent study on disparities in health-related internet content that focused on the noninstitutionalized population of the US assessed health information seeking behavior in three domains relevant to health communication (health care, health information-seeking, and user-generated content/sharing). The study indicated age, gender, race/ethnicity, education, and income related disparities across multiple domains of health communication [29].

Several studies have examined potential explanations for low EHR engagement, and HRIU more broadly, within racial and ethnic minority populations. A recent study which examined disparities in trust in sources of cancer-related health information among Hispanics in the US observed that older Hispanics had higher odds of trusting cancer information from a religious organization compared to younger Hispanics [30]. Another study of men with chronic conditions developed an eHealth usage score using seven domains of eHealth communication, which included EHR use [31]. The eHealth questions used to create the score asked whether respondents had done the following: used a computer, smartphone or other electronic means to (1) look for health information or medical information for yourself, (2) look for health or medical information for someone else, (3) buy medicine or vitamins online, (4) look for assistance for the care you provided someone else, (5) use email or the internet to communicate with a doctor or doctor's office, (6) track health care costs/changes, or (7) look up medical tests. This study identified disparities in eHealth usage across social and demographic characteristics. Particularly, education and income were positively correlated with eHealth score, with participants with higher levels of education and those with higher incomes having increased scores for eHealth usage. However, the same study observed that age and Hispanic ethnicity were negatively correlated with eHealth score, such that older patients had lower scores of eHealth usage, and Hispanics also had lower scores of eHealth usage [31]. Low eHealth usage among individuals of Hispanic ethnicity may be due to low English proficiency and lower levels of health literacy [32,33]. Among older individuals, lower levels of health literacy and technological skill have been found to be associated with lower eHealth usage [34,35]. A previous study examining associations between health literacy and health information seeking found that participants with chronic conditions were more likely to be engaged in health information seeking and higher instances of seeking care based on information found on the Web. Additionally, participants with chronic conditions had higher eHealth literacy scores compared to participants without chronic conditions [36]. When compared to patients without chronic health conditions, patients with a history of chronic conditions reported frequent use of patient portals for different aspects of health care delivery such as checking lab reports, messaging their doctors, and setting up appointments [36]. However, this study failed to adjust for social and demographic factors that are associated with both chronic disease status and health information seeking behaviors. While many studies have looked at disparities in the usage of patient portals in patients with specific chronic conditions, no studies thus far have looked at disparities in the usage of patient portals in only people with chronic conditions in a nationally-representative sample of the noninstitutionalized US population. As EHR usage has been associated with significant improvements in patient self-management of chronic diseases, as well as improved quality of care given by providers, understanding disparities in EHR use may provide important insights into health care disparities among adults living with chronic health conditions in the US [37].

2. Materials and Methods

The National Cancer Institute's Health Information National Trends Survey (HINTS) is a publicly available national representative survey of the noninstitutionalized adult population that collects data about Americans' use of cancer related information. Data for this study came from the HINTS 5 cycle 1 (N = 3335), collected from January 2017 to May 2017, and HINTS 5 cycle 2 (N = 3504), collected from January 2018 to May 2018. The sampling design for the HINTS survey has been described extensively [22,38]. The response rate was 32.4% for HINTS 5 cycle 1 and 32.985% for HINTS 5 cycle 2.

The goal of this study was to assess differences in EHR usage among respondents with chronic diseases conditions. Using self-reported data, the study was restricted to respondents with any of the following conditions: diabetes, hypertension, lung disease, heart conditions, depression, cancer, and arthritis. The final analytic sample was further restricted to respondents who reported accessing their online medical record at least once in the past 12 months for various reasons (N = 736 and 816 respectively) for a total sample size of 1552. The outcomes of EHR usage were assessed using 4 HINTS questions relating to the purposes of accessing the online medical record to (1) securely message their health care provider, (2) request a refill of medications, (3) make a decision on how to treat illness or condition, and (4) securely share it with another provider.

Primary predictors of interest were gender, race/ethnicity, age, education, income, multimorbidity, and nativity status. The covariates of interest were smoking status, employment status, regular access to a health care provider, insurance status, general health status, and family history of cancer. To measure the change in EHR use across the two HINTS releases, we used a dummy-coded variable to represent the survey year.

We used multivariable regression models to find patterns of associations of sociodemographic characteristics with domains of eHealth usage in the population of US patients with chronic conditions who have access to their online records. The use of online records to securely message health care providers in the past 12 months and the use of online records to request prescription refills in the past 12 months were analyzed using Poisson regression with a log link and robust estimates of standard errors [39]. The use of online records to make decisions on how to treat an illness or condition in the past 12 months and the use of online records to securely share health records with another provider in the past 12 months were analyzed using logistic regression. To account for the complex survey design used to collect the data, we used jackknife replicate weights to compute accurate standard errors, with all analyses weighted to provide nationally representative estimates. We conducted all statistical analyses using SAS 9.4® and Stata 16®. The threshold for the significance of the p-value was set to ≤ 0.05.

We summarized the data using appropriate descriptive statistics such as frequency (percent) and weighted percent (standard error). We presented the multivariable regression models using incident rate ratios (IRR) and odds ratio (OR) with a 95% confidence interval (CI).

3. Results

3.1. Descriptive Results

The final analytic sample comprising respondents with at least one chronic condition that accessed their online medical record at least once in the past 12 months resulted in

1552 participants from the two HINTS cycles. We have presented the summary of the sociodemographic characteristics of the respondents in Table 1. The analytic sample was 56% female, 42% 18–49 years old, 70% non-Hispanic White, 78.2% with some college education, 58% employed, and 47.2% with an income of $75,000 or more. We combined Asian and other races for the purpose of the multivariable analysis. The sample analyzed was 88.8% US born, and 93% reported speaking English very well. Additionally, 96.4% had access to health insurance, 83% had a regular health care provider, and over 73% of the participants reported having a family history of cancer.

Table 1. Sociodemographic characteristics of HINTS respondents with at least one chronic condition who accessed their patient portal at least once in the past 12 months (*N* = 1552).

Sociodemographic Characteristics	*n*	Unweighted Percent	Weighted Percent (SE)
Age			
18–34	120	7.73	14.53 (1.88)
35–49	301	19.39	28.20 (2.01)
50–64	579	37.31	35.10 (1.78)
65–74	376	24.23	13.91 (0.84)
≥75	147	9.47	6.98 (0.72)
Missing	29	1.87	1.27 (0.29)
Gender			
Male	554	35.70	40.46 (1.79)
Female	917	59.09	55.59 (1.80)
Missing	81	5.22	3.95 (0.71)
Race/Ethnicity			
Non-Hispanic White	1036	66.75	70.13 (1.59)
Non-Hispanic Black	178	11.47	7.92 (0.77)
Hispanic	136	8.76	11.48 (1.24)
Asian	51	3.29	3.63 (0.77)
Other	60	3.87	2.95 (0.57)
Missing	91	5.86	3.88 (0.65)
Education			
High school or Less	226	14.56	20.64 (1.58)
Some College or More	1307	84.21	78.24 (1.61)
Missing	19	1.22	1.12 (0.39)
Employment			
Employed	811	52.26	58.18 (1.92)
Unemployed	722	46.52	41.12 (1.93)
Missing	19	1.22	0.71 (0.21)
Income			
Less than $20,000	131	8.44	8.53 (1.11)
$20,000 to <$35,000	144	9.28	8.05 (1.07)
$35,000 to <$50,000	174	11.21	11.51 (1.28)
$50,000 to <$75,000	296	19.07	18.23 (1.43)
$75,000 or More	682	43.94	47.19 (1.98)
Missing	125	8.05	6.49 (0.74)

Table 1. *Cont.*

Sociodemographic Characteristics	*n*	Unweighted Percent	Weighted Percent (SE)
Born in the United States			
Yes	1373	88.47	88.81 (1.25)
No	158	10.18	10.41 (1.27)
Missing	21	1.35	0.78 (0.22)
Health Insurance			
Yes	1512	97.42	96.35 (1.01)
No	30	1.93	3.34 (1.01)
Missing	10	0.64	0.31 (0.13)
Regular Provider			
Yes	1325	85.37	82.89 (1.56)
No	212	13.66	15.81 (1.53)
Missing	15	0.97	1.30 (0.54)
Family History of Cancer			
Yes	1174	75.64	73.89 (1.91)
No	348	22.42	24.76 (1.88)
Missing	30	1.93	1.35 (0.32)
HINTS 5 Survey			
Cycle 1	736	47.42	47.95 (1.92)
Cycle 2	816	52.58	52.05 (1.92)

SE: Standard Error.

Table 2 presents the clinical characteristics of the respondents. About 59% of the analytic sample were never smokers. Overall estimates suggest that about 56% of respondents reported having more than one chronic condition. Within the analytic sample, the most prevalent chronic condition was high blood pressure, with 55% of the respondents reporting having high blood pressure. However, about 45% reported being in very good or excellent health, and less than 20% reported being in fair or poor health.

Table 2. The Clinical characteristics of HINTS respondents with at least one chronic condition who accessed their patient portal at least once in the past 12 months (*N* = 1552).

Clinical Characteristics	*n*	Unweighted Percent	Weighted Percent (SE)
General Health			
Excellent	132	8.51	8.12 (0.89)
Very Good	607	39.11	37.13 (1.97)
Good	557	35.89	39.09 (1.98)
Fair	207	13.34	13.46 (1.18)
Poor	40	2.58	2.62 (0.57)
Missing	9	0.58	0.57 (0.22)
Smoking Status			
Current	152	9.79	11.67 (1.21)
Former	471	30.35	29.45 (1.52)
Never	912	58.76	57.90 (1.94)
Diabetes			
Yes	423	27.26	24.70 (1.52)
No	1119	72.10	74.66 (1.53)
Missing	10	0.64	0.65 (0.29)
High Blood Pressure			
Yes	884	56.96	54.87 (1.95)
No	654	42.14	44.45 (1.99)
Missing	14	0.90	0.68 (0.25)

Table 2. *Cont.*

Clinical Characteristics	n	Unweighted Percent	Weighted Percent (SE)
Lung Disease			
Yes	272	17.53	15.95 (1.30)
No	1273	82.02	83.76 (1.29)
Missing	7	0.45	0.29 (0.13)
Heart Conditions			
Yes	195	12.56	11.26 (1.37)
No	1349	86.92	88.43 (1.37)
Missing	8	0.52	0.31 (0.12)
Depression			
Yes	547	35.24	39.88 (1.76)
No	997	64.24	59.87 (1.76)
Missing	8	0.52	0.25 (0.11)
Arthritis			
Yes	621	40.01	33.84 (1.45)
No	925	59.60	65.83 (1.46)
Missing	6	0.39	0.33 (0.19)
Cancer			
Yes	357	23.00	14.96 (1.08)
No	1192	76.80	84.98 (1.09)
Missing	3	0.19	0.05 (0.03)
Multimorbidity			
Yes	958	61.73	56.03 (1.72)
No	594	38.27	43.97 (1.72)

SE: Standard Error.

We have presented the summary for the outcome variables in Table 3. Of all the respondents who accessed their medical records in the past 12 months, 47.3% of the participants used it 1–2 times, while less than 10% accessed their medical record 10 or more times. In the EHR communication domains, 48% of the respondents used the online medical record system to securely message their health care provider, 43.8% of the participants used the online medical record system to request a refill of medications, 21.1% of the respondents used their online medical record system to make a decision on how to treat an illness or condition, and 12% securely shared their medical record with another provider.

Table 3. Summary statistics of patient portal-related communication ($N = 1552$).

Use of Online Medical Record	n	Unweighted Percent	Weighted Percent (SE)
Number of times you accessed your record online in the past 12 months?			
1–2 times	695	44.78	47.27 (1.92)
3–5 times	523	33.70	32.94 (1.90)
6–9 times	172	11.08	10.09 (0.94)
10 times or more	162	10.44	9.70 (1.05)
Used online record to securely message a health care provider in the past 12 months			
Yes	750	48.32	50.05 (2.03)
No	742	47.81	46.47 (2.04)
Missing	60	3.87	3.48 (0.63)

Table 3. *Cont.*

Use of Online Medical Record	*n*	Unweighted Percent	Weighted Percent (SE)
Used online record to request a refill of medications in the past 12 months			
Yes	680	43.81	42.07 (1.97)
No	820	52.84	54.75 (1.96)
Missing	52	3.35	3.17 (0.61)
Used online record to make a decision on how to treat illness or condition in the past 12 months			
Yes	327	21.07	20.81 (1.62)
No	1170	75.39	75.89 (1.70)
Missing	55	3.54	3.30 (0.61)
Used online record to securely share it with another provider in the past 12 months			
Yes	186	11.98	12.02 (1.15)
No	1314	84.66	84.79 (1.21)
Missing	52	3.35	3.19 (0.62)

SE: Standard Error.

3.2. Multivariable Model

We have presented the results of the multivariable model in Table 4. Among HINTS respondents with chronic conditions who accessed their online medical record at least once in the past 12 months, respondents 65 to 74 years and those 75 years or older were significantly less likely to use the system to securely message their health care providers compared to respondents 18 to 34 years (IRR = 0.73, 95% CI = 0.57, 0.94; and IRR = 0.54, 95% CI = 0.35, 0.83) respectively.

Table 4. Multivariable Model for Patient Portal-related Communication (*N* = 1552).

Sociodemographic Characteristics	Securely Message Provider in the Past 12 Months		Request PrescriptionRefills in the Past 12 Months		Make a Decision on How to Treat Condition in the Past 12 Months		Securely Share it with Other Providers in the Past 12 Months	
	IRR	95% CI	IRR	95% CI	OR	95% CI	OR	95% CI
Race/Ethnicity								
Non-Hispanic White	1.00	1.00, 1.00	1.00	1.00, 1.00	1.00	1.00, 1.00	1.00	1.00, 1.00
Non-Hispanic Black	0.94	0.71, 1.25	1.11	0.84, 1.49	1.46	0.77, 2.79	1.85	0.84, 4.07
Hispanic	0.84	0.62, 1.13	0.99	0.71, 1.37	1.08	0.41, 2.88	0.79	0.25, 2.45
Other	1.09	0.79, 1.50	1.10	0.75, 1.60	2.64	1.12, 6.24 *	3.61	1.25, 10.42 *
Age								
18–34	1.00	1.00, 1.00	1.00	1.00, 1.00	1.00	1.00, 1.00	1.00	1.00, 1.00
35–49	0.98	0.77, 1.26	1.08	0.68, 1.72	0.93	0.37, 2.35	0.80	0.30, 2.13
50–64	0.77	0.58, 1.01	1.08	0.70, 1.68	0.49	0.19, 1.25	0.48	0.18, 1.28
65–74	0.73	0.57, 0.94 *	1.06	0.64, 1.74	0.30	0.11, 0.84 *	0.34	0.09, 1.22
≥ 75	0.54	0.35, 0.83 **	1.14	0.65, 2.01	0.70	0.22, 2.16	0.17	0.03, 0.99 *
Gender								
Male	1.00	1.00, 1.00	1.00	1.00, 1.00	1.00	1.00, 1.00	1.00	1.00, 1.00
Female	0.99	0.83, 1.18	0.83	0.68, 1.02	1.02	0.61, 1.71	0.99	0.55, 1.79

Table 4. *Cont.*

Sociodemographic Characteristics	Securely Message Provider in the Past 12 Months		Request PrescriptionRefills in the Past 12 Months		Make a Decision on How to Treat Condition in the Past 12 Months		Securely Share it with Other Providers in the Past 12 Months	
	IRR	95% CI	IRR	95% CI	OR	95% CI	OR	95% CI
Education								
High School or Less	0.87	0.67, 1.14	0.87	0.64, 1.19	1.03	0.47, 2.27	0.67	0.26, 1.73
Some College	1.00	0.82, 1.21	0.95	0.75, 1.22	0.91	0.50, 1.65	1.02	0.50, 2.07
College Graduate or More	1.00	1.00, 1.00	1.00	1.00, 1.00	1.00	1.00, 1.00	1.00	1.00, 1.00
Income								
Less than $20,000	0.87	0.59, 1.29	0.73	0.43, 1.25	1.14	0.39, 3.31	0.37	0.12, 1.18
$20,000 to <$35,000	0.75	0.52, 1.10	1.14	0.76, 1.71	0.85	0.34, 2.12	0.72	0.27, 1.93
$35,000 to <$50,000	0.62	0.44, 0.86 **	1.00	0.73, 1.35	1.56	0.69, 3.53	1.39	0.45, 4.35
$50,000 to <$75,000	0.99	0.78, 1.26	0.98	0.74, 1.31	1.03	0.51, 2.06	0.79	0.36, 1.73
$75,000 or More	1.00	1.00, 1.00	1.00	1.00, 1.00	1.00	1.00, 1.00	1.00	1.00, 1.00
Born in the United States								
Yes	1.00	1.00, 1.00	1.00	1.00, 1.00	1.00	1.00, 1.00	1.00	1.00, 1.00
No	0.92	0.68, 1.23	0.74	0.50, 1.08	0.59	0.27, 1.28	0.81	0.30, 2.16
Multimorbidity								
Yes	0.93	0.78, 1.11	1.20	0.97, 1.48	1.48	0.86, 2.57	2.04	1.16, 3.59 *
No	1.00	1.00, 1.00	1.00	1.00, 1.00	1.00	1.00, 1.00	1.00	1.00, 1.00
HINTS 5 Survey								
Cycle 1	1.00	1.00, 1.00	1.00	1.00, 1.00	1.00	1.00, 1.00	1.00	1.00, 1.00
Cycle 2	1.17	1.00, 1.36 *	1.07	0.88, 1.30	1.99	1.23, 3.20 **	1.28	0.77, 2.11

The multivariable regression model (logistic and Poisson) was adjusted for insurance status, employment status, having a regular provider, general health status, smoking status, and family history of cancer. We present the results of the Poisson regression models as incident rate ratios (IRR) and the results of the logistic regression models as odds ratio (OR) with 95% confidence interval (CI). * p-value $\leq$ 0.05, ** p-value $\leq$ 0.01.

The results of the multivariable model indicate that race, age, and survey years are associated with respondents who reported the use of the online medical record to make a decision about treating a condition or an illness. Respondents in the 65 to 74 age range had reduced odds of using the medical record to make a decision regarding treating a condition or illness (OR = 0.30, 95% CI = 0.11, 0.84). There was a significant increase in the use of the online medical record to make a decision on how to treat a condition or illness from HINTS 5 cycle 1 to HINTS 5 cycle 2 (OR = 1.99, 95% CI = 1.23, 3.20). Additionally, respondents in the other racial category had increased odds of using the online medical record to make decisions on how to treat a condition or illness in the past 12 months (OR = 2.64, 95% CI = 1.12, 6.24).

Those in the 75 years and older age range had reduced odds of securely sharing their online record with another provider compared to respondents aged 18 to 34 years (OR = 0.17, 95% CI = 0.03, 0.99). Respondents with multimorbidity (2 or more diagnoses) had significantly increased odds of securely sharing their online record with another provider in the past 12 months (OR = 2.04, 95% CI = 1.16, 3.59). Lastly, respondents in the other racial category had increased odds of securely sharing their online medical record with another provider in the past 12 months compared to Non-Hispanic Whites (OR = 3.61, 95% CI = 1.25, 10.42).

The sociodemographic characteristics of the respondents were not associated with their likelihood of requesting prescription refills using the online medical record. No other predictors of interest were significantly associated with any of the outcomes after adjustment for covariates.

4. Discussion

To our knowledge, this is the first study to examine disparities in accessing patient portals for disease management among chronically ill noninstitutionalized adults using nationally representative data. Our study examined a broad range of reasons for the use patient portals for disease management, including secure messaging with providers, requesting medications refills, and sharing medical records with other providers.

Our study is consistent with previous studies examining eHealth communication in US adults, revealing that age disparities exist in the use of eHealth communication methods, with older participants having a lower rate of eHealth communication across multiple domains [26,27,29,40]. However, our study yielded specific insights into the use of the patient portals, while the other studies were focused on the more general use of the eHealth domain of communication, such as personal email, searching for health information, buying medications online, and sharing health content though social networking sites [26,27,29,40]. Other studies of seniors 65–79 years old revealed that patients aged 70 years and older were less likely to register to use web portals [24,25]. Some potential explanations as to why older respondents were less likely to be engaged in use of the patient portals include the accessibility of the patient portals, safety concerns, and the usefulness of the portal for communication as opposed to face-to-face meetings with providers. This result, however, is concerning, as the likelihood of using the patient portal may be reflective of the ability to use telehealth services, and because the COVID-19 pandemic has accelerated the expansion of telehealth services [41]. Since a recent study of older adults' readiness to engage in eHealth and mHealth indicated that over 80% of the respondents reported having access to the internet at home, and 44% of those using the internet reported doing so on a mobile device [42], further studies should be done to understand the digital divide for older patients and how to engage them in eHealth.

In a prior HINTS study of HRIU use among cancer survivors, there has been an increasing trend in the use of HRIU [21,43]. In our study, we similarly found an increase in the use of EHR to make clinical decisions about how to treat an illness or condition. However, no increasing trend was noted in any of the other domains analyzed. This result underscores the importance of encouraging and promoting the use of the portal for other aspects of chronic disease management, particularly communicating with providers and sharing medical records with other providers. Despite the increasing trend in HRIU among cancer survivors, in a prior HINTS study of cancer survivors, there were age, race, education, and geographic disparities in the use of HRIU [27,44]. This study is in line with our result of lower use of online medical records for secure communication, to treat medical conditions, and to share with other providers among older participants. With the recent COVID-19 pandemic, health care providers have shifted to the use of telehealth for primary and specialty care for disease management [45]. The shift to telehealth during the pandemic could further widen the gap of health outcomes for patients with low and limited access to technology and those who are not ready to adopt new or emerging technologies.

Our study indicates that respondents with two or more chronic conditions were 2.04 times more likely to share the EHR with another provider. This result may be attributed to the need for different specialists to be involved in the care of patients with multimorbidity. This current finding is somewhat in line with other study that indicated that patients with chronic conditions were more engaged in care-seeking behavior based on health-related internet searches [36]. Patients with multimorbidity should be encouraged to participate in other forms of eHealth communication through the patient portal for disease management.

Prior studies have found sociodemographic disparities in the use of HRIU [23,28,29,46]. In our study, those identified as other race (Asian and others) were significantly more likely than those identified as White to use the portal for clinical decision making and transferring medical records to another provider. However, only 3.63% of the participants self-identified as Asian, and only 2.95% identified as other, limiting any form of generalization of the results. Although sociodemographic disparities exist in who is being offered access to EHR [47], in our study, which focused on only respondents who reported accessing their EHR at least once in the preceding 12 months, characteristics such as income and education level were not associated with the different domain of using EHR. We suspect that the result is due to our restriction of the analytic sample to respondents who reported having accessed the web portal at least once in the past 12 months. This subset of respondents is important because those respondents were not only offered access to their EHR, but have indicated using it. This restriction on the inclusion criteria for the final analytic sample could potentially control for some of the disparities that exist in accessing EHR, such as access to and use of the internet [26], and being offered its use [47]. For that reason, we did not adjust for the access/use of internet because using the web portal is an indication of internet access/usage.

5. Study Strengths and Limitations

There are several strengths to this study. First, the HINTS survey is a national survey of adults in the United States 18 and older. Thus, the subsample used in this study represents adults in the United States living with chronic conditions who have accessed their online health portal at least once in the past 12 months. We also used two HINTS cycles, allowing us to see changes in the use of EHR over time. Additionally, the study included an exploration of different reasons for using the EHR in disease management.

Despites the strengths, our study has some limitations. Given that HINTS is a cross-sectional study, we cannot make conclusions about causality. Furthermore, the survey has participants with a limited number of chronic conditions. Therefore, we can assess the association of respondents' characteristics with reasons for using EHR only in a subset of patients with chronic conditions. Future iterations of the survey should ask participants about other prevalent chronic conditions. In addition, future iterations of HINTS should include questions that address barriers to using the online medical records.

6. Conclusions

A recent HINTS brief indicated that the proportion of US adults accessing their online medical records increased from 27% in 2014 to 40% in 2018 [48]. Despite the increase in the access and use of EHR, our study reveals age as a factor of disparity in assessing the EHR for health care management. However, past studies have shown that over 80% of older adults are already using the internet, and 44% have access to smartphones. Therefore, health care providers should develop strategies to inform older patients and their proxies about the accessibility of the EHR as a secure means of communication to providers. Some strategies can include the development of user-friendly patient portals for multiple platforms that encourage greater use. Other strategies should include eHealth literacy programs that address patients' concerns regarding safeguarding their protected health information. Additional technological improvements may include the use of integrated displays to decrease user cognitive load [49–51], integration of linked EHR records at the household level to facilitate delivery of services, and embedding of health literacy tools (e.g., embedded medical search engines, integrated AI voice chatbots for on-demand self-care advice) to facilitate meaningful patient engagement.

The use of electronic patient portals raises challenges and ethical issues regarding older adults. Notably, disparities in internet access, a key factor for the use of e-health, persist among underserved populations such as older adults and individuals of lower socioeconomic status [52,53]. Older adults may encounter more challenges in the use of EHR and electronic patient portals because older generations must learn and acquire the necessary skills needed to navigate the internet and are less comfortable using technology compared to their counterparts. Furthermore, some older adults with certain illnesses may require support to navigate the complexities of eHealth portals [54]. Our findings are consistent with prior work using HINTS data, which showed that older US adults were less likely to engage in eHealth, and point to a need for additional support to ensure equitable access to e-health for older adults [22]. Other ethical aspects that are of importance are autonomy, privacy, confidentiality, consent, and beneficence [55]. Those ethical issues can be extended to the adaptability and accessibility of patient portals for personal use by patients. With older participants being less likely to engage in eHealth communication through the patient portal, adopters of EHR should consider the issue of autonomy, privacy, confidentiality, and equality of access as they encourage older patients and their proxies to make full use of the system for disease management.

Author Contributions: Conceptualization, R.C. and M.C.-R.; methodology, R.C and M.C.-R.; software, R.C.; formal analysis, R.C.; investigation, R.C., S.I., Z.T.O., A.R., and M.C.-R.; resources, R.C., S.I., and M.C.-R.; data curation, R.C.; writing—original draft preparation, R.C., S.I., Z.T.O., A.R., and M.C.-R.; writing—review and editing, R.C., S.I., Z.T.O., A.R., and M.C.-R.; visualization, R.C. and S.I.; supervision, R.C. and M.C.-R.; project administration, R.C. All authors have read and agreed to the published version of the manuscript.

Funding: This research received no external funding.

Institutional Review Board Statement: Not applicable.

Informed Consent Statement: Not applicable.

Data Availability Statement: The data that support the findings of this study are available from https://hints.cancer.gov/data/download-data.aspx.

Conflicts of Interest: The authors declare no conflict of interest.

References

1. Jhamb, M.; Cavanaugh, K.L.; Bian, A.; Chen, G.; Ikizler, T.A.; Unruh, M.L.; Abdel-Kader, K. Disparities in electronic health record patient portal use in nephrology clinics. *Clin. J. Am. Soc. Nephrol.* **2015**, *10*, 2013–2022. [CrossRef]
2. Rathert, C.; Mittler, J.N.; Banerjee, S.; McDaniel, J. Patient-centered communication in the era of electronic health records: What does the evidence say? *Patient Educ. Couns.* **2017**, *100*, 50–64. [CrossRef]
3. Kim, E.; Rubinstein, S.M.; Nead, K.T.; Wojcieszynski, A.P.; Gabriel, P.E.; Warner, J.L. The Evolving Use of Electronic Health Records (EHR) for Research. *Semin. Radiat. Oncol.* **2019**, *29*, 354–361. [CrossRef]
4. De Angelis, C.D. The electronic health record: Boon or bust for good patient care? *Milbank Q.* **2014**, *92*, 442. [CrossRef]
5. Tome, J.; Ahmed, S.; Fagerlin, A.; Powell, C.; Mourao, M.; Chen, E.; Harrison, S.; Segal, J.; Abdel-Kader, K.; Nunes, J.W. Patient electronic health record portal use and patient-centered outcomes in CKD. *Kidney Med.* **2021**, *3*, 231–240.e1. [CrossRef] [PubMed]
6. World Health Organization. *Global Status Report on Noncommunicable Diseases 2014*; World Health Organization: Geneva, Switzerland, 2014.
7. Freid, V.M.; Bernstein, A.B.; Bush, M.A. Multiple chronic conditions among adults aged 45 and over: Trends over the past 10 years. *NCHS Data Brief.* **2012**, *100*, 1–8.
8. van Oostrom, S.H.; Gijsen, R.; Stirbu, I.; Korevaar, J.C.; Schellevis, F.G.; Picavet, H.S.J.; Hoeymans, N. Time Trends in Prevalence of Chronic Diseases and Multimorbidity Not Only due to Aging: Data from General Practices and Health Surveys. *PLoS ONE* **2016**, *11*, e0160264. [CrossRef] [PubMed]
9. Ward, B.W.; Schiller, J.S. Peer reviewed: Prevalence of multiple chronic conditions among US adults: Estimates from the National Health Interview Survey, 2010. *Prev. Chronic. Dis.* **2013**, *10*, E65. [CrossRef] [PubMed]
10. Erdem, E. Prevalence of chronic conditions among Medicare Part A beneficiaries in 2008 and 2010: Are Medicare beneficiaries getting sicker? *Prev. Chronic. Dis.* **2014**, *11*, 130118. [CrossRef]
11. Cassell, A.; Edwards, D.; Harshfield, A.; Rhodes, K.; Brimicombe, J.; Payne, R.; Griffin, S. The epidemiology of multimorbidity in primary care: A retrospective cohort study. *Br. J. Gen. Pract.* **2018**, *68*, e245–e251. [CrossRef]
12. Reeves, D.; Pye, S.; Ashcroft, D.M.; Clegg, A.; Kontopantelis, E.; Blakeman, T.; van Marwijk, H. The challenge of ageing populations and patient frailty: Can primary care adapt? *BMJ* **2018**, *362*, k3349. [CrossRef]
13. Mollica, R.; Gillespie, J. *Care Coordination for People with Chronic Conditions*; Partnership for Solutions: Portland, ME, USA, 2003.
14. Shakib, S.; Dundon, B.K.; Maddison, J.; Thomas, J.; Stanners, M.; Caughey, G.E.; Clark, R.A. Effect of a Multidisciplinary Outpatient Model of Care on Health Outcomes in Older Patients with Multimorbidity: A Retrospective Case Control Study. *PLoS ONE* **2016**, *11*, e0161382. [CrossRef] [PubMed]

15. Barbabella, F.; Melchiorre, M.G.; Quattrini, S.; Papa, R.; Lamura, G.; Richardson, E.; van Ginneken, E. *How Can. eHealth Improve Care for People with Multimorbidity in Europe*; World Health Organization, Regional Office for Europe: Copenhagen, Denmark, 2017.

16. Wright, E.; Darer, J.; Tang, X.; Thompson, J.; Tusing, L.; Fossa, A.; Delbanco, T.; Ngo, L.; Walker, J. Sharing Physician Notes Through an Electronic Portal is Associated with Improved Medication Adherence: Quasi-Experimental Study. *J. Med. Internet Res.* **2015**, *17*, e226. [CrossRef] [PubMed]

17. Osborn, C.Y.; Mayberry, L.S.; Mulvaney, S.A.; Hess, R. Patient web portals to improve diabetes outcomes: A systematic review. *Curr. Diab. Rep.* **2010**, *10*, 422–435. [CrossRef] [PubMed]

18. Kalicki, A.V.; Moody, K.A.; Franzosa, E.; Gliatto, P.M.; Ornstein, K.A. Barriers to telehealth access among homebound older adults. *J. Am. Geriatr. Soc.* **2021**. Epub ahead of print. [CrossRef]

19. Camacho-Rivera, M.; Islam, J.Y.; Rivera, A.; Vidot, D.C. Attitudes toward Using COVID-19 mHealth Tools Among Adults with Chronic Health Conditions: Secondary Data Analysis of the COVID-19 Impact Survey. *JMIR Mhealth Uhealth* **2020**, *8*, e24693. [CrossRef]

20. Chou, W.Y.; Hunt, Y.M.; Beckjord, E.B.; Moser, R.P.; Hesse, B.W. Social media use in the United States: Implications for health communication. *J. Med. Internet Res.* **2009**, *11*, e48. [CrossRef]

21. Chou, W.Y.; Liu, B.; Post, S.; Hesse, B. Health-related Internet use among cancer survivors: Data from the Health Information National Trends Survey, 2003–2008. *J. Cancer Surviv.* **2011**, *5*, 263–270. [CrossRef]

22. Kontos, E.; Blake, K.D.; Chou, W.Y.; Prestin, A. Predictors of eHealth usage: Insights on the digital divide from the Health Information National Trends Survey 2012. *J. Med. Internet Res.* **2014**, *16*, e172. [CrossRef]

23. Prestin, A.; Vieux, S.N.; Chou, W.Y. Is Online Health Activity Alive and Well or Flatlining? Findings from 10 Years of the Health Information National Trends Survey. *J. Health Commun.* **2015**, *20*, 790–798. [CrossRef]

24. Thackeray, R.; Crookston, B.T.; West, J.H. Correlates of health-related social media use among adults. *J. Med. Internet Res.* **2013**, *15*, e21. [CrossRef] [PubMed]

25. Gordon, N.P.; Hornbrook, M.C. Differences in Access to and Preferences for Using Patient Portals and Other eHealth Technologies Based on Race, Ethnicity, and Age: A Database and Survey Study of Seniors in a Large Health Plan. *J. Med. Internet Res.* **2016**, *18*, e50. [CrossRef] [PubMed]

26. Hong, Y.A.; Jiang, S.; Liu, P.L. Use of Patient Portals of Electronic Health Records Remains Low from 2014 to 2018: Results from a National Survey and Policy Implications. *Am. J. Health Promot.* **2020**, *34*, 677–680. [CrossRef]

27. Jiang, S.; Hong, Y.A.; Liu, P.L. Trends of online patient-provider communication among cancer survivors from 2008 to 2017: A digital divide perspective. *J. Cancer Surviv.* **2019**, *13*, 197–204. [CrossRef]

28. Hong, Y.A.; Cho, J. Has the Digital Health Divide Widened? Trends of Health-Related Internet Use among Older Adults From 2003 to 2011. *J. Gerontol. B Psychol. Sci. Soc. Sci.* **2017**, *72*, 856–863. [CrossRef]

29. Calixte, R.; Rivera, A.; Oridota, O.; Beauchamp, W.; Camacho-Rivera, M. Social and Demographic Patterns of Health-Related Internet Use Among Adults in the United States: A Secondary Data Analysis of the Health Information National Trends Survey. *Int. J. Environ. Res. Public Health* **2020**, *17*, 6856. [CrossRef]

30. Camacho-Rivera, M.; Gonzalez, C.J.; Morency, J.A.; Blake, K.D.; Calixte, R. Heterogeneity in Trust of Cancer Information among Hispanic Adults in the United States: An Analysis of the Health Information National Trends Survey. *Cancer Epidemiol. Biomark. Prev.* **2020**, *29*, 1348–1356. [CrossRef]

31. Sherman, L.D.; Patterson, M.S.; Tomar, A.; Wigfall, L.T. Use of digital health information for health information seeking among men living with chronic disease: Data from the health information national trends survey. *Am. J. Mens. Health* **2020**, *14*, 1557988320901377. [CrossRef] [PubMed]

32. Millar, R.J.; Sahoo, S.; Yamashita, T.; Cummins, P.A. Literacy skills, language use, and online health information seeking among Hispanic adults in the United States. *Patient Educ. Couns.* **2020**, *103*, 1595–1600. [CrossRef] [PubMed]

33. Aponte, J.; Nokes, K.M. Electronic health literacy of older Hispanics with diabetes. *Health Promot. Int.* **2017**, *32*, 482–489. [CrossRef] [PubMed]

34. Yamashita, T.; Bardo, A.R.; Cummins, P.A.; Millar, R.J.; Sahoo, S.; Liu, D. The Roles of Education, Literacy, and Numeracy in Need for Health Information during the Second Half of Adulthood: A Moderated Mediation Analysis. *J. Health Commun.* **2019**, *24*, 271–283. [CrossRef] [PubMed]

35. Choi, N.G.; Dinitto, D.M. The digital divide among low-income homebound older adults: Internet use patterns, eHealth literacy, and attitudes toward computer/Internet use. *J. Med. Internet Res.* **2013**, *15*, e93. [CrossRef]

36. Madrigal, L.; Escoffery, C. Electronic Health Behaviors among US Adults with Chronic Disease: Cross-Sectional Survey. *J. Med. Internet Res.* **2019**, *21*, e11240. [CrossRef] [PubMed]

37. Kruse, C.S.; Argueta, D.A.; Lopez, L.; Nair, A. Patient and provider attitudes toward the use of patient portals for the management of chronic disease: A systematic review. *J. Med. Internet Res.* **2015**, *17*, e40. [CrossRef]

38. Finney Rutten, L.J.; Davis, T.; Beckjord, E.B.; Blake, K.; Moser, R.P.; Hesse, B.W. Picking up the pace: Changes in method and frame for the health information national trends survey (2011–2014). *J. Health Commun.* **2012**, *17*, 979–989. [CrossRef]

39. Zou, G. A modified poisson regression approach to prospective studies with binary data. *Am. J. Epidemiol.* **2004**, *159*, 702–706. [CrossRef]

40. Anthony, D.; Campos-Castillo, C. Why Most of Your Patients Aren't Using an Online Portal, and What You Can Do About It. *Rhode Isl. Med. J.* **2020**, *103*, 32–34.

41. Bosworth, A.; Ruhter, J.; Samson, L.W.; Sheingold, S.; Taplin, C.; Tarazi, W.; Zuckerman, R. *Medicare Beneficiary Use of Telehealth Visits: Early Data from the Start of COVID-19 Pandemic*; Office of the Assistant Secretary for Planning and Evaluation, U.S. Department of Health and Human Services: Washington, DC, USA, 28 July 2020. Available online: https://aspe.hhs.gov/system/files/pdf/263866/HP_IssueBrief_MedicareTelehealth_final7.29.20.pdf (accessed on 5 August 2020).
42. Gordon, N.P.; Hornbrook, M.C. Older adults' readiness to engage with eHealth patient education and self-care resources: A cross-sectional survey. *BMC Health Serv. Res.* **2018**, *18*, 1–13. [CrossRef] [PubMed]
43. Fareed, N.; Swoboda, C.M.; Jonnalagadda, P.; Huerta, T.R. Persistent digital divide in health-related internet use among cancer survivors: Findings from the Health Information National Trends Survey, 2003–2018. *J. Cancer Surviv.* **2021**, *15*, 87–98. [CrossRef]
44. Jiang, S.; Liu, P.L. Digital divide and Internet health information seeking among cancer survivors: A trend analysis from 2011 to 2017. *Psycho Oncol.* **2020**, *29*, 61–67. [CrossRef] [PubMed]
45. Koonin, L.M.; Hoots, B.; Tsang, C.A.; Leroy, Z.; Farris, K.; Jolly, B.T.; Antall, P.; McCabe, B.; Zelia, C.B.R.; Tong, I.; et al. Trends in the Use of Telehealth During the Emergence of the COVID-19 Pandemic—United States, January–March 2020. *MMWR Morb. Mortal. Wkly. Rep.* **2020**, *69*, 1595–1599. [CrossRef]
46. Spooner, K.K.; Salemi, J.L.; Salihu, H.M.; Zoorob, R.J. eHealth patient-provider communication in the United States: Interest, inequalities, and predictors. *J. Am. Med. Inform. Assoc.* **2017**, *24*, e18–e27. [CrossRef] [PubMed]
47. Anthony, D.L.; Campos-Castillo, C.; Lim, P.S. Who Isn't Using Patient Portals and Why? Evidence and Implications from A National Sample of US Adults. *Health Aff.* **2018**, *37*, 1948–1954. [CrossRef] [PubMed]
48. Health Information National Trend Survey. Trends and Disparities in Patient Portal Use. *HINTS Briefs.* **2021**, *45*, 1–2. Available online: https://hints.cancer.gov/docs/Briefs/HINTS_Brief_45.pdf (accessed on 21 June 2021).
49. Curran, R.L.; Kukhareva, P.V.; Taft, T.; Weir, C.R.; Reese, T.J.; Nanjo, C.; Rodriguez-Loya, S.; Martin, D.K.; Warner, P.B.; Shields, D.E.; et al. Integrated displays to improve chronic disease management in ambulatory care: A SMART on FHIR application informed by mixed-methods user testing. *J. Am. Med. Inform. Assoc.* **2020**, *27*, 1225–1234. [CrossRef]
50. Brady, K.J.S.; Legler, A.; Adams, W.G. Exploring Opportunities for Household-Level Chronic Care Management Using Linked Electronic Health Records of Adults and Children: A Retrospective Cohort Study. *Matern. Child. Health J.* **2020**, *24*, 829–836. [CrossRef]
51. Chen, M.; Decary, M. Embedding Health Literacy Tools in Patient EHR Portals to Facilitate Productive Patient Engagement. *Stud. Health Technol. Inform.* **2019**, *257*, 59–63.
52. Serrano, K.; Thai, C.; Greenberg, A.; Blake, K.; Moser, R.; Hesse, B. Progress on Broadband Access to the Internet and Use of Mobile Devices in the United States. *Public Health Rep.* **2017**, *132*, 27–31. [CrossRef]
53. Okoye, S.M.; Mulcahy, J.F.; Fabius, C.D.; Burgdorf, J.G.; Wolff, J.L. Neighborhood Broadband and Use of Telehealth among Older Adults: Cross-sectional Study of National Survey Data Linked with Census Data. *J. Med. Internet Res.* **2021**, *23*, e26242. [CrossRef] [PubMed]
54. Lam, K.; Lu, A.D.; Shi, Y.; Covinsky, K.E. Assessing Telemedicine Unreadiness Among Older Adults in the United States During the COVID-19 Pandemic. *JAMA Intern. Med.* **2020**, *180*, 1389. [CrossRef] [PubMed]
55. Miesperä, A.; Ahonen, S.-M.; Reponen, J. Ethical aspects of eHealth—Systematic review of open access articles. *Finn. J. EHealth EWelfare* **2013**, *5*, 165–171.

International Journal of
Environmental Research and Public Health

Article

How Can We Develop an Efficient eHealth Service for Provision of Care for Elderly People with Balance Disorders and Risk of Falling? A Mixed Methods Study

Andréa Gomes Martins Gaspar [1,2,*,†], Pedro Escada [3] and Luís Velez Lapão [1,†]

1 Instituto de Higiene e Medicina Tropical (IHMT), Universidade NOVA de Lisboa (UNL), 1349-008 Lisbon, Portugal; luis.lapao@ihmt.unl.pt
2 Hospital Beatriz Ângelo, 2674-514 Lisbon, Portugal
3 Hospital Egas Moniz, 1349-019 Lisbon, Portugal; pedroalbertoescada@gmail.com
* Correspondence: andreamartinsbr@hotmail.com
† These authors contributed equally to this work.

Abstract: This study aimed to identify relevant topics for the development of an efficient eHealth service for elderly people with balance disorders and risk of falling, based on input from physicians providing healthcare to this patient group. In the quantitative part of the study, an open multiple-choice questionnaire was made available on the website of the Portuguese General Medical Council to assess the satisfaction with electronic medical records regarding clinical data available, the time needed to retrieve data and the usefulness of the data. Of the 118 participants, 55% were dissatisfied/very dissatisfied with data availability and 61% with the time spent to access and update data related to the focused patient group. Despite this negative experience, 76% considered future e-Health solutions as pertinent/very pertinent. Subsequently, these findings were further explored with eight semi-structured interviews. The physicians confirmed the reported dissatisfactions and pointed out the lack of comprehensive data and system interoperability as serious problems, causing inefficient health services with an overlap of emergency visits and uncoordinated diagnostics and treatment. In addition, they discussed the importance of camera and audio monitoring to add significant value. Our results indicate considerable potential for e-Health solutions, but substantial improvements are crucial to achieving such future solutions.

Keywords: balance disorders; falls; elderly care; eHealth; mixed methods

Citation: Gaspar, A.G.M.; Escada, P.; Lapão, L.V. How Can We Develop an Efficient eHealth Service for Provision of Care for Elderly People with Balance Disorders and Risk of Falling? A Mixed Methods Study. *Int. J. Environ. Res. Public Health* **2021**, *18*, 7410. https://doi.org/10.3390/ijerph18147410

Academic Editors: Irene Torres-Sanchez and Marie Carmen Valenza

Received: 23 May 2021
Accepted: 8 July 2021
Published: 11 July 2021

Publisher's Note: MDPI stays neutral with regard to jurisdictional claims in published maps and institutional affiliations.

1. Introduction

1.1. The Burden of an Ageing Population: Portugal and the World

As is observed in other health systems [1–7], Portugal's increasing life expectancy in recent decades has not been followed by an increase in healthy life years [4,6]. The prevalence of chronic diseases, comorbidities, disabilities and falls have increased with aging [1–7]. In fact, there are many causes for elderly people's falls to happen, including age, environmental factors, inappropriate clothing and shoes, risky behavior, medications, and balance disorders [1,7–9].

Elderly falls represent an important public health problem, being the main cause of accidental death in the population over 65 years of age [1,8–10]. Although Portugal has one of the lowest rates of fall-related mortality in the elderly population of the Western European region, this issue has received attention from the Portuguese government [11].

The burden of aging with balance disorders and falls, and the insufficient access to healthcare data by health professionals, have led to additional medical visits, overdiagnosis, repeated diagnostic tests and multiple prescriptions [12–15]. This misuse of healthcare provision is costlier and unsustainable for the current healthcare provision model and considered unsuitable for responding to elderly population demand [4,12–15]. In order to relieve this pressure, new strategies have been recommended, including person-centered health systems and the utilization of devices and systems supported by Information Systems and Technologies (IST) [2,5,16] (Figure 1).

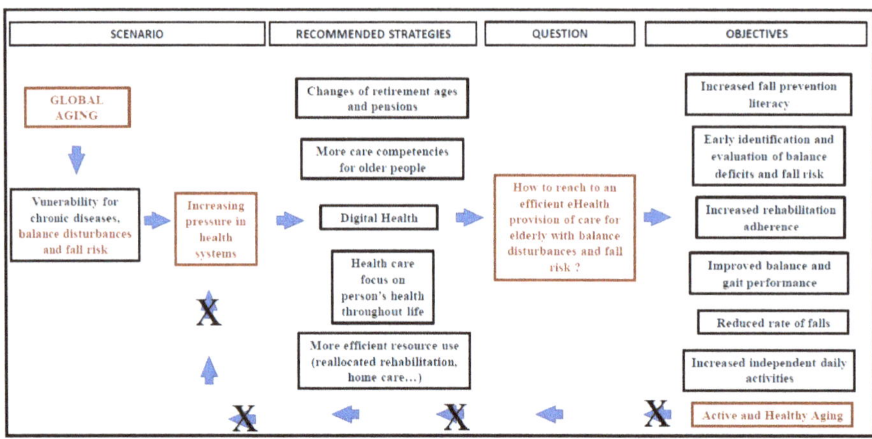

Figure 1. eHealth Framework for the elderly with balance disorders care provision. (Authors own elaboration).

These healthcare solutions have revealed the potential to provide quality health services with complete, interoperable data in near real-time [17–20], and eHealth services with the engagement of patients and families for self and remote management of chronic conditions and prevention of risky behaviors [2,20–22]. Indeed, many researchers have studied the potential of digital sensors to identify early balance deficit and identify fallers among elderly people, improving the data quality of clinical tests and functional scales as the Timed Up and Go Test (TUGT) and the Berg Balance Scale [23–29]. Other researchers have evaluated the benefits of eHealth devices in balance training, reducing the risk of falling [29–31]. The eHealth services seem to have the potential to be a complementary method for preventive monitoring of falls, telerehabilitation, and monitoring of effective rehabilitation for elderly with balance disorder and risk of falling [23–31], aligned with the eHealth definition: *"an emerging field in the intersection of medical informatics, public health and business, referring to health services and information delivered or enhanced through the Internet and related technologies. In a broader sense, the term characterizes not only a technical development, but also a state-of-mind, a way of thinking, an attitude, and a commitment for networked, global thinking, to improve health care locally, regionally, and worldwide by using information and communication technology"* [32]. However, there are still constraints to be overcome: technological obsolescence; unsuitable technological devices; regulation, standardization, auditing, inspection and quality control; lack of interoperability; health professional resistance; low organizational capability for new ways of working and organization; shortage of digital training [33–38]. In Europe, both the skill development in old age and the aging of younger generations of technology users have led to a growing number of elderly people able to use health and care services online, although to a lesser extent in Portugal [39].

1.2. Portugal and the Strategies Supported by IST

To build up a shared ecosystem of health information, the Portuguese Ministry of Health created an Electronic Health Record (EHR) called "Sclínico," which is unfortunately not yet available in all health units of the Portuguese National Health Service (NHS) [40]. More recently, another digital service was made available on the NHS' digital platform to allow the sharing of clinical information between all levels of health care and to promote the interaction between the citizen and the family health unit [40]. In addition, the Portuguese elderly people can use the current telephone and digital service of the NHS Call

Center, known as "SNS 24," which is responsible for the triage of first-level emergencies and guiding the population about health problems [40]. Another NHS telephone service, known as Senior Proximity Project, was implemented to identify the risks and needs of elderly people to reduce morbidity and promote more autonomy and health literacy [41]. Several public health units have provided retinal examination by teleradiology, teleconsultation, telediagnostic-telepathology, telemonitoring of cardiac and pulmonary diseases and telerehabilitation of osteoarticular disease of shoulder and knee [40]. Additionally, the electronic prescription system and the treatment guide for the user have allowed patients, including elderly people with chronic diseases, to obtain their medication without ever going to a health care unit [40].

However, in recent reviews [23–31], it was pointed out that the clinical applicability of eHealth devices and services in screening, assessing and treating elderly people with balance disorders and the risk of falling in Portugal is still unknown. Therefore, we aim at studying how to obtain an efficient eHealth service for the provision of care for elderly people with balance disorders and the risk of falling.

The purposes of this explanatory sequential mixed methods approach [42] were: (a) to identify and understand how to overcome the medical difficulties about availability of clinical data in the electronic medical record (EMR) relatively to the context of healthcare provision for elderly with balance disorders and risk of falling; (b) to know and understand the medical relevance about eHealth services to support health care for elderly people with balance disorders and risk of falls; (c) to understand how to develop an efficient eHealth service to support health care for elderly people with balance disorders and risk of falls.

The increasing interest of elderly people in medical digital devices and the eHealth potential to enhance health promotion and physician–patient interaction to mitigate care access inequities and to allow remote management of balance disorders and risk of falling are opportunities that should be further explored for active and healthy aging [29]. This could be viewed as an opportunity to mitigate the aging pressure in the health systems.

2. Materials and Methods

2.1. Study Design

From June to August 2019, the authors performed a quantitative observational descriptive study [42] to identify the difficulties about clinical data and the relevance of eHealth (Figure 2).

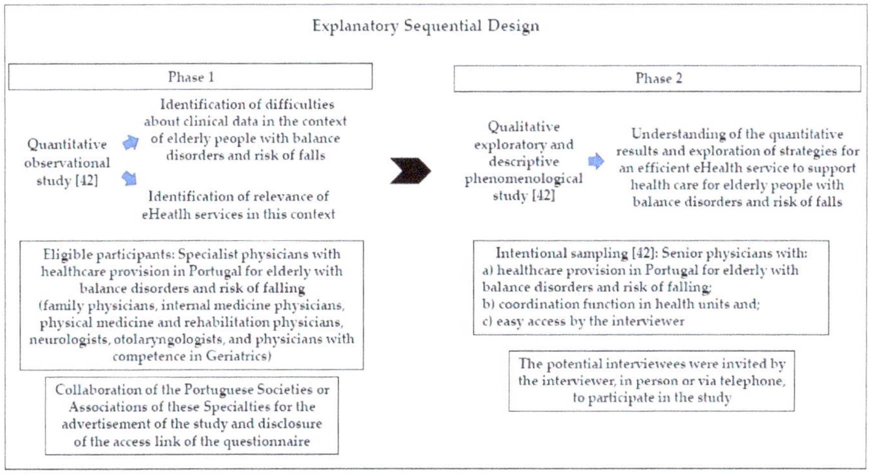

Figure 2. Design of the mixed methods study performed.

This first phase aimed at responding to the following specific research questions: "Do physicians have difficulties accessing current clinical data in EMR relatively to the context of healthcare provision for elderly with balance disorders and risk of falling?"; "In this context, what is the medical satisfaction level with the use (e.g., time spent to access and fill in clinical data) and quality (e.g., availability of sufficient and understandable clinical data) of the current clinical data in the electronic medical record (EMR)?", and "Could eHealth services be relevant to improve healthcare?".

From December 2019 to April 2020, a qualitative exploratory and descriptive phenomenological study [42] was performed to understand the quantitative results and how to obtain an efficient eHealth service to support health care for elderly people with balance disorders and risk of falls. This second phase explored the following research questions: "What are the medical difficulties related with current clinical data in the context of health care provision for elderly with balance disorders and risk of falling?"; "What strategies can be implemented to improve clinical data?"; "What do you think about the contribution of eHealth?"; "How can an eHealth service be suitable? What are the necessary strategies? What difficulties must be overcome?".

2.2. Materials

The questionnaire, entitled "Health contribution to the provision of health care for the elderly at risk of falling due to balance disorders" (in the original: "A contribuição do eHealth na prestação de cuidados de saúde ao idoso com risco de queda por distúrbios do equilíbrio") was developed with 18 multiple choice questions [42]. It included socio-demographic data of the participants, availability of data in the EMR and relevance of eHealth in the context of health care for the elderly with balance disorder and risk of falling. Except for the demographic questions, alternative responses on quantity, frequency and evaluation were used, and a non-response ("Do not know/Do not answer") was provided [42] (see Table A1). The usability, technical functionality and time to complete the questionnaire were tested. The access link was available through the website of the Portuguese General Medical Council (Ordem dos Médicos de Portugal), the entity that regulates medical practice in Portugal (https://ordemdosmedicos.pt/inquerito-a-contribuicao-do-ehealth-na-prestacao-de-cuidados-de-saude-ao-idoso/ (accessed on 25 June 2019)). The information regarding this open survey was provided online. The eligible participants were specialist physicians who provide healthcare in Portugal for the elderly with balance disorders and risk of falling, including family physicians, internal medicine physicians, physical medicine and rehabilitation (PMR) physicians, neurologists, otolaryngologists and physicians with competence in Geriatrics. For the advertisement of the study and disclosure of the access link of the questionnaire, the authors requested, via email, the collaboration of the Portuguese Society of Physical Medicine and Rehabilitation, Portuguese Association of General and Family Medicine, Portuguese Society of Internal Medicine, Center for Geriatric Studies of the Portuguese Society of Internal Medicine, Portuguese Society of Otorhinolaryngology and Portuguese Otoneurology Association. The Portuguese Society of Neurology was also contacted through this institution's website. The questionnaire was distributed online using the survey software SurveyMonkey® [43]. Each question was made available in turn, with the possibility of returning to the previous questions. All questions had one mandatory answer [44]. During the study time, the IP address of the participants was used to eliminate potential duplicate responses from the same user [44].

Regarding the qualitative study, the same interviewer (one of the authors) conducted individual semi-structured interviews [42]. Four primary thematic categories were discussed: current clinical data in the context of health provision for elderly with balance disorders and risk of falling (i.e., understanding of quantitative results), interventions to improve the clinical data, understanding of eHealth relevance pointed out by physicians in the quantitative research, and strategies to improve the use of eHealth services (see Table A2). The sampling was intentional [42], with a purposeful search for physicians with: (a) healthcare provision for elderly with balance disorders and risk of falling; (b) coordination function in health units and; (c) easy access by the interviewer. The potential interviewees were invited by the interviewer, in person or via telephone, to participate in the study. The participant number was defined after saturation or redundancy of responses; that is, the sampling process was completed when no new information emerged from the new interviews [42]. Respecting the anonymous participation of the quantitative study, the interviewer did not ask if the interviewee had participated in the previous study.

2.3. Data Analysis

Firstly, to determine the quantitative frequency tables [42], a descriptive and exploratory statistics of the data from the questionnaires were performed. The software IBM SPSS Statistics version 26 (IBM Corporation, Armonk, NY, USA) was used [45]. Secondly, a descriptive analysis of demographic data [42] of interviewed was performed. All the interviews were manually coded and transcribed by the interviewer, allowing content analysis of interviews [42]. For a better comprehension of the quotes, the authors entered words in round brackets.

2.4. Ethical Considerations

The survey's aim was clearly identified in both the website of the Portuguese General Medical Council and on the SurveyMonkey® link [SURVEY PREVIEW MODE] A contribuição do eHealth na prestação de cuidados de saúde ao idoso com risco de queda por distúrbios do equilíbrio Survey (surveymonkey.com, accessed on 25 June 2019). The physicians could voluntarily participate and leave the study until the submission of the questionnaire. The information of the quantitative study was treated confidentially and anonymously by using respondent e-mails confidentiality and anonymous responses features of the software SurveyMonkey® [43].

Regarding the interviews, the participants signed a consent form and received a copy of this and information about the study. They could leave the study until one month after the interview's date. The audio recording was authorized by the participants. To guarantee confidentiality, all interviews were manually coded. The transcriptions omitted information to avoid identifying respondents. All data were kept anonymous [42]. The information from the questionnaires and interviews and the audio records were kept in a safe place (external disk with access code) within the period provided by the Portuguese law [46], always safeguarding the confidentiality of the information obtained.

3. Results

3.1. First Phase: Quantitative Research

The online questionnaire had a total of 118 responses. This represents 1% of the total universe of 12,214 [47] family physicians, internal medicine physicians, PMR physicians, neurologists and otolaryngologists registered in Portugal (Table 1).

Table 1. Study quantitative: Demographic data of the participants. PMR: physical medicine and rehabilitation.

Demographic Data/Specialty	Family Physician	Internal Medicine Physician	PMR Physician	Neurologist	Otolaryngologist	Total
Physician number according to Portuguese General Medical Council—year 2019 [47]	7451	2847	691	549	676	12,214
Number of participants of the study according to specialty (%)	18 (15.3%)	46 (39.0%)	5 (4.2%)	4 (3.4%)	45 (38.1%)	118 (100%)

Table 1. *Cont.*

Demographic Data/Specialty	Family Physician	Internal Medicine Physician	PMR Physician	Neurologist	Otolaryngologist	Total
Participation according to specialty total number of physicians (%)	18/7451 (0.2%)	46/2847 (1.6%)	5/691 (0.7%)	4/549 (0.7%)	45/676 (6.7%)	118/12 214 (1.0%)
Participant's gender (M/F)	3/15	15/31	3/2	1/3	24/21	46/72
Participant age $\leq$ 50 years old/Total physician number age $\leq$ 50 years old [a]	15/2390	34/1299	5/271	3/234	30/233	87/4427
Participant age >51 years old/Total physician number age > 51 years old [a]	3/5061	12/1548	0/420	1/315	15/443	31/7787
Regional Health Administration of Portugal						
-North	2	13	1	0	8	24
-Center	0	7	1	0	12	20
-Lisbon and Tejo Valey	16	19	3	4	22	64
-Alentejo	0	3	0	0	1	4
-Algarve	0	4	0	0	2	6
-Madeira	0	0	0	0	0	0
-Azores	0	0	0	0	0	0
Main job—Public sector	17	43	4	4	29	97
Main job—Private sector	1	3	1	0	16	21

[a] According to Portuguese General Medical Council ("Ordem dos Médicos de Portugal")—year 2019 [47].

There was no duplicate response found with the same IP address. A relevant proportion of the respondent activity was directed to provide care to elderly people in the context studied. About the elderly people observed by physicians, 19% of the participants said that their monthly appointment time was more than 50% occupied with elderly patients with balance disorders, while 9% of physicians had their monthly appointment time more than 50% occupied with elderly patients with complaints related to consequent falls. A total of 86% of the physicians recognized the relevance of data about the previous provision of health care to the elderly with balance disorders and risk of falling. However, A total of 43% of all physicians responded that they need to access data from previous care consultations for elderly patients with balance disorder and risk of falling in more than half of cases. The majority of the participants (84%) had access to this information through the hospital or health center electronic medical record. Most respondents (60%) reported that more than half of the medical consultation time had been spent on IST-related activities. Moreover, 50% of participants were dissatisfied or very dissatisfied with the use of IST (e.g., usefulness, quality) in the context of balance disorders and the risk of falling in the elderly.

3.1.1. Socio-Demographic Participant Data

Most of the participants were female, accounting for 72 responses (61%). Younger physicians adhered more to the study: most participants (74%) had 50 years old or less. Although 64% of the eligible physicians were over 50 years old (7787 out of 12,214), the participants over 50 years old represented only 26% (31 out of 118) of the responses.

About 39% of the participants were specialists in Internal Medicine, 38% in Otolaryngology, 15% were family physicians, 4% PMR physicians, and 3% were neurologists. Only 4% were enrolled in the College of Competence in Geriatrics. Comparing the numbers, the family physicians had weak participation (18 out of 7451 family physicians), although being the specialty most represented among the eligible physicians.

Most physicians (82%) had the main job in public healthcare units, and 54% were from the larger Portuguese health region, the Lisbon and Tejo Valey (LTV) Regional Health Administration. (Table 1)

3.1.2. Difficulties and Medical Satisfaction Level Related to Current Clinical Data in the EMR (Electronic Medical Registration)

61% of all respondents were dissatisfied or very dissatisfied with the time spent accessing clinical data in the EMR, rising to 65% when considering only professionals who have the main job in a public healthcare institution. Regarding the availability of sufficient and understandable clinical data in the EMR, 55% of the physicians revealed dissatisfaction or a lot of dissatisfaction, with values of 59% for public health professionals as their main job. Again, 61% of all participants also expressed dissatisfaction or great dissatisfaction with time spent to fill in new data in the EMR, reaching 64% among professionals with the main job in a public health institution (Table 2).

Table 2. Questionnaire: Satisfaction degree with clinical data in the EMR—Context of health care provision for elderly with balance disorders and risk of falling.

Satisfaction Degree/Specialty	Family Physician	Internal Medicine Physician	PMR Physician	Neurologist	Otolaryngologist	Total
Time to data access (public and private main job)						
-S	1	12	3	1	25	42 (36%)
-D	17	34	1	3	17	72 (61%)
-Others	0	0	1	0	3	4 (3%)
TOTAL	18	46	5	4	45	118 (100%)
Time to data access (public main job)						
-S	1	11	3	1	15	31 (32%)
-D	16	32	0	3	12	63 (65%)
-Others	0	0	1	0	2	3 (3%)
TOTAL	17	43	4	4	29	97 (100%)
Sufficient/understandable data (public and private main job)						
-S	3	18	3	1	25	50 (42%)
-D	15	27	2	3	18	65 (55%)
-Others	0	1	0	0	2	3 (3%)
TOTAL	18	46	5	4	45	118 (100%)
Sufficient/understandable data (public main job)						
-S	3	16	3	1	15	38 (39%)
-D	14	26	1	3	13	57 (59%)
-Others	0	1	0	0	1	2 (2%)
TOTAL	17	43	4	4	29	97 (100%)
Time to fill data (public and private main job)						
-S	2	13	2	0	22	39 (33%)
-D	16	29	3	4	20	72 (61%)
-Others	0	4	0	0	3	7 (6%)
TOTAL	18	46	5	4	45	118 (100%)

Table 2. *Cont.*

Satisfaction Degree/Specialty	Family Physician	Internal Medicine Physician	PMR Physician	Neurologist	Otolaryngologist	Total
Time to fill data (public main job)						
-S	2	12	2	0	13	29 (30%)
-D	15	27	2	4	14	62 (64%)
-Others	0	4	0	0	2	6 (6%)
TOTAL	17	43	4	4	29	97 (100%)

S: Satisfied or very satisfied. D: Dissatisfied or very dissatisfied. Others: Did not use or Did not answer or Did not know.

3.1.3. Relevance of the Use of eHealth

The possibility of using eHealth for elderly patients with balance disorder and risk of falling was considered pertinent or very pertinent by 76% of all physicians and also by professionals with public healthcare as the main job. Regarding the medical specialties with more than 30 responses, 72% (33 out of 46) and 82% (37 out of 45) of internal medicine physicians and otolaryngologists, respectively, considered remote services as pertinent or very pertinent. If we consider only the participants of Internal Medicine and Otolaryngology working in the public sector as their main job, the percentages remain at 72% (31 out of 43) and rise to 86% (25 out of 29), respectively (Table 3).

Table 3. Questionnaire: Relevance degree about the use of eHealth in the context of care provision for the elderly with balance disorders and the risk of falling.

Relevance of eHealth/Specialty	Family Physician	Internal Medicine Physician	PMR Physician	Neurologist	Otolaryngologist	Total
Public and private main job						
-Pertinent	13	33	5	2	37	90 (76%)
-No pertinent	2	3	0	1	3	9 (8%)
-Indifferent	2	5	0	0	3	10 (8%)
-Others	1	5	0	1	2	9 (8%)
TOTAL	18	46	5	4	45	118 (100%)
Public main job						
-Pertinent	12	31	4	2	25	74 (77%)
-No pertinent	2	3	0	1	1	7 (7%)
-Indifferent	2	4	0	0	2	8 (8%)
-Others	1	5	0	1	1	8 (8%)
TOTAL	17	43	4	4	29	97 (100%)

Others: Did not answer or Did not know.

3.2. Second phase: Qualitative Research

The same interviewer conducted a total of seven face-to-face semi-structured interviews and one semi-structured interview by mobile phone due to coronavirus pandemic limitations. This interview phase was limited to senior physicians who provided healthcare to the elderly, with different training in technology and medical experience.

3.2.1. Socio-Demographic Participant Data

Five male and three female physicians, aged 47–66 years old, participated in the study. Two were family physicians, two internal medicine physicians, one neurologist and three otolaryngologists. All of them were either graduated or senior consultants. Six physicians were coordinators in their public health units, and two were coordinators of private otoneurology units. One physician was from the Regional Health Administration of the center (Center) of Portugal, and the others were from Lisbon and Tejo Valey (LTV) (Table 4).

Table 4. Study qualitative: Socio-demographic data of the participants and interview features.

Participant	Gender	Age	Specialty	Regional Health Administration of Portugal	Main Job	Interview	Audio Recording
1	M	59	Otolaryngology	LTV	Public sector	Face-to-face	Y
2	M	55	Neurology	LTV	Public sector	Face-to-face	Y
3	M	53	Internal Medicine	LTV	Public sector	Face-to-face	Y
4	M	59	Internal Medicine	LTV	Public sector	Face-to-face	Y
5	M	47	Family Medicine	LTV	Public sector	Face-to-face	Y
6	F	66	Family Medicine	Center	Public sector	Face-to-face	Y
7	F	49	Otolaryngology	LTV	Private sector	Face-to-face	Y
8	F	55	Otolaryngology	LTV	Private sector	Mobile phone	Y

M: Male; F: Female; LTV: Lisbon and Tejo Valey.

3.2.2. Content Analysis

As mentioned previously, four primary thematic categories were discussed. Twelve subthemes emerged from data analysis (Table 5).

Table 5. Thematic categories of the qualitative research.

Thematic Categories
1. Current clinical data in the context of health provision for elderly with balance disorders and risk of falling: understanding of the medical dissatisfaction identified in the quantitative research
1.1.   Availability 1.2.   Barriers
2. Interventions to improve the clinical data
2.1.   Interoperability of computer health systems 2.2.   New work organization
3. eHealth contribution in this context: understanding of the relevance observed in the quantitative research
3.1.   eHealth benefits
4. Strategies to improve eHealth services for a more effective healthcare provision
4.1.   Clinical and interactive data 4.2.   Audiovisual technology 4.3.   eHealth management 4.4.   Security of eHealth use 4.5.   Motivation and training of patient 4.6.   Caregiver involvement 4.7.   Medical training

The interviewees pointed out some misuse of healthcare provision by the elderly people in Portugal, meaning using above what is necessary of emergency visits, drug therapies and complementary diagnostic tests: *The elderly Portuguese population has no specific education on how to access healthcare services properly* (Participant 5). They also agreed on the need to access complete data: ... *the elderly people often represent complex patients ... the intervention ... requires multiple specialties ...* (Participant 1).

Relative to the medical dissatisfaction with available clinical data identified in the quantitative research, all interviewees highlighted the lack of a comprehensive data set and the lack of interoperability of computer systems: *We* (Physicians) *get to know more or less the drugs that are prescribed ... We don't know more ...* (Participant 4); *I have asked them* (family physicians) *to send me information. So, I can get a sense of what is going on with the patient.* (Participant 8); *... the data records are, sometimes, incomplete, they are not very explicit* (Participant 1); *The computer systems ... have great incompatibilities with each other. ... because the operating systems are different, or because the internet browser is different.* (Participant 5).

Some interventions to improve clinical data were pointed out: more investments in the interoperability of health information systems and in the organization of work with time for remote interaction and consultation: *Medicine will have to be a Medicine of shared information.* (Participant 1); *We should have some time allocated for this* (remote consultations). *So that we can keep our head on it and we will be really effective.* (Participant 4). Regarding the use of eHealth services for elderly people with balance disorders and the risk of falling, only Interviewee 2 questioned its proper applicability, justifying the doubts due to his lack of experience: ... *I don't know how this* (remote health care provision) *is done at a distance ... I don't even see myself doing a thing like that ... I think the physician-patient relationship is something that is impossible to be computerized.* (Participant 2). The other physicians, similar to most of the participants of the quantitative research, considered the eHealth contribution to be beneficial: ... *it* (eHealth) *could be a great help because vertigo has many decompensations ...* (and) *they* (patients) *are afraid of being ... without connection to the physician.* (Participant 7).

The interviewees pointed out the potential benefits of eHealth as a complementary channel to healthcare: rational use of resources with lower pressure on hospital resources, more healthcare access, better communication between medical specialties, closer physician–patient relation and more participation of patient and caregiver at home. However, they only agreed with the eHealth use for a follow-up consultation. The age was not considered a limitation for eHealth use. About the improvement of the use of eHealth services, the interviewees mentioned the need for more discussions to address the essential parameters for remote interaction and the need for involving eHealth system managers and programmers: ... *it is crucial the collaboration between the technology and those responsible for the technology ...* (Participant 6). In addition, the need for availability of clinical and interactive data and the motivation for human involvement were mentioned.

As essential strategies, the participants considered the inclusion of medications in use, the analyzes and imaging tests, and the registration of activities of daily living [48]. The use of questionnaires, calendars and graphics of trends on the occurrence of balance disorders and falls, supported by physician and patient's records, was also considered as a closer way of managing the disease: ... *simple questions like "Have you had a fall last year?", "Was there any injury? Yes, are you afraid of falling due to this injury?* (Participant 5); ... *interactive questionnaires ... , for example, in the recurrent vertigo ... to have documented how many episodes ... what kind of triggers ...* (Participant 7). Warning messages for adverse effects of medications or falls were classified as beneficial. The availability of individual balance exercises with a checklist and the possibility of uploading patient videos for clinical follow-up were other issues discussed: ... *the patient could record what they are feeling, for example, eye movements and then they uploaded the recording ... We* (physicians) *could include some exercises on the platform ...*

The interviewees mentioned the relevance of camera and audio for monitoring of balance rehabilitation and closer interaction, especially in cases of gait assessment and depression: ... *a phone call is one thing. If there is a camera it may even allow you* (physician) *to see, for example, the patient's gait* ... (Participant 3); ... *to monitor through videos, through cameras, as long as the patient gives his consent, of course* ... (Participant 7).

The eHealth service management by a physician was considered essential: *When there are changes* (in the health), *the physician can also be consulted.* (Participant 3); ... *always a physician.* (Participant 7).

In addition, the security of using eHealth was discussed: ... what type of password one (physician) should use ... encrypted ... addressed to the clinical team with security code (Participant 5).

The participants also referred to the relevance of active motivation and involvement of patient and caregiver in disease prevention and control: ... *Patients cannot continue to think that the responsibility of their health belongs to the physician* ... *the patients have to be involved and responsible for their health* ... (Participant 7); ... *we often think that our elderly people do not have the ability to manage new technologies* ... *but we can have a caregiver who can contact us remotely* ... (Participant 5) Finally, the investment in medical awareness and eHealth training was highlighted, allowing better physician involvement: ... *this is a work to be done in medical education* ... *after some time of implementation* (of medical education), *I am convinced that it* (eHealth) *will be the future of medicine* ... (Participant 1).

4. Discussion

A mixed-methods study was performed to know how to develop an efficient eHealth service for the provision of care for elderly people with balance disorders and the risk of falling, i.e., the research problem. Our findings revealed negative experiences with EMR, contributing to a misuse of the health care system. Despite this, the highlighted relevance of eHealth in this matter is an incentive for the development of future solutions.

Unfortunately, we had a low response number in the quantitative research, as described in other studies [49,50]. This limited the comparison between the specialties and between public or private health provision groups. Despite this, we could confirm the presence of constraints regarding the data availability in Portugal. We observed medical dissatisfaction with the information systems in general. In total, 50% of the participants of our study were dissatisfied or very dissatisfied with the current usefulness and quality of IST in the context of balance disorders and the risk of falling in the elderly. The physicians were dissatisfied or very dissatisfied with available data in EMR and time spent to access and update clinical data. According to the interviews, the incomplete or not understandable information about medical consultations and the lack of integration of clinical data between the health units have contributed to the misuse of healthcare provisions by elderly people, with multiple consultations, repeated prescriptions, polypharmacy and increasing costs that could be minimized with an appropriate digital service. These limitations and consequent costs have been reported by other authors [4,12–15]. The participants pointed out the need for investments in the interoperability of health information systems and in the organization of work to overcome this situation, as previously proposed in other studies [19–21]. In fact, data are essential for healthcare provision, monitoring of population health status and decision making. To reach real-time universal data in healthcare, interoperability issues should be addressed.

As in previous studies [2,5,16], our findings also confirmed the relevance of the use of eHealth services. However, the current way of working and interacting with patients should be restructured, including dedicated time to interact digitally with patients. eHealth can be leveraged as a complementary method to provide healthcare services, including preventive monitoring of falls and telerehabilitation with evaluation and monitoring of balance diseases and falls. For the interviewed participants, the remote consultation or management should be only for follow-up consultations, and it cannot fully replace the face-to-face clinical evaluations. As mentioned by Catan et al. [34], face-to-face consultations reduce anxiety whenever people need a physician. Only one interviewed revealed skepticism about digital solutions due to the lack of eHealth clinical experience. Several studies have already revealed the influence of limited knowledge about telemedicine on the perception of the potential of eHealth [19,35].

Finally, we confirmed the need for improvement of eHealth services for a more effective healthcare provision. The qualitative research allowed exploring interventions to achieve an efficient eHealth service to support healthcare for elderly people with balance disorders and risk of falling. Several suggestions were pointed out: the inclusion of complete clinical data, the possibility for interactive communication, message alerts and remote availability of balance exercises. Camera and audio were considered essential elements for closer interaction, allowing remote viewing of the gait, as well as the balance exercises performed. All the parameters should be aligned between technology experts and physicians to design suitable technological services. eHealth services and devices should be user-friendly and suitable for both the health professional and for the patient.

For the participants, the physician emerged as the main manager of eHealth service, but not necessarily the only one [51]. As in other studies [35,37], physicians also highlighted the need for investment for confidentiality and security of data. This should always be ensured. Relatively to human resources, strategies to motivate, educate and train the elderly patient and caregiver were also discussed. Self-care and self-management of health and disease (e.g., promotion of health and prevention of disease) should be further encouraged [2,5,16]. The need for health professional awareness and training to use all of the potential of digital solutions were mentioned, including the investment in professional health education. Thus, the potential of digital health could be widely used with motivated and trained human resources [2,20]. The Portuguese health system should be adjusted to tackle aging demand, overcoming the constraints of the EHR and the lack of interoperability of the information systems. The implementation of a universal digital health coverage system supported by comprehensive digital tools with camera and audio resources can better contribute to active and healthy aging [52] with more efficient management of health care for elderly with balance disorder and risk of falling. The design and the development of a balance disorder-related remote service, with the recommended functionalities, is an opportunity worth to be explored. The strategies identified and discussed in this study will be fed into a Design Science [50] process to design and implement a future eHealth service for a more effective provision of healthcare for elderly with balance disorders at a distance.

Limitations

Regarding the participants of the web-based questionnaire, we should acknowledge possible selection biases [53] of the quantitative research. Despite the intention to recruit physicians of different specialties, a small participant size was observed as in other online studies. This made it difficult to compare the results between specialties and to know if there is a difference between the public or private health provision groups. Another consideration, already highlighted in other web-based surveys, is the age of the respondents. Most participants of this research were younger physicians who seem to have more technological resources and online interests and to be more receptive to web-based questionnaires. In addition, the higher percentage of responses from otolaryngologists can be explained by their focus of interest in inner ear diseases that can promote balance disorders. The family physicians had weak participation (only 18 out of 7451), although this specialty represented most of the eligible participants. Due to the reduced or null number of responses from geographically more remote areas (e.g., Azores), we could not analyze and, in the second phase of the mixed methods study, explore more deeply the data of these participants that could most benefit from the potential of eHealth.

Additionally, the multiple-choice questions of the questionnaire allowed an easier analysis, but this approach did not allow the inclusion and discussion of supplementary opinions of the respondents.

Relatively to the qualitative research, the sampling was intentional. The last interview was conducted via mobile phone due to the limitation of the coronavirus pandemic.

The population targeted in both studies was limited to physicians with the provision of outpatient health care to the elderly in Portugal.

All these facts limited the generalizability of the findings.

5. Conclusions

Despite significant obstacles in existing digital solutions, 76% of the Portuguese physicians included in this study considered future e-Health services as highly relevant for complementary healthcare for elderly people with balance disorders and the risk of falling. The use of eHealth services comprised of digital technologies such as cameras, sensors and audio monitoring may reinforce such solutions. Additionally, these services may represent considerable potential for reducing the excess of emergency visits, and the overlap of drug therapies and diagnostic procedures and improved treatment. This may increase both health care efficiency and quality and contribute to relieving pressure in the escalating health care costs. However, significant constraints regarding the current availability of clinical data in EHR care systems were described. The insufficient quality of both available data in EMR and in the time needed to access such data and to register new clinical data was stated. We would like to highlight that our study group's description of incomplete, or not understandable, information about medical consultations and the lack of integration of clinical data indicate serious challenges to overcome. More research about this topic is also required to further enhance the knowledge about the use of digital tools in this field of health care.

Author Contributions: Conceptualization, A.G.M.G. and L.V.L.; methodology, A.G.M.G. and L.V.L.; formal analysis, A.G.M.G. and L.V.L.; investigation, A.G.M.G. and L.V.L.; data curation, A.G.M.G.; writing—original draft preparation, A.G.M.G., P.E. and L.V.L.; writing—review and editing, A.G.M.G., P.E. and L.V.L.; funding acquisition, L.V.L. All authors have read and agreed to the published version of the manuscript.

Funding: This work was partially supported by Fundação para a Ciência e a Tecnologia (FCT) for funds to Global Health and Tropical Medicine (GHTM) (UID/04413/2020 to LVL).

Institutional Review Board Statement: The study was conducted according to the guidelines of the Declaration of Helsinki and approved by the Scientific Council and the Ethics Council of Instituto de Higiene e Medicina Tropical of Universidade NOVA de Lisboa, Portugal (date of approval: 21 September 2018).

Informed Consent Statement: Informed consent was obtained from all subjects involved in the qualitative study.

Data Availability Statement: The data that support the findings of this study are available on request from the corresponding author, AGMG.

Acknowledgments: Portuguese General Medical Council (Ordem dos Médicos), for the approval and availability of the link on the website of the Institution; Portuguese Societies and Associations who contributed to the dissemination of the link with the partners.

Conflicts of Interest: The authors declare no conflict of interest. The funders had no role in the design of the study; in the collection, analyses, or interpretation of data; in the writing of the manuscript, or in the decision to publish the results.

Appendix A

Table A1. Quantitative study: answer choices of the questionnaire.

Number Question	Subject	Answer Choices
Q1	Sex	Male Female
Q2	Age	= or <30 31–40 41–50 51–60 = or >61
Q3	Specialty	Physical Medicine and Rehabilitation General and Family Medicine Internal Medicine Otorhinolaryngology Neurology
Q4	Competence in Geriatrics	Yes No
Q5	Main job	Personalized public health care unit Family health unit Public hospital Hospital in public-private partnership University hospital Private health unit
Q6	Regional Health Administration	North Center Lisbon and Tejo Valey Alentejo Algarve Madeira Azores
Q7	Monthly frequency of health care provision for elderly with balance disorders	= or <25% 26 a 50% 51 a 75% =or >76% Do not answer/Do not know
Q8	Monthly frequency of health care provision for elderly with consequent falls	= or <25% 26 a 50% 51 a 75% = or >76% Do not answer/Do not know
Q9	Need to access data from previous care consultations of elderly people with balance disorders and risk of falling	= or <25% 26 a 50% 51 a 75% = or >76% Do not answer/Do not know
Q10	Access to data from previous care consultations for elderly people with balance disorders and risk of falling	Clinical paper process Electronic medical record (EMR) Paper information provided by the patient Do not answer/Do not know

Table A1. *Cont.*

Number Question	Subject	Answer Choices
Q11	Relevance of data about previous health care to the elderly with balance disorders and risk of falling for a new provision of healthcare in this context	Never Rarely Sometimes Often Always Do not answer/Do not know
Q12	Estimated time spent on Information Systems and Technologies (IST)-related activities	= or >76% 51 a 75% 26 a 50% = or <25% Do not use Do not answer/Do not know
Q13	General usefulness of clinical data in the EMR	Excellent Very good Good Bad Very bad Do not use Do not answer/Do not know
Q14	Satisfaction with time spent to access clinical data, in the EMR, from previous care consultations for elderly people with balance disorders and risk of falling	Very dissatisfied Dissatisfied Satisfied Very Satisfied Do not use Do not answer/Do not know
Q15	Satisfaction with availability of sufficient and understandable clinical data, in the EMR, from previous care consultations for elderly people with balance disorders and risk of falling	Very dissatisfied Dissatisfied Satisfied Very Satisfied Do not use Do not answer/Do not know
Q16	Satisfaction with time spent to fill-in new data, in the EMR, related to the provision of health care to the elderly with balance disorders and risk of falling	Very dissatisfied Dissatisfied Satisfied Very Satisfied Do not use Do not answer/Do not know
Q17	General satisfaction with the use of IST (usefulness, quality) in the context of elderly with balance disorders and risk of falling	Very dissatisfied Dissatisfied Satisfied Very Satisfied Do not use Do not answer/Do not know
Q18	Relevance of eHealth in the context of elderly with balance disorders and risk of falling	Very relevant Relevant No difference Irrelevant Very irrelevant Do not answer/Do not know

Appendix B

Table A2. Qualitative study: Interview guide.

Thematic Categories	Questions
Current clinical data in Portugal	"What are the medical difficulties related with current clinical data in the context of health care provision for elderly with balance disorders and risk of falling?"
Interventions to improve the clinical data	What strategies can be implemented to improve clinical data?
eHealth contribution	"What do you think about the contribution of eHealth?"
Strategies to improve the use of eHealth services	How can eHealth services be suitable? What are the necessary strategies? What difficulties must be overcome?

References

1. European Innovation Partnership on Active and Healthy Ageing (EIP-AHA). Action Plan on Specific Action on Innovation in Support of Personalized Health Management, Starting with a Falls Prevention Initiative. 2013. Available online: https://ec.europa.eu/eip/ageing/library/action-plan-specific-action-innovation-support-personalized-health-management-starting-falls_en.html (accessed on 15 January 2020).
2. The International Society for Quality in Health Care (ISQua). Health Systems and Their Sustainability: Dealing with the Impending Pressures of Ageing, Chronic and Complex Conditions, Technology and Resource Constraints. Whitepaper. 2016. Available online: https://isqua.org/resources-blog/resources?page=1&search=Health%20Systems%20and%20their%20Sustainability&date_range_start=&date_range_end= (accessed on 15 January 2020).
3. National Institute for Health and Care Excellence (NICE). Multimorbidity: Clinical Assessment and Management. 2016. Available online: https://www.nice.org.uk/guidance/ng56 (accessed on 15 January 2020).
4. Simões, J.; Augusto, G.F.; Fronteira, I.; Hernández-Quevedo, C. Portugal: Health system review. *Health Syst. Transit.* **2017**, *19*, 1–184. Available online: http://www.healthobservatory.eu (accessed on 17 January 2020).
5. Amalberti, R.; Vincent, C.; Nicklin, W.; Braithwaite, J. Coping with more people with more illness. Part 1: The nature of the challenge and the implications for safety and quality. *Int. J. Qual. Health Care* **2019**, *31*, 154–158. [CrossRef] [PubMed]
6. OECD. *Health at a Glance 2019: OECD Indicators*; OECD Publishing: Paris, France, 2019. [CrossRef]
7. European Union. Ageing Europe—Looking at the Lives of Older People in the EU. 2020. Available online: https://ec.europa.eu/eurostat (accessed on 26 November 2020).
8. Salzman, B. Gait and balance disorders in older adults. *Am. FAM Physician* **2010**, *82*, 61–68. [PubMed]
9. Health Evidence Network (HEN)—WHO. What are the Main Risk Factors for Falls Amongst Older People and What are the Most Effective Interventions to Prevent These Falls? 2004. Available online: http://www.euro.who.int/document/E82552.pdf (accessed on 20 October 2020).
10. Ha, V.A.T.; Nguyen, T.N.; Nguyen, T.X.; Nguyen, H.T.T.; Nguyen, T.T.H.; Nguyen, A.T.; Pham, T.; Thanh Vu, H.T. Prevalence and Factors Associated with Falls among Older Outpatients. *Int. J. Environ. Res. Public Health* **2021**, *18*, 4041. [CrossRef]
11. Haagsma, J.A.; Olij, B.F.; Majdan, M.; van Beeck, E.F.; Vos, T.; Castle, C.D.; Dingels, Z.V.; Fox, J.T.; Hamilton, E.B.; Liu, Z.; et al. Falls in older aged adults in 22 European countries: Incidence, mortality and burden of disease from 1990 to 2017. *INJ Prev.* **2020**, *26*, i67–i74. [CrossRef] [PubMed]
12. Kerber, K.A. Vertigo and Dizziness in the Emergency Department. *Emerg. Med. Clin. N. Am.* **2009**, *27*, 39–50. [CrossRef]
13. Heinrich, S.; Rapp, K.; Rissmann, U.; Becker, C.; König, H.H. Cost of falls in old age: A systematic review. *Osteoporos. Int.* **2010**, *21*, 891–902. [CrossRef]
14. Tehrani, A.S.S.; Coughlan, D.; Hsieh, Y.H.; Mantokoudis, G.; Korley, F.K.; Kerber, K.A.; Frick, K.D.; Newman-Toker, D.E. Rising Annual Costs of Dizziness Presentations to U.S. Emergency Departments. *Acad. Emerg. Med.* **2013**, *20*, 689–696. [CrossRef] [PubMed]
15. Reis, L.; Lameiras, R.; Cavilhas, P.; Escada, P. Epidemiology of Vertigo on Hospital Emergency. *Acta Med. Port.* **2016**, *29*, 326–331. [CrossRef]
16. Nolte, E.; Merkur, S.; Anell, A. *Achieving Person-Centred Health Systems*; Cambridge University Press: Cambridge, UK, 2020; pp. 1–396.
17. Winter, A.; Takabayashi, K.; Jahn, F.; Kimura, E.; Engelbrecht, R.; Haux, R.; Honda, M.; Hübner, U.H.; Inoue, S.; Kohl, C.D.; et al. Quality Requirements for Electronic Health Record Systems. A Japanese-German Information Management Perspective. *Methods Inf. Med.* **2017**, *56*, e92–e104.
18. Ammenwerth, E.; Duftschmid, G.; Al-Hamdan, Z.; Bawadi, H.; Cheung, N.T.; Cho, K.H.; Goldfarb, G.; Gülkesen, K.H.; Harel, N.; Kimura, M.; et al. International Comparison of Six Basic eHealth Indicators Across 14 Countries: An eHealth Benchmarking Study. *Methods Inf. Med.* **2020**, *59*, e46–e63. [PubMed]

19. Lapão, L.V.; Dussault, G. The contribution of eHealth and mHealth to improving the performance of the health workforce: A review. *Public Health Panorama.* **2017**, *3*, 463–471.
20. Blandford, A. HCI for health and wellbeing: Challenges and opportunities. *Int J. Hum. Comput. Stud.* **2019**, *131*, 41–51. [CrossRef]
21. World Health Organization (WHO). mHealth Use of Appropriate Digital Technologies for Public Health—EB142/20. 2017. Available online: https://apps.who.int/gb/ebwha/pdf_files/EB142/B142_20-en.pdf (accessed on 15 January 2020).
22. Uei, S.L.; Kuo, Y.M.; Tsai, C.H.; Kuo, Y.L. An Exploration of Intent to Use Telehealth at Home for Patients with Chronic Diseases. *Int. J. Environ. Res. Public Health* **2017**, *14*, 1544. [CrossRef]
23. Sun, R.; Sosnoff, J.J. Novel sensing technology in fall risk assessment in older adults: A systematic review. *BMC Geriatr.* **2018**, *18*, 14. [CrossRef] [PubMed]
24. Rucco, R.; Sorriso, A.; Liparoti, M.; Ferraioli, G.; Sorrentino, P.; Ambrosanio, M.; Baselice, F. Type and Location of Wearable Sensors for Monitoring Falls during Static and Dynamic Tasks in Healthy Elderly: A Review. *Sensors* **2018**, *18*, 1613. [CrossRef] [PubMed]
25. Nguyen, H.; Mirza, F.; Naeem, M.A.; Baig, M.M. Falls management framework for supporting an independent lifestyle for older adults: A systematic review. *Aging Clin. Exp. Res.* **2018**, *30*, 1275–1286. [CrossRef]
26. Leirós-Rodríguez, R.; García-Soidán, J.L.; Romo-Pérez, V. Analyzing the Use of Accelerometers as a Method of Early Diagnosis of Alterations in Balance in Elderly People: A Systematic Review. *Sensors* **2019**, *19*, 3883. [CrossRef]
27. Montesinos, L.; Castaldo, R.; Pecchia, L. Wearable Inertial Sensors for Fall Risk Assessment and Prediction in Older Adults: A Systematic Review and Meta-Analysis. *IEEE Trans. Neural Syst. Rehabil. Eng.* **2018**, *26*, 573–582. [CrossRef]
28. Bet, P.; Castro, P.C.; Ponti, M.A. Fall detection and fall risk assessment in older person using wearable sensors: A systematic review. *Intern. J. Med. Inform.* **2019**, *130*, 103946. [CrossRef]
29. Gaspar, A.G.M.; Lapão, L.V. eHealth for Addressing Balance Disorders in the Elderly: Systematic Review. *J. Med. Internet Res.* **2021**, *23*, e22215. [CrossRef] [PubMed]
30. Skjæret, N.; Nawaz, A.; Morat, T.; Schoene, D.; Helbostad, J.L.; Vereijken, B. Exercise and rehabilitation delivered through exergames in older adults: An integrative review of technologies, safety and efficacy. *Int. J. Med. Inf.* **2016**, *85*, 1–16. [CrossRef] [PubMed]
31. Choi, S.D.; Guo, L.; Kang, D.; Xiong, S. Exergame technology and interactive interventions for elderly fall prevention: A systematic literature review. *Appl. Erg.* **2017**, *65*, 570–581. [CrossRef]
32. Eysenbach, G. What is e-health? *J. Med. Internet Res.* **2001**, *3*, E20. [CrossRef]
33. Poenaru, C.; Poenaru, E.; Vinereanu, D. Current Perception of Telemedicine in an EU Country. *Maedica* **2014**, *9*, 367–374.
34. Catan, G.; Espanha, R.; Mendes, R.V.; Toren, O.; Chinitz, D. Health information technology implementation—Impacts and policy considerations: A comparison between Israel and Portugal. *ISR J. Health Policy Res.* **2015**, *4*, 41. [CrossRef]
35. Ayatollahi, H.; Sarabi, F.Z.P.; Langarizadeh, M. Clinicians' Knowledge and Perception of Telemedicine Technology. *Perspect Health Inf. Manag.* **2015**, *12*, 1c.
36. Radhakrishnan, K.; Xie, B.; Berkley, A.; Kim, M. Barriers and Facilitators for Sustainability of Tele-Homecare Programs: A Systematic Review. *Health Serv. Res.* **2016**, *51*, 48–75. [CrossRef]
37. Albarraka, A.I.; Mohammedb, R.; Almarshoudc, N.; Almujalli, L.; Aljaeed, R.; Altuwaijiri, S.; Albohairy, T. Assessment of physician's knowledge, perception and willingness of telemedicine in Riyadh region, Saudi Arabia. *J. Infect. Public Health* **2021**, *14*, 97–102. [CrossRef]
38. Wynn, R.; Gabarron, E.; Johnsen, J.K.; Traver, V. Special Issue on E-Health Services. *Int. J. Environ. Res. Public Health* **2020**, *17*, 2885. [CrossRef] [PubMed]
39. Gil, H. The elderly and the digital inclusion: A brief reference to the initiatives of the European union and Portugal. *MOJ Gerontol. Ger.* **2019**, *4*, 213–221. [CrossRef]
40. SPMS, Portuguese National Centre of Telehealth (CNTS). National Strategic Telehealth Plan (PENTS) 2019–2022. Available online: https://www.spms.min-saude.pt/wp-content/uploads/2019/11/PENTS_Tradu%C3%A7%C3%A3o.pdf (accessed on 15 February 2021).
41. SPMS. SNS 24 Projeto Proximidade Senior. Available online: https://www.spms.min-saude.pt/pesquisa-geral/?_sf_s=idoso (accessed on 15 August 2020).
42. Creswell, J.W.; Creswell, J.D. Research design. In *Qualitative, Quantitative, and Mixed Methods Approaches*, 5th ed.; SAGE Publications Inc.: Los Angeles, CA, USA, 2018.
43. SurveyMonkey. Available online: https://help.surveymonkey.com/articles/en_US/kb/How-do-I-make-surveys-anonymous (accessed on 15 January 2019).
44. Eysenbach, G. Improving the quality of Web surveys: The Checklist for Reporting Results of Internet E-Surveys (CHERRIES). *J. Med. Internet Res.* **2004**, *6*, e34. [CrossRef]
45. IBM. SPSS Statistics. Available online: https://www.ibm.com/products/spss-statistics/details (accessed on 12 September 2019).
46. Assembleia da República. Lei nº 58/2019 de 8 de Agosto de 2019. Diário da República n.º 151/2019, Série I de 2019-08-08. 3-40. Available online: https://data.dre.pt/eli/lei/58/2019/08/08/p/dre (accessed on 22 February 2021).
47. Ordem dos Médicos de Portugal. Estatísticas 2019. Available online: https://ordemdosmedicos.pt/wp-content/uploads/2020/01/ESTATISTICAS_ESPECIALIDADES_2019.pdf (accessed on 17 July 2020).

48. WHO. Active Ageing, A Policy Framework. A Contribution of the WHO to the Second United Nations World Assembly on Ageing, Madrid, Spain, April, 2002. Available online: https://extranet.who.int/agefriendlyworld/wp-content/uploads/2014/06/WHO-Active-Ageing-Framework.pdf (accessed on 17 January 2020).
49. Aitken, C.; Power, R.; Dwyer, R. A very low response rate in an on-line survey of medical practitioners. *Aust. N. Z. J. Public Health* **2008**, *32*, 288–289. [CrossRef] [PubMed]
50. Lapão, L.V.; Da Silva, M.M.; Gregório, J. Implementing an online pharmaceutical service using design science research. *BMC Med. Inform. Decis. Mak.* **2017**, *17*, 1–4. [CrossRef] [PubMed]
51. Lapão, L.V.; Pisco, L. Primary health care reform in Portugal, 2005–2018: The future and challenges of coming of age. *Cad. Saude Publica* **2019**, *35* (Suppl. 2), e00042418.
52. Liotta, G.; Canhão, H.; Cenko, F.; Cutini, R.; Vellone, E.; Illario, M.; Kardas, P.; Poscia, A.; Sousa, R.D.; Palombi, L.; et al. Active Ageing in Europe: Adding Healthy Life to Years. *Front. Med.* **2018**, *5*. [CrossRef]
53. Eysenbach, G.; Wyatt, J. Using the Internet for surveys and health research. *J. Med. Internet Res.* **2002**, *4*, E13. [CrossRef] [PubMed]

International Journal of
Environmental Research and Public Health

MDPI

Article

Cross Sectional E-Health Evaluation Study for Telemedicine and M-Health Approaches in Monitoring COVID-19 Patients with Chronic Obstructive Pulmonary Disease (COPD)

Abdullah H. Alsharif

Department of Management Information Systems, College of Business Administration-Yanbu, Taibah University, Yanbu, Saudi Arabia; alsharifa@taibahu.edu.sa

Abstract: Monitoring COVID-19 patients with COPD has become one of the major tasks in preventing transmission and delivering emergency healthcare services after vaccination in case of any issues. Most COVID-19-affected patients are suggested to self-quarantine at home or in institutionalized quarantine centers. In such cases, it is essential to provide remote healthcare services. For remote healthcare monitoring, two approaches are being considered in this study, which include mHealth and Telehealth. A mixed-methods approach is adopted, where survey questionnaires are used for collecting information from 108 patients and semi-structured interviews are used with seven physicians regarding mHealth and Telehealth approaches. Survey results indicated that mHealth is rated to be slightly more effective than Telehealth, and interview results indicated that Telehealth is identified to be slightly more effective than mHealth in relation to parameters including usefulness, ease of use and learnability, interface and interaction quality, reliability, and satisfaction. However, both physicians and patients opined that both mHealth and Telehealth have a promising future with increasing adoption. Based on the findings, it can be concluded that both mHealth and Telehealth are considered to be effective in delivering remote care for COPD patients infected with COVID-19 at home. Implications of the study findings are discussed.

Keywords: mHealth; Telehealth; COPD; usefulness; evaluation; satisfaction

Citation: Alsharif, A.H. Cross Sectional E-Health Evaluation Study for Telemedicine and M-Health Approaches in Monitoring COVID-19 Patients with Chronic Obstructive Pulmonary Disease (COPD). *Int. J. Environ. Res. Public Health* **2021**, *18*, 8513. https://doi.org/10.3390/ijerph18168513

Academic Editors: Marie Carmen Valenza and Irene Torres-Sanchez

Received: 25 May 2021
Accepted: 14 July 2021
Published: 12 August 2021

Publisher's Note: MDPI stays neutral with regard to jurisdictional claims in published maps and institutional affiliations.

1. Introduction

COVID-19, since its identification in December 2019, has been affecting many people across the world, resulting in different waves of rising infections and deaths. As of 9 May 2021, there are 1.5 billion confirmed COVID-19 cases, including 3 million deaths reported globally [1]. The vast majority COVID-19 cases are mainly found in the West, including the Americas (63 million), followed by Europe (52 million). However, a steep rise in the number of cases was identified in the past 2 months in Southeast Asia (25 million), especially India (22 million) [1]. The number of deaths related to COVID-19 has been on the surge, with more than 570,000 reported deaths in the USA [2], 419,114 deaths in Brazil [3], and 242,362 deaths in India [4]. The rise in the number of COVID-19-related deaths were attributed to older age of patients and previous health complications such as diabetes, blood pressure, asthma and other critical health conditions which increase the risk of death among patients [5–9]. One such health condition wherein COVID-19 can have a serious impact on people is chronic obstructive pulmonary disease (COPD). It is identified that COVID-19 patients with COPD have a high risk of admission to an intensive care unit, mechanical ventilation, or death [10]. In addition, levels of angiotensin-converting enzyme 2 (ACE2), the reported host receptor of the virus responsible for COVID-19 (SARS-CoV-2), have been observed to be increased in patients with COPD [11]. Given the severity associated with COVID-19-infected COPD patients, there is increasing pressure on the health system to provide additional care. There is a need to maintain and reinforce follow-up and close management for these patients, with the aim of limiting collateral effects that

67

could be induced by non-optimal management of COPD during and after the pandemic [12]. Strategies such as increased vaccination can help in preventing the mortality rate of COVID-19 infected COPD patients. However, in spite of administering over 11 billion vaccine doses globally [1], there is still a considerable global population that needs access to vaccines, especially in low- and middle-income group countries [13,14]. However, despite vaccination and other implementation of preventive and mitigation strategies such as lockdowns, increasing vaccination, etc. the new cases continue to rise globally. The risk of drug shortage due to COVID-19 restrictions is another major factor that can affect the treatment of COVID-19-infected patients with COPD and other critical respiratory conditions such as asthma, diabetes, etc. [15]. Moreover, the complications after vaccination were identified to be mild to moderate, which include pain, fever, headache, diarrhea, etc. [16,17]. Recently, black fungal infection (mucormycosis) is on the rise in Southeast Asia, resulting in a rising number of deaths and blindness among the vaccinated [18,19]. Therefore, there is an increasing need to monitor COVID-19 patients after treatment and vaccinated people, which can significantly increase the burden on healthcare systems.

Considering these factors, the best available options for managing the increased burden of managing COVID-19 patients in home quarantine and vaccinated people with additional health conditions such as COPD is to adopt effective remote monitoring of the patients' conditions and adopt effective health information management techniques [20]. Effective information management techniques including real-time monitoring, data storing, transfer, retrieval, and update, are essential for improved clinical decision-making. Timely response to chronic conditions can help in preventing mortalities and provide quality care to the patients [21,22]. It has been identified that medically necessary, time-sensitive procedures can efficiently manage resources and improve clinical decision-making for treating COVID-19 patients [23]. Advances in technology have helped in developing various monitoring systems such as smart watches and diabetes monitors that can be integrated with mobile applications and transfer real-time data to hospital servers, improving the effectiveness of remote monitoring by efficiently managing the health information of the patients. eHealth, for instance, is an emerging field at the intersection of medical informatics, public health, and business, referring to the health services and information delivered or enhanced through the internet and related technologies [24]. mHealth and Telehealth are two major eHealth approaches that are being extensively used during the COVID-19 pandemic to deliver remote healthcare services. mHealth is a medical and public health practice supported by mobile devices, such as mobile phones, patient monitoring devices, personal digital assistants (PDAs), and other wireless devices [25], whereas telemedicine is the use of medical information exchanged from one site to another via electronic communications to improve a patient's clinical health status [26]. While Telehealth is mainly used for information exchange, mHealth can be used for both information exchange and remote monitoring and diagnosis. The difference between both approaches lies in the modes of information transmission. In mHealth, the information is transferred through mobile applications, while in Telehealth, information has to be communicated by the patients to the medical representative.

In relation to chronic diseases, such as COPD, associated with COVID-19, parameters such as oxygen levels, electrocardiograms (ECGs), carbon dioxide partial pressure (PaCO2), etc. can be monitored remotely (mHealth) using sensors and devices linked to the patients mobile, which transmits information to hospitals through the mobile applications [27]. The cost of care assistance for chronic diseases such as COPD are dramatically increasing, as a result of which remote healthcare management systems such as Telehealth can be used to reduce costs and increase healthcare efficiency [28]. Therefore, remote monitoring solutions such as mHealth and Telehealth can be considered as an alternative to traditional healthcare operations, which not only decreases the burden on the already strained healthcare sector due to COVID-19, but also improves healthcare information management and delivers quality care for patients. In this context, it was observed that mHealth technologies could help in mitigating the effects of the COVID-19 pandemic [29] by increasing the reach of local guidance for healthcare professionals in managing COVID-19 outbreak and treatment procedures [30], and in coordinating mHealth infrastructure for managing the COVID-19 pandemic [31]. Studies [32–35] have evaluated mHealth applications in relation to various parameters, including usefulness, ease of use, learnability, interface and interaction quality, reliability, satisfaction, and future use, etc., which reflected positive outcomes. Similarly, Telehealth approaches were also identified to be effective in relation to the above-mentioned parameters [36].

Both mHealth and Telehealth approaches have some advantages and drawbacks. For instance, patients at home may not regularly monitor their health condition and may not enter the health data into the mobile applications, which may lead to ineffective remote monitoring. However, the Telehealth approach may address this issue, as healthcare practitioners may call the patients and regularly collect the data and provide instant feedback. However, repeated calls from the healthcare professionals may create discomfort for the patients. However, for effective healthcare management, the need for accurate and daily health data is essential, and both mHealth and Telehealth can serve this purpose. However, there is a lack of research on comparing both approaches in remote monitoring of COPD patients infected with COVID-19 in home quarantine or vaccinated patients at home. Therefore, the purpose of this study is aimed to evaluate Telehealth and mHealth approaches in monitoring (remote monitoring of exercise tolerance, comorbidity, and smoking habits, oxygen levels, blood pressure, sugar levels, and other factors, as prescribed by the patients' respective hospitals) COVID-19 patients with COPD after the treatment and vaccination.

2. Materials and Methods

The purpose of this study is to compare and evaluate the mHealth and Telehealth approaches in monitoring COVID-19 patients with COPD after treatment and vaccination while at home. A mixed-methods approach employing both qualitative (semi-structured interviews) and quantitative (questionnaire-based survey) were adopted for collecting the data regarding mHealth and Telehealth approaches from physicians and patients, respectively, in Saudi Arabia.

2.1. Questionnaire Design

As discussed in the introduction section, there are various parameters used for evaluating both mHealth and Telehealth applications [32–36]. To cover different contexts of evaluation, Telehealth Usability Questionnaire (TUQ) is adopted in this study, including items related to various parameters (usefulness, ease of use, interface quality, interaction quality, reliability, satisfaction, and future use) from [36]. TUQ is preferred in this study because it covers wide range of parameters that are used individually in various studies [29–32], making a comprehensive list of items to be used in evaluating Telehealth applications. Furthermore, items from (MAUQ) [37], including ease of use, satisfaction, usefulness, are considered along with TUQ for developing questionnaires for mHealth and Telehealth evaluation, respectively, as shown in Appendix A. Both questionnaires (mHealth and Telehealth) have same set of questions with parameters including usefulness (three items), ease of use and learnability (three items), interface quality (three items), interaction quality (three items), reliability (three items), and satisfaction and future use (four items). Multiple-choice answers and five-point Likert scale ratings [38] were used for answering the questions by the participants. The questionnaire was initially designed in English (a copy of survey questionnaire is presented in Appendix A), which was then translated to Arabic using two professional Arabic translators. The Arabic version questionnaire was designed using QuestionPro application, and a survey link was generated for accessing the survey. A pilot study was conducted with 12 randomly selected patients from Saudi Arabia for evaluating the questionnaire. Based on the feedback from pilot study participants, few changes were made in relation to the questions' formulation and grammatical errors in Arabic. In addition, Cronbach's alpha for all items in in the questionnaire was identified to be greater than 0.81, revealing good consistency and reliability. In addition, the interview questionnaire included eight questions reflecting the interviewees' experiences with Telehealth and mHealth, ease of use, satisfaction and future use, usefulness, learnability, interface quality, interaction quality, and reliability (as shown in Appendix B). Thus, semi-structured interviews were adopted in order to evaluate both mHealth and Telehealth from the perspectives of healthcare practitioners, and a survey instrument was adopted to evaluate both approaches from perspectives of patients, reflecting the two main actors in remote monitoring in healthcare system.

2.2. Recruitment

COPD patients diagnosed with COVID-19 and vaccinated COPD patients were recruited for the survey using the survey link generated using QuestionPro application. The survey link was forwarded to the patients through emails and other social media platforms such as Facebook and WhatsApp. The survey was conducted for a period of 5 weeks from 15 April to 20 May 2021. Physicians were contacted through emails and over phone, requesting them to participate in the interviews. Interviews were scheduled from 16 April to 5 May 2021, and were conducted online. On average, each interview lasted for 35 min.

2.3. Sampling

Considering the purpose and objective of the study, which is to collect the data from a specific group of population (COPD patients diagnosed with COVID-19 and vaccinated COPD patients at home involved in remote monitoring), a purposive sampling approach was adopted [39]. Accordingly, a nonprobability sample was obtained based on the objective of the study, which mainly focused on analyzing the patients' and physicians' perceptions of mHealth and Telehealth. The survey link was initially forwarded 142 patients using various online channels. A total of 108 students participated in the survey, reflecting a response rate of 76.05%. While 19 physicians were contacted for semi-structured interviews, 7 accepted invitations and took part in the interviewees. Low response rate in interviews is due to the busy schedules of the physicians owing to the COVID-19 outbreak.

2.4. Data Analysis

The survey was developed using Google forms and conducted for a period of 5 weeks. Both survey and interview data are analyzed and discussed using nine themes, which included experiences, usefulness, ease of use, learnability, interface quality, interaction quality, reliability, satisfaction, and future use. In relation to survey data, relative frequencies for each item under these themes and statistical tests (t-tests) are used for analyzing the data, which are presented in the following section.

3. Results and Discussion

3.1. Survey Results

The final sample achieved for the study was 108. The demographic information of the participants is presented in Table 1. Among the total participants, 58.3% were male (63/108), and 41.7% were female (45/108). Considering the age groups, 43.5% were aged between 35 and 44 years (47/108), followed by 30.6% between 25 and 34 years (114/479), 12% between 18 and 24 years (13/108), 12% between 45 and 54 years (13/108), and 1.9% participants aged 55 or more than 55 years (2/108). Focusing on the education of the participants, 38.9% were bachelor's degree graduates (42/108), followed by 30.6% master's graduates (33/108), 15.7% high school graduates or diploma graduates (17/108), and 14.8% doctorates (16/108). Demographics of the participants reflected good participation levels by both genders. Moreover, the majority of the participants were aged between 25 and 44 years, reflecting the population who is better equipped with the skills of using health information technologies [40–42], and has good education levels.

Table 1. Frequency distribution of demographic variables.

Variables	*n* (%)
Gender	
Male	63 (58.3%)
Female	45 (41.7%)
Age	
18–24	13 (12%)
25–34	33 (30.6%)
35–44	47 (43.5%)
45–54	13 (12%)
55 and above	2 (1.9%)
Education	
High school graduate, diploma, or equivalent	17 (15.7%)
Bachelor's degree	42 (38.9%)
Master's degree	33 (30.6%)
Doctorate	16 (14.8%)

All the participants were having an experience of using both Telehealth and mHealth approaches. In relation to the experience of using Telehealth (via telephone/mobile), 65.7% of the participants had 2 or less than 2 years of experience (71/108), followed by 17.6% having 2 to 5 years of experience (19/108), 9.3% having 10 or more years of experience (10/108), and 7.4% having 5 to 10 years of experience (8/108). In relation to the experience of using mHealth (via mobile application/smart sensors), 63.9% of the participants had 2 or less than 2 years of experience (69/108), followed by 25.9% having 2 to 5 years of experience (28/108), 6.5% having 10 or more years of experience (7/108), and 3.7% having 5 to 10 years of experience (4/108). The experience levels of the participants in relation to mHealth and Telehealth reflected almost similar statistics, with the majority of them having 2 years or less and 2 to 5 years of experience.

Participants' opinions on the usefulness of both mHealth and Telehealth are presented in Table 2. Improved access to healthcare was identified to be a highly rated factor related to usefulness, followed by the time-saving factor, and approaches meeting the healthcare needs of the users. Both approaches were identified to be similarly rated by the participants in relation to the usefulness parameter.

Table 2. Usefulness of mHealth and Telehealth.

Items	mHealth		Telehealth	
	Mean	Std.Dev	Mean	Std.Dev
Access to healthcare services	3.75	1.10	3.67	1.08
Saves time	3.68	1.17	3.65	1.13
Meets healthcare needs	3.49	1.09	3.53	1.03

Furthermore, to identify the differences of opinions in relation to mHealth and Telehealth, a *t*-test was conducted, as shown in Table 3. The mean scores of mHealth (Mean = 3.64, SD = 1.12) and Telehealth (Mean = 3.61, SD = 1.08), identified in the analysis, reflected that participants found that both mHealth and Telehealth approaches to be effective in terms of usefulness. *t*-value, as shown in Table 3, was found to be (t = 0.2004) at 0.05 confidence interval, and was identified as not statistically significant ($p > 0.05$). Therefore, no significant differences of opinions in relation to the usefulness of mHealth and Telehealth can be observed. The findings are similar to [32], reflecting the usefulness of both approaches. As both approaches are aimed at improving access to healthcare and save time in accessing healthcare needs, it is possible that both approaches are identified to be useful by the participants.

Table 3. Difference in usefulness of mHealth and Telehealth.

Variable	Approach	*n*	Mean	Std.Dev	*df*	*t*-Value	*p*-Value
Usefulness	mHealth	108	3.64	1.12	214	*0.2004*	*0.8414*
	Telehealth	108	3.61	1.08			

Participants' opinions on the ease of use and learnability parameters are presented in Table 4. It can be observed that simple to use and easy to learn factors of mHealth are slightly greater than that of Telehealth; no major differences were identified in relation to the ability of the approach for enhancing productivity. While the mHealth application is mobile-based, and easy to use, Telehealth completely relies on calls in providing care and monitoring health information. Therefore, slight differences in terms of ease of use and learnability factors can be expected.

Table 4. Ease of use and learnability of mHealth and Telehealth.

Items	mHealth		Telehealth	
	Mean	Std.Dev	Mean	Std.Dev
Simple to use	3.70	1.09	3.62	1.15
Easy to learn	3.75	1.04	3.73	1.11
Productivity	3.72	1.13	3.73	1.04

Furthermore, to identify the differences of opinions in relation to mHealth and Telehealth, a *t*-test was conducted, as shown in Table 5. The mean scores of mHealth (Mean = 3.72, SD = 1.12) and Telehealth (Mean = 3.69, SD = 1.1), identified in the analysis, reflected that participants found that both mHealth and Telehealth approaches to be effective in terms of ease of use and learnability. *t*-value, as shown in Table 5, was found to be ($t = 0.1986$) at 0.05 confidence interval, and was identified as not statistically significant ($p > 0.05$). Therefore, no significant differences of opinions in relation to ease of use and learnability parameters of mHealth and Telehealth can be observed. These findings can be related to [33], in which ease of use and learnability were rated to be effective by the majority of the participants.

Table 5. Difference in ease of use and learnability of mHealth and Telehealth.

Variable	Approach	*n*	Mean	Std.Dev	*df*	*t*-Value	*p*-Value
Ease of use and Learnability	mHealth	108	3.72	1.12	214	*0.1986*	*0.8428*
	Telehealth	108	3.69	1.1			

While mHealth applications have a mobile interface, Telehealth does not have any physical interface, but the quality of the interface can be identified from the communication between patients and healthcare practitioners over telephone or mobiles. In relation to the interface quality, both mHealth and Telehealth approaches were rated slightly above average (Mean = 2.5), as shown in Table 6. It is interesting to note that only 50.9% of the participants either strongly agreed or agreed that the Telehealth approach meets their healthcare needs, and only 52.7% of the participants either strongly agreed or agreed that the mHealth approach meets their healthcare needs, reflecting that there is a considerable number of participants who are not happy with interface quality in both approaches.

Table 6. Interface quality of mHealth and Telehealth.

Items	mHealth		Telehealth	
	Mean	**Std.Dev**	**Mean**	**Std.Dev**
Pleasant	3.5	1.04	3.48	1.07
Simple and easy to understand	3.55	1.03	3.61	1.06
Ability to do according to users' needs	3.22	1.10	3.20	1.05

Furthermore, to identify the differences of opinions in relation to mHealth and Telehealth, a *t*-test was conducted, as shown in Table 7. The mean scores of mHealth (Mean = 3.42, SD = 1.05) and Telehealth (Mean = 3.43, SD = 1.06), identified in the analysis, reflected that participants found that both mHealth and Telehealth approaches to be effective in terms of interface quality. *t*-value, as shown in Table 7, was found to be (t = 0.0697) at 0.05 confidence interval, and was identified as not statistically significant (p > 0.05). Therefore, no significant differences of opinions in relation to interface quality parameters of mHealth and Telehealth can be observed. These findings related to mHealth are similar to [34], indicating good interface quality of mHealth applications.

Table 7. Difference in interface quality of mHealth and Telehealth.

Variable	Approach	*n*	Mean	Std.Dev	*df*	*t*-Value	*p*-Value
Usefulness	mHealth	108	3.42	1.05	214	*0.0697*	*0.9445*
	Telehealth	108	3.43	1.06			

In relation to interaction quality, no significant differences were identified between mHealth and Telehealth, as identified from Table 8. In relation to the ability of the approaches reflecting in similar to personal interactions, less than 50% of the participants reflected the opinion that these approaches are similar to personal interactions. Moreover, more than 50% of the participants stated that they are not able to express their opinions effectively on mHealth and Telehealth applications.

Table 8. Interaction quality of mHealth and Telehealth.

Items	mHealth		Telehealth	
	Mean	**Std.Dev**	**Mean**	**Std.Dev**
Easy to talk to clinician	3.46	1.01	3.48	0.98
Able to express effectively	3.5	1.02	3.50	1.07
Similar to personal interaction	3.28	1.13	3.23	1.01

Furthermore, to identify the differences of opinions in relation to mHealth and Telehealth, a *t*-test was conducted, as shown in Table 9. The mean scores of mHealth (Mean = 3.41, SD = 1.05) and Telehealth (Mean = 3.40, SD = 0.99), identified in the analysis, reflected that participants found that both mHealth and Telehealth approaches to be effective in terms of interaction quality. *t*-value, as shown in Table 9, was found to be (t = 0.0720) at 0.05 confidence interval, and was identified as not statistically significant (p > 0.05). Therefore, no significant differences of opinions in relation to interface quality parameters of mHealth and Telehealth can be observed. These findings related to mHealth are similar to [33,34], indicating average interaction quality of mHealth and Telehealth approaches.

Table 9. Difference in interaction quality of mHealth and Telehealth.

Variable	Approach	*n*	Mean	Std.Dev	*df*	*t*-Value	*p*-Value
Usefulness	mHealth	108	3.41	1.05	214	0.0720	0.9427
	Telehealth	108	3.40	0.99			

In relation to reliability (Table 10), the mHealth approach was rated as slightly better than the Telehealth approach, stating that the approach was similar to hospital visits in delivering the care, and also in the ability to fix issues by receiving messages through the application (in comparison to messages received through calls). The majority of the participants (>50%) were identified to be neutral (neither agree nor disagree) in relation to all the factors listed in the reliability parameter.

Table 10. Reliability of mHealth and Telehealth.

Items	mHealth		Telehealth	
	Mean	Std.Dev	Mean	Std.Dev
Approach was similar to in-person visits	3.16	1.03	2.96	1.09
Ability to recover from the mistakes in the system	3.37	1.03	3.28	1.03
Ability of the system in sending messages to fix issues	3.28	1.04	3	1.02

Furthermore, to identify the differences of opinions in relation to mHealth and Telehealth, a *t*-test was conducted, as shown in Table 11. The mean scores of mHealth (Mean = 3.27, SD = 1.03) and Telehealth (Mean = 3.08, SD = 1.05), identified in the analysis, reflected that participants found that both mHealth and Telehealth approaches to be effective in terms of reliability. *t*-value, as shown in Table 11, was found to be ($t = 1.3424$) at 0.05 confidence interval, and was identified as not statistically significant ($p > 0.05$). Therefore, no significant differences of opinions in relation to the reliability parameter of mHealth and Telehealth can be observed. Considering the less acceptance of eHealth in Saudi Arabia, due to various factors of influence, it may be possible that low responses were identified in relation to reliability factor, which can be compared to the findings in [43].

Table 11. Difference in Reliability of mHealth and Telehealth.

Variable	Approach	*n*	Mean	Std.Dev	*df*	*t*-Value	*p*-Value
Usefulness	mHealth	108	3.27	1.03	214	1.3424	0.1809
	Telehealth	108	3.08	1.05			

In relation to satisfaction (Table 12), it can be identified that participants were slightly more satisfied with mHealth compared to Telehealth across all the factors related to satisfaction and future use. Considering the overall satisfaction, there is no difference of opinions expressed in relation to both approaches. However, while 60% of the participants agreed that they would use mHealth in the future, 64% of the participants stated they would use Telehealth in the future, indicating a slightly more preference towards Telehealth over mHealth.

Table 12. Satisfaction and future use of mHealth and Telehealth.

Items	mHealth		Telehealth	
	Mean	Std.Dev	Mean	Std.Dev
Comfortability	3.50	1.08	3.46	1.04
Acceptable way to receive healthcare services	3.60	1.06	3.55	1.03
I would use in future	3.71	1.06	3.57	1.05
Overall satisfaction	3.71	1.05	3.57	1.05

Furthermore, to identify the differences of opinions in relation to mHealth and Telehealth, a *t*-test was conducted, as shown in Table 11. The mean scores of mHealth (Mean = 3.63, SD = 1.06) and Telehealth (Mean = 3.53, SD = 1.04), identified in the analysis, reflected that participants found that both mHealth and Telehealth approaches to be effective in terms of satisfaction and future use. *t*-value, as shown in Table 13, was found to be (t = 0.6998) at 0.05 confidence interval, and was identified as not statistically significant ($p > 0.05$). Therefore, no significant differences of opinions in relation to the reliability parameter of mHealth and Telehealth can be observed.

Table 13. Difference in satisfaction and future use of mHealth and Telehealth.

Variable	Approach	*n*	Mean	Std.Dev	*df*	*t*-Value	*p*-Value
Usefulness	mHealth	108	3.63	1.06	214	0.6998	0.4848
	Telehealth	108	3.53	1.04			

These findings may be compared to [31,32,40] in relation to satisfaction levels. Moreover, preference over these approaches may be influenced by the recent COVID-19 outbreak, which has led to increased adoption of eHealth approaches due to the surge in COVID-19 cases and preventive measures such as lockdowns and curfews. However, lack of reliability as identified in [43] can be one of the reasons for leaning more towards Telehealth rather than mHealth.

3.2. Interview Results

A total of seven healthcare practitioners were interviewed, and all of them were males. Among them, three participants belonged to the age group of 35–44 years; another three in 45–54 years; and one participant in 25–34 years. Four participants were general physicians, one participant was a dentist, one was a surgery specialist, and another was a medical specialist. Three participants had experience of 2 or less years in using mHealth and Telehealth approaches, two had an experience of 2 to 5 years, and another two had an experience of 5 to 10 years. The participants' experience levels and roles reflect a good sample for collecting the information about mHealth and Telehealth.

Focusing on the opinions expressed in relation to Telehealth, all the participants identified good usefulness levels for Telehealth. One of the interviewees identified it to be an easier approach to reach patients without any difficulty or requirement to learn new technologies, reflecting the edge over mHealth. Another interviewee identified Telehealth to be an effective approach in providing distant care. In relation to mHealth, only one participant identified its usefulness to be poor, while the rest of them indicated good levels of usefulness. One of the interviewees identified it to be very useful, providing real-time information at any time. The findings in relation to usefulness indicated that the majority of the participants identified both approaches to be of good usefulness. However, Telehealth was identified to be slightly more useful compared to mHealth.

In relation to ease of use, all the interviewees mentioned Telehealth to be very easy to use and learn. However, one of the interviewees mentioned that additional training relating to compliances and standards is required in using Telehealth. One of the interviewees mentioned that Telehealth may not be easy to learn in the beginning, but as handling of it improves, one can learn effectively. Similarly, all the interviewees identified mHealth to be easy to use, with little knowledge of computers and technology. However, one of the interviewees mentioned that experience is required in using mHealth, and another interviewee mentioned that mHealth is hard to use. Another interviewee mentioned that mHealth may be effective for learning for only those who use mobiles more frequently, and have experience of using applications. The overall analysis of responses indicated that Telehealth was rated slightly more than mHealth in terms of ease of use and learnability.

In relation to interface and interaction quality, all the interviewees reflected Telehealth to be good. Similarly, in relation to mHealth, only one interviewee identified it with poor interface and interaction quality. Using Telehealth was identified to be effective because interaction using Telehealth takes comparatively less time than visits, which can save time; while using mHealth, it was indicated that a lot of time could be saved as it is one-way messaging. Both approaches indicated time-saving as an outcome of effective interface and interaction quality. However, Telehealth was slightly rated more than mHealth.

Focusing on the reliability parameter, all the interviewees indicated Telehealth to be reliable but extended their statement that reliability may depend on many factors, including the physician and patient and the technology used to connect them. While Telehealth mainly relies on calls, it is sometimes possible that internet technologies such as voice over internet protocol or applications such as skype or zoom may be used, which may raise concerns over security and privacy. However, in relation to mHealth, five interviewees stated it to be reliable. One of the interviewees stated that more research is needed to assess its reliability, while others raised privacy and security concerns over mHealth.

Findings relating to usefulness, ease of use, learnability, interface and interaction quality, and reliability from interviewees' perspectives reflected a slightly greater preference towards Telehealth compared to mHealth in contrast to patients' perspectives. The differences of opinions among the participants may be related to their experience and understanding of these approaches, and also the features and design of the applications they have been using, which can influence their perspectives [44–46].

In relation to the satisfaction parameter, all the interviewees reflected good satisfaction levels about the Telehealth approach. However, focusing on the mHealth approach, one interviewee stated moderate satisfaction and another interviewee stated poor satisfaction. Findings reflected that the interviewees are slightly more satisfied with Telehealth compared to mHealth. These results regarding satisfaction contrasted with survey results, where participants identified with being slightly more satisfied with mHealth compared to Telehealth, supporting findings from [47–50]. Furthermore, in relation to future use prospects, Telehealth was identified to be a promising approach in reducing clinical visits, improving quality healthcare. Similarly, mHealth was also identified to be having a promising future where electronic health records can be integrated with daily monitoring systems, providing 24×7 remote healthcare services which can significantly improve effectiveness and efficiency of care. The results have indicated that both mHealth and Telehealth would be increasingly adopted in the future, similar to survey results.

4. Conclusions

This study has compared and evaluated mHealth approaches in remote monitoring of COPD patients diagnosed with COVID-19. The importance of this study arises from the rising complications and effects of COVID-19 after the recovery and during the home quarantine, associated with the additional complications of COPD condition. To evaluate the approaches, both patients' and physicians' perspectives are considered. The findings have indicated that patients' views were in contrast to physicians' views. While patients leaned towards mHealth, physicians leaned towards Telehealth. However, both approaches were identified to be effective in terms of their usefulness, ease of use and learnability, interface and interaction quality, reliability, satisfaction, and future use, with minor differences. This study has few limitations. As different patients use different mHealth applications and adopt different practices in Telehealth, there could be a certain bias in the results. In addition, the number of participants in both survey and interviews was lower due to the impact of the COVID-19 outbreak. Therefore, generalizations should be made with care. This study has both theoretical and practical implications. First, this study addresses the gaps in the literature in evaluating the remote monitoring approaches in the context of COVID-19 pandemic. Second, the findings can be used to improve the mHealth and Telehealth approaches in relation to the needs of the patients and physicians. Moreover, as the study has been conducted in Saudi Arabia, the findings can only be compared to the population with similar demographics. Therefore, future research may focus on the evaluation of different remote monitoring approaches in different regions reflecting varying demographics.

Funding: This research received no external funding.

Institutional Review Board Statement: Not applicable.

Informed Consent Statement: Informed consent was obtained from all subjects involved in the study.

Data Availability Statement: Not applicable.

Conflicts of Interest: The author declares no conflict of interest.

Appendix A

Evaluating Telehealth and mHealth approaches in monitoring the COVID-19 patients with chronic obstructive pulmonary disease (COPD): Survey Questionnaire (Telehealth)

1. Name.
2. Gender: Male/Female.
3. Education: Associate's degree, Bachelor's degree, Completed some postgraduate, Master's degree, Ph. D.
4. Experience with Telehealth: None, Less than 3 months, 3–6 months, 6 months–1 year, more than 1 year.
5. Please rate the following aspects of the health system on a scale of one to five (1: Strongly Disagree; 2: Disagree; 3: Neutral; 4: Agree; 5: Strongly Agree).

Components	Factors
Usefulness	
1	Telehealth improves my access to healthcare services
2	Telehealth saves me time traveling to a hospital or specialist clinic
3	Telehealth provides for my healthcare needs
Ease of Use & Learnability	
1	It was simple to use this system
2	It was easy to learn to use the system
3	I believe I could become productive quickly using this system
Interface Quality	
1	The way I interact with this system is pleasant
2	The system is simple and easy to understand
3	This system is able to do everything I would want it to be able to do
Interaction Quality	
1	I could easily talk to the clinician using the telehealth system
2	I felt I was able to express myself effectively
3	Using the telehealth system, I can see the clinician as well as if we met in person

Reliability	
1	I think the visits provided over the telehealth system are the same as in-person visits
2	Whenever I made a mistake using the system, I could recover easily and quickly
3	The system gave error messages that clearly told me how to fix problems
Satisfaction and Future Use	
1	I feel comfortable communicating with the clinician using the telehealth system
2	Telehealth is an acceptable way to receive healthcare services
3	I would use telehealth services again
4	Overall, I am satisfied with this telehealth system

Appendix B

Evaluating Telehealth and mHealth approaches in monitoring COVID-19 patients with chronic obstructive pulmonary disease (COPD): Survey Questionnaire (mHealth)

1. Name.
2. Gender: Male/Female.
3. Education: Associate's degree, Bachelor's degree, Completed some postgraduate, Master's degree, Ph. D.
4. Experience with Telehealth: None, Less than 3 months, 3–6 months, 6 months–1 year, more than 1 year.
5. Please rate the following aspects of the health system on a scale of one to five (1: Strongly Disagree; 2: Disagree; 3: Neutral; 4: Agree; 5: Strongly Agree).

Components	Factors
Usefulness	
1	mHealth improves my access to healthcare services
2	mHealth . saves me time traveling to a hospital or specialist clinic
3	mHealth provides for my healthcare needs
Ease of Use & Learnability	
1	It was simple to use this system
2	It was easy to learn to use the system
3	I believe I could become productive quickly using this system
Interface Quality	
1	The way I interact with this system is pleasant
2	The system is simple and easy to understand
3	This system is able to do everything I would want it to be able to do

Interaction Quality

1	I could easily talk to the clinician using the telehealth system
2	I felt I was able to express myself effectively
3	Using the mHealth system, I can see the clinician as well as if we met in person

Reliability

1	I think the visits provided over the mHealth system are the same as in-person visits
2	Whenever I made a mistake using the system, I could recover easily and quickly
3	The system gave error messages that clearly told me how to fix problems

Satisfaction and Future Use

1	I feel comfortable communicating with the clinician using the mHealth system
2	mHealth is an acceptable way to receive healthcare services
3	I would use mHealth services again
4	Overall, I am satisfied with this mHealth system

Appendix C

Interview Questionnaire
Telehealth

1. Please reflect your opinions on the usefulness of Telehealth.
2. Please reflect your opinions on the ease of use of Telehealth.
3. Please reflect your opinions on the learnability of Telehealth.
4. Please reflect your opinions on the interface quality of Telehealth.
5. Please reflect your opinions on the interaction quality of Telehealth.
6. Please reflect your opinions on the reliability of Telehealth.
7. Please reflect your opinions on the overall satisfaction of Telehealth.
8. Please reflect your opinions on the future use of Telehealth.

mHealth

1. Please reflect your opinions on the usefulness of mHealth.
2. Please reflect your opinions on the ease of use of mHealth.
3. Please reflect your opinions on the learnability of mHealth.
4. Please reflect your opinions on the interface quality of mHealth.
5. Please reflect your opinions on the interaction quality of mHealth.
6. Please reflect your opinions on the reliability of mHealth.
7. Please reflect your opinions on the overall satisfaction of mHealth.
8. Please reflect your opinions on the future use of mHealth.

References

1. World Health Organization. WHO Coronavirus (COVID-19) Dashboard. Available online: https://covid19.who.int/ (accessed on 10 May 2021).
2. World Health Organization. United States of America. Available online: https://covid19.who.int/region/amro/country/us (accessed on 10 May 2021).
3. World Health Organization. Brazil. Available online: https://covid19.who.int/region/amro/country/br (accessed on 10 May 2021).
4. World Health Organization. India. Available online: https://covid19.who.int/region/searo/country/in (accessed on 10 May 2021).

5.   Yao, Y.; Wang, H.; Liu, Z. Expression Of ACE2 In Airways: Implication For COVID-19 Risk And Disease Management In Patients With Chronic Inflammatory Respiratory Diseases. *Clin. Exp. Allergy* **2020**, *50*, 1313–1324. [CrossRef] [PubMed]

6.   Wang, Q.; Xu, R.; Volkow, N. Increased Risk Of COVID-19 Infection And Mortality In People With Mental Disorders: Analysis From Electronic Health Records In The United States. *World Psychiatry* **2020**, *20*, 124–130. [CrossRef] [PubMed]

7.   Driggin, E.; Madhavan, M.V.; Chuich, T.; Laracy, J.; Biondi-Zoccai, G.; Brown, T.S.; Der Nigoghossian, C.; Zidar, D.A.; Haythe, J.; Brodie, D.; et al. Cardiovascular Considerations for Patients, Health Care Workers, and Health Systems During the COVID-19 Pandemic. *J. Am. Coll. Cardiol.* **2020**, *75*, 2352–2371. [CrossRef]

8.   Chen, S.; Jones, P.; Underwood, B.; Moore, A.; Bullmore, E.; Banerjee, S.; Osimo, E.; Deakin, J.; Hatfield, C.; Thompson, F.; et al. The Early Impact Of COVID-19 On Mental Health And Community Physical Health Services And Their Patients' Mortality In Cambridgeshire And Peterborough, UK. *J. Psychiatr. Res.* **2020**, *131*, 244–254. [CrossRef] [PubMed]

9.   Williamson, E.J.; Walker, A.J.; Bhaskaran, K.; Bacon, S.; Bates, C.; Morton, C.E.; Curtis, H.J.; Mehrkar, A.; Evans, D.; Inglesby, P.; et al. Factors associated with COVID-19-related death using OpenSAFELY. *Nature* **2020**, *584*, 430–436. [CrossRef]

10.   Wu, F.; Zhou, Y.; Wang, Z.; Xie, M.; Shi, Z.; Tang, Z.; Li, X.; Li, X.; Lei, C.; Li, Y.; et al. Clinical Characteristics Of COVID-19 Infection In Chronic Obstructive Pulmonary Disease: A Multicenter, Retrospective, Observational Study. *J. Thorac. Dis.* **2020**, *12*, 1811–1823. [CrossRef]

11.   Lippi, G.; Henry, B. Chronic Obstructive Pulmonary Disease Is Associated With Severe Coronavirus Disease 2019 (COVID-19). *Respir. Med.* **2020**, *167*, 105941. [CrossRef]

12.   Deslée, G.; Zysman, M.; Burgel, P.; Perez, T.; Boyer, L.; Gonzalez, J.; Roche, N. Chronic Obstructive Pulmonary Disease And The COVID-19 Pandemic: Reciprocal Challenges. *Respir. Med. Res.* **2020**, *78*, 100764. [CrossRef]

13.   Wouters, O.; Shadlen, K.; Salcher-Konrad, M.; Pollard, A.; Larson, H.; Teerawattananon, Y.; Jit, M. Challenges In Ensuring Global Access To COVID-19 Vaccines: Production, Affordability, Allocation, And Deployment. *Lancet* **2021**, *397*, 1023–1034. [CrossRef]

14.   Torjesen, I. Covid-19 Vaccine Shortages: What Is The Cause And What Are The Implications? *BMJ* **2021**, *372*, n781. [CrossRef]

15.   Badreldin, H.; Atallah, B. Global Drug Shortages Due To COVID-19: Impact On Patient Care And Mitigation Strategies. *Res. Soc. Adm. Pharm.* **2021**, *17*, 1946–1949. [CrossRef] [PubMed]

16.   World Health Organization. Side Effects of COVID-19 Vaccines. Available online: https://www.who.int/news-room/feature-stories/detail/side-effects-of-covid-19-vaccines#:~{}:text=Reported%20side%20effects%20of%20COVID,muscle%20pain%2C%20chills%20and%20diarrhoea. (accessed on 10 May 2021).

17.   Menni, C.; Klaser, K.; May, A.; Polidori, L.; Capdevila, J.; Louca, P.; Sudre, C.; Nguyen, L.; Drew, D.; Merino, J.; et al. Vaccine Side-Effects And SARS-Cov-2 Infection After Vaccination In Users Of The COVID Symptom Study App In The UK: A Prospective Observational Study. *Lancet Infect. Dis.* **2021**. [CrossRef]

18.   Biswas, S. Mucormycosis: The 'Black Fungus' Maiming Covid Patients in India. Available online: https://www.bbc.com/news/world-asia-india-57027829 (accessed on 10 May 2021).

19.   Debroy, S. Covid-Sparked Fungal Infection Assuming Epidemic Proportions | India News—Times of India. Available online: https://timesofindia.indiatimes.com/india/covid-sparked-fungal-infection-assuming-epidemic-proportions/articleshow/82473382.cms (accessed on 10 May 2021).

20.   Watson, A.; Wah, R.; Thamman, R. The Value Of Remote Monitoring For The COVID-19 Pandemic. *Telemed. e-Health* **2020**, *26*, 1110–1112. [CrossRef] [PubMed]

21.   Wu, J.; Chang, L.; Yu, G. Effective Data Decision-Making And Transmission System Based On Mobile Health For Chronic Disease Management In The Elderly. *IEEE Syst. J.* **2020**, 1–12. [CrossRef]

22.   Odei-Lartey, E.; Prah, R.; Anane, E.; Danwonno, H.; Gyaase, S.; Oppong, F.; Afenyadu, G.; Asante, K. Utilization Of The National Cluster Of District Health Information System For Health Service Decision-Making At The District, Sub-District And Community Levels In Selected Districts Of The Brong Ahafo Region In Ghana. *BMC Health Serv. Res.* **2020**, *20*, 514. [CrossRef]

23.   Prachand, V.; Milner, R.; Angelos, P.; Posner, M.; Fung, J.; Agrawal, N.; Jeevanandam, V.; Matthews, J. Medically Necessary, Time-Sensitive Procedures: Scoring System To Ethically And Efficiently Manage Resource Scarcity And Provider Risk During The COVID-19 Pandemic. *J. Am. Coll. Surg.* **2020**, *231*, 281–288. [CrossRef] [PubMed]

24.   Uribe-Toril, J.; Ruiz-Real, J.L.; Nievas-Soriano, B.J. A Study of eHealth from the Perspective of Social Sciences. *Healthcare* **2021**, *9*, 108. [CrossRef]

25.   Global Observatory for Ehealth. Mhealth New Horizons for Health through Mobile Technologies. 2020. Available online: https://www.who.int/goe/publications/goe_mhealth_web.pdf (accessed on 5 July 2021).

26.   American Telemedicine Association. What Is Telemedicine, Exactly? 2021. Available online: https://www.americantelemed.org/ata-news/what-is-telemedicine-exactly/#.VquyglLMbbo (accessed on 5 July 2021).

27.   Tomasic, I.; Tomasic, N.; Trobec, R.; Krpan, M.; Kelava, T. Continuous Remote Monitoring Of COPD Patients—Justification And Explanation Of The Requirements And A Survey Of The Available Technologies. *Med Biol. Eng. Comput.* **2018**, *56*, 547–569. [CrossRef]

28.   Ambrosino, N.; Vagheggini, G.; Mazzoleni, S.; Vitacca, M. Telemedicine in chronic obstructive pulmonary disease. *Breathe* **2016**, *12*, 350–356. [CrossRef]

29.   Adans-Dester, C.; Bamberg, S.; Bertacchi, F.; Caulfield, B.; Chappie, K.; Demarchi, D.; Erb, M.; Estrada, J.; Fabara, E.; Freni, M.; et al. Can Mhealth Technology Help Mitigate The Effects Of The COVID-19 Pandemic? *IEEE Open J. Eng. Med. Biol.* **2020**, *1*, 243–248. [CrossRef]

30. Windisch, O.; Zamberg, I.; Zanella, M.; Gayet-Ageron, A.; Blondon, K.; Schiffer, E.; Agoritsas, T. Using Mhealth To Increase The Reach Of Local Guidance To Health Professionals As Part Of An Institutional Response Plan To The COVID-19 Outbreak: Usage Analysis Study. *JMIR mHealth uHealth* **2020**, *8*, e20025. [CrossRef] [PubMed]

31. van der Velden, R.; Hermans, A.; Pluymaekers, N.; Gawalko, M.; Vorstermans, B.; Martens, H.; Buskes, S.; Crijns, H.; Linz, D.; Hendriks, J. Coordination Of A Remote Mhealth Infrastructure For Atrial Fibrillation Management During COVID-19 And Beyond: Telecheck-AF. *Int. J. Care Coord.* **2020**, *23*, 65–70. [CrossRef]

32. Liew, M.; Zhang, J.; See, J.; Ong, Y. Usability Challenges For Health And Wellness Mobile Apps: Mixed-Methods Study Among Mhealth Experts And Consumers. *JMIR mHealth uHealth* **2019**, *7*, e12160. [CrossRef] [PubMed]

33. Costa, P.; de Jesus, T.; Winstein, C.; Torriani-Pasin, C.; Polese, J. An Investigation Into The Validity And Reliability Of Mhealth Devices For Counting Steps In Chronic Stroke Survivors. *Clin. Rehabil.* **2019**, *34*, 394–403. [CrossRef]

34. Verhagen, S.; Berben, J.; Leue, C.; Marsman, A.; Delespaul, P.; van Os, J.; Lousberg, R. Demonstrating The Reliability Of Transdiagnostic Mhealth Routine Outcome Monitoring In Mental Health Services Using Experience Sampling Technology. *PLoS ONE* **2017**, *12*, e0186294.

35. Cobelli, N.; Chiarini, A. Improving Customer Satisfaction And Loyalty Through Mhealth Service Digitalization. *TQM J.* **2020**, *32*, 1541–1560. [CrossRef]

36. Parmanto, B.; Lewis, A., Jr.; Graham, K.; Bertolet, M. Development Of The Telehealth Usability Questionnaire (TUQ). *Int. J. Telerehabilitation* **2016**, *8*, 3–10. [CrossRef] [PubMed]

37. Zhou, L.; Bao, J.; Setiawan, I.; Saptono, A.; Parmanto, B. The Mhealth App Usability Questionnaire (MAUQ): Development And Validation Study. *JMIR mHealth uHealth* **2019**, *7*, e11500. [CrossRef]

38. Likert, R. A Technique for the Measurement of Attitudes. *Arch. Psychol.* **1932**, *140*, 1–55.

39. Etikan, I. Comparison of Convenience Sampling and Purposive Sampling. *Am. J. Theor. Appl. Stat.* **2016**, *5*, 1. [CrossRef]

40. Ivanitskaya, L.; O'Boyle, I.; Casey, A. Health Information Literacy And Competencies Of Information Age Students: Results From The Interactive Online Research Readiness Self-Assessment (RRSA). *J. Med. Internet Res.* **2006**, *8*, e6. [CrossRef]

41. Banas, J. A Tailored Approach To Identifying And Addressing College Students' Online Health Information Literacy. *Am. J. Health Educ.* **2008**, *39*, 228–236. [CrossRef]

42. Eriksson-Backa, K.; Ek, S.; Niemelä, R.; Huotari, M. Health Information Literacy In Everyday Life: A Study Of Finns Aged 65–79 Years. *Health Inform. J.* **2012**, *18*, 83–94. [CrossRef] [PubMed]

43. Alharbi, A.S. Evolution of E-Health in Saudi Arabia: Mobile Technology and MHealth. *Int. Multiling. Acad. J.* **2021**, *3*, 41–47.

44. Tistad, M.; Lundell, S.; Wiklund, M.; Nyberg, A.; Holmner, Å.; Wadell, K. Usefulness And Relevance Of An Ehealth Tool In Supporting The Self-Management Of Chronic Obstructive Pulmonary Disease: Explorative Qualitative Study Of A Cocreative Process. *JMIR Human Factors* **2018**, *5*, e10801. [CrossRef]

45. Klocek, A.; Šmahelová, M.; Knapová, L.; Elavsky, S. Gps' Perspectives OnEhealth Use In The Czech Republic: A Cross-Sectional Mixed-Design Survey Study. *BJGP Open* **2019**, *3*, bjgpopen19X101655. [CrossRef]

46. Wattanapisit, A.; Wattanapisit, S.; Tuangratananon, T.; Amaek, W.; Wongsiri, S.; Petchuay, P. Primary Health Care Providers' Perspectives On Developing An Ehealth Tool For Physical Activity Counselling: A Qualitative Study. *J. Multidiscip. Healthc.* **2021**, *14*, 321–333. [CrossRef]

47. Alwashmi, M.; Fitzpatrick, B.; Davis, E.; Gamble, J.; Farrell, J.; Hawboldt, J. Perceptions Of Health Care Providers Regarding A Mobile Health Intervention To Manage Chronic Obstructive Pulmonary Disease: Qualitative Study. *JMIR mHealth uHealth* **2019**, *7*, e13950. [CrossRef]

48. Lee, H.; Uhm, K.; Cheong, I.; Yoo, J.; Chung, S.; Park, Y.; Lee, J.; Hwang, J. Patient Satisfaction With Mobile Health (Mhealth) Application For Exercise Intervention In Breast Cancer Survivors. *J. Med. Syst.* **2018**, *42*, 254. [CrossRef] [PubMed]

49. Kirby, D.; Fried, J.; Buchalter, D.; Moses, M.; Hurly, E.; Cardone, D.; Yang, S.; Virk, M.; Rokito, A.; Jazrawi, L.; et al. Patient And Physician Satisfaction With Telehealth During The COVID-19 Pandemic: Sports Medicine Perspective. *Telemed. e-Health* **2021**. [CrossRef]

50. Darcourt, J.; Aparicio, K.; Dorsey, P.; Ensor, J.; Zsigmond, E.; Wong, S.; Ezeana, C.; Puppala, M.; Heyne, K.; Geyer, C.; et al. Analysis Of The Implementation Of Telehealth Visits For Care Of Patients With Cancer In Houston During The COVID-19 Pandemic. *JCO Oncol. Pract.* **2021**, *17*, e36–e43. [CrossRef] [PubMed]

International Journal of
Environmental Research and Public Health

Description of e-Health Initiatives to Reduce Chronic Non-Communicable Disease Burden on Brazilian Health System

Daniela Laranja Gomes Rodrigues [1,*], Gisele Silvestre Belber [1], Igor da Costa Borysow [1], Marcos Aurelio Maeyama [2] and Ana Paula Neves Marques de Pinho [1]

[1] Social Responsibility Department, Hospital Alemão Oswaldo Cruz, São Paulo 01323-020, Brazil; gsbelber@haoc.com.br (G.S.B.); iborysow@haoc.com.br (I.d.C.B.); ampinho@haoc.com.br (A.P.N.M.d.P.)

[2] Telehealth Center, Florianópolis 88040-900, Brazil; marcos.aurelio@univali.br

* Correspondence: dlrodrigues@haoc.com.br; Tel.: +55-11-985421342

Citation: Rodrigues, D.L.G.; Belber, G.S.; Borysow, I.d.C.; Maeyama, M.A.; Pinho, A.P.N.M.d. Description of e-Health Initiatives to Reduce Chronic Non-Communicable Disease Burden on Brazilian Health System. *Int. J. Environ. Res. Public Health* **2021**, *18*, 10218. https://doi.org/10.3390/ijerph181910218

Academic Editors: Irene Torres-Sanchez and Marie Carmen Valenza

Received: 31 August 2021
Accepted: 15 September 2021
Published: 28 September 2021

Abstract: Chronic non-communicable diseases (NCD) account for 72% of the causes of death in Brazil. In 2013, 54 million Brazilians reported having at least one NCD. The implementation of e-Health in the Unified Health System (SUS) could fill gaps in access to health in primary health care (PHC). Objective: to demonstrate telehealth strategies carried out within the scope of the Institutional Development Support Program of the Unified Health System (PROADI-SUS) and developed by Hospital Alemão Oswaldo Cruz, between 2018 and 2021, on evaluation, supply, and problem-solving capacity for patients with NCDs. Methodology: a prospective and descriptive study of three projects in the telehealth areas, using document analysis. The Brasil Redes project used availability, implementation, and cost-effectiveness analysis, TELEconsulta Diabetes is a randomized clinical trial, and Regula Mais Brasil is focused on the waiting list for regulation of specialties. All those strategies were developed within the scope of the SUS. Results: 161 patients were attended by endocrinology teleconsultation in one project and another two research projects, one evaluating Brazil's Telehealth Network Program, and another evaluating effectiveness and safety of teleconsultation in patients with diabetes mellitus referred from primary care to specialized care in SUS. Despite the discrepancy in the provision of telehealth services in the country, there was an increase in access to specialized care on the three projects and especially on the Regula Mais Brasil Collaborative project; we observed a reduction on waiting time and favored distance education processes. Conclusion: the three projects offered subsidies for decision-making by the Ministry of Health in e-Health and two developed technologies that could be incorporated into SUS.

Keywords: chronic non-communicable diseases; unified health system; primary health care; Brazil

1. Introduction

With the change in the epidemiological and demographic profile of the population, in addition to the growing increase in chronic pathologies and people in vulnerable situations, the need for an integrated approach to the individual concerning health care emerges [1–3]. Chronic non-communicable diseases (CNCD) are described as diseases that involve the circulatory system, diabetes, cancer, and chronic respiratory diseases. This group of diseases usually occurs more commonly in developing countries, accounting for 63% of the causes of death in the world [4,5] and corresponding to one-third of deaths in people under 60 years of age [6]. The last national health survey conducted by the Brazilian government in 2013 [7] showed that 72% of the causes of death were related to CNCD, that is, by diseases that could have their impact reduced if health strategies focused mainly on primary care were implemented.

Digital health, or e-Health, can be defined as "the safe and cost-effective use of information and communication technologies in support of health and health-related fields",

83

with the World Health Organization considering making e-Health a global priority for health systems development [8]. The implementation of e-Health can substantially contribute to filling some gaps in the performance of primary health care (PHC), especially in developing countries such as Brazil [9]. One of these strategies is to expand the use of telemedicine to provide diagnostic and consulting support to professionals in distant locations, supporting the physician in such locations and improving the skills and knowledge of remote healthcare providers [10].

The distribution of income and access of the Brazilian population to technological resources is characterized by geographical and social inequality. Even so, the report "Digital 2020", published by *We Are Social* in January 2020, identified 150.4 million internet users in Brazil [11], accounting for an increase of 6.0% (8.5 million) when compared to the same month of the previous year, and by June 2020 there were identified 234 million active smartphones, a number that has been larger than the Brazilian population since 2017, according to the 31st edition of the Getúlio Vargas Foundation (FGV) Annual Research on the IT Usage [12]. Considering the public health area, there is a partnership between the Ministry of Health and the Ministry of Science, Technology, and Communications to provide internet access in 100% of health care units in Brazil [13]. That way, we can recognize that there is an increasingly favorable context for the wide deployment of telemedicine in the country. Despite this, there are still challenges and important aspects to be considered, for example, a Brazilian resolution that allows greater use of telehealth strategies throughout the country [14,15].

This article aims to describe telehealth strategies carried out in the years 2018 to 2021 by the Hospital Alemão Oswaldo Cruz, in partnership with the Brazilian Ministry of Health through the Support Program for Institutional Development of the Unified Health System (PROADI-SUS) [16,17].

2. Materials and Methods

Between 2018 and 2021, Hospital Alemão Oswaldo Cruz, in partnership with the Ministry of Health, through PROADI-SUS, developed projects and actions in Telehealth aimed at improving management in various spheres of the SUS. Of these, we review three strategies aimed at reducing the impact of CNCD on health using different strategies in e-Health for this purpose. As these are three different projects, we will describe the methodology of each one separately.

2.1. Brasil Redes Project

The Brasil Redes Project aimed to carry out a diagnostic evaluation of the National Telehealth Brazil Networks Program (PNTBR–in Portuguese), which completed 10 years of existence in 2017 [18]. This program aims to promote strategies to support primary health care through distance communication actions that are carried out by telehealth centers. As of 2011, the main activities to be carried out by the centers were as defined: teleconsulting, telediagnosis, tele-education, and training second opinion [19].

Five studies were carried out for this evaluation: evaluability study, implementation evaluation, and cost-effectiveness evaluation of two services offered in the list of activities foreseen by the program and program cost analysis [20]. The evaluability study was the first stage, aiming to identify the resources, the stakeholders, the processes carried out, the evaluative questions, and the expectations regarding the results of the referred program. This study included data collection work between the months of August and December 2018. The second study was the implementation evaluation, which had two major stages: socio-historical analysis of the genesis of the program and the implementation evaluation itself. The economic evaluations addressed the teledermatology service of the Telehealth Center of Santa Catarina [21] and the telecardiology service of the Telehealth Center of the Hospital das Clínicas of the Federal University of Minas Gerais [22] (Figure 1). The last study evaluated the unit value of the electrocardiogram and Holter report to be implemented as a procedure made available to the population by the Unified Health System.

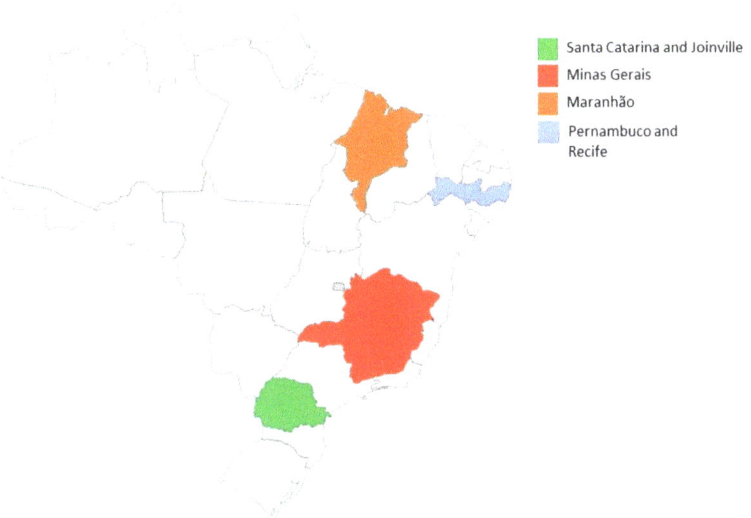

Figure 1. Brazil's states with telehealth initiatives included in the current article.

Table 1 outlines a systematization of the five studies carried out.

2.2. Regula Mais Brasil Colaborativo

Regula Mais Brasil Collaborative [17] is a project developed to qualify the outpatient care regulation process (evaluation of referrals to specialized care) using telehealth technologies, with an evaluation of cases in regulation, use of access protocols to specialties, and teleconsulting support for case resolution. Furthermore, this project monitors the waiting period process, where the patient is registered in the online system where he awaits the emergence of a vacancy for the requested specialty. Usually, this waiting time is 1 to 2 years.

Motivated by the COVID-19 pandemic [15], the Ministry of Health of Brazil authorized teleconsultations to face the pandemic, which had been prohibited in Brazil until then, and invited the Hospitals to participate in the Regula Mais Brasil Collaborative to offer teleconsultations, thus expanding the scope of the project. The purpose of this offer was to reduce the risk of transmission of the virus, both in displacement and in the environment of health services, to avoid contact with possible contaminants, and to maintain the care of acute and chronic diseases in an off-site manner. In addition, the importance of this offer was enhanced by the closing of several elective outpatient care services at the beginning of the pandemic, which could lead to the worsening of cases due to a lack of assistance.

2.3. Teleconsulta Diabetes

This project's main objective, until in recruitment phase, is to conduct clinical research to test the hypothesis that teleconsultation is non-inferior to the face-to-face care of patients with diabetes mellitus (DM) type II referred from primary health care (PHC) to specialized care in SUS [23].

This is a randomized, pragmatic, phase-2, single-center, open, non-inferiority clinical trial, with central randomization, allocation confidentiality, and data analyst blinding, to assess the efficacy and safety of specialized care by teleconsultation compared to face-to-face care. A total of 250 participants of both sexes over 18 years old, with type 2 diabetes mellitus, will be included. This sample size will allow evaluating the non-inferiority of up to 0.5% between groups, assuming a standard deviation of 1.30.

Table 1. Systematization of Brazil's Rede project.

Analysis	Objective	Methodology	Institutions
Evaluability	• Check the feasibility of the process • Evaluative, exploring expectations and needs of the interest groups involved and the degree of organization and implementation of the initiative	• Participant observation, members interviews, elaboration of a Logical Model	• State telehealth centers in Maranhão, Pernambuco, Santa Catarina and 2 Telehealth Centers in Minas Gerais (Figure 1)
Implementation	• Check the necessary adaptations of the proposals of the program by the centers, according to their realities (region, population characteristics, political context, history of disputes) and conformation. Identify limitations of the service network and the articulation with the main actors of the program in question.	• Socio-historical study • 17 semi-structured members interviews were carried out • Implementation evaluation • Participant observation and interviews with members of the telehealth centers and network agents • Implementation degree evaluation matrix	• Socio-historical study: representatives of telehealth centers across the country and former employees of the Ministry of Health
Teledermatology's Economic Evaluation	• Carry out cost-effectiveness analysis of the teledermatology service • Identify the specific activities and vocations of the Nucleus, aiming at raising expenses and results obtained to feed the economic evaluation models • Define the effectiveness indicators and economic evaluation models • Develop economic assessment	• Cost-effectiveness analysis comparing the teledermatology service with the provision of conventional care, from the society's perspective, considering the period from January 2016 to December 2018. The analysis was developed using a tree decision, considering the possible transition to alternatives: teledermatology or conventional service.	• Santa Catarina's Telehealth Center (Figure 1)
Telecardiology's Economic Evaluation	• Cost-effectiveness analysis of the telecardiology service.	• Cost-effectiveness analysis comparing the telecardiology service with the provision of conventional care, from the perspective of society, considering the period between July 2018 and June 2019 • The analysis was developed using a decision tree, considering the possibilities transition to alternatives: telecardiology or conventional care	• Federal University of Minas Gerais Telehealth Center
Analysis of costs and budgetary impact of the inclusion of the telecardiology service in the list of Unified Public Health System (SUS) procedures	• Identify the unit payment amount, by the SUS, of an electrocardiogram and Holter report, via telehealth, with a view to its incorporation into the list of SUS procedures	• Description, measurement, and analysis of the costs involved in providing the telecardiology service. Calculation of the total monthly cost and the average cost per report. Analysis of the budgetary impact arising from the incorporation of the service into the SUS.	

The outcomes that will be analyzed in this study are: the mean difference in the percentage of glycated hemoglobin (HbA1c) post- and pre-intervention in patients diagnosed with type 2 DM in 3 and 6 months, fasting blood glucose measurements and blood count, measurements of serum urea and creatinine, lipid profile measurements, systolic pressure measurements, in-office measurements, in-office diastolic pressure measurements, measurements of body weight and body mass index (BMI), the incidence of any adverse events, and quality measurements of patients' lives using the diabetes quality of life measurement questionnaire (DQOL) [24].

In the micro-costing analysis [25], this study performs the definition of the consumption time equations of each resource per product/service and extrapolation of the findings to define the productive capacities of the basic health units (UBS–in Portuguese) through real care data by the flow of regulation in the city of Joinville.

3. Results

The main target audience of the Telessaúde Brasil Redes National Program is primary health care teams, and with that many of the actions offered by the centers were focused on caring for people with CNCD. In this sense, the project results highlighted opportunities for improvement of the telehealth program for CNCD at the federal level. In Brazil's public health system, there is a necessity to homogenize the financing values considering the production potential of the centers, mainly for telediagnosis, and monitoring and evaluation methods of results. For the spheres of state centers aiming to value their actions, there is a need to reflect on challenges in the use of these services by health workers and on regional partnerships. The set of methods offered a broad vision of the program and articulated more detailed analyses of certain services offered by the telehealth centers. The purpose was to offer subsidies to the Ministry of Health in improving the telehealth program and using the expertise of certain centers as inspiring the formulation of procedures used to be offered by SUS.

Since the creation of the Brazil Redes program in 2007, 33 telehealth centers have been identified in various regions of the country (Figure 1), with very diversified health access strategies, as described in Table 2.

Table 2. Contributions of the evaluative studies of the National Telehealth Brazil Networks Program to MS decision-making, with a focus on CNCD.

Analysis	Results	Contributions to the Ministry of Health in Decision Making for Chronic Non-Communicable Diseases (CNCD)
Evaluability study	• Management mechanisms identified and clear definition of work proposals and expected results. Heterogeneity between health centers, including activities offered and receipt of funds. • There is institutional weakness in the program's national coordination.	• In-depth analysis of the heterogeneity of the centers and their regional contexts is necessary to better plan the actions to be offered by each one, and regularize the funds transferred by the federal entity. This could improve the telehealth strategy aimed at primary care.
Implementation assessment	• Through the analysis of four cores, heterogeneity in the provision of telehealth services, differences in the receipt and use of funds, and diversity in articulations with state decision-making bodies were confirmed. • The federal administration did not develop adequate strategies for managing the funding and productivity of the centers. The centers raise funds from different sources and seek to keep their services running.	• The actions of teleconsultation, tele-education and training second opinion are aimed at primary care and help in the training of workers in the care of CNCD. Despite the fragile measurement of results, these actions are recognized by workers as relevant to improvement. • Telediagnosis in cardiology, dermatology, spirometry, stomatology, and ophthalmology demonstrate an impact on the CNCD healthcare network. Support to regulation centers organize and optimize queues for specialists and promote greater resolution in the primary care.

Table 2. *Cont.*

Analysis	Results	Contributions to the Ministry of Health in Decision Making for Chronic Non-Communicable Diseases (CNCD)
Teledermatology's economic evaluation	• The cost-effectiveness analysis shows that the service, using the teledermatology strategy, costs USD 191.38 per patient, while the use of the conventional service costs USD 220.68 per patient. • As the performance of the service was used as an effectiveness parameter, there are no significant differences between the alternatives. If the services have equal patient care capacity, there are only differences in costs, which result in USD 59.89 more for the care of a patient in the conventional strategy compared to teledermatology.	• It is concluded that conventional care is dominated by teledermatology, presenting itself as a good strategy to be implemented and/or financed by the public administration. This service will be able to contribute with greater agility in scheduling and carrying out consultations, preventing dermatological problems from evolving in severity, as well as helping people from regions farther away from the places where the specialists are located to have easier access to the dermatological examination.
Telecardiogy's economic evaluation	• The cost-effectiveness analysis shows that the service, using the telecardiology strategy, considering that all patients have their problems solved in secondary care, cost USD 47.35 purchasing power parity (PPP) per patient, while the use of the conventional service costs USD 99.94 PPP dollars per patient. As the performance of patient care was used as a parameter of effectiveness, there is no significant difference in effectiveness between the alternatives. If the services have equal patient care capacity, there are only differences in costs, which result in USD 52.59 PPP more for the care of a patient in the conventional strategy compared to telecardiology.	• It is then concluded that conventional care is dominated by telecardiology, presenting itself as a good strategy to be implemented and/or financed by the public administration. This service will be able to contribute with greater agility in scheduling and carrying out consultations, preventing cardiovascular problems from evolving in severity, as well as helping people from regions farther away from the places where the specialists are located to have easier access to the related exams.

In this sense, the centers sought to articulate different fronts of activities to contemplate permanent education and the improvement of primary care professionals, whether through courses, virtual lectures, teleconsulting, production of training second opinions, and telediagnosis. Regarding the processes evaluated in the economic component, the two studies carried out with the Telehealth Center at Clinic's Hospital from the Federal University of Minas Gerais clarify the cost and budget impact forecast for the implementation of telecardiology services in the public service network, which can greatly contribute to the improvement of care for people with CNCD.

The Hospital Alemão Oswaldo Cruz carried out consolations in the Regula Mais Brasil Collaborative Project via teleconsultation of users, in the specialties of neurology, endocrinology, orthopedics, and mental health, totaling 1097 teleconsultations in this period. Specific actions for CNCDs are concentrated in the specialty of endocrinology, in which 161 teleconsultations were carried out, all with SUS users in the city of Recife, city located in the state of Pernambuco, in northeastern Brazil (Figure 1). The results obtained in this project are displayed in Table 3.

Depending on the prevalence of endocrinological diseases, diabetes mellitus appears as the most prevalent reason for teleconsultations performed. Regarding the technology used for contact between the doctor and the patient, video calling (audio and video) was mostly used, since it represents the best possibility of remote interaction. The exclusive audio resource was only used when it was not possible to use video calling.

Table 3. Results of Regula Mais Brazil Project.

Teleconsultation	Diseases	N (%)
Reason for teleconsultation	Insulin-dependent diabetes	80
	Non-insulin dependent diabetes	67
	Hypothyroidism/myxedema	4
	Hypertension with complications	3
	Thyreoid cancer	2
	Goiter	1
	Hypertension without complications	1
	Hyperthyroidism	1
	Other diseases	2
Technology used	Phone	46 (28.6%)
	Video calling	115 (71.4%)
Outcomes after the last teleconsultation	Follow up at primary care	40.3%
	Follow up at specialized care	33.8%
	Teleconsultation follow up	23.4%
	Urgent follow up at specialized care	2.6%
Net promoter score (NPS)	0–100	93

The outcomes of the last teleconsultation were performed to allow an assessment of the resolvability potential of this assistive technology. The follow-up outcomes in PHC (40.3%) represent cases in which the specialist, based on the assessment in teleconsultation, considers that, based on the complexity of the case, the patient can be followed up in the PHC, close to their residence, avoiding unnecessary referral and displacement of the patient to the specialized center. In addition, he receives the first guidance and conduct by the specialist via teleconsultation. Likewise, the reassessment in teleconsultations (23.4%) also represents the cases in which digital technology allows for adequate follow-up with a specialist, without the need to travel. On the other hand, cases in which the outcomes indicate the need for a face-to-face consultation with the specialist, whether immediate or elective, represent the limitation of teleconsultation in the assessment or follow-up of cases. It is worth mentioning that all cases evaluated did not have a first face-to-face consultation which, in some cases, limits a better diagnosis.

What corroborates the potential of teleconsultations is the result of the Net Promoter Score [26], which assesses user satisfaction concerning the service, and which, on a scale from 0 to 100, presented an index of 93 points in teleconsultations in endocrinology carried out and evaluated by the project patients.

Regarding the results of the teleconsultation diabetes research project, it was not yet possible to demonstrate aspects of the research results related to the results of complementary exams, questionnaires measuring the quality of life in diabetes (DQOL), and satisfaction assessment with telemedicine, due to the small number of participants included so far. Of the patients included in the study, the mean age was 59.4 years, with a predominance of 78% of white people, followed by black (12%) and mixed race (10%), and a slight predominance of males (54%) of the included participants.

In addition, within the scope of the teleconsultation project, the time-driven activity-based costing (TDABC) micro-costing methodology, or costs based on time and activity, was adopted because it allows the identification of the unit cost of the service within the expected efficiency conditions, also acting as a comparison metric. In this scenario, the first hypothesis considered the duration of the teleconsultation and the face-to-face consultation to be identical, as well as the duration of the pre-consultation activities. This is a preliminary conclusion, considering that recruitment, until this period, was minimal and with low statistical power to detect differences in times. The second hypothesis considered that the costs with materials and structure of the UBS's and the Polyclinic are also identical since the Health Department of Joinville is in the middle of the process of changing the methodology for this type of cost and must provide these data in a manner more accurate by mid-2021. In this hypothetical scenario, the cost of teleconsultation was 5% higher than that of a face-to-face consultation. As in both cases, the consultation is responsible for more than 90% of the total cost, the difference in duration between the two types of consultation will be the determining factor for the difference in costs between them. More details that relate the methodologic and statistic description are outlined in the design paper published before [23].

4. Discussion

Teleconsultation is a medical act and must abide by the Brazilian Medical Code of Ethics (MCE) [27–29]. As provided in Article 37 of the MCE, the physician is prohibited from "prescribing treatment or other procedures without direct examination of the patient, except in cases of urgency or emergency and proven impossibility of performing it, and, in such circumstances, must do so immediately after ceasing the impediment" [30]. The State of Emergency in Public Health of National Importance (ESPIN) [31] triggered during the SARS-CoV-2 pandemic fits into this prerogative, thus supporting teleconsultations during this exceptional period where the Brazilian society had the opportunity to experience remote medical care more extensively.

Critics of telemedicine approval warn that the possible overuse and careless use of telemedicine could turn doctors into "telemarketing operators", which could lead to poor quality in appointments, medical errors, and unemployment by reducing the number of face-to-face doctors [32]. Although the movements of criticism to teleconsultation caused a fanfare in the mainstream media, they do not seem to reproduce what the Brazilian medical professional wants. A survey carried out in February 2020 by the São Paulo Physicians Association (APM) with over 2200 physicians from 55 different specialties revealed that 64.39% of doctors wanted a regulation that would allow the expansion of services and assistance to the population, including direct doctor-to-patient teleconsultation [33].

This whole context evidences the lack of alignment between the federal government, the Brazilian Medical Federal Council, health plan operators, medical associations, and medical professionals, especially concerning teleconsultations. The challenge is how to expand access to medical services mainly to specialists for populations in remote regions, reduce healthcare costs and the displacement of patients, and on the other hand, minimize the fear of damage to the medical profession [27]. Other challenges now faced by specialists are data insecurity, trivialization of teleconsultations, and the production of misdiagnoses and prescriptions, in addition to avoiding possible unemployment of doctors. The country should take advantage of the current situation and promote a wider discussion on the benefits and limitations of permanent and full permission to use telemedicine, bringing to the agenda socioeconomic, cultural, and technological issues.

The Brazil Redes program showed the presence of decentralized and disseminated strategies throughout the national territory, with actions in telehealth, focused not only on CNCDs but also other health promotion strategies. The actions of teleophthalmology with identification of diabetic retinopathy based in Mato Grosso [34,35] confirm the efficiency and potential of supporting the health of the population using the Unified Health System if the use of similar strategies were more widespread in the country. They also serve as a model as strategies to be implemented in other developing countries.

The Regula Mais Brasil project, through the implementation of teleconsultation in the Unified Health System, demonstrated, in an unprecedented way, how strategies that allow the population's access to different medical specialties can be feasible, in addition to allowing the capillarization of care, through the strategy to take the specialist wherever the patient is and not the other way around. Especially in the period of the SARS-CoV-2 pandemic, this project can maintain access to health care, especially for users with diabetes and sequelae after a stroke, ensuring the maintenance of care [36].

Regarding the teleconsultation diabetes research project, if the efficacy and safety for patients assisted by teleconsultation are confirmed, once the normative issues are resolved by the Federal Council of Medicine (CFM), this will be able to effectively contribute to the promotion access of patients to the public health system, including specialist physicians. Likewise, there may be an increase in the resolvability of the population's health needs, breaking the geographical barrier that a country with continental dimensions, such as Brazil, imposes on the provision and standardization of health, in addition to the potential savings for health systems, which may be used safely and with quality.

This paper has several limitations: one of then is the fact we make a compilation of three different strategies implemented at SUS in a tentative way to delivery medical assistance to the general population. As the initial projects were not designed to analyze the impact on outcomes, for example, the incidence of myocardial infarct or stroke, the authors only could initiate a discussion, showing that e-Health strategies can be performed in a public health system.

5. Conclusions

Although there is evidence in favor of the use of telehealth strategies to deal with CNCDs, in Brazil, due to difficulties in accessing technology in addition to important care gaps and legal impediments, we observe a lack of strategies in this area for health promotion within the scope of the SUS.

Therefore, given the above, there is a need to generate reliable scientific evidence of the effectiveness and safety of applicability within the context of regulation and access to the SUS. In this article, the authors demonstrate how specific initiatives in telehealth, through partnerships between the Ministry of Health and Hospitals of Excellence, can foster the development of new research and assistance strategies within the scope of e-Health in Brazil.

Author Contributions: Conceptualization, D.L.G.R.; methodology, D.L.G.R., G.S.B., I.d.C.B. and M.A.M.; validation, D.L.G.R., G.S.B., I.d.C.B., M.A.M. and A.P.N.M.d.P.; formal analysis, D.L.G.R., G.S.B., I.d.C.B. and M.A.M.; investigation, D.L.G.R., G.S.B., I.d.C.B. and M.A.M.; resources, A.P.N.M.d.P.; data curation, D.L.G.R., G.S.B., I.d.C.B. and M.A.M.; writing—original draft preparation, D.L.G.R., G.S.B., I.d.C.B. and M.A.M.; writing—review and editing, D.L.G.R., G.S.B., I.d.C.B., M.A.M. and A.P.N.M.d.P.; supervision, D.L.G.R.; funding acquisition, A.P.N.M.d.P. All authors have read and agreed to the published version of the manuscript.

Funding: This research was funded by Brazil's Ministry of Health, by the Institutional Development Support Program of the Unified Health System (PROADI-SUS).

Institutional Review Board Statement: The study TELEconsulta Diabetes was conducted according to the guidelines of the ResNo466 of the National Health Council, is the basis for the definition of ethical precepts in research involving humans in Brazil, since we are not signatories to the Declaration of Helsinki and approved by the Ethics Committee of (CAAE: 03434218.1.0000.0070, protocol code 3.623.207 and approval on 5 October 2019).

Informed Consent Statement: Informed consent was obtained from all subjects involved in the study.

Conflicts of Interest: The authors declare no conflict of interest and the funders had no role in the design of the study; in the collection, analyses, or interpretation of data; in the writing of the manuscript, or in the decision to publish the results.

References

1. Freire Filho, J.R.; Silva, C.B.G.; Costa, M.V.; da Forster, A.C. Educação Interprofissional nas políticas de reorientação da formação profissional em saúde no Brasil. *Saúde Debate* **2019**, *43*, 86–96. [CrossRef]
2. Peduzzi, M.; Norman, I.J.; Germani, A.C.C.G.; da Silva, J.A.M.; de Souza, G.C. Educação interprofissional: Formação de profissionais de saúde para o trabalho em equipe com foco nos usuários. *Rev. Esc. Enferm. USP* **2013**, *47*, 977–983. [CrossRef]
3. Tziraki, C.; Grimes, C.; Ventura, F.; O'Caoimh, R.; Santana, S.; Zavagli, V.; Varani, S.; Tramontano, D.; Apóstolo, J.; Geurden, B.; et al. Rethinking palliative care in a public health context: Addressing the needs of persons with non-communicable chronic diseases. *Prim. Health Care Res. Dev.* **2020**, *21*, e32. [CrossRef] [PubMed]
4. World Health Organization. Global Status Report on Noncommunicable Diseases 2010. Available online: https://www.who.int/nmh/publications/ncd_report_full_en.pdf (accessed on 1 September 2021).

5. Malta, D.C.; Andrade, S.S.; Oliveira, T.P.; Moura, L.D.; Prado, R.R.; Souza, M.D. Probabilidade de morte prematura por doenças crônicas não transmissíveis, Brasil e regiões, projeções para 2025. *Rev. Bras. Epidemiol.* **2019**, *22*, e190030. [CrossRef]
6. Malta, D.C.; de Azeredo Passos, V.M.; Machado, Í.E.; Souza, M.D.; Ribeiro, A.L. The GBD Brazil network: Better information for health policy decision-making in Brazil. *Popul. Health Metr.* **2020**, *18*, 23. [CrossRef] [PubMed]
7. Instituto Brasileiro de Geografia e Estatística. (Ed.). *Pesquisa Nacional de Saúde, 2013: Acesso e Utilização dos Serviços de Saúde, Acidentes e Violências: Brasil, Grandes Regiões e Unidades da Federação*; Instituto Brasileiro de Geografia e Estatística-IBGE: Rio de Janeiro, Brazil, 2015; 98p.
8. World Health Organization. *Resolution WHA58.28: eHealth*; WHO: Geneva, Swizterland, 2005. Available online: http://apps.who. int/iris/bitstream/10665/20378/1/WHA58_28-en.pdf?ua=1) (accessed on 27 August 2021).
9. Lima-Toivanen, M.; Pereira, R.M. The contribution of eHealth in closing gaps in primary health care in selected countries of Latin America and the Caribbean. *Rev. Panam. Salud Pública* **2018**, *42*, 1–11. [CrossRef]
10. Silva, A.B.; da Silva, R.M.; Ribeiro, G.D.; Guedes, A.C.; Santos, D.L.; Nepomuceno, C.C.; Caetano, R. Three decades of telemedicine in Brazil: Mapping the regulatory framework from 1990 to 2018. *PLoS ONE* **2020**, *15*, e0242869.
11. Digital in 2020. Available online: https://wearesocial.com/digital-2020 (accessed on 1 September 2021).
12. Uso da TI-Tecnologia de Informação nas Empresas Pesquisa Anual do FGVcia. Available online: https://eaesp.fgv.br/sites/ eaesp.fgv.br/files/u68/fgvcia2021pesti-relatorio.pdf (accessed on 22 August 2021).
13. Unidades de Saúde Serão 100% Conectadas à Internet até o Fim de Abril, Garante Marcos Pontes. Available online: https://www.gov.br/pt-br/noticias/saude-e-vigilancia-sanitaria/2020/04/unidades-de-saude-serao-100-conectadas-a-internet-ate-o-fim-de-abril (accessed on 27 August 2021).
14. Conselho Federal de Medicina. RESOLUÇÃO CFM no2.228/2019. Sect. I 6 March 2019. p. 58. Available online: https: //sistemas.cfm.org.br/normas/visualizar/resolucoes/BR/2019/2228 (accessed on 8 September 2021).
15. Carvalho, C.R.R.; Scudeller, P.G.; Rabello, G.; Gutierrez, M.A.; Jatene, F.B. Use of telemedicine to combat the COVID-19 pandemic in Brazil. *Clinics* **2020**, *75*, e2217. [CrossRef]
16. Decreto n° 7237 de 20 de Julho de 2010. Available online: https://legislacao.presidencia.gov.br/atos/?tipo=DEC&numero=7237 &ano=2010&ato=299c3ZU5EMVpWT10d (accessed on 22 August 2021).
17. PROADI-SUS. Available online: https://hospitais.proadi-sus.org.br/ (accessed on 27 August 2021).
18. Ministério da Saúde. *Telessaúde Brasil Redes*; Ministério da Saúde: Brasília, Brazil, 2015. Available online: http://189.28.128.100 /dab/docs/portaldab/documentos/manual_tecnico_telessaude_preliminar.pdf (accessed on 22 August 2021).
19. Brasil. Ministério da Saúde. Gabinete do Ministro. Portaria no 2.546, de 27 de Outubro de 2011. Diário Oficial da União, Brasília: Ministério da Saúde. 1.html. 2011. Available online: http://bvsms.saude.gov.br/bvs/saudelegis/gm/2011/prt2546_27_10_2011 .html (accessed on 27 August 2021).
20. Hospital Alemão Oswaldo Cruz. Avaliação Diagnóstica: Relatório de Avaliabilidade do Programa Nacional Telessaúde Brasil Redes. 2019. Available online: https://hospitais.proadi-sus.org.br/projetos/96/telessaude-brasil-redes (accessed on 27 August 2021).
21. Ferreira, I.G.; Godoi, D.F.; Perugini, E.R.; de Bastiani Lancini, A.; Zonta, R. Teledermatologia: Uma interface entre a atenção primária e atenção especializada em Florianópolis. *Rev. Bras. Med. Fam. E Comunidade* **2019**, *14*, 2003. [CrossRef]
22. Ribeiro, A.L.; Alkmim, M.B.; Cardoso, C.S.; Carvalho, G.G.; Caiaffa, W.T.; Andrade, M.V.; Cunha, D.F.; Antunes, A.P.; Resende, A.G.; Resende, E.S. Implantação de um sistema de telecardiologia em Minas Gerais: Projeto Minas Telecardio. *Arq. Bras. Cardiol.* **2010**, *95*, 70–78. [CrossRef]
23. Rodrigues, D.L.; Belber, G.S.; Padilha, F.V.; Spinel, L.F.; Moreira, F.R.; Maeyama, M.A.; Pinho, A.P.; Júnior, Á.A. Impact of Teleconsultation on Patients with Type 2 Diabetes in the Brazilian Public Health System: Protocol for a Randomized Controlled Trial (TELEconsulta Diabetes Trial). *JMIR Res. Protoc.* **2021**, *10*, e23679. [CrossRef] [PubMed]
24. Pereira, E.V.; Tonin, F.S.; Carneiro, J.; Pontarolo, R.; Wiens, A. Evaluation of the application of the Diabetes Quality of Life Questionnaire in patients with diabetes mellitus. *Arch Endocrinol. Metab.* **2020**, *64*, 59–65. [CrossRef]
25. Da Silva Etges, A.P.; Paixão Schlatter, R.; Lavanholi Neyeloff, J.; Vianna Araújo, D.; Bahia, L.R.; Cruz, L.; Regina Godoy, M.; da Silva Bittencourt, O.N.; da Rosa, P.R.; Anne Polanczyk, C. Estudos de Microcusteio aplicados a avaliações econômicas em saúde: Uma proposta metodológica para o Brasil. *J. Bras. Econ. Saúde* **2019**, *11*, 87–95.
26. Hamilton, D.F.; Lane, J.V.; Gaston, P.; Patton, J.T.; Macdonald, D.J.; Simpson, A.H.; Howie, C.R. Assessing treatment outcomes using a single question: The Net Promoter Score. *Bone Jt. J.* **2014**, *96-B*, 622–628. [CrossRef] [PubMed]
27. Maldonado, J.M.S.V.; Marques, A.B.; Cruz, A. Telemedicine: Challenges to Dissemination in Brazil. *Cad. Saúde Pública* **2016**, *32* (Suppl. S2). Available online: http://www.scielo.br/scielo.php?script=sci_arttext&pid=S0102-311X2016001402005&lng=en& tlng=en (accessed on 27 August 2021). [CrossRef]
28. A Página Não Foi Encontrada. Available online: https://www.iess.org.br/cms/rep/Telemedicina_Chao.pdf (accessed on 1 September 2021).
29. Ferrari, D.V.; Lopez, E.A. A review of hearing aid teleconsultation in brazil. *J. Hear. Sci.* **2017**, *7*, 9–24.
30. Código de Ética Médica: Resolução CFM no 2.217, de 27 de Setembro de 2018, Modificada Pelas Resoluções CFM no 2.222/2018 e 2.226/2019/Conselho Federal de Medicina–Brasília: Conselho Federal de Medicina. 2019. Available online: https://portal.cfm. org.br/images/PDF/cem2019.pdf (accessed on 27 August 2021).

31. Dall'Alba, R.; Rocha, C.F.; de Pinho Silveira, R.; da Silva Costa Dresch, L.; Vieira, L.A.; Germanò, M.A. COVID-19 in Brazil: Far beyond biopolitics. *Lancet* **2021**, *397*, 579–580. [CrossRef]
32. Caetano, R.; Silva, A.B.; Guedes, A.C.; Paiva, C.C.; Ribeiro, G.D.; Santos, D.L.; Silva, R.M. Desafios e oportunidades para telessaúde em tempos da pandemia pela COVID-19: Uma reflexão sobre os espaços e iniciativas no contexto brasileiro. *Cad. Saúde Pública* **2020**, *36*, e00088920. [CrossRef] [PubMed]
33. Connectivity and Digital Health on the Brazilian Physician Life. Associação Paulista de Medicina, February 2020. Available online: http://associacaopaulistamedicina.org.br/assets/uploads/textos/Pesquisa-APM-2020.pdf (accessed on 10 September 2021).
34. Ribeiro, A.G.; Rodrigues, R.A.M.; Guerreiro, A.M.; Regatieri, C.V.S. A teleophthalmology system for the diagnosis of ocular urgency in remote areas of Brazil. *Arq. Bras. Oftalmol.* **2014**, *77*, 214–218. [CrossRef] [PubMed]
35. Grisolia, A.B.D.; Abalem, M.F.; Lu, Y.; Aoki, L.; Matayoshi, S. Teleophthalmology: Where are we now? *Arq. Bras. Oftalmol.* **2017**, *80*, 401–406. [CrossRef]
36. Mantese, C.E.; da Aquino, E.R.S.; Figueira, M.D.; Rodrigues, L.; Basso, J.; Raupp Da Rosa, P. Telemedicine as support for primary care referrals to neurologists: Decision-making between different specialists when guiding the case over the phone. *Arq. Neuropsiquiatr.* **2021**, *79*, 299–304. [CrossRef] [PubMed]

International Journal of
**Environmental Research
and Public Health**

Article

An eCoach-Pain for Patients with Chronic Musculoskeletal Pain in Interdisciplinary Primary Care: A Feasibility Study

Cynthia Lamper [1,*], Ivan Huijnen [1,2], Maria de Mooij [1], Albère Köke [1,2], Jeanine Verbunt [1,2] and Mariëlle Kroese [3]

[1] Department of Rehabilitation Medicine, Functioning, Participation & Rehabilitation, Faculty of Health, Medicine and Life Sciences, Care and Public Health Research Institute (CAPHRI), Maastricht University, 6200 MD Maastricht, The Netherlands; ivan.huijnen@maastrichtuniversity.nl (I.H.); m.demooij@maastrichtuniversity.nl (M.d.M.); albere.koke@maastrichtuniversity.nl (A.K.); jeanine.verbunt@maastrichtuniversity.nl (J.V.)

[2] Centre of Expertise in Rehabilitation and Audiology, Adelante, 6432 CC Hoensbroek, The Netherlands

[3] Department of Health Services Research, Faculty of Health, Medicine and Life Sciences, Care and Public Health Research Institute (CAPHRI), Maastricht University, 6200 MD Maastricht, The Netherlands; marielle.kroese@maastrichtuniversity.nl

* Correspondence: cynthia.lamper@maastrichtuniversity.nl

Citation: Lamper, C.; Huijnen, I.; de Mooij, M.; Köke, A.; Verbunt, J.; Kroese, M. An eCoach-Pain for Patients with Chronic Musculoskeletal Pain in Interdisciplinary Primary Care: A Feasibility Study. *Int. J. Environ. Res. Public Health* **2021**, *18*, 11661. https://doi.org/10.3390/ijerph182111661

Academic Editors: Irene Torres-Sanchez and Marie Carmen Valenza

Received: 8 October 2021
Accepted: 4 November 2021
Published: 6 November 2021

Abstract: eHealth could support cost-effective interdisciplinary primary care for patients with chronic musculoskeletal pain. This study aims to explore the feasibility of the eCoach-Pain, comprising a tool measuring pain complexity, diaries, pain education sessions, monitoring options, and chat function. Feasibility was evaluated (June–December 2020) by assessing learnability, usability, desirability, adherence to the application, and experiences from patients and general practitioners, practice nurses mental health, and physiotherapists. Six primary healthcare professionals (PHCPs) from two settings participated in the study and recruited 29 patients (72% female, median age 50.0 years (IQR = 24.0)). PHCPs participated in a focus group. Patient data was collected by evaluation questionnaires, individual interviews, and eCoach-Pain-use registration. Patients used the eCoach during the entire treatment phase (on average 107.0 days (IQR = 46.0); 23 patients completed the pain complexity tool and used the educational sessions, and 12 patients the chat function. Patients were satisfied with the eCoach-Pain (median grade 7.0 (IQR = 2.8) on a 0–10 scale) and made some recommendations for better fit with patient-specific complaints. According to PHCPs, the eCoach-Pain is of added value to their treatment, and patients also see treatment benefits. However, the implementation strategy is important for successful use of the eCoach-Pain. It is recommended to improve this strategy and involve a case-manager per patient.

Keywords: chronic musculoskeletal pain; primary care; eHealth; blended care; interdisciplinary care; feasibility; mixed-methods design

1. Introduction

Chronic musculoskeletal pain (CMP) is a significant public health problem occurring in 19–28% of the European population [1,2]. It is expected that this number will increase in the next years, in line with an aging population [3]. The current health system for patients with CMP is fragmented, leading to high societal and healthcare costs [4–6]. Therefore, the World Health Organization (WHO) calls for a change in health systems focusing on interdisciplinary rehabilitation care and the improvement of self-management skills of patients on long term [7]. However, in order to reach this, there is a need for changes in knowledge, skills, and attitudes of healthcare professionals, as well as changes in the organization of healthcare.

Challenges in this change are accessibility and cost-effectiveness of rehabilitation care, for which eHealth can be a solution [8]. eHealth is defined as the use of information and communication technology for health [9,10]. A wide range of eHealth tools (such as mobile

95

applications and online interventions) have been developed to improve self-management for acute and chronic pain, with promising results regarding their effectiveness [11–13]. Several reasons for the additional value of eHealth in the treatment for patients with CMP can be mentioned.

First, current care for chronic pain is fragmented and continuity of care for the individual patient is often lacking. eHealth can improve healthcare organization as it can facilitate communication and collaboration between healthcare professionals of different disciplines [14]. Accordingly, the WHO advises integrating rehabilitation care within and between primary (general practice), secondary (general hospital), and tertiary care (specialized care centers) [7]. They advise to implement eHealth to facilitate continuity of care in integrated health systems by stimulating daily activities and participation of patients, which are rehabilitation goals [15].

Second, currently, healthcare professionals receive training on diagnosis and treatment, primarily focused on knowledge within their own discipline [16]. This ranges from biomedical oriented care focusing on attempts to solve the pain, toward biopsychosocial oriented care which focuses on optimizing functioning despite pain [17,18]. However, the recommended approach by the WHO requires an integral biopsychosocial vision applied by all healthcare professionals. Currently, patients receive various treatment approaches causing confusion, resulting in unsuccessful organization of integrated care. An eHealth application can facilitate an integral vision on pain and a common language, which are components of integrated care [19]. In this way, it supports the treatment program of all participating healthcare professionals.

Third, earlier studies indicated that eHealth improves self-care support and improves daily activities for people with chronic illnesses [20–22]. eHealth, consisting of a combination of tools, might be of added value and useful as part of a blended care intervention. This is studied previously with separate tools for online pain education or keeping track of daily activities and participation in combination with face-to-face consultations [23,24]. The combination of tools is not studied previously and might lead to better informed and more actively involved patients with increased autonomy, as well as a shift of the role of the healthcare professional into adviser or coach [25]. Moreover, it is assumed that this blended care can stimulate integrated care in the long term and decrease healthcare costs [26,27].

To study the additional value of eHealth in an interdisciplinary network of healthcare professionals for patients with CMP, we implemented an electronic Coach (eCoach-Pain) to facilitate pain rehabilitation within the South East of the Netherlands [28,29]. Based on feedback in this earlier performed implementation, the eCoach-Pain is further improved into its current version. The eCoach-Pain aims to support the provision of integrated rehabilitation care with a shared biopsychosocial vision on health within the Network Pain Rehabilitation Limburg. Within this network, patients and Primary Health Care Professionals (PHCPs), existing of general practitioners (GPs), physiotherapists (PTs), and practice nurses mental health (PNMHs) use the eCoach-Pain. It comprises a measurement tool for assessing complexity of the pain problem, diaries, pain education sessions, monitoring options, and a chat function. Whether it is feasible to use in clinical practice is currently unknown. Therefore, this study aims to explore the feasibility of the eCoach-Pain for patients and PHCPs.

2. Materials and Methods

This study (June 2020 and December 2020) had a mixed-methods design. Feasibility was evaluated with a focus on learnability, usability, desirability, adherence to the application, and experiences from patients and PHCPS. These were measured by use of patient questionnaires, data about eCoach-Pain-use, a focus group with PHCPs, and interviews with patients. Ethical approval was obtained from the Medical Ethics Committee Z, the Netherlands (METCZ20190037). Patients did not have to pay for participation in Network Pain Rehabilitation Limburg or the eCoach-Pain. During a patient's first login in the eCoach-Pain, an electronic informed consent for the use of the eCoach-Pain and consent for transferability of their contact details to the researcher were registered. Additionally, for the telephonic interview, patients were asked for informed consent and for recording the interview. PHCPs were asked for informed consent at the start of the focus group.

2.1. Sample and Setting

PHCPs (GPs, PNMHs, and PTs) of two interdisciplinary primary care practices were recruited to participate in this feasibility study ($n = 6$). They all participated in the Network Pain Rehabilitation Limburg (situated in the South East region of the province Limburg, The Netherlands). This network within and between primary, secondary, and tertiary care aims to support a shared biopsychosocial vision regarding CMP, early recognition of patients with subacute complaints, and a person-centered referral and treatment.

For the current project, patients were recruited by the participating PHCPs. They were eligible if they were $\geq$18 years at the start of the study, had CMP or musculoskeletal pain at increased risk of becoming chronic (based among criteria on the STarT MSK tool [30,31]), were willing to improve their functioning despite the pain, and had adequate Dutch literacy to use the eCoach-Pain. Exclusion criteria were pregnancy or any medical (orthopedic, rheumatic, or neurological) or psychiatric disease which could be treated by a more appropriate therapy, according to the expert opinion of the GP.

Once a new patient with CMP, or with an increased risk of developing chronic pain, consulted a PHCP, the patient was asked to use the eCoach-Pain. The PHCP gave the main instructions and sent a manual by email.

2.2. The eCoach-Pain

The eCoach-Pain has been designed by Sananet Care B.V., based on earlier developed eCoaches, such as for Inflammatory Bowel Disease and heart failure [29,32,33]. It contains different goals or opportunities for both patients with CMP and their PHCPs. For patients, the goal is to improve and maintain self-management in coping with pain. For PHCPs the goal is to facilitate biopsychosocial assessment for treatment planning and to monitor the treatment progress of patients with CMP. The eCoach-Pain has been developed in an iterative co-creative development process with the collaboration of researchers, technical experts, patients, and PHCPs. The results of this study will be published elsewhere [28]. The eCoach-Pain can be used on mobile phones, tablets, laptops, and PCs with internet connections.

2.2.1. Application for the Patient

Each patient has an own account for the eCoach-Pain (Figure 1), which could be created in two different ways: First, the PHCP could create a patient's account by filling in the patient's contact details after which the patient receives two emails, one with an account name and one with a password. Consecutively, the patients' account is automatically linked to that of his/her treating PHCP. Second, patients could do a self-subscription throughout a webpage. In this way, a patient is invited to complete contact details, to create a password, and to connect him/herself to his/her own PHCP. Subsequently, the patient's username is sent to the patient by email.

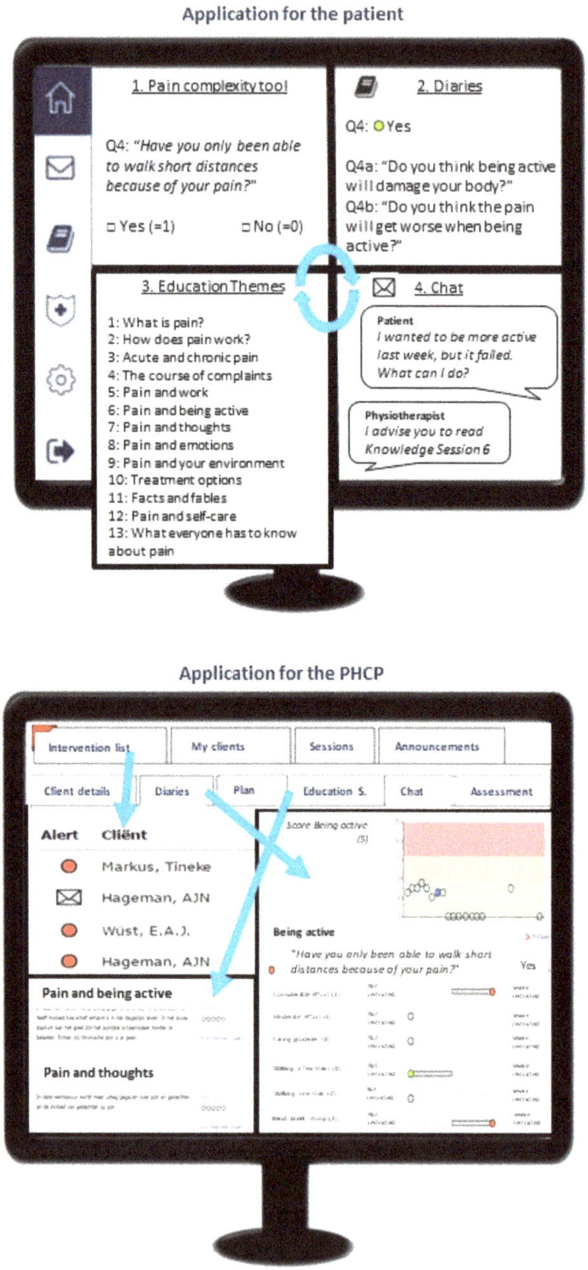

Figure 1. Content of the eCoach-Pain. At the top is the application for the patient, with the pain complexity tool, diaries, educational sessions, and chat function. At the bottom is the application for the primary healthcare professional (PHCP) with an intervention list, an overview of the diaries, and an overview of educational sessions.

After login by the patient, a home screen is presented, which contains four different elements (Figure 1).

1.   The pain complexity tool:

The pain complexity tool supports the PHCPs in their decision-making for problem-mapping and treatment selection. It consists of two parts:

(A)   The STarT MSK Tool assessing the complexity of the pain problem for referral within primary care. The patient's first action in the eCoach-Pain is completing this questionnaire. The Dutch version of the STarT MSK Tool is translated and validated [30,31]. The STarT MSK Tool exists of nine Yes (=1) or No (=0) questions regarding activity level, anxiety, depression, and thoughts about CMP and one Visual Analogue Scale (0–10) to assess pain intensity (0–4 = 0 points, 5–6 = 1 point, 7–8 = 2 points, 9–10 = 3 points). All scores are summed, and a total score of 0–4 indicated a low risk, a total score from 5–8 indicated a moderate risk, and a total score from 9–12 indicated a high risk of developing CMP.

(B)   To further differentiate within the range of primary, secondary and tertiary rehabilitation care an additional set of questions about the biopsychosocial complaints and background of the patient was added to be filled in by the PHCP. After completion of both parts of the complexity tool, the eCoach-Pain calculates the score and assigns the best-fitting referral option to assist the PHCP. The PHCP discusses the results with the patient and refers him/her to the most appropriate treatment via shared decision making.

2.   Diaries:

The eCoach-Pain also contains the possibility to use diaries. The PHCP decides, together with the patient, if and how often diaries will be sent to the patient. Diaries can automatically be sent every week, every two weeks, or once a month. However, the diary option can also be neglected. The automatic setting of these diaries is once a week. The diaries exist of the pain complexity tool with additional questions. This extension exists of additional questions based on the questions in the STarT MSK tool scored with "Yes". The answers to these additional questions could be discussed with the PHCP during consultation and used to adjust the treatment or to provide additional educational material to the patient.

3.   Education sessions:

The educational sessions provide patients background information about topics related to pain and pain-related disability, such as the difference between acute and chronic pain, treatment of pain, biopsychosocial influences on their pain, information about work and pain, and treatment options. The educational sessions are interactive (YouTube videos and quiz questions with feedback on answers), and they are integrated to stimulate learning and improving knowledge about (chronic) pain. The educational materials are presented in 13 themes and per theme subdivided over several sessions (Figure 1).

4.   Chat function:

The chat function is used to send bidirectional messages containing questions or treatment material between patient and PHCP. All communication between patient and PHCP remains accessible in the eCoach-Pain to enable patients to reread answers, advice, or treatment exercise at later moments and times.

2.2.2. Application for the PHCPs

PHCPs could access the eCoach-Pain via a secured webpage on their own device. The PHCPs were instructed to monitor and analyze the patient's situation within a few working days after the patient had completed the pain complexity tool or diary, and to respond as quickly as possible to messages from the patients. To facilitate interpretation of the pain complexity tool and diaries and to save PHCPs' time, information within the application was supportively presented using overviews, graphs, and colored risk flags. Based on the results of the pain complexity tool, different flags appeared on the intervention list: a red flag for a high risk, an orange flag for a medium risk, and a green flag for a low risk for developing CMP (Figure 1).

PHCPs had only access to data of patients treated by themselves. It was possible that more PHCPs, for example, a PT and GP, have access to the data of the same patient in case it was a joint patient. When the PHCP sent a message to a patient or another PHCP, respectively, the other PHCPs and patient were able to read this message in the chat function.

An instruction meeting of one hour to become familiar with the possibilities of the eCoach-Pain was provided to all PHCPs before the start of the study. The software developers and research team facilitated this meeting. Afterward, a paper-copy instruction manual was provided. Moreover, during the pilot, the PHCPS could contact the service desk of the software developers when help was needed or technical issues occurred.

2.3. Data Collection and Analysis

2.3.1. Learnability, Usability, and Desirability

In September 2020, when the PHCPs had used the eCoach-Pain already for approximately three months, the researcher sent a questionnaire to participating patients. The questionnaire assessed learnability (5-items), usability (5-items), and desirability (6-items) and was an adjusted version of a questionnaire used in a study by Hochstenbach et al. (2016) [34]. Usability was defined as 'the extent to which the application could be used by patients with CMP to monitor their pain, physical activity, and participation level effectively, efficiently, and satisfactorily in everyday practice'. Learnability was defined as 'the time and effort required for these patients to use the application'. Desirability was defined as 'the extent to which the application was fun and engaging to use for these patients. Patients rated each item on a 1–5 Likert scale (completely disagree–completely agree); higher scores indicated better learnability, usability, and desirability. A separate item about the recommendation of the eCoach-Pain to family and friends on a 5-point scale, and a separate item about the overall acceptance of the eCoach-Pain for treatment purposes on a 10-point scale, were added.

Before data analysis, negatively-keyed items were reversed-scored using Microsoft Excel, version Professional Plus 2016, the Microsoft Corporation, Santa Rosa, CA, USA. Median scores with interquartile ranges per item and category were calculated. To identify differences between the PHCPs disciplines, a sub-analysis with discipline as dependent variable was performed.

2.3.2. Adherence to the Application

To assess the patients' adherence, process data from the pain complexity tool, diaries (filled out or not, time of fill out, answers), and from the educational sessions (opened or not, time of opening, how often opened) were logged on the server. The data collected between June 2020 and September 2020 were exported in October 2020.

Median scores and interquartile ranges were calculated using Microsoft Excel, for the number of days patients were active in the eCoach-Pain, number of completed pain complexity tools, diaries, educational sessions, and chat messages used.

Moreover, data about the PHCPs was collected. The median and interquartile ranges of the number of log-ins of the PHCPs was registered overall and per PHCP discipline.

2.3.3. Experiences of Patients

Based on stratified probability sampling on sex, age, and PHCP, patients were contacted for a telephonic interview by the researcher to gain more insight into the experiences with the eCoach-Pain in September 2020. It was intended to ask approximately 16 patients, of which eight agreed, until data-saturation would be reached. However, as data saturation was not reached after this number of interviews, five additional interviews were performed in December 2020. Topics discussed in the semi-structured interviews included: the use and acceptance of the pain complexity tool, diaries, educational sessions, and chat function, the supportiveness of the application regarding self-management, and technological functioning of the application. Interviews were audio-recorded.

The audio recordings of the interviews were transcribed verbatim. These written interviews were independently analyzed with inductive and deductive thematic analysis by two researchers (C.L. and M.d.M.) using QSR International Pty Ltd. (Melbourne, Australia) (2018) NVivo (Version 12), https://www.qsrinternational.com/nvivo-qualitative-data-analysis-software/home (accessed on 4 November 2021) [35–37]. Data were sorted based on pre-defined themes of the semi-structured interview-guide. Within these themes, sub-categories were created based on the data. After the first two interviews, main themes and codes were discussed and finalized. Thereafter, all other interviews were analyzed and discussed by adding additional codes under the predefined themes.

2.3.4. Experiences of PHCPs

In September 2020, the researcher (C.L.) and observer (A.K.) held an online focus group interview with all participating PHCPs via Zoom [38]. Technical experts of Sananet B.V. (manufacturers of the eCoach-Pain) were available to take notes for future improvements. Before the start of the focus group, participating PHCPs completed questions about the topics on the agenda for the focus group. This encouraged them to formulate an individual opinion before the focus group started and to share this during the meeting. Topics discussed included: use and acceptance of the application, supportiveness of the application in monitoring, advising and treating patients, fit with daily care, and technical functioning of the application. The focus group was audio-recorded.

During the focus group with the PHCPs, the observer (A.K.) made notes and gave a summary per topic discussed. These summaries were asked to be confirmed by the PHCPs during the focus group. Before analysis, the audio recording was used to add additional notes to the summaries by the researchers (C.L. and M.d.M.). These summaries were independently analyzed with thematic analysis on the topics discussed by two researchers (C.L. and M.d.M.) using QSR International Pty Ltd. (2018) NVivo (Version 12), https://www.qsrinternational.com/nvivo-qualitative-data-analysis-software/home (accessed on 4 November 2021) [35].

3. Results

Data was collected from 29 patients in total; see Table 1. They were eight male and 21 female participations aged between 24–71 years (Median = 50, IQR = 24). The GPs were the primary contact person for 14 patients, the PTs for 13 patients, and the PNMH for two patients. Sixteen patients used the self-registration webpage, while 13 patients were registered by their PHCP. The GP and PNMH of primary care practice one recruited, together, eight patients. The PT of primary care practice one did not recruit any patients. In primary care practice two, 21 patients were recruited by the GP, PNMH, and PT.

Table 1. Patient characteristics.

Participant Number	Sex (%)	Age (y) (Median, IQR)	Days Active in the Coach Till Export (Median, IQR)	First Contact Person	Registration via		Primary Care Practice		Data Available		
					PHCP (*n*, %)	Webpage (*n*, %)	1 (*n*, %)	2 (*n*, %)	Questionnaire (*n*, %) September 2020	Export (*n*, %) October 2020	Interview (*n*, %) September + December 2020
R01	M	41	103	GP	X		X		X	X	
R02	M	61	86	GP		X	X		X	X	X
R03	M	36	124	GP		X	X		X	X	
R04	F	69	109	PT		X		X	X	X	
R05	M	70	54	GP		X		X	X	X	
R06	M	71	115	GP		X		X	X	X	X
R07	F	67	133	PT	X			X	X	X	X
R08	M	70	61	GP		X	X			X	
R09	F	43	150	PT	X			X	X	X	
R10	F	65	7	PT		X		X	X	X	X
R11	F	66	112	GP		X		X	X	X	
R12	F	47	68	PT	X			X	X	X	
R13	F	46	117	PT	X			X		X	X
R14	F	60	92	GP		X	X			X	X
R15	M	62	117	GP		X		X		X	X
R16	F	32	83	GP		X		X		X	X
R17	F	50	n.a.	PNMH	X			X			X
R18	F	41	n.a.	PT	X			X			X
R19	F	45	n.a.	PT	X			X			X
R20	M	50	65	PT	X			X		X	
R21	F	57	98	GP		X	X			X	
R22	F	35	56	PT	X			X		X	
R23	F	44	105	GP		X		X		X	
R24	F	24	127	GP		X	X			X	
R25	F	29	118	PNMH	X		X			X	
R26	F	64	115	GP		X		X		X	
R27	F	40	59	PT	X			X		X	
R28	F	38	144	PT	X			X		X	
R29	F	65	127	PT		X		X		X	
Total:	F:21 72%	50.0—24.0	107.0—46.0	GP: 14, PT: 13, PNMH: 2	13 45%	16 55%	8 28%	21 72%	11 38%	26 90%	11 38%

F: female, M: male, GP: general practitioner, PT: physiotherapist, PNMH: practice nurse mental health. n.a.: not applicable. IQR: interquartile range, *n*: total number, %: percentage of total.

3.1. Learnability, Usability, and Desirability

Twenty-three patients received the invitation for the evaluation questionnaire (September 2020), of whom 11 patients responded (48%). The responders were older (65 (IQR = 23) years old) than the non-responders (48.0 (IQR = 24) years old), and there were less females than males (female responders: 55% (6 out of 11); female non-responders: 83% (10 out of 12)) compared to the total sample.

Six patients started using the eCoach-Pain after the questionnaire was sent and were therefore not invited. The patients who filled in the evaluation questionnaire were an average of 65 (IQR = 23) years old, 55% (6 out of 11) were female, and, on average, active in the eCoach-Pain for 109 (IQR = 41) days. Ten patients answered all questions, and one patient answered only the questions regarding learnability and usability.

Table 2 presents the overall median score (GP and PT), as well as the median score of the categories, and items separately per discipline (by a GP or PT). The scores show that patients learned quickly how to manage the application (Median = 5.0, IQR = 1.0) and could easily use the different components of the eCoach-Pain (Median = 5.0, IQR = 1.5). The desirability was scored with a median score of 4.0, IQR = 2.0. The overall acceptance, rated by the question "I would like to recommend the application to other patients" was scored with a median of 4.0 (IQR = 2.0). Patients gave the eCoach-Pain a total overall score of 7.0 (IQR = 2.8) on a 0–10 Numeric Rating Scale. Patients subscribed by GPs ($n = 6$) scored 5.0 (IQR = 0.0) for learnability, 5.0 (IQR = 1.0) for usability, and 4.5 (IQR = 2.0) for desirability, while patients subscribed by PTs ($n = 5$) scored 5.0 (IQR = 1.0), 5.0 (IQR = 2.0), and 4.0 (IQR = 1.0), respectively.

Table 2. Median (IQR) learnability, usability, and desirability scores for the total patient group, patients subscribed by GPs, and patients subscribed by PTs.* (1–5).

		Subscribed by	
	GP and PT ($n = 11$)	GP ($n = 6$)	PT ($n = 5$)
Learnability	5.0 (1.0)	5.0 (0.0)	5.0 (1.0)
It was easy to learn how to use the application.	5.0 (0.5)	5.0 (0.0)	5.0 (1.0)
I think the application was very complicated. [a]	5.0 (0.5)	5.0 (0.5)	5.0 (1.0)
I needed a lot of help to learn using the application. [a]	5.0 (0.5)	5.0 (0.0)	5.0 (3.0)
I quickly caught on how I could use the application.	5.0 (1.0)	5.0 (0.8)	4.0 (1.0)
I am confident that I used the application in the right way.	5.0 (1.5)	5.0 (0.8)	4.0 (0.8)
Usability	5.0 (1.5)	5.0 (1.0)	5.0 (2.0)
I could easily login into the application.	5.0 (1.0)	5.0 (0.8)	4.0 (2.0)
I could easily report my pain, activities, feelings, thoughts, and emotions.	5.0 (1.5)	5.0 (0.8)	5.0 (3.0)
I understood the information in the educational sessions about my pain, activities, feelings, thoughts, and emotions.	5.0 (1.0)	5.0 (0.8)	5.0 (2.0)
I could easily search for information about pain with the application.	5.0 (2.0)	5.0 (1.5)	4.0 (2.0)
I could easily leave a message for the PHCP via the application.	5.0 (1.5)	4.5 (1.8)	5.0 (0.0)
Desirability	4.0 (2.0) ($n = 10$)	4.5 (2.0)	4.0 (1.0) ($n = 4$)
I liked using the application.	4.0 (1.0)	5.0 (1.5)	4.0 (1.8)
I liked using the pain diary for reporting my pain, activities, feelings, thoughts, and emotions.	4.0 (1.0)	5.0 (0.8)	4.0 (1.3)
I liked using the educational sessions.	4.0 (1.8)	4.5 (2.0)	3.5 (1.0)
I liked using the chat function.	3.0 (1.8)	3.5 (1.8)	3.0 (1.0)
I liked the idea that my PHCP monitors my pain, activities, feelings, thoughts, and emotions.	4.5 (1.0)	4.5 (1.0)	4.5 (1.5)
I liked the idea that my PHCP could adjust my treatment based on my answers in the eCoach-Pain.	4.0 (2.0)	4.0 (1.8)	4.0 (2.3)
I would like to recommend the application to other patients.	4.0 (2.0)	4.5 (1.8)	3.5 (1.0)
Total overall score (0–10)	7.0 (2.8)	8.5 (2.3)	6.5 (0.5)

Scores: 0—totally disagree, 5 or 10—totally agree. [a] Negatively-keyed items were reversed-scored before data analyses but the original question is presented in the table with the reversed-score. * PNMHs have no patients subscribed which completed the questionnaire.

3.2. Adherence to the Application

At the end of October 2020, for 26 of the 29 patients (median age 53.5 years (IQR = 24.75), 69% female (18 out of 26)), the export data about adherence to the application was available. Three patients were asked to participate in the interviews in December 2020. At that moment, exports were already performed; therefore, no export data of them were available. At the moment of data export, the included 26 patients were, on average, 107.0 (IQR = 46.0) days active in the eCoach-Pain. Ten of them stopped using the eCoach-Pain prematurely because they finished treatment ($n = 3$) or did not want to use it anymore ($n = 7$). The other 16 patients were still active in the eCoach-Pain at that time.

Twenty-three patients completed the pain complexity tool (Median = 7.0, IQR = 3.0, 1x low risk, 16x medium risk, 7x high risk). For 21 patients, their PHCPs also answered the second part of the pain complexity tool. On average, the diaries were 6.0 (IQR = 3.5) times filled ($n = 23$).

The educational sessions were opened by 23 patients, and they read, on average, 12.0 (IQR = 5.0) educational sessions per person. On average, each separate educational session in the eCoach-Pain was started by 19.5 (IQR = 6.3) individual patients. In total, 224 unique educational sessions were opened by these patients. As there were 13 sessions, this means that some patients read (a part of) the educational sessions several times. Fourteen patients completed all education sessions.

Twelve messages were sent from seven (27%) unique patients to their PHCP, and five messages were sent from the PHCPs to the patients by the chat function. The patients started all conversations. They often elaborated on their diary answers, technical dysfunction of the eCoach-Pain, or they explained why they were not able to fill in the diaries.

The six PHCPs together logged in on average 6 times (IQR = 16.75), the GPs on average 16 times (IQR = 8), the PTs on average 35.5 times (IQR = 32.5), and the PNMHs on average 2.5 times (IQR = 1.5).

3.3. Experiences of Patients and PHCPs

At the end of September 2020, 16 patients were asked to participate in a telephonic interview, to which eight agreed. To reach data saturation, five additional patients had to be asked, of which three agreed to be interviewed in December 2020. This led to 11 patients participating in the telephonic interviews (mean duration 15 min). The participants had a median age of 60.0 (IQR = 2) years, 73% was female (8 out of 11), and they were active in the eCoach-Pain for 100.5 (IQR = 31.75) days.

Two GPs (one male and one female), two PTs (one male and one female), and one PNMH (one female) participated in a focus group. In addition, one other PNMH (one female) participated in an individual telephonic interview, as she was not able to participate in the focus group.

3.3.1. Overall Opinion and Usage

Patients stated that they were positive about the eCoach-Pain because the functionality worked well for their treatment, it was easy to use, and text was written in clear and understandable language. The content was perceived as informative concerning their pain complaints and knowledge about pain pathophysiology. The interaction between patient and PHCP in the diaries and the quiz questions in the educational sessions of the eCoach-Pain were experienced of added value. Some patients appreciated the reminders for diaries and educational sessions as it gave them structure and control. However, for other patients, these automatic reminders were perceived as somewhat stressful.

Before patients ($n = 6$) started using the eCoach-Pain, they expected the content would be more tailored to their own medical complaints and history. Furthermore, patients expected that the eCoach-Pain would motivate them for treatment compliance to improve their complaints. Although PHCPs are able to change diary frequency, six patients expected less frequent diaries and repetition of information. Besides, some patients indicated that they had preferred to receive more information (by their PHCP or a pamphlet) about the content, frequency of questions, and expected duration of the eCoach-Pain program when they started to use it. Some patients expected more feedback from the eCoach-Pain itself about their answers or an automatic end-session in the eCoach-Pain to close it. Seven patients found it frustrating that, in their opinion, "non-relevant" questions kept returning. The option to indicate a holiday leave and stop sending reminders during this leave was felt to be missing.

R16: "Basically, I think it is a good app. However, the questions appear too frequent, too standard."

Among the PHCPs, the eCoach-Pain was most often used by the PTs. One PT used it to structure the content of the treatment sessions and to deliver additional information to the patient.

PT2: "I like the idea that every week new educational sessions about pain are open for the patient. And, that I can see what the patient answered, which information they have read, and that I can use that during the treatment session. This causes more structure in my treatments."

Another PT used it for educational purposes for the patient, as well, but did not use the results to guide or adjust treatment as the other PT did. In this feasibility study, the PNMHs hardly used the eCoach-Pain because it was not clear for them how to integrate it in their treatment. PNMHs perceived the eCoach-Pain options offered as specifically PTs treatment options. PTs registered most patients by themselves, which gave them control over the number of patients in the eCoach-Pain. Furthermore, controlling this registration facilitated the ability to inform patients before the start. In addition, the GPs used the eCoach-Pain to score the pain complexity assessment and to support the referral of the patient to the PT. They did not use it to offer treatment purposes or pain education to the patient. The patients that entered the study by a GP most often used the self-registration webpage. GPs indicated that this route was timesaving for them. GPs mentioned that the eCoach-Pain provided them an extra treatment option above the current treatment when they referred a patient to the PT.

GP2: "The eCoach-Pain is an extra treatment option above the existing options. As a GP, it is important to know the content of the treatment options when referring to a PT, and it is great that we can offer something extra."

3.3.2. Pain Complexity Tool and Diaries

Patients indicated difficulties in distinguishing the pain complexity tool from the diaries, as the tool and the diaries both were presented as a questionnaire in the eCoach-Pain. Therefore, in this paragraph, the tool and diaries are presented together. Eight of the 11 patients perceived the usability of the pain complexity tool and diaries as good. They indicated that the pain complexity tool and diaries were easy to use, not too time-consuming, easy to understand, and that the reminders by email were of added value. In addition, patients perceived the content as easy to keep track of their pain complaints and the amount of questions as good. However, most patients ($n = 8$) indicated that the repetition and frequency of the questions were too high. They also missed background information of the questions in the introduction of the eCoach-Pain.

R02: "I thought that I had to fill in some questions a few times. However, the questions came every day or week for two or three months. And this was not explained to me beforehand."

Overall, most patients indicated that they perceived the questions in the pain complexity tool and diaries as less applicable in their situation. As several patients had co-morbidities besides CMP, it was difficult for them to know how to interpret the questions. For some questions, it was unclear for them whether the answers should be given with the perspective of having CMP, or from the perspective as a person having pain and other co-morbidities. For example, it was not always possible to indicate exactly their own pain complaints or to adjust answers to questions properly when their situation changed. Sometimes, the eCoach-Pain gave more insight into patients' complexity and impact of their own complaints, which was perceived as heavy to encounter for some patients. Patients without difficulties in daily social participation or psychosomatic problems perceived answer options as less applicable in their specific situation. However, they understood that general questions were formulated for all different kinds of CMP.

The pain complexity tool was the most important tool for GPs in the eCoach-Pain. GP1 indicated that he used it to objectify referral and to get more insight into the complexity of the pain problem. However, GP1 indicated that the digital version directed the referral more than the paper-version. The eCoach-Pain automatically calculates the score and assigns the best-fitting referral option, while, with the paper-version, this can easily be overruled when necessary, according to the opinion of the GP.

GP1: "When using the paper-version, you have more freedom in the choice of the treatment. As you can overrule the score of the patient easier. In the eCoach-Pain, the treatment options are more limited based on the answers of the patients. Which is a strength of the eCoach-Pain."

The two PTs used the pain complexity tools in combination with the diaries. For GP1, the graphical display of the results was especially of added value as it gave insight into the effect of the treatment. As improvement, all PHCPs indicated that the graphical displays of the diaries could be upgraded, as it was not always immediately clear for them if the patient's score was positive or negative.

3.3.3. Educational Sessions

The educational sessions were perceived as interesting with clarifying quiz questions and links to YouTube videos. The sessions about 'What is pain' and 'Pain and being active' were perceived as the most useful sessions. The sessions about work and work disability were not appropriate for every patient as some were retired or did not have a job. Two patients indicated that they desired more subjects and educational sessions, for example, about general health.

R10: "It was a revelation for me, because through the information in the educational sessions, in addition with information on the same topic given by my PHCP, I understand how my brain controls the pain".

The usability and comprehensibility of the educational sessions were perceived as good as the language used was easy to understand. However, three patients found the language level even too easy and the repetition of subjects in the text as too much. One patient indicated that it was more useful for her when the sessions were not divided over several days, but that all sessions can be followed at once.

Overall, five of the eleven patients indicated that they did not receive new information in the educational sessions in comparison with what they already knew about pain (out of earlier treatments). Some other patients indicated that they perceived recognition and acceptance of their CMP during the sessions due to explanations about the pathophysiology of pain. One patient indicated the sessions as confronting as she/he recognized her/himself for the first time as a patient with chronic pain.

R13: "I have read all sessions and the total overview was good for me. But at the same time it was also confronting, maybe that was good, as well."

The educational sessions were most often used by the PTs, and sometimes by the PNMH. PT1 used it to guide the content of his treatment, and PT2 and both PNMHs used it as additional education material for the patient. They indicated that patients were satisfied with the content of the educational sessions and that it gave them more insight into their pain problem. However, they perceived the educational sessions as less applicable for patients with a lower IQ-score or restricted literacy.

3.3.4. Chat Function and Communication with PHCP

Two patients used the chat function, while nine patients indicated they did not. Those two patients were positive about its usability.

Four patients indicated that they had contact with their PT about the diaries and educational sessions they performed in the eCoach-Pain and rated these of added value. For at least one patient, the physiotherapy treatment was adjusted based on the results in the eCoach-Pain. Moreover, some patients discussed the diary questions about their psychosocial status. Furthermore, patients indicated that the pain education received by their PT fitted well with the information in the educational sessions. The eCoach-Pain resulted in a better patient-PT treatment relationship. Three patients had questions about the eCoach-Pain and needed extra support from their PHCP, for example, about the content of the eCoach-Pain, when to finish using the eCoach-Pain, or extra practical tips regarding their pain complaints. Moreover, some patients mentioned that they had discussed technical issues with their PHCP, such as logging in, bugs in the sessions, or difficulties with data exchange between the PHCP and patient.

The other seven patients mentioned no contact with their PHCP about their activities in the eCoach-Pain. Reasons for this ranged from patients' holidays and sick leave periods, technical issues which limited eCoach-Pain-use, or the fact that the patient had not filled in the pain complexity tools and diaries before the next contact with the PHCP could take place.

Patients did not bother with the fact that their PHCP was able to track their activity in the eCoach-Pain, while some of them did not know this option before the interview. Two patients mentioned that they felt no need to discuss their activity in the eCoach-Pain with their PHCP. Most patients indicated that the possibility to discuss their activity online with the eCoach-Pain was of added value, especially in the situation of COVID-19 they were in during the pilot period, as live contact with PHCPs was only limited to emergency consultations.

PHCPs indicated that they did not use the chat function of the eCoach-Pain often as they preferred other ways to communicate with the patient, such as email, chat functions of other applications, or a real-life contact. Furthermore, the fact that they had to log in again to answer these messages was another reason not to use the chat function. GPs mentioned that they did not always communicate with the patient about the results of the eCoach-Pain themselves, but, instead, they asked the PT or PNMH to respond to the patient.

> GP1: *"Because of our work-flow, it is the easiest way that the PT communicates with the patient and has a prominent role in the follow-up."*

They checked if a patient scored a red flag, and only then did they contact the patient or the PT. PT1 discussed the results during nearly each treatment session, while PT2 and the PNMH used a less frequent basis or when the patient had questions about it.

3.3.5. Technical Issues

Six patients did not report any technical issues using the eCoach-Pain. Others mentioned problems in finding how to use all functions of the eCoach-Pain, bugs in sessions, or difficulties connecting their eCoach-Pain to the PHCP's profile. Two patients perceived difficulties with logging in into the eCoach-Pain because they had to renew their password more than once or had to log in several times in a row. Two patients had help from family or friends with logging in, use of a computer, or receiving reminders. There were no problems mentioned with the instruction manual, and nobody contacted the Helpdesk of the software developer during the pilot period. Four patients registered themselves with the self-subscription option via a website without any problems. The others were registered by their PHCP; in one case, the connection between the application of the patient and the application of the PHCP failed.

Overall, PHCPs indicated that the eCoach-Pain is easy to use. However, all PHCPs reported having difficulties with the two-way factor identification for logging-in, which is obligated by the General Data Protection Regulation (GDPR). They perceived a delay in receiving the codes by email or the email is marked as spam. The fact that there is an extra step for logging-in hindered them in using the eCoach-Pain more often. They also indicated that it is difficult for them to combine the eCoach-Pain with other existing applications in daily practice, as each application has its own login system, function, and layout.

3.3.6. Future Usage and Recommendations of the eCoach-Pain

Most patients were satisfied with the eCoach-Pain. Some patients indicated that the eCoach-Pain supports to increase insight in how pain impacts daily activities and participation and that it answers questions about their pain complaints. Moreover, they recommend it for the use of the chat function with their PHCP. Some patients would recommend the eCoach-Pain because they were satisfied with it themselves. Most of them would recommend it to patients with other complaints than their own, as they indicated that the content did not fit perfectly based on their own situation. They would especially recommend it to patients who are recently diagnosed, have problems in daily activities and participation, are low literate, or who want to use an eCoach-Pain frequently.

> *R16: "I would recommend it to people who get acquainted with pain complaints, or who have not so much knowledge yet, for them it is useful to get to know more about pain. But for people who have complaints for years, like me, I would not recommend it."*

They would not recommend it to patients with pain complaints for years, elderly who are not familiar with eHealth, or patients who do not want to use the eCoach-Pain frequently. However, some patients who are recently diagnosed would recommend it for patients with chronic complaints.

Most PHCPs indicated that they will keep on using the eCoach-Pain in the future as they find it important to offer the patient something extra besides usual care. However, due to time constraints in daily practice, GPs hope that PNMHs can get a more prominent role in the follow-up of patients and contact other PHCPs about the results in the eCoach-Pain. In this case, the PNMH has to contact the GP when expertise or referral of the patient is needed.

> *GP1: "Because of the high work-load in primary care, it would be of added value when someone as a PNMH can get a more prominent role in follow-up of patients. They will also be able to keep track of the eCoach-Pain activities. We as GPs have not enough time to do this properly."*

PTs think they will keep on using the eCoach-Pain in the same way as they did during the feasibility study. However, all PHCPs indicated that the costs of the eCoach-Pain concerning their patient volume are important indicators for future usage. During this feasibility study, these costs were covered by the project budget of NPRL.

4. Discussion

The current study provides insight into the feasibility of an eCoach-Pain for patients with CMP or a high risk of becoming chronic, and for PHCPs in interdisciplinary primary care. In general, patients and PHCPs had positive experiences using the eCoach Pain. The answers to questions/statements about learnability, usability, desirability, and adherence to the application confirm that the eCoach Pain has sufficient quality for further use. However, some further adjustments for successful implementation and use are needed.

Some patients mentioned that the content of the eCoach-Pain does not fit with their situation, such as multi-morbidities and previous experiences with treatment. An explanation why patients do not find the eCoach-Pain suitable could be that patients with CMP often experience multi-morbidities, such as depression, anxiety disorders, obesity, hypertension, and diabetes [39–42]. It has been shown that these patients with multi-morbidities need a personalized treatment [43,44]. The eCoach-Pain has not enough attention for these multi-morbidities. Some patients mentioned that the eCoach-Pain was more suitable for patients with other complaints than they had. Remarkably, the patients with severe complaints for several years mentioned that the eCoach-Pain was better suitable for patients in a subacute phase or those recently diagnosed. Patients with complaints for several years indicated that the information about CMP in the eCoach-Pain was not new and perceived the education sessions as too basic for them. They indicated that the pain education was given in earlier treatments. However, patients with subacute complaints mentioned that the eCoach-Pain is better suited to patients with a clear diagnosis or, in contrast to the comment of patient with long-term complaints, patients with a longer duration of complaints. A possible explanation for this finding could be that subacute patients are still searching for an explanation and solution for their complaints and are, therefore, more biomedically oriented and not yet focused on a biopsychosocial treatment. [45]. As accepting of CMP is an ongoing process, it could be that the patients with subacute musculoskeletal pain do not see themselves as patients with CMP [46]. Therefore, further research is needed to discover for which patients group(s) the eCoach-Pain can be used in primary care and, accordingly, how the eCoach-Pain can be aligned for personalized treatment.

The eCoach-Pain is well integrated into the treatment of the PTs. All PHCPs perceived advantages of the use of the eCoach-Pain during physiotherapy treatment. Patients indicated that eCoach-Pain connects to the treatment of the PT. Positive thoughts about blended rehabilitation care for other diseases are also seen in several other studies [47,48]. The integration of an eCoach as blended physiotherapy care for patients with temporomandibular disorders lead to an increase in self-efficacy, support of data collection and personalization of the application in the Netherlands [47]. The review of Orlando et al. (2019) showed an overall positive impact on patient and caregivers' satisfaction and it appears to enhance communication and engagement between healthcare professionals for different kinds of telehealth in rural settings [48]. However, questions about the integration of an eCoach in the treatment, such as duration of the treatment, fit in each consultation, and the frequency of the consultations remain [49]. Tilburg et al. recommend to integrate an eCoach into the total treatment and not to implement it as a separate component to the treatment. Further research needs to design and evaluate the integration of the eCoach-Pain into the treatment to deliver blended care.

eCoaches can stimulate and influence interdisciplinary collaboration in primary care [50]. Based on the findings that PHCPs indicated suboptimal collaboration during treatment, it can be concluded that interdisciplinary collaboration between the PHCPs was a point of attention. Accordingly, it seems that the eCoach-Pain did not contribute to interdisciplinary care. GPs indicated that they preferred to refer the patient automatically to a PT as they had not enough time to contact patients and discuss the treatment plan with the PTs, as purposed in interdisciplinary care. The preference of GPs, due to their lack of time, for treatment of these patients by a PT is in line with another study with eCoaches in primary care in the Netherlands [51]. In this study, the PHCPs perceived advantages of the eCoach-Pain in referring a patient to a PT or adding an extra role of a case-manager (for instance, a PNMH or specialized nurses for mental health) in the future. Previous successful implemented eCoaches used a case-manager as first contact for patients [32,34]. In addition, the Standard of Care for Chronic Pain in the Netherlands advices the use of a case-manager for patients with CMP in primary care [14]. However, the eCoaches in these earlier studies were all implemented in secondary care. Therefore, the role for case-manager in primary care needs to be optimized before implementation. Currently, there is no regular financing of a case-manager for patients with CMP in primary care in the Netherlands yet. However, it is crucial to have a case-manager when focusing on integrated and interdisciplinary primary care to stimulate a common vision and treatment plan [19].

Some patients mentioned technical problems that limited the use of diaries or education sessions, even though they received a reminder. Although the developers could not find an explanation for this, it could have influenced the adherence rates for the diaries and education sessions. Other patients indicated that they received too many reminders for diaries. Therefore, in future use of the eCoach-Pain, attention must be given to the communication between the PHCPs and patients. The PHCPs must discuss in advance the number of diaries and reminders offered, based on the preferences of the patients. Research has shown that shared-decision making for chronic illness with treatments containing more than one session leads to treatment agreement [52]. Therefore, shared-decision making in eCoach-Pain adjustments could lead to increased treatment adherence. Connection to the electronic patient file is another technical problem mentioned. PHCPs, and especially

GPs, experienced barriers in the use of the eCoach-Pain in daily practice which was not connected with their electronic patient file. This caused double registration steps, which was a reason to restrict use of the eCoach-Pain. Therefore, it would be favored to find a possibility to integrate the eCoach-Pain in the electronic patient file (Dutch: Huisarts Informatie Systeem) to avoid extra registration steps [53].

A major strength of this study is the use of qualitative and quantitative data (mixed-methods) alongside objective data on use of the eCoach-Pain from both patients' and PHCPs' perspectives. These data gave a broad overview of the usability of the eCoach-Pain, as well as the experiences. Moreover, the content of the eCoach-Pain was developed together with the PHCPs before the start of the study with a user-centered design [28]. Higgins et al. (2018) recommend user-centered designs and implementation science methods to improve the availability of eHealth tools. [54]. However, the GPs and PNMH rarely used the eCoach-Pain despite their influence in the development process. Reasons for this are the login-facility and lack of time during and after consultations, which are also seen as barriers in the study of Daniëls et al. (2019) in primary care [51].

Some limitations of this study need to be acknowledged. First, the small sample of patients that were available for this study and the limited use of the eCoach-Pain could have introduced selection bias. It could be that patients with, for example, low literacy or co-morbidities, were not asked for participation by the PHCPs with a risk for selection bias. However, despite the small sample, patients differed in demographic characteristics, resulting in sufficient confidence to have studied a representative group of users. Second, not all patients performed all measurements, so the completeness of available data per measurement differed. Six patients did not respond on the evaluation questionnaires, and, for three patients export data of eCoach-Pain-use is missing. This could have led to information bias which could have influenced the data. Third, the sample of primary care practices and PHCPs was small, which could have influenced the results. As for primary care practice 1, all patients are subscribed by a GP or PNMH, and, for primary care practice 2, the PT also subscribed patients, besides the GP and PNMH. Most patients were recruited by primary care practice 2 ($n = 21$), and most of the patients participating in the interviews were also recruited by this practice (9 out of 11). Therefore, limited results were available about the recruitment of GPs in the interviews.

5. Conclusions

In conclusion, the eCoach-Pain seems to be promising in primary care: the patients, as well as the PHCPs, experienced advantages for treatment of patients with CMP. However, adjustments to the content have to be made for better fit with patient-specific CMP complaints. Moreover, the implementation strategy seems to be an important factor for successful use among PHCPs. This should be improved for successful use in interdisciplinary primary care settings. The involvement of a case-manager for CMP should be further explored when implementing the eCoach-Pain. Thereby, it is important to use user-centered designs and implementation science methods to evaluate adjustments resulting in a successful implementation.

Author Contributions: C.L. was involved in data curation, formal analysis, investigation, methodology, project administration, supervision, validation, visualization, and writing—original draft. I.H. was involved in supervision, methodology, validation, and writing—review & editing. M.d.M. was involved in formal analysis, investigation, validation, and writing—review & editing. A.K. was involved in investigation, methodology, supervision, and writing—review & editing. M.K. and J.V. were involved in supervision and writing—review & editing. All authors have read and agreed to the published version of the manuscript.

Funding: Health Insurance Companies CZ, VGZ, and Zilveren Kruis Achmea funded the study, but they had no role in the design of the study and collection, analysis, and interpretation of data and in writing the manuscript.

Institutional Review Board Statement: The study was conducted according to the guidelines of the Declaration of Helsinki, and approved by the Medical Ethics Committee Z, the Netherlands (METCZ20190037, latest date approved 19/05/2020). Informed consent will be obtained from study participants.

Informed Consent Statement: Informed consent was obtained from all subjects involved in the study.

Data Availability Statement: The data presented in this study are available on request from the corresponding author. The data are not publicly available due to the small number of healthcare professionals and patients included. This could interfere with the privacy of the participants.

Conflicts of Interest: C.L., M.d.M. and M.K. have nothing to disclose. I.H., A.K. and J.V. report grants from Health Insurance Companies CZ, VGZ, and Zilveren Kruis Achmea, during the conduct of the study.

References

1. Breivik, H.; Collett, B.; Ventafridda, V.; Cohen, R.; Gallacher, D. Survey of chronic pain in Europe: Prevalence, impact on daily life, and treatment. *Eur. J. Pain* **2006**, *10*, 287–333. [CrossRef] [PubMed]
2. Cieza, A.; Causey, K.; Kamenov, K.; Hanson, S.W.; Chatterji, S.; Vos, T. Global estimates of the need for rehabilitation based on the Global Burden of Disease study 2019: A systematic analysis for the Global Burden of Disease Study. *Lancet* **2020**, *396*, 2006–2017. [CrossRef]
3. Fayaz, A.; Croft, P.; Langford, R.M.; Donaldson, L.J.; Jones, G.T. Prevalence of chronic pain in the UK: A systematic review and meta-analysis of population studies. *BMJ Open* **2016**, *6*, e010364. [CrossRef] [PubMed]
4. Bekkering, G.E.; Bala, M.M.; Reid, K.; Kellen, E.; Harker, J.; Riemsma, R.; Huygen, F.J.; Kleijnen, J. Epidemiology of chronic pain and its treatment in The Netherlands. *Neth J. Med.* **2011**, *69*, 141–153. [PubMed]
5. Breivik, H.; Eisenberg, E.; O'Brien, T.; Openminds. The individual and societal burden of chronic pain in Europe: The case for strategic prioritisation and action to improve knowledge and availability of appropriate care. *BMC Public Health* **2013**, *13*, 1229. [CrossRef] [PubMed]
6. Gaskin, D.J.; Richard, P. The economic costs of pain in the United States. *J. Pain* **2012**, *13*, 715–724. [CrossRef]
7. World Health Organization. *Rehabilitation in Health Systems*; World Health Organization: Geneva, Switzerland, 2017.
8. Roberts, A.; Philip, L.; Currie, M.; Mort, A. Striking a balance between in-person care and the use of eHealth to support the older rural population with chronic pain. *Int. J. Qual Stud. Health Well-Being* **2015**, *10*, 27536. [CrossRef] [PubMed]
9. World Health Organization. *Global Observatory for eHealth. Building Foundations for eHealth: Progress of Member States: Report of the Global Observatory for eHealth*; World Health Organization: Geneva, Switzerland, 2006.
10. Eysenbach, G. What is e-health? *J. Med. Internet Res.* **2001**, *3*, E20. [CrossRef] [PubMed]
11. Özden, F.; Sarı, Z.; Karaman, Ö.N.; Aydoğmuş, H. The effect of video exercise-based telerehabilitation on clinical outcomes, expectation, satisfaction, and motivation in patients with chronic low back pain. *Ir. J. Med. Sci.* **2021**, 1–11. [CrossRef]
12. van de Graaf, D.L.; Trompetter, H.R.; Smeets, T.; Mols, F. Online Acceptance and Commitment Therapy (ACT) interventions for chronic pain: A systematic literature review. *Internet Interv.* **2021**, *26*, 100465. [CrossRef]
13. McGuire, B.E.; Henderson, E.M.; McGrath, P.J. Translating e-pain research into patient care. *Pain* **2017**, *158*, 190–193. [CrossRef]
14. Dutch Pain Society en het Samenwerkingsverband Pijnpatiënten. Zorgstandaard Chronische Pijn. Available online: https://www.zorginzicht.nl/kwaliteitsinstrumenten/chronische-pijn (accessed on 1 November 2021).
15. World Health Organization. *Digital Health Investements should be Coordinated to Support Continuity of Care*; World Health Organization: Geneva, Switzerland, 2021.
16. Niv, D.; Devor, M.; European Federation of, I.C. Position paper of the European Federation of IASP Chapters (EFIC) on the subject of pain management. *Eur J. Pain* **2007**, *11*, 487–489. [CrossRef]
17. Committee on Education of the EFIC (European Federation of IASP Chapters). *The Pain Management Core Curriculum for European Medical Schools*; European Federation of IASP: Washington, DC, USA, 2013; p. 33.
18. Domenech, J.; Sanchez-Zuriaga, D.; Segura-Orti, E.; Espejo-Tort, B.; Lison, J.F. Impact of biomedical and biopsychosocial training sessions on the attitudes, beliefs, and recommendations of health care providers about low back pain: A randomised clinical trial. *Pain* **2011**, *152*, 2557–2563. [CrossRef]
19. Bronstein, L.R. A Model for Interdisciplinary Collaboration. *Soc. Work* **2003**, *48*, 297–306. [CrossRef]
20. Puntillo, F.; Giglio, M.; Brienza, N.; Viswanath, O.; Urits, I.; Kaye, A.D.; Pergolizzi, J.; Paladini, A.; Varrassi, G. Impact of COVID-19 pandemic on chronic pain management: Looking for the best way to deliver care. *Best Pract. Res. Clin. Anaesthesiol.* **2020**, *34*, 529–537. [CrossRef]
21. Slattery, B.W.; Haugh, S.; O'Connor, L.; Francis, K.; Dwyer, C.P.; O'Higgins, S.; Egan, J.; McGuire, B.E. An Evaluation of the Effectiveness of the Modalities Used to Deliver Electronic Health Interventions for Chronic Pain: Systematic Review With Network Meta-Analysis. *J. Med. Internet Res.* **2019**, *21*, e11086. [CrossRef]

22. Lindberg, B.; Nilsson, C.; Zotterman, D.; Söderberg, S.; Skär, L. Using Information and Communication Technology in Home Care for Communication between Patients, Family Members, and Healthcare Professionals: A Systematic Review. *Int. J. Telemed. Appl.* **2013**, *2013*, 461829. [CrossRef]

23. Moseley, G.L.; Nicholas, M.K.; Hodges, P.W. A randomized controlled trial of intensive neurophysiology education in chronic low back pain. *Clin. J. Pain* **2004**, *20*, 324–330. [CrossRef]

24. Malfliet, A.; Kregel, J.; Meeus, M.; Roussel, N.; Danneels, L.; Cagnie, B.; Dolphens, M.; Nijs, J. Blended-Learning Pain Neuroscience Education for People With Chronic Spinal Pain: Randomized Controlled Multicenter Trial. *Phys. Ther.* **2017**, *98*, 357–368. [CrossRef]

25. Ekman, I.; Swedberg, K.; Taft, C.; Lindseth, A.; Norberg, A.; Brink, E.; Carlsson, J.; Dahlin-Ivanoff, S.; Johansson, I.-L.; Kjellgren, K.; et al. Person-Centered Care—Ready for Prime Time. *Eur. J. Cardiovasc. Nurs.* **2011**, *10*, 248–251. [CrossRef]

26. Koppenaal, T.; Arensman, R.M.; van Dongen, J.M.; Ostelo, R.W.J.G.; Veenhof, C.; Kloek, C.J.J.; Pisters, M.F. Effectiveness and cost-effectiveness of stratified blended physiotherapy in patients with non-specific low back pain: Study protocol of a cluster randomized controlled trial. *BMC Musculoskelet. Dis.* **2020**, *21*, 265. [CrossRef] [PubMed]

27. Keogh, E.; Rosser, B.A.; Eccleston, C. e-Health and chronic pain management: Current status and developments. *Pain* **2010**, *151*, 18–21. [CrossRef] [PubMed]

28. Cynthia, L.; Ivan, P.J.H.; Mariëlle, E.A.L.K.; Albère, J.K.; Gijs, B.; Dirk, R.; Jeanine, A.V. Exploring the feasibility of a network of organizations for pain rehabilitation: What are the lessons learned? *Res. Square* **2021**. [CrossRef]

29. Sananet Care BV. Sananet: Specialist in eHealth. Available online: https://www.sananet.nl/ (accessed on 27 August 2021).

30. Dunn, K.; Campbell, P.; Afolabi, E.; Lewis, M.; van der Windt, D.; Hill, J.; Mallen, C.; Protheroe, J.; Hay, E.; Foster, N. Refinement and validation of the keele start msk tool for musculoskeletal pain in primary care. *Rheumatology* **2017**, *56*. [CrossRef]

31. van den Broek, A.G.; Kloek, C.J.J.; Pisters, M.F.; Veenhof, C. Construct validity and test-retest reliability of the Dutch STarT MSK tool in patients with musculoskeletal pain in primary care physiotherapy. *PLoS ONE* **2021**, *16*, e0248616. [CrossRef]

32. de Jong, M.J.; van der Meulen-de Jong, A.E.; Romberg-Camps, M.J.; Becx, M.C.; Maljaars, J.P.; Cilissen, M.; van Bodegraven, A.A.; Mahmmod, N.; Markus, T.; Hameeteman, W.M.; et al. Telemedicine for management of inflammatory bowel disease (myIBDcoach): A pragmatic, multicentre, randomised controlled trial. *Lancet* **2017**, *390*, 959–968. [CrossRef]

33. Amin, H.; Weerts, J.; Brunner-La Rocca, H.P.; Knackstedt, C.; Sanders-van Wijk, S. Future perspective of heart failure care: Benefits and bottlenecks of artificial intelligence and eHealth. *Future Cardiol.* **2021**, *17*, 917–921. [CrossRef]

34. Hochstenbach, L.M.; Zwakhalen, S.M.; Courtens, A.M.; van Kleef, M.; de Witte, L.P. Feasibility of a mobile and web-based intervention to support self-management in outpatients with cancer pain. *Eur. J. Oncol. Nurs.* **2016**, *23*, 97–105. [CrossRef]

35. QSR International Pty Ltd. (2018) NVivo (Version 12). Available online: https://www.qsrinternational.com/nvivo-qualitative-data-analysis-software/home (accessed on 4 November 2021).

36. Vaismoradi, M.; Turunen, H.; Bondas, T. Content analysis and thematic analysis: Implications for conducting a qualitative descriptive study. *Nurs. Health Sci.* **2013**, *15*, 398–405. [CrossRef]

37. Kiger, M.E.; Varpio, L. Thematic analysis of qualitative data: AMEE Guide No. *Med. Teach.* **2020**, *42*, 846–854. [CrossRef]

38. Yuan, E.S. *Zoom*; Zoom Video Communications, Inc.: San Jose, CA, USA, 2011.

39. Bair, M.J.; Robinson, R.L.; Katon, W.; Kroenke, K. Depression and pain comorbidity: A literature review. *Arch. Intern. Med.* **2003**, *163*, 2433–2445. [CrossRef] [PubMed]

40. Tsang, A.; Von Korff, M.; Lee, S.; Alonso, J.; Karam, E.; Angermeyer, M.C.; Borges, G.L.; Bromet, E.J.; Demytteneare, K.; de Girolamo, G.; et al. Common chronic pain conditions in developed and developing countries: Gender and age differences and comorbidity with depression-anxiety disorders. *J. Pain* **2008**, *9*, 883–891. [CrossRef] [PubMed]

41. Walsh, T.P.; Arnold, J.B.; Evans, A.M.; Yaxley, A.; Damarell, R.A.; Shanahan, E.M. The association between body fat and musculoskeletal pain: A systematic review and meta-analysis. *BMC Musculoskelet. Dis.* **2018**, *19*, 1–13. [CrossRef] [PubMed]

42. Williams, A.; Kamper, S.J.; Wiggers, J.H.; O'Brien, K.M.; Lee, H.; Wolfenden, L.; Yoong, S.L.; Robson, E.; McAuley, J.H.; Hartvigsen, J.; et al. Musculoskeletal conditions may increase the risk of chronic disease: A systematic review and meta-analysis of cohort studies. *BMC Med.* **2018**, *16*, 167. [CrossRef]

43. Rijken, M.; Korevaar, J. *Goede Zorg Voor Mensen met Multimorbiditeit: Handvatten Voor de Ontwikkeling en Evaluatie van Zorg Voor Mensen met Meerdere Chronische Aandoeningen*; Nivel: Utrecht, The Netherlands, 2021.

44. Smeets, R.G.M.; Kroese, M.; Ruwaard, D.; Hameleers, N.; Elissen, A.M.J. Person-centred and efficient care delivery for high-need, high-cost patients: Primary care professionals' experiences. *BMC Fam. Pract.* **2020**, *21*, 106. [CrossRef]

45. Nijs, J.; Roussel, N.; Paul van Wilgen, C.; Köke, A.; Smeets, R. Thinking beyond muscles and joints: Therapists' and patients' attitudes and beliefs regarding chronic musculoskeletal pain are key to applying effective treatment. *Man. Ther.* **2013**, *18*, 96–102. [CrossRef]

46. Pietila Holmner, E.; Stalnacke, B.M.; Enthoven, P.; Stenberg, G. "The acceptance" of living with chronic pain—an ongoing process: A qualitative study of patient experiences of multimodal rehabilitation in primary care. *Pain Med.* **2018**, *50*, 73–79.

47. van der Meer, H.A.; de Pijper, L.; van Bruxvoort, T.; Visscher, C.M.; Nijhuis-van der Sanden, M.W.G.; Engelbert, R.H.H.; Speksnijder, C.M. Using e-Health in the physical therapeutic care process for patients with temporomandibular disorders: A qualitative study on the perspective of physical therapists and patients. *Disabil. Rehabil.* **2020**, 1–8. [CrossRef]

48. Orlando, J.F.; Beard, M.; Kumar, S. Systematic review of patient and caregivers' satisfaction with telehealth videoconferencing as a mode of service delivery in managing patients' health. *PLoS ONE* **2019**, *14*, e0221848. [CrossRef]

49. van Tilburg, M.L.; Kloek, C.J.J.; Staal, J.B.; Bossen, D.; Veenhof, C. Feasibility of a stratified blended physiotherapy intervention for patients with non-specific low back pain: A mixed methods study. *Physiother. Theory Pract.* **2020**, 1–13. [CrossRef]
50. Zorgimpuls. e-Health in de Eerste Lijn, Wat Kunnen We Ermee? Available online: https://www.zorgimpuls.nl/upload/files/Artikelen%20ZorgImpuls/ZorgImpuls%20whitepaper%20e-health%20totaal_feb%202018.pdf (accessed on 1 November 2021).
51. Daniels, N.E.M.; Hochstenbach, L.M.J.; van Bokhoven, M.A.; Beurskens, A.; Delespaul, P. Implementing Experience Sampling Technology for Functional Analysis in Family Medicine—A Design Thinking Approach. *Front. Psychol.* **2019**, *10*, 2782. [CrossRef] [PubMed]
52. Joosten, E.A.; DeFuentes-Merillas, L.; de Weert, G.H.; Sensky, T.; van der Staak, C.P.; de Jong, C.A. Systematic review of the effects of shared decision-making on patient satisfaction, treatment adherence and health status. *Psychother. Psychosom.* **2008**, *77*, 219–226. [CrossRef] [PubMed]
53. Nederlands Huisartsen Genootschap. NHG-Standpunt: E-Health Voor Huisarts en Patiënt. Available online: https://www.nhg.org/nhg-e-health (accessed on 31 March 2021).
54. Higgins, K.S.; Tutelman, P.R.; Chambers, C.T.; Witteman, H.O.; Barwick, M.; Corkum, P.; Grant, D.; Stinson, J.N.; Lalloo, C.; Robins, S.; et al. Availability of researcher-led eHealth tools for pain assessment and management: Barriers, facilitators, costs, and design. *PAIN Rep.* **2018**, *3*, e686. [CrossRef] [PubMed]

International Journal of
*Environmental Research
and Public Health*

Article

Patient Experience and Satisfaction with an e-Health Care Management Application for Inflammatory Bowel Diseases

Aria Zand [1,2,*], Audrey Nguyen [1], Courtney Reynolds [1], Ariela Khandadash [1], Eric Esrailian [1] and Daniel Hommes [1,2]

[1] UCLA Center for Inflammatory Bowel Diseases, Vatche and Tamar Manoukian Division of Digestive Diseases, David Geffen School of Medicine, University of California at Los Angeles, Los Angeles, CA 90095, USA; audrey.d.nguyen@ucla.edu (A.N.); courtney.reynolds@gmail.com (C.R.); akhandad@uci.edu (A.K.); EEsrailian@mednet.ucla.edu (E.E.); d.w.hommes@gmail.com (D.H.)
[2] Department of Digestive Diseases, Leiden University Medical Center, 2333 ZA Leiden, The Netherlands
* Correspondence: azand89@gmail.com

Citation: Zand, A.; Nguyen, A.; Reynolds, C.; Khandadash, A.; Esrailian, E.; Hommes, D. Patient Experience and Satisfaction with an e-Health Care Management Application for Inflammatory Bowel Diseases. *Int. J. Environ. Res. Public Health* **2021**, *18*, 11747. https://doi.org/10.3390/ijerph182211747

Academic Editors: Marie Carmen Valenza, Irene Torres-Sanchez and Paul B. Tchounwou

Received: 19 August 2021
Accepted: 2 November 2021
Published: 9 November 2021

Publisher's Note: MDPI stays neutral with regard to jurisdictional claims in published maps and institutional affiliations.

Abstract: Background: Rising healthcare expenditures have been partially attributed to suboptimal management of inflammatory bowel diseases (IBD). Electronic health interventions may help improve care management for IBD patients, but there is a need to better understand patient perspectives on these emerging technologies. Aims: The primary aim was to evaluate patient satisfaction and experience with the UCLA eIBD mobile application, an integrative care management platform with disease activity monitoring tools and educational modules. The secondary objective was to capture patient feedback on how to improve the mobile application. Methods: We surveyed IBD patients treated at the UCLA Center for Inflammatory Bowel Diseases. The patient experience survey assessed the patients' overall satisfaction with the application, perception of health outcomes after participation in the program, and feedback on educational modules as well as areas for application improvement. Results: 50 patients were included. The responses indicated that the patients were greatly satisfied with the ease of patient–provider communication within the application and appointment scheduling features (68%). A majority of respondents (54%) also reported that program participation resulted in improved perception of disease control and quality of life. Lastly, a majority of participants (79%) would recommend this application to others. Conclusions: Mobile tools such as UCLA eIBD have promising implications for integration into patients' daily lives. This patient satisfaction study suggests the feasibility of using this mobile application by patients and providers. We further showed that UCLA eIBD and its holistic approach led to improved patient experience and satisfaction, which can provide useful recommendations for future electronic health solutions.

Keywords: electronic health; mobile health; mobile applications; chronic disease management; patient experience; inflammatory bowel disease

1. Introduction

Value-based healthcare (VBHC) can be described as the systematic pursuit of the triple aim in healthcare: to improve the individual's experience, improve health outcomes, and reduce costs [1]. The concept of VBHC is particularly ready for application to long-term management of chronic illnesses since rising healthcare expenditures have been partially attributed to suboptimal management of chronic illnesses, including IBD [2]. The estimated annual disease-attributable cost of IBD is $6.3 billion [3]. Hospitalizations represented over a third of costs, outpatient services—one third. Reducing hospitalization and readmission rates, therefore, continues to be a challenge in chronic disease management. There is clearly an opportunity to reduce costs by increasing the efficiency of outpatient care and preventing hospitalizations.

Electronic health (e-health) interventions are one solution for more effective IBD care management beyond the clinical setting, both in terms of patient outcomes and cost

115

reduction. Smartphone applications are widely available for consumers, and the large population of smartphone users make applications useful tools to manage chronic illnesses like IBD [4]. In fact, smartphone devices with mobile applications and short message reminders have been used effectively by patients with IBD of mild or moderate severity [5].

Furthermore, mobile health technologies have been shown to improve patient outcomes and quality of life [6]. Patient satisfaction with mobile technologies has been observed for many chronic diseases, including asthma [7], HIV [8], diabetes [9], atrial fibrillation [10], and IBD [11]. IBD patients generally have positive views on mobile applications, but there are desirable improvements. A study of Con et al. [11] surveying 86 IBD patients found that 98.8% of the participants were willing to use communication technologies for IBD management, with mobile applications being one of the top two preferred forms. These previous IBD mobile technologies were often created to assess a major single aspect such as quality of life [5], education curriculum [12], or diets [13]. Additional features that patients seek in their chronic disease management applications include easy user interface [14], tracking of disease symptoms [11], and easy access to medical data and services [11].

A systematic assessment of 26 IBD mobile applications found that applications offered a variety of features including diary functionality, pain tracking, bowel movement tracking, and reminders, with application's content playing a major role in driving patient behavior change [4]. The MyIBD Coach telemedicine tool, which monitors adherence, disease activity, quality of life, and mental health among other measures through validated questionnaires, was shown to be successful, with high rates of patient satisfaction and compliance [15]. It involves collaboration among healthcare providers but is not synchronized with electronic medical records and lacks educational application features on alternative medicine, behavioral health, and physical activity.

To enhance VBHC in IBD, we developed UCLA eIBD to integrate various successful features of previous applications (e.g., appointment reminders, medication trackers) in addition to a healthcare provider portal. UCLA eIBD seeks to provide patients more agency in managing their IBD by increasing their access to healthcare professionals and providing self-help educational modules. Access to care providers through a messaging application provides patients with fast feedback on their conditions and streamlines patient care [16]. The application also contains disease activity, quality of life, and work productivity surveys that facilitate interactions between patients and providers. These tools allow healthcare providers to monitor patients' disease activity and give direct feedback. This comprehensive application therefore seeks to enhance patient outcomes by including direct connections to the healthcare team and extensive module options.

We previously conducted a pilot study of UCLA eIBD, which found significantly fewer endoscopies and decreases in healthcare utilization, long-term steroid use and IBD-related costs [17]. While it is important to evaluate the efficacy and outcomes of IBD management platforms, it is just as crucial to understand patients' satisfaction with these platforms to inform their feasibility. Gathering user feedback is necessary to develop the next generation of applications, improve product design, and reduce the gap between application developers and consumers [18–20]. This study therefore aims to provide an evaluation of perceived patient satisfaction and experience with the UCLA eIBD mobile application.

2. Materials and Methods

2.1. Objectives

The primary objective was to measure patient satisfaction and experience with the UCLA eIBD mobile application for care management. The secondary objective was to capture patient feedback on how to improve the mobile application.

2.2. Design and Population

We surveyed IBD patients treated at the UCLA Center for Inflammatory Bowel Diseases from October 2017 to October 2018. Included patients were at least 18 years old; diagnosed with Crohn's disease (CD) or ulcerative colitis (UC) either by means of endoscopy, imaging, or pathology; and had objectively logged into the application in the past year (assessed on the platform). Patients with intestinal cancer, active chemotherapy, or a known intestinal infection were excluded.

All the eligible patients who had logged into the application in the past year were emailed and asked to complete a patient experience survey. Those who did not complete the survey in response to the initial email were followed up and interviewed via phone. No sample size estimation was performed.

2.3. Description of UCLA eIBD

UCLA eIBD is a mobile application that administers a clinic-centered, care management program to its users (Figure 1). It was designed to be a comprehensive tool for patients' long-term disease management in the IBD outpatient setting. The features of this application include disease activity monitoring, messaging, educational modules, lifestyle modules, and electronic cognitive behavioral therapy (eCBT). The platform is also integrated with UCLA Health's electronic medical records, allowing patients to view their testing and laboratory results within the application.

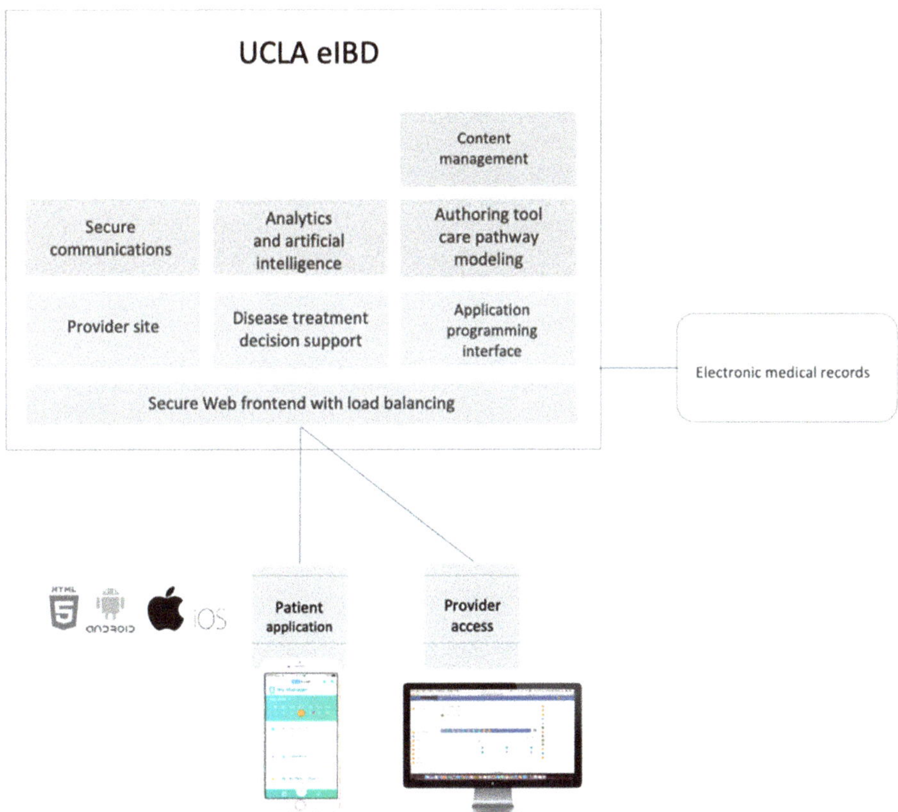

Figure 1. The UCLA eIBD mobile application is an integrative care management platform for patients and providers.

For disease activity monitoring, a previously validated tool called the Mobile Health Index was integrated to assess the patients' disease activity, quality of life, and work productivity [21]. If the surveys indicated poor disease control or a significant change from prior surveys, a message was automatically generated through the application to clinic staff. The enrolled patients could also elect to take these surveys in their own time if they felt they were experiencing a sudden change in their health.

Lastly, the application provided education through several optional interactive modules designed to promote healthy lifestyle habits, including nutrition (My Menu), exercise (My Yoga, My Fitness), relaxation (My Acupressure, My Meditation), and mental health (My Coach). My Menu teaches patients about specific foods to eat and avoid and includes recipes (breakfast, snack, lunch and dinner) designed for IBD patients. My Yoga provides a 6-week program promoting relaxation and flexibility for users. My Acupressure teaches patients about different pressure points for alleviating IBD pain via instructional videos and pictures. My Meditation is a self-guided mindfulness therapy tool that aims to reduce stress-related health issues. My Coach is a personalized mental life coaching program (6-week mental support program) aimed at improving mental well-being and stress management through a cognitive behavioral therapy method.

2.4. Data Collection and Outcomes

Patient demographic data were acquired via chart review. The data from the patient experience survey were collected via REDCap [22]. The patient experience survey (Table 1) consisted of 24 items aimed at assessing the patients' overall satisfaction with the application and their perception of health outcomes after participation in the program. Responses were provided either via a Likert scale or open text. Questionnaire items addressing the application's features and interface requested feedback on the ease of application use, ability to communicate with staff, and informativeness of modules. Questionnaire items pertaining to the patient's outcomes asked the patients how effective they felt the application was at improving disease control, work productivity, and quality of life. Lastly, the patients could provide optional open-ended feedback via free text input on ways to improve the application.

Table 1. Patient experience survey.

No.	Question	$n = 50$
1	How easy was it to communicate with program staff overall?	26 (52%)—very easy 8 (16%)—somewhat easy 13 (26%)—neutral 3 (6%)—somewhat difficult
2	How easy was it to schedule appointments?	26 (52%)—very easy 8 (16%)—somewhat easy 13 (26%)—neutral 3 (6%)—somewhat difficult
3	How satisfied were you with program staff's response rate to messages and questions?	22 (44%)—very satisfied 18 (36%)—satisfied 3 (6%)—somewhat dissatisfied 7 (14%)—neutral
4	How did participating in the program affect your disease control?	15 (30%)—significant improvement 12 (24%)—some improvement 21 (42%)—no change 2 (4%)—somewhat worse
5	How participating in the program affect your quality of life?	13 (26%)—significant improvement 15 (30%)—some improvement 20 (40%)—no change 2 (4%)—somewhat worse
6	How did participating in the program affect your work productivity?	11 (22%)—significant improvement 14 (28%)—some improvement 24 (48%)—no change 1 (2%)—somewhat worse
7	Did you participate in the cognitive behavioral therapy modules?	6 (12%)—yes 44 (88%)—no

Table 1. *Cont.*

No.	Question	*n* = 50
8	How did participating in the program affect your mental health?	8 (16%)—significant improvement 6 (12%)—some improvement 25 (50%)—no change 1 (2%)—somewhat worse 10 (20%)—unknown
9	Were your clinic visits scheduled too often, just right or not often enough?	44 (88%)—just right 6 (2%)—not often enough
10	Did you feel you were having lab tests done too often, just right or not often enough?	44 (88%)—just right 1 (2%)—not often enough 5 (10%)—too often
11	Did you feel you had to fill out surveys too often, just right or not often enough?	39 (78%)—just right 4 (8%)—not often enough 7 (14%)—too often
12	How accurately do you feel the survey results reflected your opinion of your disease activity and well-being?	17 (34%)—very accurately 20 (40%)—somewhat accurately 11 (22%)—neutral 2 (4%)—somewhat inaccurate
13	How easy was it to navigate the mobile application?	18 (36%)—very easy 19 (38%)—somewhat easy 6 (12%)—neutral 4 (8%)—somewhat difficult 3 (6%)—very difficult
14	Did you find the graphics and overall 'look' of the application appealing?	40 (81.63%)—yes 9 (18.37%)—no
15	Overall, how informative was the application, particularly My Academy?	12 (24%)—very informative 11 (22%)—somewhat informative 24 (48%)—neutral 3 (6%)—not informative
16	Which of the following modules did you complete? (choice = My Fitness)	17/50 (34%)
17	Which of the following modules did you complete? (choice = My Meditation)	13/50 (26%)
18	Which of the following modules did you complete? (choice = My Menu)	17/50 (34%)
19	Which of the following modules did you complete? (choice = My Yoga)	10/50 (20%)
20	Which of the following modules did you complete? (choice = My Accupressure)	5/50 (10%)
21	Is there a topic you would like to see added to My Academy or My Wellness? If so, what topic?	Displayed in Table 4.
22	Did you need to access technical support at any time during this study?	7 (14%)—yes 43 (86%)—no
23	If so, how many times did you need to access technical support? *	4 (1 time) 5 (2–5 times)
24	How reliably were you able to reach technical support? *	3 (27%)—somewhat reliable 7 (64%)—neutral 1 (9%)—very unreliable

* Optional question.

3. Results

3.1. Patient Demographics

In total, 151 patients had been active on the mobile application in the past year, of whom 50 patients responded and completed the survey and thus were included in this study. Regarding the type of IBD, 44% were diagnosed with CD (*n* = 22), 56%—with UC (*n* = 28). Our inclusion cohort had a mean age of 43 years (SD, 14 years) and an average BMI of 25.3 (SD, 6.6). Of the patients, 44% were female, and the majority were White (42%) and non-Hispanic (90%) (Table 2). Most of the patients were non-smokers (78%), and 28% of the patients reported alcohol use. The patients stated use of the following medications:

anti-TNF (34%), ASA (16%), combination therapy (32%), IMM (10%), and steroids (6%). Previous abdominal surgeries were reported in 36% of the participants.

Table 2. Patient demographics.

Variable	All (*n* = 50)
Gender	22 (44%)—female
Disease Type	22 (44%)—Crohn's disease
	28 (56)—ulcerative colitis
Race	21 (42%)—White
	4 (8%)—Black
	3 (6%)—Asian
	1 (2%)—Armenian
	21 (42%)—unknown
Ethnicity	4 (8%)—Hispanic
	45 (90%)—non-Hispanic
	1 (2%)—unknown
Current smoker	3 (6%)—current smoker
	8 (16%)—former smoker
	39 (78%)—never smoker
Age (mean SD)	42.58 (SD, 13.6)
Alcohol use	14 (28%)—yes
	36 (42%)—no
BMI (mean SD)	25.3 (SD, 6.6)
Disease duration (mean SD)	14.6 (SD, 11.2)
Disease activity	29 (58%)—clinical remission
	11 (22%)—mild disease activity
	6 (12%)—moderate disease activity
	3 (6%)—severe disease activity
	1 (2%)—unknown
Medications	
- Anti-TNF	17 (34%)—anti-TNF
- ASA	8 (16%)—ASA
- Combination of any medications	16 (32%)—combo
- IMM	5 (10%)—IMM
- Steroids	3 (6%)—steroids
- No Meds	1 (2%)—no meds
Abdominal surgeries (%)	18 (36%)

3.2. Patient Satisfaction

Fifty participants out of the 151 users responded and completed the patient experience survey to provide feedback on the mobile application (Table 1). Responses to the Likert scale questions indicated that the patients were overall satisfied with the patient–provider communication interface of the application. When asked how easy it was to communicate with the program's staff overall, 52% of the participants responded with "very easy" and 16% responded with "somewhat easy". A majority of the participants also found it easy to schedule appointments through the application, with 52% and 16% responding with "very easy" and "somewhat easy", respectively. In addition, a large majority (88%) of the participants reported that the frequency of completing laboratory tests and surveys and scheduling clinic visits was "just right" (Table 1). Regarding the ease of application use, 74% of the participants indicated the application was either "very easy" or "somewhat easy" to navigate.

Additionally, a majority of the participants reported an improved perception of disease control and QoL; 54% of the participants indicated significant or some improvement in their disease control. When asked how program participation affected QoL, 26% indicated significant improvement, 30%—some improvement. Regarding work productivity, 44% indicated significant or some improvement.

When the participants were asked whether they would recommend this application to their friends, family, or other patients on a ten-point scale, with 10 being most likely, the median score was 8, and 79% indicated a score greater than 5. When asked about how informative the application was, 46% of the patients felt that the application was "somewhat" or "very" informative.

3.3. Patient Usage of Educational Modules

A majority of the patients completed modules as part of their participation in the program. The most used modules were My Fitness and My Menu (Table 1). Among the patients who participated in the CBT modules (12%), 28% indicated significant or some improvement in their mental health.

When asked about what they liked and disliked about the modules, the patients identified positive aspects to be the modules' informative content, ease of use, and support of overall well-being (Table 3). For example, one patient said, "They're easy and I feel great afterwards." Another patient expressed liking the modules because they "encourage me to take care of my whole self instead of the focus just being on taking my meds".

Table 3. The patients' optional feedback on the modules (*n* = 50). The patients provided open-ended feedback about the educational modules. Their responses were grouped into categories based on the common themes identified across the responses.

What the Patients Liked about the Modules	Count	Examples of Patient Feedback
Informative content	7	"Modules contained useful information." "My Meditation provided helpful tips."
Ease of use	3	"Very user friendly."
Ease of communication with the provider	1	"Liked the VQ visual display. The app gave me comfort because it gave me access to the doctors especially when you have this disease."
Supports overall well-being	2	"I like that the modules encourage me to take care of my whole self instead of the focus just being on taking my meds."
Reminders to complete the modules	1	"I like to get reminded to complete the modules, they're easy and I feel great afterwards."
The yoga module was simple and effective	1	"I liked the yoga app because it was simple and effective..."
Total	15	
What the patients disliked about the modules	**Count**	**Examples of patient feedback**
Not informative	1	"Modules need to contain information that is more specialized."
Difficult to use	2	"Hard to navigate."
Unresponsiveness from the staff	1	"Not responsive from staff."
Did not know about the modules	2	"I did not know about the modules."
Takes too long to complete	1	"Liked overall content and goal that IBD trying to aim for. Time issue for completing the module."
Problem with a specific module (My Yoga, My Acupuncture, etc.)	1	"Yoga portion could contain an audio aspect... stopping and reading about doing the yoga was counter-productive to my relaxation."
Unsure of the purpose or need for them	4	"Didn't feel like they applied to me, personally."
Total	12	

The most common reason for not liking the modules was being unsure of the purpose or need for them (8%), particularly for the modules where patients already had their own interventions in place. For example, one patient said they "didn't feel [the modules] applied to me" while another expressed that they "thought [the module] was good but [I have my] own routine for working out [with regards to My Fitness]".

3.4. Patient Feedback

In the patient experience survey, the patients could provide optional suggestions about additional topics and functionalities they would like the application to cover which were not presently included (Table 4). One participant, for instance, suggested adding a subsection about nutritional advice related to veganism within the My Menu module. Other recommendations included adding a "symptoms tracker", allowing patients to indicate what symptoms or lack thereof they were experiencing and generating in-application reminders for blood draws or laboratory orders. The other patient-recommended categories to add were the ability to chart laboratory results, side effects of their consequent medications, and gender-specific health topics (Table 4).

Table 4. The patients' optional feedback on UCLA eIBD. The patients provided open-text suggestions to improve the application in general. These suggestions were grouped into categories of comment types, including improvements in application content such as possible additional topics and features, as well as miscellaneous critiques.

Comment Types	Total Count	Examples of Patient Feedback (Count)
Suggestions for new application articles and topics	8	A module on acupuncture (1) A module on veganism (1) Side effects of medications (1) Female health topics (1) Blood draw instructions (1) Resources for the recommended pathways (e.g., local places to get nutritional advice, do yoga, fitness) (1) FAQ for family and friends (1)
Suggestions for new application features and tools	3	Ability to chart laboratory results (1) Symptoms tracker (2)
Suggestions for better technical aspects of the application	3	Touch ID for signing in (1) No automatic logoff (1) Different languages (1)
Miscellaneous improvement suggestions	4	Staff response rate faster at the beginning of the program (1) Poor wording of some in-application questionnaires (2) - e.g., *"I don't like the wording of the questionnaires. i felt they lacked nuance. none asked if i felt overwhelmed, anxious, or preoccupied by disease things. just 'angry' or 'depressed' which i think are really different experiences."* - e.g., *"Sometimes i feel just saying on a scale from 1 to 10, how my disease affects my work or social life is too broad a question."* Lacks in-depth, longer-term information about IBD (1) - e.g., *"app is good for people new to ibd but doesnt offer as much for people who have had ibd for a while and want more in depth information."*

The patients' feedback regarding general comments about the application is also shown in Table 4 (miscellaneous improvement suggestions). One patient stated, "I think this is a great idea and will be very helpful to future patients. I really like being able to communicate with the office without always having to call." Most patients who provided comments also highlighted aspects that could be improved, such as the application's interface (e.g., adding a touch ID option to log in; preventing automatic logoff from the application). Other participants reported critical feedback on the application's content. For instance, one patient stated that the application "is good for people new to IBD, but doesn't offer as much for people who have had IBD for a while and want more in depth information".

3.5. Summary of Principal Findings

The outcomes suggest that the patients strongly favored the ease of patient–provider communication, with 78% being satisfied. Beneficial outcomes were also seen in patient-reported measures, with 54% reporting a perceived improvement in disease control and 56% reporting a perceived improvement in QoL, indicating that a majority of patients felt the platform positively impacted their health. Additionally, the participants rated this application with a median score out of 10 (10 being most likely) to recommend this application to friends, family, or other patients. My Fitness and My Menu were the two most used optional wellness modules, each reaching the 34% completed status.

4. Discussion

4.1. Strengths and Comparisons

Our study collected feedback on patient experiences with the UCLA eIBD application after one year of use. Our results could provide guidance for further application development and provide critical feedback for other e-health applications like this one. In fact, mobile tools such as UCLA eIBD have been shown to have promising implications in improving healthcare delivery and integrating into patients' daily lives. Earlier comparison studies of UCLA eIBD found impacts on costs and healthcare utilization and identified its unique features, such as automated messaging to care coordinators [17,23–25]. To complement the previous outcome studies, this study aimed to understand patients' satisfaction and feedback to help elucidate gaps in the current e-health technologies and inform future designs.

For instance, GI Buddy is a mobile application developed by the Crohn's Colitis Foundation which enables patients to self-monitor their disease and receive reminders about clinical appointments; however, users cannot directly interact with their providers [26]. Similarly, while the current applications for IBD may be useful for patient monitoring and self-management, many lack professional medical involvement and adherence to clinical guidelines [4]. UCLA eIBD addressed this gap by allowing users to make appointments and message their providers via the platform, in which a majority of users found it "easy" or "very easy" to communicate with their providers. Another self-management tool, myIBD Coach, showed feasibility among patients and providers [15]. As many as 79% of UCLA eIBD users would recommend this application to others (indicated by a score of greater than 5 on the recommendation score item), compared to the 93% found in the myIBD Coach feasibility study [15].

The findings of this novel patient satisfaction study demonstrate the feasibility of UCLA eIBD as a home monitoring tool and some advantages it can provide for both patients and providers. In addition to the patient–provider communication features, the platform's educational modules are more diverse than the previous tools and provide patients with more alternatives to aid traditional medicine, such as acupuncture, cognitive behavioral therapy, and meditation. These optional modules may improve IBD patients' well-being and productivity beyond the scope of their disease. For providers, tracking the various modules that the patients use can also provide guidance for tailoring treatment and counseling to the patients' interests, including nutrition, exercise, and mental health. Lastly, the holistic nature of the application, including features about alternative medicine and assessments for work productivity via the Mobile Health Index, can more completely address the complex, multidimensional factors of chronic disease management.

While the integration of mobile health in IBD management is rapidly expanding, our study also presents novel data from the patient perspective and emphasizes a patient-centered approach towards mobile application development. For instance, the patients' suggestions to improve the application were centered on specific content interests and the need for additional educational categories (e.g., female health topics), rather than technical problems or lack of need in an application. The fact that the suggestions were less focused on design features could be explained by the overall satisfaction rate of 74% of the participants finding the application easy to navigate. The current and future applications can thus utilize these methods and/or findings to adapt their platforms to address patients' specific needs, improve satisfaction with their product, and better engage patients in their medical care beyond a doctor's office visit.

4.2. Limitations

Some study limitations should be noted. As the selected patients were individuals who use smartphones, they may be more adept at the usage of applications. Participants were also actively recruited and agreed to participate in this study; thus, a selection bias may have impacted the study results due to the participants being predisposed to wanting to improve their health via e-health solutions. We further acknowledge the sample size was small and relatively homogenous; however, we feel it was adequate for the purpose of directing the future development of this UCLA application and other healthcare applications.

Additionally, the fact that we invited participants to evaluate the application's feasibility rather than making it mandatory during application usage may explain the response rate of 33%. The response rate should further be considered in the context of challenges associated with adopting e-health technologies into the healthcare space. The obstacles to widespread long-term integration of e-health technologies (e.g., loss of interest, data entry burdens) are still being investigated [27,28]. Despite the growing population of individuals who use mobile health applications, many stop using them over time [29]. Our findings help provide insight to consumer perspectives on application usability and possible explanations to circumvent these challenges.

4.3. Future Directions

In an era where the use of mobile technology has become irreplaceable in daily life, there is undoubted benefit of incorporating e-health applications in the management of chronic conditions. Studies have shown proven effect of mobile applications, but also that patients still desire improvements to the existing solutions. We showed that UCLA eIBD and its holistic approach led to greater patient experience and satisfaction, which can provide useful recommendations for healthcare providers and application developers. However, larger and controlled studies are recommended to assess its efficacy at a larger scale and its impact on costs.

Author Contributions: A.Z.: study concept and design; analysis and interpretation of data; drafting of the manuscript; critical revision of the manuscript for important intellectual content; statistical analysis; study supervision; A.N.: study concept and design; acquisition of data; analysis and interpretation of data; drafting of the manuscript; critical revision of the manuscript for important intellectual content; C.R.: study concept and design; drafting of the manuscript; critical revision of the manuscript for important intellectual content; A.K.: acquisition of data; administrative, technical, or material support, drafting of the manuscript; critical revision of the manuscript for important intellectual content; E.E.: revision of the manuscript for important intellectual content; D.H.: study concept and design; analysis and interpretation of data; drafting of the manuscript; critical revision of the manuscript for important intellectual content; statistical analysis; study supervision. All authors have read and agreed to the published version of the manuscript.

Funding: This research received no external funding.

Institutional Review Board Statement: This study was approved by the Institutional Review Board at UCLA with IRB protocol number 17-001208. All the procedures performed in the studies involving human participants were in accordance with the ethical standards of the institutional and/or national research committee and the 1964 Helsinki Declaration and its later amendments or comparable ethical standards.

Informed Consent Statement: Informed consent was obtained from all the individuals included in the study.

Conflicts of Interest: The authors declare that they have no conflict of interest.

References

1. Berwick, D.M.; Nolan, T.W.; Whittington, J. The Triple Aim: Care, Health, and Cost. *Health Aff.* **2008**, *27*, 759–769. [CrossRef]
2. Thorpe, K.E. The Rise in Health Care Spending and What To Do About It. *Health Aff.* **2005**, *24*, 1436–1445. [CrossRef] [PubMed]
3. Kappelman, M.D.; Rifas–Shiman, S.L.; Porter, C.Q.; Ollendorf, D.A.; Sandler, R.S.; Galanko, J.A.; Finkelstein, J.A. Direct Health Care Costs of Crohn's Disease and Ulcerative Colitis in US Children and Adults. *Gastroenterology* **2008**, *135*, 1907–1913. [CrossRef]
4. Con, D.; De Cruz, P.; Van Mierlo, T.; Aramaki, E. Mobile Phone Apps for Inflammatory Bowel Disease Self-Management: A Systematic Assessment of Content and Tools. *JMIR mHealth uHealth* **2016**, *4*, e13. [CrossRef] [PubMed]

5. Stunkel, L.; Karia, K.; Okoji, O.; Warren, R.; Jean, T.; Jacob, V.; Swaminath, A.; Scherl, E.; Bosworth, B. Impact on quality of life of a smart device mobile application in patients with inflammatory bowel disease. *Am. J. Gastroenterol.* **2012**, *107*, S635–S636. [CrossRef]

6. Atreja, A.; Khan, S.; Szigethy, E.; Otobo, E.; Shroff, H.; Chang, H.; Keefer, L.; Rogers, J.; Deorocki, A.; Ullman, T.; et al. Improved quality of care for IBD patients using healthPROMISE App: A randomized, control trial. *Am. J. Gastroenterol.* **2018**, *113*, S1.

7. Geryk, L.L.; Roberts, C.A.; Sage, A.J.; Coyne-Beasley, T.; Sleath, B.L.; Carpenter, D.M.; Hall, A.; Rhee, H.; Morrison, C.; Morrison, D. Parent and Clinician Preferences for an Asthma App to Promote Adolescent Self-Management: A Formative Study. *JMIR Res. Protoc.* **2016**, *5*, e229. [CrossRef]

8. Swendeman, D.; Farmer, S.; Mindry, D.; Lee, S.-J.; Medich, M. HIV Care Providers' Attitudes regarding Mobile Phone Applications and Web-Based Dashboards to support Patient Self-Management and Care Coordination: Results from a Qualitative Feasibility Study. *J. HIV AIDS* **2016**, *2*, 1310–1318.

9. Lyles, C.R.; Harris, L.T.; Le, T.; Flowers, J.; Tufano, J.; Britt, D.; Hoath, J.; Hirsch, I.B.; Goldberg, H.I.; Ralston, J.D. Qualitative Evaluation of a Mobile Phone and Web-Based Collaborative Care Intervention for Patients with Type 2 Diabetes. *Diabetes Technol. Ther.* **2011**, *13*, 563–569. [CrossRef]

10. Hirschey, J.; Bane, S.; Mansour, M.; Sperber, J.; Agboola, S.; Kvedar, J.; Jethwani, K. Evaluating the Usability and Usefulness of a Mobile App for Atrial Fibrillation Using Qualitative Methods: Exploratory Pilot Study. *JMIR Hum. Factors* **2018**, *5*, e13. [CrossRef]

11. Con, D.; Jackson, B.; Gray, K.; De Cruz, P. eHealth for inflammatory bowel disease self-management—The patient perspective. *Scand. J. Gastroenterol.* **2017**, *2017*, 1–8. [CrossRef]

12. Hommel, K.A.; Gray, W.N.; Hente, E.; Loreaux, K.; Ittenbach, R.F.; Maddux, M.; Baldassano, R.N.; Sylvester, F.A.; Crandall, W.; Doarn, C.; et al. The Telehealth Enhancement of Adherence to Medication (TEAM) in pediatric IBD trial: Design and methodology. *Contemp. Clin. Trials* **2015**, *43*, 105–113. [CrossRef] [PubMed]

13. Vlahou, C.H.; Cohen, L.L.; Woods, A.M.; Lewis, J.D.; Gold, B.D. Age and Body Satisfaction Predict Diet Adherence in Adolescents with Inflammatory Bowel Disease. *J. Clin. Psychol. Med. Settings* **2008**, *15*, 278–286. [CrossRef]

14. Liu, Y.; Wang, L.; Chang, P.; Lamb, K.V.; Cui, Y.; Wua, Y. What features of smartphone medication applications are patients with chronic diseases and caregivers looking for? *Stud. Health Technol. Inform.* **2016**, *225*, 515–519.

15. De Jong, M.; Van Der Meulen-De Jong, A.; Romberg-Camps, M.; Degens, J.; Becx, M.; Markus, T.; Tomlow, H.; Cilissen, M.; Ipenburg, N.; Verwey, M.; et al. Development and Feasibility Study of a Telemedicine Tool for All Patients with IBD: MyIBDcoach. *Inflamm. Bowel. Dis.* **2017**, *23*, 485–493. [CrossRef] [PubMed]

16. Ventola, C.L. Mobile Devices and Apps for Health Care Professionals: Uses and Benefits. P T: A peer-reviewed. *J. Formul. Manag.* **2014**, *39*, 356–364.

17. Van Deen, W.K.; Spiro, A.; Burak Ozbay, A.; Skup, M.; Centeno, A.; Duran, N.E.; Lacey, P.N.; Jatulis, D.; Esrailian, E.; van Oijen, M.G.H.; et al. The impact of value-based healthcare for inflammatory bowel diseases on healthcare utilization: A pilot study. *Eur. J. Gastroenterol. Hepatol.* **2017**, *29*, 331–337. [CrossRef]

18. Jake-Schoffman, D.E.; Silfee, V.J.; Waring, M.E.; Boudreaux, E.D.; Sadasivam, R.S.; Mullen, S.P.; Carey, J.L.; Hayes, R.B.; Ding, E.Y.; Bennett, G.G.; et al. Methods for Evaluating the Content, Usability, and Efficacy of Commercial Mobile Health Apps. *JMIR mHealth uHealth* **2017**, *5*, e190. [CrossRef] [PubMed]

19. Liew, M.S.; Zhang, J.; See, J.; Ong, Y.L. Usability Challenges for Health and Wellness Mobile Apps: Mixed-Methods Study Among mHealth Experts and Consumers. *JMIR mHealth uHealth* **2019**, *7*, e12160. [CrossRef]

20. Maramba, I.; Chatterjee, A.; Newman, C. Methods of usability testing in the development of eHealth applications: A scoping review. *Int. J. Med. Inform.* **2019**, *126*, 95–104. [CrossRef]

21. Deen, W.K.; Van Jong, A.E.V.D.M.; Parekh, N.K.; Kane, E.; Zand, A.; Dinicola, C.A.; Hall, L.; Inserra, E.K.; Choi, J.M.; Ha, C.Y.; et al. Development and Validation of an Inflammatory Bowel Diseases Monitoring Index for Use with Mobile Health Technologies. *Clin. Gastroenterol. Hepatol.* **2016**, *14*, 1742–1750.e7. [CrossRef] [PubMed]

22. Harris, P.A.; Taylor, R.; Thielke, R.; Payne, J.; Gonzalez, N.; Conde, J.G. Research electronic data capture (REDCap)-A metadata-driven methodology and workflow process for providing translational research informatics support. *J. Biomed. Inform.* **2009**, *42*, 377–381. [CrossRef] [PubMed]

23. Walsh, A.; Travis, S. What's app? Electronic health technology in inflammatory bowel disease. *Intest. Res.* **2018**, *16*, 366–373. [CrossRef] [PubMed]

24. Yin, A.L.; Hachuel, D.; Pollak, J.P.; Scherl, E.J.; Estrin, D. Digital Health Apps in the Clinical Care of Inflammatory Bowel Disease: Scoping Review. *J. Med. Int. Res.* **2019**, *21*, e14630. [CrossRef]

25. Kelso, M.; Feagins, L.A. Can Smartphones Help Deliver Smarter Care for Patients with Inflammatory Bowel Disease? *Inflamm. Bowel. Dis.* **2018**, *24*, 1453–1459. [CrossRef] [PubMed]

26. Ehrlich, O.; Atreja, A.; Markus-kennell, S.; Frederick, K. CCFA GI Buddy provides patient reported outcomes and IBD symptoms evaluation: P-50. *Inflamm. Bowel. Dis.* **2012**, *18*, 35–36. [CrossRef]

27. Gutiérrez-Ibarluzea, I.; Chiumente, M.; Dauben, H.-P. The Life Cycle of Health Technologies. Challenges and Ways Forward. *Front. Pharmacol.* **2017**, *8*, 14. [CrossRef] [PubMed]

28. Drake, C.; Cannady, M.; Howley, K.; Shea, C.; Snyderman, R. An evaluation of mHealth adoption and health self-management in emerging adulthood. *AMIA Annu. Symp. Proc. AMIA Symp.* **2020**, *2019*, 1021–1030. [PubMed]
29. Krebs, P.; Duncan, D.T. Health App Use among US Mobile Phone Owners: A National Survey. *JMIR mHealth uHealth* **2015**, *3*, e101. [CrossRef]

International Journal of *Environmental Research and Public Health*

MDPI

Building a Research Roadmap for Caregiver Innovation: Findings from a Multi-Stakeholder Consultation and Evaluation

Kieren J. Egan [1,*], Kathryn A. McMillan [1,2], Marilyn Lennon [1], Lisa McCann [1] and Roma Maguire [1]

[1] Digital Health and Wellness Research Group, Department of Computer and Information Science, Faculty of Science, University of Strathclyde, Glasgow G1 1XH, UK; katie.mcmillan3@nhs.scot (K.A.M.); marilyn.lennon@strath.ac.uk (M.L.); lisa.mccann@strath.ac.uk (L.M.); roma.maguire@strath.ac.uk (R.M.)

[2] Information and Digital Technology Department, NHS Shetland, Lerwick, Shetland ZE1 0LB, UK

* Correspondence: kieren.egan@strath.ac.uk; Tel.: +44-0141-548-3138

Citation: Egan, K.J.; McMillan, K.A.; Lennon, M.; McCann, L.; Maguire, R. Building a Research Roadmap for Caregiver Innovation: Findings from A Multi-Stakeholder Consultation and Evaluation. *Int. J. Environ. Res. Public Health* **2021**, *18*, 12291. https://doi.org/10.3390/ijerph182312291

Academic Editors: Irene Torres-Sanchez and Marie Carmen Valenza

Received: 28 October 2021
Accepted: 14 November 2021
Published: 23 November 2021

Publisher's Note: MDPI stays neutral with regard to jurisdictional claims in published maps and institutional affiliations.

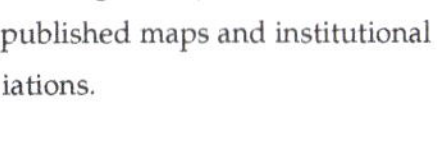

Abstract: Across the world, informal (unpaid) caregiving has become the predominant model for community care: in the UK alone, there are an estimated 6.5 million caregivers supporting family members and friends on a regular basis, saving health and social care services approximately £132 billion per year. Despite our collective reliance on this group (particularly during the COVID-19 pandemic), quality of life for caregivers is often poor and there is an urgent need for disruptive innovations. The aim of this study was to explore what a future roadmap for innovation could look like through a multi-stakeholder consultation and evaluation. An online survey was developed and distributed through convenience sampling, targeting both the informal caregiver and professionals/innovators interested in the caregiver demographic. Data were analysed using both quantitative (summary statistics) and qualitative (inductive thematic analysis) methods in order to develop recommendations for future multi-stakeholder collaboration and meaningful innovation. The survey collected 174 responses from 112 informal caregivers and 62 professionals/innovators. Responses across these stakeholder groups identified that there is currently a missed opportunity to harness the value of the voice of the caregiver demographic. Although time and accessibility issues are considerable barriers to engagement with this stakeholder group, respondents were clear that regular contributions, ideally no more than 20 to 30 min a month could provide a realistic route for input, particularly through online approaches supported by community-based events. In conclusion, the landscape of digital health and wellness is becoming ever more sophisticated, where both industrial and academic innovators could establish new routes to identify, reach, inform, signpost, intervene and support vital and vulnerable groups such as the caregiver demographic. Here, the findings from a consultation with caregivers and professionals interested in informal caring are presented to help design the first stages of a roadmap through identifying priorities and actions that could help accelerate future research and policy that will lead to meaningful and innovative solutions.

Keywords: caregivers; innovation; research; co-design; interdisciplinary; digital health; participatory design; collaboration at distance

1. Introduction

Informal (unpaid) carers, also termed "caregivers", are family members and friends who support a loved one who needs help due to illness, frailty, disability, mental health problems or addiction. In the UK, it is estimated that there are at least 6.5 million informal caregivers, a workforce substantially larger than the National Health Service (NHS) [1] and the collective saving to the health and social care services is estimated to be £132 billion per year. This situation is similar across Europe, and beyond, where 80% of all care is delivered by informal caregivers [2]. As the tide of an ageing global population continues to advance in tandem with a shrinking health and social care workforce [3], there is a public

127

health emergency looming, whereby the pressures on caregivers across the world are set to significantly escalate.

While the recent events during the COVID-19 pandemic have demonstrated the societal value of caring, it also highlighted that many will face short, and long-term, health and wellness consequences. Informal caregivers share many of the same challenges of a professional workforce, such as NHS employees, but lack much of the associated infrastructure. For example, there is little to no: support, training, pre-agreed workload such as hours per week or a working role definition, and caregivers often must balance caring responsibilities around work, families and their own health and wellness needs. While some caregivers do enjoy benefits from caring roles, such as a sense of fulfilment from caring [4], there is now compelling evidence that caregiving adversely impacts on health and wellness, both in the short and long term [5–7]. Crisis points, such as hospitalizations, significant worsening of mental and/or physical health and irreversible changes to caring circumstances are commonplace (even in the absence of COVID-19) and frequently cause deterioration in health for caregivers, and those being cared for [8,9].

Given the considerable number of unmet needs described above, there have been many different innovations developed for the caregiver demographic. Although successful evidence has emerged [10–12], there still remains a striking number of caregivers who remain "out of reach" [8] through a combination of factors, including (but not limited to): digital and health literacy levels, socioeconomic status, health, mobility, and level of dependency required from the person cared for. Thus, there remains a continued need to consult, innovate and evaluate for, and with, caregivers and to develop research excellence in this area. Emerging from the evidence in this field, there are several areas of particular interest, which include: methodologies to improve identification of caregivers (including in hard to reach groups such as ethnic minorities [13], the development and adaption of theoretical/conceptual frameworks, techniques for cultural adaptation [14] and addressing implementation and abandonment issues [15].

Digital health and wellness solutions are a core part of the current drive to deliver equity in health to all—including informal caregivers. In 2020, for example, the WHO (World Health Organization) produced a long-term strategy for the use and scale-up of digital health, highlighting the positive impact digital health can have on the access to, and provision of healthcare as well as the health and wellbeing of the population [16]. Such findings are paralleled in recent review work [17]. Much of this work focuses on caregiver health and wellbeing or signposting to sources of support and are not restricted to one or two types of technology (e.g., mobile app, internet-based, integrate platforms, sensors) [17]. Despite such progress, the market reality is that most startups fail and many successful research studies are challenging to implement. Reducing uncertainties and increasing the availability of market knowledge, such as bettering current understanding of ongoing caregiver needs, would be considerably advantageous.

Taken collectively, innovation for informal caregivers is sorely needed but the landscape still falls short of meeting the everyday needs. Identifying both better ways to collaborate across all stakeholders and methodologies to rapidly, but robustly, develop and test innovations could help improve the translational hit rate. Therefore, here we took a first step to improve engagement with caregivers by undertaking a multi-stakeholder consultation to design a future roadmap for innovation in caregiver research.

2. Materials and Methods

Initially, this study was due to involve face-to-face interviews and focus groups, however, due to the COVID-19 pandemic we had to be agile and to achieve our aim of engaging with as many caregivers and other stakeholders as we could, we changed methodology and conducted the multi-stakeholder consultation via survey. We developed our own survey approach as we did not find a suitable existing one. Although the same content of surveys was delivered to both caregivers and professional groups (e.g., healthcare professionals, innovators), we tailored wording on occasion to represent each stakeholder group.

The surveys were iteratively designed to collate participant information regarding: (i) demographics, (ii) previous research involvement, (iii) feelings towards research and innovation, (iv) barriers to taking part in research, (v) future participation in research and, (vi) focus/outcomes of future research. No questions were mandatory and, thus, the total number of responses for each question varied.

Ethical approval was granted from the Computer and Information Science (CIS) Department Ethics Committee at the University of Strathclyde, Glasgow. The surveys, aimed at caregivers and those with a professional role interested in caregivers, were designed to take no more than 10 min to complete, and to be used across the UK. See Appendix A for example questions.

Inclusion criteria and survey design: Participants 18 years and over were invited to take part. A broad definition of an informal (unpaid) caregiver was shared with participants, *"People that provide unpaid care by looking after an ill, older or disabled family member, friend or partner"*: hereafter referred to as caregivers. A minimum number of hours per week that caregivers needed to be caring for was not specified, leaving it up to participants to self- identify with the term. For professional groups, our participant information sheet stated that we were interested to find *"Professionals (not employed by Universities) who have an interest in innovating around carer health and wellbeing (e.g., health and social care professionals or those working in the technology industry/third sector)"*. Where we state professionals, hereafter, this implies professionals/innovators interested in the caregiver demographic.

Survey distribution: Consent was implied after participants read, acknowledged, and accepted the initial terms of the anonymised survey. The distribution of the survey involved sharing the online version of our questionnaire through social media channels (e.g., Twitter), alongside email distribution through networks accessible to partners, such as the Digital Health & Care Innovation Centre (DHI) and Carers UK (e.g., Healthy Ageing Innovation Group and the "Digital Health & Care" Mailing list). The survey was open from 15 June 2020 until 30 September 2020.

Data handling: The survey was constructed using Qualtrics Software. Participants were reminded that any data entered must not contain any identifiable information. An integrated mixed-methods approach was taken for the data analysis. All quantitative analyses (frequencies and summary statistics) were completed using R studio (version 1.1.456). Qualitative analyses were undertaken using a content analysis approach by two researchers (KM and KE) [18]. Free text responses to questions were collated and the first question coded by KM. The coding structure cross-checked and agreed by KE as a measure of inter-rater reliability. Once agreed, the rest of the coding was completed by KM, before being cross-checked by KE. Data were frequently referred back to throughout this process to ensure the coding framework developed was appropriate. The flexible nature of content analysis tolerates the use of a combination of deductive and inductive creation of themes and patterns from the data. Deductive analysis was used to create the key themes and inductive analysis was used to create sub-themes. More specifically, the six key themes were identified from the survey structure (deductive analysis), sub-themes were identified through researcher (KM and KE) interpretation and coding of free text responses (inductive analysis). Data gathered from informal caregivers and professionals were analysed, and are presented, separately in the results section of this paper. Major themes were identified where at least half of respondents aligned with a specific finding. Where respondents identified as both a caregiver and a professional/innovator interested in caregiving, we classed these individuals as part of the caregiver demographic.

Data analysis: This work was developed without an a priori hypothesis but instead was conducted as an observational/explorative piece used to generate a future roadmap/hypothesis. Therefore, our statistical analysis is limited to frequencies, presented as stacked bars throughout the manuscript (we interpreted priorities where >50% of the respondents suggested an item was important/very important or equivalent). Qualitative analysis included the identification and discussion of key themes, such as clear priorities outlined by participants, throughout the narrative. For the purposes of readability and figure simplicity, long statements have been abbreviated and a full list of statements presented to participants can be found in Appendix B.

3. Results

3.1. Description of Our Sample

3.1.1. Caregiver Demographics

The survey received 112 responses from informal caregivers (see Table 1). Caregiver respondents were aged 25 to up to 75 to 84 years old, with the mode age within the 45 to 54 years category. Over 85% (96/112) of caregiver respondents identified as a woman/female. Caregivers were highly educated, 60.7% were educated to degree level or equivalent, and 111/112 identified as white (Table 1). Just over half (50.9%, 57/112) of caregivers had been caring for 10 or more years and 6% (7/112) had been caring for 1 year or less. Caregivers varied in their ability to undertake work/study alongside their caring role: 35% were able to continue to work/study, 34% had been forced to reduce their work/study hours and 19.6% had been forced to give up work/study. Almost all (96%) of caregiver respondents were using digital technologies, such as smartphones and laptops on a daily basis. The vast majority of caregiver responses were from Scotland (96%, 108/112), with 1 response from Northern Ireland (<1%) and 3% (3/112 responses) from England.

Table 1. Participant Demographics. * n = 61 for professionals' gender question.

Variable	Group	Caregivers (n= 112, [%])	Professionals (n = 62 *, [%])
Age Group	18 to 24	0 [0%]	0 [0%]
	25 to 34	4 [3.6%]	16 [26%]
	35 to 44	21 [18.8%]	13 [21%]
	45 to 54	47 [42%]	22 [35%]
	55 to 64	30 [26.8%]	11 [18%]
	65 to 74	7 [6.3%]	0 [0%]
	75 to 84	2 [1.8%]	0 [0%]
	Prefer not to say	1 [0.9%]	0 [0%]
Gender	Man/Male (including trans man)	12 [10.7%]	13 [21%]
	Woman/Female (including trans woman)	96 [85.7%]	48 [79%]
	In another way	2 [1.8%]	0 [0%]
	I prefer not to answer	2 [1.8%]	0 [0%]
Ethnicity	White	111 [99.1%]	58 [94%]
	Asian/Asian British	0	2 [3%]
	Other ethnic group	0	2 [3%]
	Mixed/multiple ethnic groups	1 [0.9%]	0 [0%]
Education level	Degree or equivalent	68 [60.7%]	46 [74%]
	Higher education	27 [24.1%]	10 [16%]
	Other qualifications	5 [4.5%]	5 [8%]
	School qualifications	10 [8.9%]	1 [2%]
	No qualifications	2 [1.8%]	0 [0%]

3.1.2. Professionals Demographics

The survey received 62 responses from professionals (Table 1). Professional respondents were aged from 25 to up to 55 to 64 years, most identified as women/female (79%) and 94% of respondents were white. The professional respondents were highly educated, 74% had a degree or equivalent level of education. When asked what their professional role was, 61 of the 62 participants responded. Most respondents (36%, 22/61) worked in healthcare, followed by 25% (15/61) working in social care, 23% (14/61) working in the 3rd sector, 11% of respondents put other, 3% worked in caregiver policy development and one respondent worked in a Small to Medium Enterprise (SME). All participants reported everyday use of technology. Similar to caregivers, 98% (60 /61) of professionals were from Scotland, with 1 response from England (2%).

3.2. Views of Informal Caregivers

Three key themes were established from the structure of our questioning and eight sub-themes were identified from the thematic analysis of the professionals free-text survey data. The three pre-specified (deductive) themes were; (1) Previous research participation, (2) Future research participation and (3) Future research aspirations. Sub-themes (identified through inductive thematic analysis) are presented in Table 2.

Table 2. Key themes and sub-themes from Caregiver Survey. (CAHMS= Child and Adolescent Mental Health Services).

Key Theme	Sub-Theme	Example Quote (s)
1. Previous research participation	1.1. Experience and connection to research/ers	*The research facilitated access to [a] psychologist and assessment whilst [I was] awaiting CAHMS appointments*
	1.2. Barriers	*I always found great difficulty finding the time with all other responsibilities, however, feel it would have been good opportunity, but struggle with other commitments to find time*
2. Future research participation	2.1. Value of the input from caregivers	*It is important to have your voice heard to give an accurate picture of caring and carers*
	2.2. Methods of participation	*Face-to-face is always preferred to build meaningful relationship. Online is next best thing and personally I would be reluctant for the final two options [telephone and post] as [it's] difficult to engage*
	2.3. Time available for participation	*There needs to be a balance between the time commitment and the formation of a relationship between the carer and the researcher*
3. Future research aspirations	3.1. Innovative technology	*The use of remote technology so that carers don't worry or have to be with the person as much*
	3.2. Improved support for carers	*Give carers a voice and some real support, there are many of us who are unpaid and dedicate ourselves to our person whilst struggling with life ourselves* *Please help us be less invisible in our communities. Please help us help the person we love and care for to be less invisible in our communities.*
	3.3. Impact on policy	*Shaping government agenda is probably where change is mostly required*

3.2.1. Previous Caregiver Research Participation

Most respondents (79%, 89/112) had not previously participated in research being conducted by a university in relation to their role as a caregiver, while 18% (20/112) of participants had previously taken part in research. For those who had previously taken part in research, the majority had completed surveys: there was no reference by any of the caregivers to taking part in other forms of research, such as focus groups or interviews. Most had taken part previously due to a professional interest in research: for example, it being part of their job, or from a personal interest whereby participating in the research enabled access to specific support or services. The main reasons for not participating in research previously was lack of awareness, being unaware of any ongoing research or not being asked to participate in research prior to the current project. The other barrier towards participating in research was the lack of free time caregivers had allowing them to commit, given the other time pressures in their life. Time was, by far (both in Likert and free text responses) the most significant barrier to taking part in research, followed by money and personal health (See Figure 1).

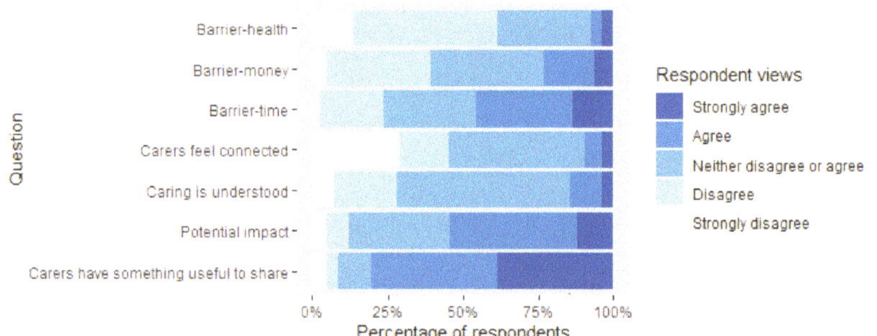

Figure 1. Caregiver responses to a range of statements according to whether they "Strongly agree", "Agree", "Neither disagree or agree", "Disagree" and "Strongly Agree". Statements focused on whether: (i) caregivers have something useful to share with researchers; (ii) researchers could make potential impact on caregiver health and wellbeing, (iii) the role of a caregiver is understood, and (iv) whether caregivers feel connected to researchers at university settings. In addition, we explored views on the barriers of (v) time, (vi) money and (vii) health to participation in research.

Closer inspection of our Likert responses around current and previous experiences of caregiver research suggests over 75% of caregivers agree/strongly agree that caregivers have something "useful to share" with researchers, yet far fewer respondents felt that researchers could make a significant impact on their health and wellness as a caregiver (Figure 1). Few caregivers responding to this work felt they were connected to researchers in universities. When asked to rate whether they felt connected to university research, 29% (24/84) of participants strongly disagreed with this statement (Figure 1). It was suggested that university researchers do not have a particularly strong grasp of the realities of a caring situation.

3.2.2. Future Research Participation

Online and community-based events, along with newsletters, were the most common ways in which caregivers felt they could contribute to university-led research in the future. However, when asked what the best way would be for caregivers to work alongside academia in the future, online and in person were the most popular options, and almost equally split among participants. The split opinion seems to be between the gold standard being face-to-face to ensure a rapport is built between caregiver and researcher, among other benefits, and the practicality and perhaps reality that online would be more suitable, given the time pressures faced by caregivers. The majority (53%, 39/73) of respondents would be willing to commit 30 min or more a month to participating in research.

3.2.3. Future Research Aspirations

In terms of future focus for the health and wellbeing of caregivers, from Likert scale responses, there was a need stated for a diverse range of areas including: information-based research, remote monitoring technology, communication technologies to enable connection between caregivers and policy related research to help shape national agendas and policies (Figure 2a). When asked about different ways in which informal caregivers could provide input in research, many placed high value on all the options presented to them, including developing new ideas and new approaches to working together. Other popular options included ensuring that fresh solutions are relevant for other caregivers. Reviewing innovative technology ideas in development was not a clear priority for caregivers compared to other solutions. Thematic analysis of participant quotations highlighted a strong need to improve the support provided to caregivers. This included financial, emotional, psychological and training and educational support. It was very apparent that participants do not currently feel heard or that they have received sufficient levels and/or quality of support in their role as a caregiver.

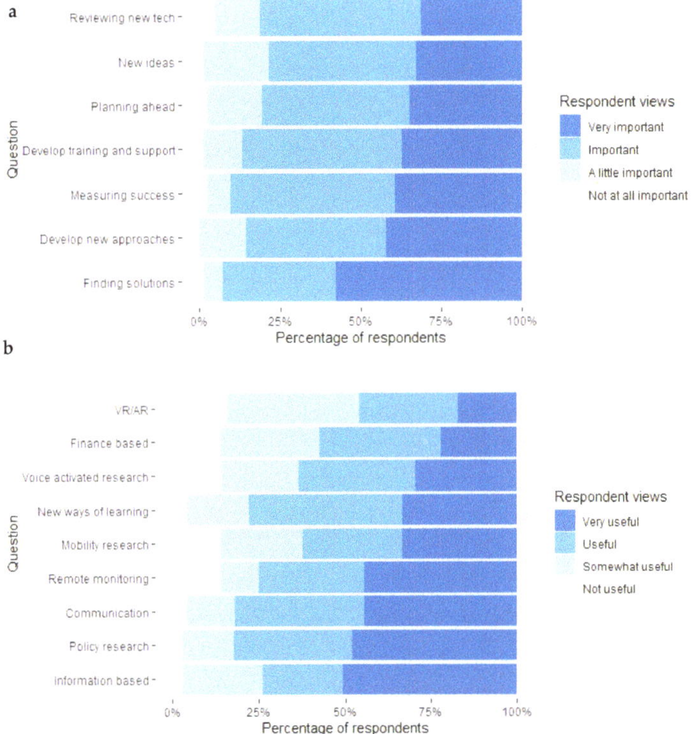

Figure 2. (**a**) Caregivers' responses to a range of statements according to whether they are "Very important", "Important", "A little important", "Not at all important". Statements included whether caregivers were interested in (i) finding solutions, (ii) developing new approaches to work with researchers, (iii) measuring success, (iv) developing training and support, (v) planning ahead, (vi) developing new ideas, and (vii) reviewing new technologies. (**b**) A separate question for caregivers focused on whether future work would be "Very useful", "Useful", "Somewhat useful" or "Not Useful". Statements included research around: (i) information (ii) policy, (iii) communication, (iv) remote monitoring, (v) mobility, (vi) new ways of learning, (vii) voice activated technologies, (viii) finance based and (ix) VR/AR. VR = Virtual Reality, AR= Augmented Reality.

Further feedback for future work was obtained around which specific areas of collaboration would make the biggest impact on caregiver health and wellness (Figure 2b). There was a wide array of different interests in this group, for example, information science was a particular area of interest highlighted where caregivers viewed research that shares key information around: caring policy, rights, and entitlements as a priority. While more than 50% of the participants viewed research around mobility, innovation and learning, finance and voice activated research as useful/very useful, fewer participants were interested in Virtual Reality/Augmented Reality as a research priority.

3.3. Professionals

Three key themes were established from the structure of our questioning and eight sub-themes were identified from the thematic analysis. The three pre-specific (deductive) themes were; (1) Previous research participation, (2) Future research participation and (3) Future research aspirations. The key themes and sub-themes are presented in Table 3.

Table 3. Key themes and sub-themes from Professionals Survey.

Key Theme	Sub-Theme	Example Quote
1. Previous research participation	1.1. Experience and connection to Research/ers	*I feel obligated to take part since we need more attention on the work we do as carers.*
	1.2. Interest in research	*Think it is really important to gather information from all walks of life and people who are doing the work at ground level and not people sitting in offices that are not meeting clients and families on a daily basis.*
	1.3. Barriers	*Time is my main commodity.*
2. Future research participation	2.1. Methods of participation	*Carers have little time for themselves, so it would make sense to provide ways in which they could contribute with their experiences at a convenient time. They are exhausted day by day so asking them to go somewhere or receiving people at home is uncomfortable and burdensome for them.*
	2.2. Involving hard to reach stakeholders	*Often when research or consultations are carried out with carers, it is the same people saying the same things and often following an organisation or organisation's agenda. Would suggest there is value in speaking with carers who haven't been supported by carer organisations.*
3. Future research aspirations	3.1. Innovative technology	*Technology for monitoring care for persons when they have memory issues that can give the carer peace of mind ... Having a life outside of caring, maybe tracking what the carer is doing and encouraging them to take a break where possible.*
	3.2. Training and support for Carers	*Training and peer support is currently a massive challenge and services are currently looking to move as much as possible onto online platforms.* *To find the best possible solutions to help people manage their own well-being as carers, which will impact on those they care for.*

3.3.1. Previous Research Participation

Professionals were asked if they had previously been involved with or collaborated with universities and just over 62% (38/61) of respondents said they had not. When asked to comment further, many of those who had previous experience with research had done so in a professional capacity as part of their role. A key reason for having participated in research or being interested in doing so in the future was the importance and want to share their knowledge and experience.

The majority of professionals suggested that caregivers do not feel connected to universities or researchers and, similarly, many did not think that those working in universities have a good understanding of the challenges faced by caregivers (Figure 3). However, the vast majority of respondents felt that caregiver experiences would be extremely valuable to researchers and that research could have a significant impact on the health and wellbeing of caregivers (Figure 3). When asked about barriers to caregivers participating in research, lack of time was the most significant barrier, followed by money and then poor caregiver health.

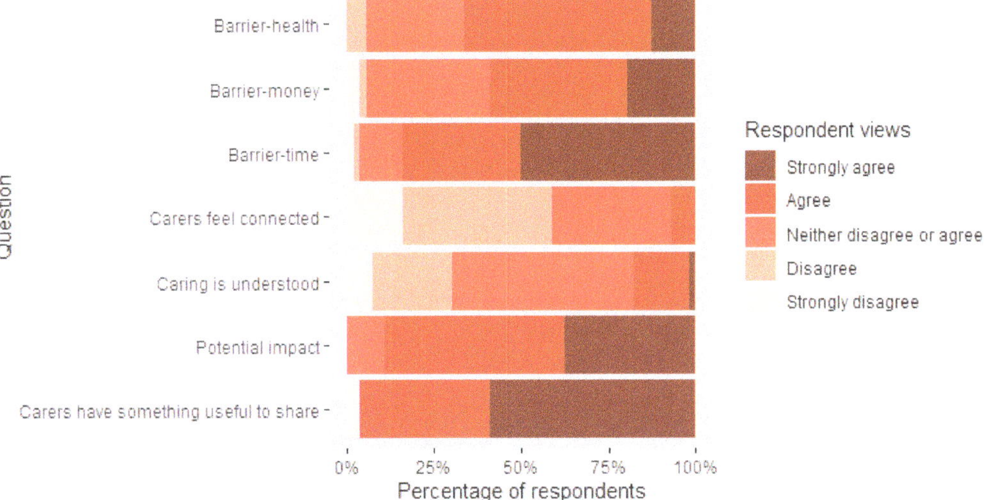

Figure 3. Professionals' responses to a range of statements according to whether they "Strongly agree", "Agree", "Neither disagree or agree", "Disagree" and "Strongly Agree". Statements focused on whether: (i) caregivers have something useful to share with researchers, (ii) researchers could make potential impact on caregiver health and wellbeing, (iii) the role of a caregiver is understood, and (iv) whether caregivers feel connected to researchers at university settings. In addition, we explored views on the barriers of (v) time, (vi) money and (vii) health to participation in research.

3.3.2. Future Research Participation

Professionals stated that online group activities and community-based events were the most popular ways which would make it easier or more likely for people to contribute to caregiver research 82% (46/56) and 66% (37/56), respectively. However, when asked to explain further, the importance of engaging those who are harder to reach and not simply using charity organisations as a means of recruiting caregivers to participate in research came through strongly in the free text answers. Perhaps, reflective of the COVID-19 situation people were living in at the time of completing the survey, online methods followed by in person face-to-face meetings were preferred with phone-based and postal methods being less popular. Many respondents, however, highlighted the benefits of face-to-face meetings.

3.3.3. Future Research Aspirations

We asked professional respondents similar questions to caregivers around research priority areas. Although there was some indication that finding available solutions and measuring success were important, training and support, planning ahead, developing new ideas and developing new approaches to work with research were all a priority that over 50% of professionals thought were "important" or "very important" (Figure 4a). In terms of specific types of technology, respondents felt research and development into all areas of technology would be useful (Figure 4b). Those respondents appeared to think that what would be particularly useful were technologies enabling remote monitoring and communication technologies to facilitate connection between caregivers and other caregivers or professionals. Interestingly, virtual reality technology was of interest, but noticeably less so than other forms of technology (Figure 4b).

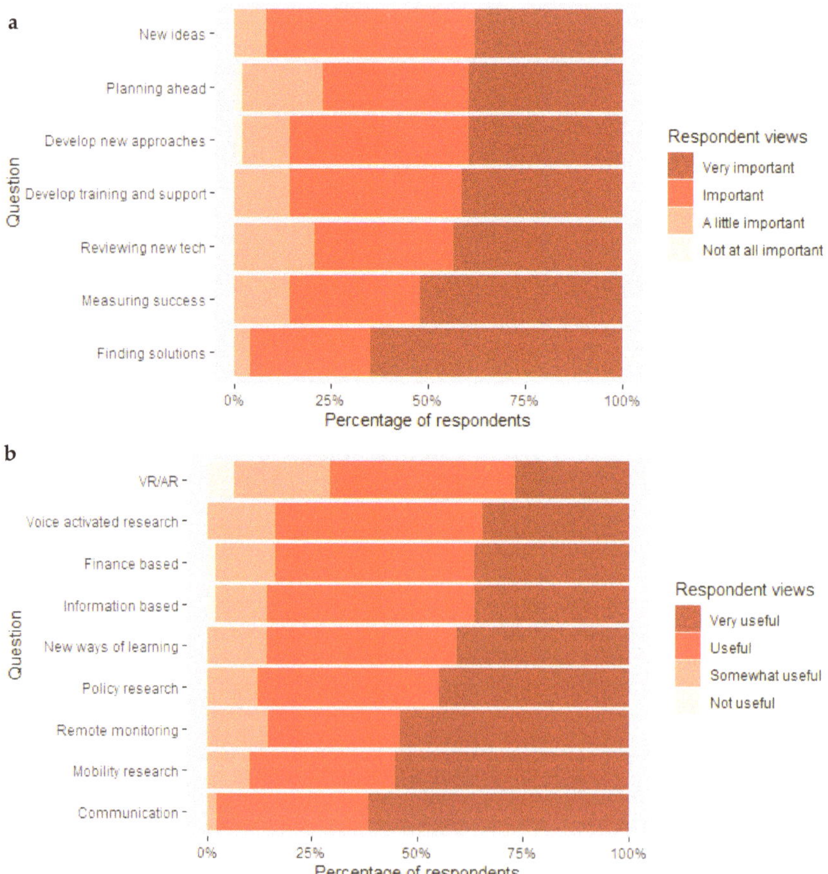

Figure 4. (**a**) Professionals' responses to a range of statements according to whether they are "Very important", "Important", "A little important", "Not at all important". Statements included whether caregivers were interested in (i) finding solutions, (ii) developing new approaches to work with researchers, (iii) measuring success, (iv) developing training and support, (v) planning ahead, (vi) developing new ideas, and (vii) reviewing new technologies. (**b**) A separate question for professionals focused on whether future work would be "Very useful", "Useful", "Somewhat useful" or "Not Useful". Statements included research around: (i) information (ii) policy, (iii) communication, (iv) remote monitoring, (v) mobility, (vi) new ways of learning, (vii) voice activated technologies, (viii) finance based and (ix) VR/AR.

When asked to elaborate on the most important outputs they felt could come from their involvement in caregiver research, the response fits into four key themes: (1) Solutions that are relevant to caregivers and truly reflective of their needs, (2) Long-term, sustainable solutions, (3) For caregivers to feel empowered and supported in their role, (4) Possible impact on policy and practice.

4. Discussion

The pressures on the caregiver demographic continue to grow. An ageing growing global population [19], the disruption caused by COVID-19 [8] and limited support for the caregivers are a perfect storm for a population already under strain. Recent UK statistics suggest that year on year, that stress outcomes have been worsening for caregivers, underlining the need for disruptive change [20]. Here, we have explored what a future roadmap for innovation could look like through a multi-stakeholder consultation and evaluation: an area of considerable importance for public health. Through incorporating the views of 174 individuals (including 112 informal caregivers and 62 professionals), there are a wide range of unmet needs, and an appetite to work closer with researchers and universities through both in person and online approaches (see Figure 5). While our responses suggest that working together to understand core needs is plausible for a sizable group of caregivers, care must be taken not to become overzealous in our interpretations—many caregivers may remain difficult to reach, including those who are delivering care on 24/7 basis. Nevertheless, this work forms some early steps to engaging the wider caregiver demographic and, taken collectively, our findings suggest a need to: (i) identify and work sustainably with caregivers, (ii) listen to, and encompass the wide range of needs and perspectives of caregivers and other stakeholder groups, (iii) improve the quality and quantity of methodologies for caregiver research and (iv) widely share research knowledge (e.g., successful and non-successful innovations for research/implementation).

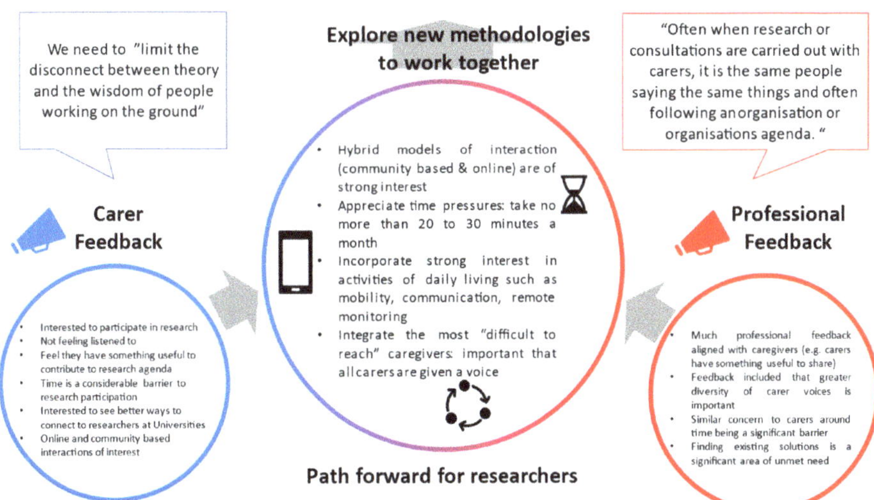

Figure 5. Summary figure summating both informal caregiver and professionals/innovators interested in the caregiver demographic.

While we achieved our main objectives, there are several limitations. Our work represents an online survey where there are biases caused by convenience sampling. For example, representation from ethnic minorities is relatively sparse, however, this was not a specific objective we aimed to address and further work is required to understand whether our findings are paralleled in such groups. We may also be inadvertently missing out on views on those with substantial caring commitments or those who do not engage in digital technologies. Second is that, while we have attempted to be as objective as possible in our analyses and interpretations, our questions were pre-specified and, therefore, our qualitative analysis should be interpreted as deductive in terms of thematic approach. Third is that there are many important barriers to innovations that we cannot comment on, such as government incentives, cost-benefit models and start up environments. Our responses are based almost exclusively in the Scottish setting: in future, it would be advantageous to better understand other parts of the UK, alongside more nuanced differences such as differences between urban vs. rural settings. Finally, this work took place during the COVID-19 pandemic, which may have skewed representation to more digitally advanced groups of caregivers and missed important viewpoints.

Our main findings highlight a need to identify and work sustainably with caregivers: a task not straightforward given the lack of emotional, physical, and monetary support readily available. Moving forward, there is a need to find better ways to recognize the value that caregivers bring. This may be achievable through developing a more continuous interaction with caregivers, for example, moving beyond individual projects to open ended or problem focused collaborations. While many caregivers would welcome routes to engage with universities, researchers must see that the process needs to be efficient and inclusive: ideally taking no more than 20 to 30 min a month. Relevant to such a trade-off between researchers aims and citizen engagement is the concept of living labs. Living labs are a methodology/approach that can allow citizens to participate in the design, development, and evaluation of innovative solutions to address societal problems [21]. There is considerable interest in using such approaches to explore health and wellness problems, including scope to carefully explore a range of different issues through participatory design, across the life course [22,23]. These approaches are helping to redefine research methodologies, including the integration of citizen generated and non-traditional data sources [24,25]. Although careful consideration will be required of how to best build upon emerging online methodologies (e.g., ethical use of data), our findings here indicate that a hybrid model of both online and in-person approaches could work well for the caregiver demographic. "Citizen science" approaches could facilitate multiple research stages, such as ideation, co-design of the approach, data-gathering and knowledge transfer of findings through "light touch" interactions (e.g., smartphone/computer-based) or through community "pop up" events. Critically, this would allow multiple routes to help different groups of caregivers influence research agendas within the earliest stages [26]. Such integration could help address known issues around translational failure [15] and be supported through a wide variety of tools such as prioritization methodology [27] alongside recognition in academic publications [28].

Another key finding from this work is that caregivers (who were able to respond to this work) would largely welcome being "championed" and provided with a channel to engage with universities. However, the practicalities of delivering this are yet to be realized—political and social landscapes change over time, as do priorities. While both we and many other research groups [29,30] consult caregivers about current unmet needs on a regular basis, there remains a risk that some specific groups (e.g., according to age, geographical location, ethnicity, socioeconomic status, number of hours caring per week) will be excluded from conversations through selection bias [31]. Further work is needed to contextualize how representative current research is of all caregivers and, critically, we must establish community links to help those in need who do not traditionally connect with research agendas or third sector partners. Lastly, while the concept of sustained engagement proposed here may be new to many university settings, there is a need to ensure efforts augment, opposed to replace existing efforts ongoing elsewhere (e.g., health and social care data, charity, and voluntary sector annual reports).

Given that much of the conversation around caregivers, needs and innovations falls out with traditional models of empirical research (i.e., Randomized Controlled Trials), there is a need to keep furthering knowledge of research methodologies—this could be to improve the value of non-traditional data sources and/or to increase capacity for more caregivers to become involved. A source of inspiration that highlights passive involvement from participants includes the "Dreamlab" project from University College London [32], developed so that individuals can engage with researchers through charging their smartphone to assist with cancer research. As the concept has already reached 1 million downloads, this is a clear indication that citizens are interested to connect and make impacts to research priorities. Further, the recent contributions of the COVID ZOE smartphone app have also underlined the power of citizen science, whereby seemingly small actions of individuals on a larger scale can accelerate current knowledge to help navigate the unchartered waters of an emerging pandemic [33]. Given such examples and our participant feedback, it is feasible to see smartphones being better used to connect caregivers (and other citizens/professionals) to ongoing research opportunities and emerging agendas, and to signpost others to recruitment calls. An immediate utility of such citizen science approaches could be to connect caregivers of rarer/under-researched conditions to highlight such needs to researchers and funding councils alike.

Another key take-home point from this work is the demonstration of the real depth of knowledge that caregivers hold for the development of future innovations and co-design approaches [34]. For example, caregivers demonstrated intimate knowledge of the daily barriers and enablers to existing technologies, and from the findings presented here, many caregivers are still actively seeking solutions around information needs and remote monitoring, problems that researchers may think are already solved. In terms of contributing specifically to ongoing research, findings from our own work and others [17,29] have highlighted key research topics for caring (e.g., access to information online), and such priorities could form the foundation of conversations with caregivers in a variety of settings. An achievable and realistic first step from here could be to pick exemplar caregiver priorities, and to run pilot projects that could facilitate the development of caregiver co-designed "principles" to help scaffold the way by which researchers engage with caregivers moving forward.

5. Conclusions

To conclude, the aims of this work were to initiate the development of a roadmap for future caregiver innovation. We found that, despite a number of barriers to participation, many caregivers are willing to regularly contribute to the research agenda and have a wide range of unmet needs that could be better addressed by existing and future research outputs. Visibility is still an issue to caregivers—we were only able to hear from a small subsample of the caregiver population here and even these participants did not feel visible and heard in research agendas. As a wider community, this needs to change. Clearly, accessible routes for engagement (e.g., integrated online and face-to-face approaches) and more formal recognition of their expertise would be welcomed alongside the delivery of real-world impacts on those who care. A next logical step would be to determine which approaches (such as living lab methodologies) are capable of delivering for caregivers not just in terms of research outputs/outcomes, but in terms of transparency, engagement, and experience.

Author Contributions: Conceptualization, K.J.E. and K.A.M.; methodology, K.J.E. and K.A.M.; writing—original draft preparation, K.J.E., K.A.M.; writing—review and editing, K.J.E., K.A.M., L.M., R.M., L.M. and M.L.; funding acquisition, K.J.E. and K.A.M. All authors have read and agreed to the published version of the manuscript.

Funding: This research received funding from the University of Strathclyde "Strathwide" Early Career Research Grant awards.

Institutional Review Board Statement: The study was conducted according to the guidelines of the Declaration of Helsinki and approved by the University of Strathclyde Computer and Information Science Ethics Committee.

Informed Consent Statement: Informed consent was obtained from all subjects involved in the study.

Data Availability Statement: The data presented in this study are available on request from the corresponding author. The full data are not publicly available to ensure anonymity of participants who took part in this work.

Acknowledgments: We would like to recognize the input of all caregivers and professionals involved in this work. We are also grateful to Patricia Clark at Carers Scotland (UK) and her colleagues to help us recruit carers to this work and to colleagues within the Digital Health & Care Innovation Centre (DHI) for assisting us to recruit professionals/innovators (including Moira MacKenzie, Kara McKenzie). We would also like to acknowledge contributions from the Centre for Lifelong Learning (CLL) at the University of Strathclyde—including Alix McDonald and Gemma Gilliland.

Conflicts of Interest: The authors declare no conflict of interest.

Appendix A

Q1 How old are you in years? Please select one.

- 18 to 24 (1)
- 25 to 34 (2)
- 35 to 44 (3)
- 45 to 54 (4)
- 55 to 64 (5)
- 65 to 74 (6)
- 75 to 84 (7)
- 85+ (8)
- Prefer not to say (9)

Q2 How do you describe you gender? Please select one.

- Man/Male (including trans man) (1)
- Woman/Female (including trans woman) (2)
- I prefer not to answer (3)
- In another way (4)

Q2.1 If you selected in another way, please share further information here:
Q3 What is your highest level of education? Please select one.

- Degree or equivalent (1)
- Higher education (2)
- School qualifications (3)
- Other qualifications (4)
- No qualifications (5)
- Don't know (6)

Q4 What is your ethnicity? Please select one.

- White (1)
- Mixed/multiple ethnic groups (2)
- Asian/Asian British (3)
- Black/African/Caribbean/Black British (4)
- Other ethnic group (5)
- Prefer not to say (6)

Q5 How long have you been a carer (in years)

- 1 year or less (1)
- Up to 2 years (2)
- Up to 4 years (3)
- Up to 10 years (4)
- 10 years or more (5)

Q6 Are you still able to work or study as you care?

- Yes, caregiving has not affected my working/studying hours (1)
- Yes, caregiving has caused me to reduce my working/studying hours (2)
- Caregiving has caused me to give up work/study (3)
- I do not currently work/study and did not have to give up work/study due to caregiving (4)

Appendix B

Table A1. Likert statements in full.

Title	Abbreviation (i.e., in Figure)
Question group: To what extent do you support the following statements	
I feel connected to researchers working at universities	I feel connected
I feel that researchers working in universities have a good understanding of the challenges of being a carer	Caring is understood
My experience (including as a carer) means that I have something useful to share with researchers working within universities	I have something useful to share
The work that researchers in universities do could make a significant impact on my health and wellness as a carer	Potential impact
Time is a significant barrier for me to become involved in research	Barrier—time
Money is a significant barrier for me to become involved in research	Barrier—money
My health is a significant barrier for me to become involved in research	Barrier—health
Question group: Where do you think your input as a carer is most important for researchers working in universities?	
Developing new ideas	New ideas
Reviewing new technology ideas in development (e.g., using a star rating)	Reviewing new tech
Developing new approaches to work together (e.g., improving the way researchers connect with carers)	Develop new approach
Developing training and support materials for development of new ideas	Develop training & support
Helping to decide how to measure the success of technologies (e.g., understanding what matters to you most)	Measure success
Making sure that other carers can find new solutions relevant to them	Finding solutions
Working with researchers in universities long term to plan ahead	Planning ahead
Other	Other
Question group: In your view, what areas of collaboration with universities would make the biggest impact on your own health and wellness?	
Information based technology research (e.g., research that shares key information around caring, your rights, entitlements etc.)	Information-based
Finance based research (e.g., innovations/applications that would help you manage any finance activities, including reminders or notifications)	Finance-based
Voice activated technology research (e.g., Alexa, Siri or other voice based technologies)	Voice activated research
Virtual reality/Augmented reality research (e.g., through the use of mobile phones or more specialised equipment)	VR/AR
Mobility research (e.g., any technologies that help within the home or getting out an about such as wheelchairs)	Mobility research
Research on technologies for remote monitoring (e.g., looking after someone at distance through self reported measures)	Remote monitoring
Communication technologies (e.g solutions to help you connect to other caregivers or professionals)	Communication
Policy related research (e.g., gathering evidence to help shape national agendas and priorities)	Policy research
Research into new/innovative ways of learning (e.g., online learning)	New ways of learning
Other	Other

Table A1. *Cont.*

Title	Abbreviation (i.e., in Figure)
To what extent do you support the following statements	
Carers are connected to researchers working at universities	I feel connected
Researchers working in universities have a good understanding of the challenges of being a carer	Caring is understood
Carer experience is something useful to share with researchers working within universities	Carers have something useful to share
The work that researchers in universities do could make a significant impact on carer health and wellness	Potential impact
Time is a significant barrier for carers to become involved in research	Barrier—time
Money is a significant barrier for carers to become involved in research	Barrier—money
Poor carer health is a significant barrier for carers to become involved in research	Barrier—health
Where do you think your input is most important for researchers working in universities to develop new ideas/solutions for carers?	
Developing new ideas	New ideas
Reviewing new technology ideas in development (e.g., using a star rating)	Reviewing new tech
Developing new approaches to work together (e.g., improving the way researchers connect with carers)	Develop new approach
Developing training and support materials for development of new ideas	Develop training & support
Helping to decide how to measure the success of technologies (e.g., understanding what matters to you most)	Measure success
Making sure that carers can find new solutions relevant to them	Finding solutions
Working with researchers in universities long term to plan ahead	Planning ahead
Other	Other
If you were to work with a carers in research on a regular basis, what technologies/innovations or areas of work do you think would make the biggest impact on carers own health and wellness?	
Information based technology research (e.g., research that shares key information around caring, your rights, entitlements etc.)	Information-based
Finance based research (e.g., innovations/applications that would help you manage any finance activities, including reminders or notifications)	Finance-based
Voice activated technology research (e.g., Alexa, Siri or other voice based technologies)	Voice activated research
Virtual reality/Augmented reality research (e.g., through the use of mobile phones or more specialised equipment)	VR/AR
Mobility research (e.g., any technologies that help within the home or getting out an about such as wheelchairs)	Mobility research
Research on technologies for remote monitoring (e.g., looking after someone at distance through self reported measures)	Remote monitoring
Communication technologies (e.g., solutions to help you connect to other caregivers or professionals)	Communication
Policy related research (e.g., gathering evidence to help shape national agendas and priorities)	Policy research
Research into new/innovative ways of learning (e.g., online learning)	New ways of learning
Other	Other

References

1. Carers UK. Missing out, the Identification Challenge. 2016. Available online: https://www.carersuk.org/for-professionals/policy/policy-library/missing-out-the-identification-challenge (accessed on 2 June 2021).
2. Hoffmann, F.; Rodrigues, R. Informal Carers: Who Takes Care of Them? Policy Brief 4/2010. Vienna: European Centre for Social Welfare and Policy Research. 2010. Available online: https://www.euro.centre.org/publications/detail/387 (accessed on 2 June 2021).
3. Maynard, A. Shrinking the state: The fate of the NHS and social care. *J. R. Soc. Med.* **2017**, *110*, 49–51. [CrossRef]
4. Pristavec, T. The burden and benefits of caregiving: A latent class analysis. *Gerontologist* **2019**, *59*, 1078–1091. [CrossRef] [PubMed]
5. Stansfeld, S.; Smuk, M.; Onwumere, J.; Clark, C.; Pike, C.; McManus, S.; Harris, J.; Bebbington, P. Stressors and common mental disorder in informal carers—An analysis of the English Adult Psychiatric Morbidity Survey 2007. *Soc. Sci. Med.* **2014**, *120*, 190–198. [CrossRef] [PubMed]
6. Metzelthin, S.F.; Verbakel, E.; Veenstra, M.Y.; van Exel, J.; Ambergen, A.W.; Kempen, G.I.J.M. Positive and negative outcomes of informal caregiving at home and in institutionalised long-term care: A cross-sectional study. *BMC Geriatr.* **2017**, *17*, 232. [CrossRef]
7. del-Pino-Casado, R.; Priego-Cubero, E.; López-Martínez, C.; Orgeta, V. Subjective caregiver burden and anxiety in informal caregivers: A systematic review and meta-analysis. *PLoS ONE* **2021**, *16*, e0247143. [CrossRef] [PubMed]
8. Egan, K. Digital technology, health and wellbeing and the COVID-19 pandemic: It's time to call forward informal carers from the back of the queue. *Semin. Oncol. Nurs.* **2020**, *36*, 151088. [CrossRef]
9. Ewing, G.; Austin, L.; Jones, D.; Grande, G. Who cares for the carers at hospital discharge at the end of life? A qualitative study of current practice in discharge planning and the potential value of using The Carer Support Needs Assessment Tool (CSNAT) Approach. *Palliat. Med.* **2018**, *32*, 939–949. [CrossRef]
10. Egan, K.J.; Pinto-Bruno, Á.C.; Bighelli, I.; Berg-Weger, M.; van Straten, A.; Albanese, E.; Pot, A.M. Online training and support programs designed to improve mental health and reduce burden among caregivers of people with dementia: A systematic review. *J. Am. Med. Dir. Assoc.* **2018**, *19*, 200–206.e201. [CrossRef]
11. Shin, J.Y.; Choi, S.W. Online interventions geared toward increasing resilience and reducing distress in family caregivers. *Curr. Opin. Supportive Palliat. Care* **2020**, *14*, 60–66. [CrossRef]
12. Pot, A.M.; Gallagher-Thompson, D.; Xiao, L.D.; Willemse, B.M.; Rosier, I.; Mehta, K.M.; Zandi, D.; Dua, T.; iSupport Development Team. iSupport: A WHO global online intervention for informal caregivers of people with dementia. *World Psychiatry* **2019**, *18*, 365–366. [CrossRef]
13. Pham, Q.; El-Dassouki, N.; Lohani, R.; Jebanesan, A.; Young, K. The future of virtual care for older ethnic adults beyond COVID-19. *arXiv* **2021**. preprint.
14. James, T.; Mukadam, N.; Sommerlad, A.; Guerra Ceballos, S.; Livingston, G. Culturally tailored therapeutic interventions for people affected by dementia: A systematic review and new conceptual model. *Lancet Healthy Longev.* **2021**, *2*, e171–e179. [CrossRef]
15. Greenhalgh, T.; Wherton, J.; Papoutsi, C.; Lynch, J.; Hughes, G.; A'Court, C.; Hinder, S.; Fahy, N.; Procter, R.; Shaw, S. Beyond adoption: A new framework for theorizing and evaluating nonadoption, abandonment, and challenges to the scale-up, spread, and sustainability of health and care technologies. *J. Med. Internet Res.* **2017**, *19*, e367. [CrossRef]
16. WHO. *Global Strategy on Digital Health 2020–2025*; WHO: Geneva, Switzerland, 2021.
17. Lindeman, D.A.; Kim, K.K.; Gladstone, C.; Apesoa-Varano, E.C. Technology and caregiving: Emerging interventions and directions for research. *Gerontologist* **2020**, *60* (Suppl. 1), S41–S49. [CrossRef]
18. Braun, V.; Clarke, V. Using thematic analysis in psychology. *Qual. Res. Psychol.* **2006**, *3*, 77–101. [CrossRef]
19. World Health Organization. *Decade of Healthy Ageing: Baseline Report*; World Health Organization: Geneva, Switzerland, 2020.
20. NHS Digital. Available online: https://digital.nhs.uk/data-and-information/publications/statistical/personal-social-services-survey-of-adult-carers/england-2018-19 (accessed on 19 August 2021).
21. Kim, J.; Kim, Y.; Jang, H.; Cho, M.; Lee, M.; Kim, J.; Lee, H. Living labs for health: An integrative literature review. *Eur. J. Public Health* **2019**, *30*, 55–63. [CrossRef] [PubMed]
22. Archibald, M.M.; Wittmeier, K.; Gale, M.; Ricci, F.; Russell, K.; Woodgate, R.L. Living labs for patient engagement and knowledge exchange: An exploratory sequential mixed methods study to develop a living lab in paediatric rehabilitation. *BMJ Open* **2021**, *11*, e041530. [CrossRef]
23. Verbeek, H.; Zwakhalen, S.M.G.; Schols, J.; Kempen, G.; Hamers, J.P.H. The living lab in ageing and long-term care: A sustainable model for translational research improving quality of life, quality of care and quality of work. *J. Nutr. Health Aging* **2020**, *24*, 43–47. [CrossRef]
24. van den Kieboom, R.C.; Bongers, I.M.; Mark, R.E.; Snaphaan, L.J. User-driven living lab for assistive technology to support people with dementia living at home: Protocol for developing co-creation-based innovations. *JMIR Res. Protoc.* **2019**, *8*, e10952. [CrossRef]
25. Okop, K.J.; Murphy, K.; Lambert, E.V.; Kedir, K.; Getachew, H.; Howe, R.; Niyibizi, J.B.; Ntawuyirushintege, S.; Bavuma, C.; Rulisa, S.; et al. Community-driven citizen science approach to explore cardiovascular disease risk perception, and develop prevention advocacy strategies in sub-Saharan Africa: A programme protocol. *Res. Involv. Engagem.* **2021**, *7*, 11. [CrossRef]
26. Sanders, E.B.N.; Stappers, P.J. Co-creation and the new landscapes of design. *CoDesign* **2008**, *4*, 5–18. [CrossRef]

27. Interaction Design Website. Making Your UX Life Easier with the MoSCoW. Available online: https://www.interaction-design.org/literature/article/making-your-ux-life-easier-with-the-moscow (accessed on 19 August 2021).
28. Staniszewska, S.; Brett, J.; Simera, I.; Seers, K.; Mockford, C.; Goodlad, S.; Altman, D.G.; Moher, D.; Barber, R.; Denegri, S.; et al. GRIPP2 reporting checklists: Tools to improve reporting of patient and public involvement in research. *BMJ* **2017**, *358*, j3453. [CrossRef] [PubMed]
29. Egan, K.J. A national U.K. survey to understand current needs and future expectations of informal caregivers for technology to support health and wellbeing. *JMIR Aging* **2021**, in press. [CrossRef]
30. Graffigna, G.; Gheduzzi, E.; Morelli, N.; Barello, S.; Corbo, M.; Ginex, V.; Ferrari, R.; Lascioli, A.; Feriti, C.; Masella, C. Place4Carers: A multi-method participatory study to co-design, piloting, and transferring a novel psycho-social service for engaging family caregivers in remote rural settings. *BMC Health Serv. Res.* **2021**, *21*, 591. [CrossRef] [PubMed]
31. Enzenbach, C.; Wicklein, B.; Wirkner, K.; Loeffler, M. Evaluating selection bias in a population-based cohort study with low baseline participation: The life-adult-study. *BMC Med Res. Methodol.* **2019**, *19*, 135. [CrossRef]
32. Veselkov, K.; Gonzalez, G.; Aljifri, S.; Galea, D.; Mirnezami, R.; Youssef, J.; Bronstein, M.; Laponogov, I. HyperFoods: Machine intelligent mapping of cancer-beating molecules in foods. *Sci. Rep.* **2019**, *9*, 9237. [CrossRef]
33. Nguyen, L.H.; Drew, D.A.; Graham, M.S.; Joshi, A.D.; Guo, C.G.; Ma, W.; Mehta, R.S.; Warner, E.T.; Sikavi, D.R.; Lo, C.H.; et al. Risk of COVID-19 among front-line health-care workers and the general community: A prospective cohort study. *Lancet Public Health* **2020**, *5*, e475–e483. [CrossRef]
34. O'Connor, S.; Bouamrane, M.M.; O'Donnell, C.A.; Mair, F.S. Barriers to co-designing mobile technology with persons with dementia and their carers. *Stud. Health Technol. Inform.* **2016**, *225*, 1028–1029.

International Journal of
***Environmental Research
and Public Health***

Article

Come for Information, Stay for Support: Harnessing the Power of Online Health Communities for Social Connectedness during the COVID-19 Pandemic

Brian M. Green [1,*], Casey A. Hribar [2], Sara Hayes [1], Amrita Bhowmick [1,3] and Leslie Beth Herbert [1]

[1] Health Union, LLC, Philadelphia, PA 19107, USA; sara.hayes@health-union.com (S.H.); amrita.bhowmick@health-union.com (A.B.); LB.herbert@health-union.com (L.B.H.)
[2] School of Medicine, University of North Carolina at Chapel Hill, Chapel Hill, NC 27599, USA; casey_hribar@med.unc.edu
[3] Department of Health Behavior, Gillings School of Global Public Health, University of North Carolina at Chapel Hill, Chapel Hill, NC 27599, USA
* Correspondence: brian.m.green@health-union.com

Abstract: The COVID-19 pandemic created a globally shared stressor that saw a rise in the emphasis on mental and emotional wellbeing. However, historically, these topics were not openly discussed, leaving those struggling without professional support. One powerful tool to bridge the gap and facilitate connectedness during times of isolation is online health communities (OHCs). This study surveyed Health Union OHC members during the pandemic to determine the degree of COVID-19 concern, social isolation, and mental health distress they are facing, as well as to assess where they are receiving information about COVID-19 and what sources of support they desire. The survey was completed in six independent waves between March 2020 and April 2021, and garnered 10,177 total responses. In the United States, OHCs were utilized significantly more during peak lockdown times, and the desire for emotional and/or mental health support increased over time. Open-ended responses demonstrated a strong desire for connection and validation, which are quintessential characteristics of OHCs. Through active moderation utilizing trained moderators, OHCs can provide a powerful, intermediate and safe space where conversations about mental and emotional wellbeing can be normalized and those in need are encouraged to seek additional assistance from healthcare professionals if warranted.

Keywords: eHealth; chronic disease; online community; social support; COVID-19

Citation: Green, B.M.; Hribar, C.A.; Hayes, S.; Bhowmick, A.; Herbert, L.B. Come for Information, Stay for Support: Harnessing the Power of Online Health Communities for Social Connectedness during the COVID-19 Pandemic. *Int. J. Environ. Res. Public Health* **2021**, *18*, 12743. https://doi.org/10.3390/ijerph182312743

Academic Editors: Irene Torres-Sanchez and Marie Carmen Valenza

Received: 29 October 2021
Accepted: 1 December 2021
Published: 3 December 2021

1. Introduction

Although social isolation, loneliness, and mental health distress have always been commonplace, societal stigma often means they are not openly discussed [1]. However, the COVID-19 pandemic created a globally shared stressor that saw a rise in the emphasis on mental and emotional wellbeing. From popular icons such as Michelle Obama freely speaking about experiencing "low-grade depression" as a result of the pandemic and societal unrest in America [2], to the World Health Organization's emphasis on increased mental health infrastructure due to an impending and critical increase in demand [3], the COVID-19 pandemic has changed the way we talk about, prioritize, and consume healthcare and services that address mental health needs.

Despite still being in the midst of the pandemic, research so far points toward COVID-19's ubiquitous negative impact on mental health [3–6]. A Kaiser Family Foundation survey from April to May of 2021 found that 30% of adults in the United States (U.S.) reported symptoms of depression or anxiety, a rise of more than 10% higher than expected based on pre-pandemic trends. In addition, nearly a quarter of those with symptoms of psychological distress who reported needing mental health services were not able to access them [7]. These COVID-19 impacts are far-reaching; however, factors such as a previous

145

lack of social support, perceived likelihood of survival, comorbid mental or physical health conditions, and challenging home demands, such as homeschooling young children, are all contributors to an increased risk for mental health distress [4,6].

In research conducted prior to the pandemic, social connectedness and social support were associated with benefits such as decreased risk of cardiovascular disease and depression, improved immune system functioning, and reduced morbidity and mortality, among others [8–11].

A metanalysis of 40 studies confirmed that comorbidities of mental health and chronic physical conditions are a burden for people, not just in the developed world where research often focuses, but also in developing and emerging countries [12]. While psychological interventions of various types can improve a patient's quality of life [13–15], there are mixed results depending upon in-patient or out-patient settings [16].

A positive state of mind as a result of strong social support, is also linked to greater medication adherence, a critical concern when it comes to living with one or more chronic health conditions [17].

Chronic health conditions are of particular interest in examining societal responses to the COVID-19 pandemic, as physical lockdowns and fears of contagion prevented the ease of access to in-person healthcare [18]. In addition, current research still suggests that those with pre-existing chronic conditions (including those with a history of malignancy) have a greater risk of developing severe COVID-19 and accompanying significant morbidity and mortality than those without [19,20]. Further, the population of chronically ill individuals tends to be older, and inherently possesses an increased susceptibility to social isolation and deteriorating mental health [21].

Especially in the context of the COVID-19 pandemic, telemedicine provides a means for patients to seek care for a variety of concerns in a convenient, accessible, and safe manner. While telemedicine for mental health is on the rise, there are still questions regarding patient willingness to embrace these new options when discussing sensitive issues. Additional concerns include payor issues with telemedicine services and the equitable access to technology required in order to undergo sessions [22–24].

One powerful tool to bridge the gap and facilitate social support and connectedness during times of isolation is online health communities (OHCs). OHCs are online platforms in which individuals with similar health conditions or experiences can share information, support, and connections [25,26]. OHCs provide a temporally flexible space that allows people to connect across geographic locations and often with anonymity [26,27]. OHCs allow patients to play as active or passive a role as they would like, while helping with identity development, self-confidence, personal validation, social interconnectedness, navigation of complex emotions, and fostering a sense of purpose [27–30]. Although OHCs are not a platform for medical care nor advice, they can provide an environment to gauge similar experiences and gain confidence to seek professional support.

In order to facilitate safe, online spaces that provide support, social connection, and information, Health Union created an Adaptive Engagement Model for OHCs [30]. At the beginning of the pandemic, Health Union had 28 active OHCs. Each community is a separate URL that matches the name of the condition, for example, Lupus.net, and has social media pages on Facebook, Instagram, and Twitter.

These OHCs are contextual and situational, and feature three core structural elements, including social support, adaptive engagement, and active moderation. Similar to physical communities, they rely on a shared identity (connection to a specific chronic health condition), social norms, and commonly an experience of societal stigma. These features, taken together in the creation and subsequent management and moderation of the OHCs, are used to facilitate the sharing of relevant health information (or content), build relationships, and harness participant interdependence through passive and active engagement opportunities [30]. Given the Adaptive Engagement Model's high emphasis on combating social isolation and creating space for normalizing conversations about life with a chronic illness, including mental health impacts, these OHCs may serve as a stepping stone for intermediate support as people struggle with considering and initiating professional services for mental health or emotional support.

The present study was conducted primarily to determine the information needs and supportive resources needed by people with chronic health conditions from Health Union's 28 online communities during the pandemic. After seeing how community members continued to engage and support each other throughout the pandemic and become less interested in specific content regarding COVID, the authors decided to conduct a secondary analysis of these data to further study social connectedness.

The aim of this study was to survey current Health Union OHC members to determine the degree of COVID-19 concern, social isolation, and mental health distress they are facing, as well as assessing where they are receiving information about COVID-19 and what sources of support they require. Although other research focused on the mental and emotional impacts of COVID-19 and the use of telemedicine or other technologies to access healthcare services, our efforts aim to investigate the novel space in between, wherein OHCs may play a critical role in fostering wellbeing, especially when access to traditional in-person services may be limited.

2. Materials and Methods

2.1. Data Collection and Survey Design

As this was a secondary analysis of a de-identified data set originally collected for quality improvement purposes, it was considered exempt from human subject review. This research was conducted according to the guidelines of the Declaration of Helsinki and other relevant laws in the U.S. Informed consent was obtained from all survey participants via an introductory email, allowing people to voluntarily click through to accept or decline the invitation to complete the Qualtrics survey. Survey responses were anonymous and not linked to individual identifiers. All email or IP addresses were stripped from the data set prior to data cleaning, storage, and access of the final data set for analysis by the authors.

Data were collected in an online survey format hosted through Qualtrics Survey Software (Qualtrics International Inc., Seattle, WA, USA). The survey was administered in six waves between 19 March 2020 and 19 April 2021. The first survey wave was sent to members of 10 separate OHCs hosted, managed, and moderated by Health Union. Subsequent waves were expanded to eventually reach a total of 28 separate OHCs managed and moderated by Health Union. Each platform provides a space for support, engagement, and education around the chronic health condition of focus [30]. In total, the survey was sent to community members via email across the six waves. Banners and other site features advertising the survey were also displayed on each platform during data collection periods. Data collection for each wave lasted between three and eight days, tailored with the aim of securing an adequate sample size, while being inclusive of people with many different chronic health conditions (Table 1).

Table 1. Survey wave sample size, number of OHCs included, dates fielded, and completed responses.

Survey Wave	Dates Fielded	Num. of OHCs	Total Completes
Wave 1	19–25 March 2020	10	991
Wave 2	14–17 April 2020	26	2214
Wave 3	12–14 May 2020	26	2210
Wave 4	21–23 July 2020	26	1777
Wave 5	23 October–2 November 2020	26	2005
Wave 6	12–19 April 2021	28	980

In order to participate in the survey, respondents needed to be at least 18 years old, live in the United States, have a chronic health condition from a pre-specified list (full list in Supplementary Materials), and be aware of COVID-19. Survey questions focused on life with a chronic condition during the pandemic. Initial topics of interest included current treatments used to manage chronic health conditions, where information was obtained, and concern regarding COVID-19. In addition, questions regarding changes in personal health behaviors as a result of the pandemic, changes in established treatment plans, desired support sources, communication with healthcare providers, and financial impacts were also included.

Given the evolving and uncertain nature of the COVID-19 pandemic, it was necessary to adjust topics slightly to most accurately reflect the current status of the pandemic for that time period. For example, waves three through six included questions about telehealth at the same time that telehealth was expanded by executive order through the Centers for Medicare and Medicaid Services during this time period [31]. Waves five and six included questions surrounding pandemic burnout, quality of life, and vaccination status. Due to this required flexibility, each wave ranged from 35 to 41 questions.

Participants who provided complete responses to the survey were entered in a drawing to receive a U.S. e-gift card for each wave. The drawing for the first wave featured USD 50 gift cards, while waves two through six featured USD 25 gift cards.

2.2. Measures

2.2.1. Perception of COVID-19

Concern for COVID-19 was assessed in each wave through the question, "At this time, how concerned do you feel about the novel coronavirus (COVID-19)?" Concern was measured on a 7-point Likert scale (1 = Not at all concerned to 7 = Very concerned). Wave six (post-vaccine availability) had a slightly modified version of this question by gauging agreement with the statement, "I am still very concerned about my risk of contracting COVID-19" (7-point Likert scale with 1 = Strongly disagree and 7 = Strongly agree). In a similar fashion, participants were asked about their concern with COVID-19 in the context of having a chronic health condition in each survey wave (Supplementary Materials). Significant concern was a response of 6 or 7, while a lack of concern was a response of 1 or 2. In several waves (waves two, three, and four), participants were also asked in an open-ended question to provide one word that described how they were feeling about COVID-19.

2.2.2. Utilized and Desired Sources of COVID-19 Information and Support

In order to determine current COVID-19 information resource use, respondents were prompted in waves one through four with the question, "What sources are you using to learn more about the novel coronavirus (COVID-19)?" Participants were prompted to select as many as applied from a list of resources, including social networking sites, Internet search engines, government websites, online blogs or support communities, and TV news reports (Supplementary Materials). Subsequently, participants were asked, "What types of information and/or support would be most helpful to you right now?" Respondents were permitted to choose up to three desired types of information or support including information from their healthcare provider about COVID-19 in relation to their health condition, emotional and/or mental health support, financial support for medications, and home delivery options (Supplementary Materials).

2.2.3. Quality of Life and Health-Related Behavioral Changes

Later waves, specifically waves five and six, incorporated questions about quality of life, mental health impacts, burnout, and perception of returning to pre-pandemic "normal" life. Changes in more tangible behaviors were asked in wave five, through a 3-option ranking of doing less, doing the same as, or doing more of a specific behavior pre-pandemic versus present. For example, participants were asked to rank their current social media use with pre-pandemic levels on this scale. Less tangible changes, such as stress, impacts on mental health, and overall quality of life were asked in wave six using a 5-point Likert-type scale, including options such as mental health being much worse during COVID-19, somewhat worse, about the same during COVID-19, somewhat better, and much better during COVID-19 (Supplementary Materials). Wave five also featured the open-ended question, "What is the biggest struggle that you're having at this point in time, as a result of (or related to) the coronavirus (COVID-19) pandemic?"

2.2.4. Demographics

Participants across all waves were asked a series of demographic questions including age, gender, annual household income, primary health insurance form, residence type (suburban, urban, rural), highest level of education attained, and employment status. Age was selected from a dropdown menu, while others allowed participants to choose from a categorical list of item responses. Participants were asked to select chronic health conditions from a list including, but not limited to, COPD, migraine, asthma, HIV, rheumatoid arthritis, hypertension, and several types of cancers (Supplementary Materials).

2.2.5. Analysis

Responses were analyzed using descriptive statistics and z-tests to explore differences across waves. This analysis was conducted in order to identify differences in the need for mental and/or emotional support and changes to quality of life relative to each wave of the survey and the corresponding time period of the pandemic. Responses to open-ended questions were reviewed for common themes and impactful quotes regarding the need for social support.

3. Results

3.1. Participants

In total, there were 10,177 responses across all six waves. The number of survey participants for each wave is shown in Table 1. Demographics of participants for each wave, including, but not limited to, mean age, gender, and employment status, were collected (Table 2). The most commonly experienced chronic health condition was hypertension, followed by asthma, migraine, rheumatoid arthritis, COPD, and multiple sclerosis. Nearly seven in ten had never been diagnosed with cancer. Of those diagnosed with cancer, skin cancer (of any form) was the most common.

Table 2. Survey participants; select demographics per wave.

Survey Wave	Gender	Mean Age	Employment Status
Wave 1	87% Female 12% Male	58.7	35% Employed 34% Retired 21% Disability
Wave 2	84% Female 16% Male	57.6	33% Employed 30% Retired 24% Disability
Wave 3	81% Female 19% Male	56.3	30% Employed 38% Retired 21% Disability
Wave 4	83% Female 17% Male	59.7	31% Employed 36% Retired 22% Disability
Wave 5	82% Female 18% Male	58.9	30% Employed 38% Retired 21% Disability
Wave 6	78% Female 21% Male	60.4	31% Employed 39% Retired 19% Disability

The majority of participants were female; however, this is in line with prior research among people who were seeking health information or support online [30]. Unsurprisingly, given the chronically ill nature of participants and the older average age, nearly half were either on disability benefits or fully retired, and roughly 30% were either employed full-time, part-time, or self-employed (Table 2).

3.2. COVID-19 Concern

The percentage of respondents reporting a 6 or 7 on the scale of general COVID-19 concern stayed relatively constant throughout waves one through five, peaking in the first wave with 71% of respondents ($n = 699$) and remaining in the mid-to-high 60% range through to wave five. In wave six, the question was adjusted slightly to inquire about the concern of contracting COVID-19 (after over one year of infections and the introduction of several vaccines). As such, significant concern for personally contracting COVID-19 in wave six was only 38% ($n = 370$). The trends in patterns of concern with regard to personal health history were the opposite for those with general chronic health conditions versus those with cancer. Those with a cancer history reported the highest levels of concern for COVID-19 in relation to their personal health at the beginning of the pandemic (wave one, 86%, $n = 105$). However, this significantly decreased to the 40% range for waves two through four ($p < 0.01$). In contrast, those with non-cancerous chronic health conditions saw a steady trend in concern for COVID-19 in relation to their personal medical history, with those reporting a strong concern hovering between 67% and 71% throughout the field of study.

3.3. Self-Directed Research and Desire for Additional Support

The top sources respondents used for COVID-19-related information throughout waves one through four were TV news reports (64%, $n = 4591$), government websites such as the CDC or NIH (61%, $n = 4393$), news websites (57%, $n = 4099$), and Internet search engines such as Yahoo or Google (47%, $n = 3353$). Social networking sites such as Facebook and Twitter were used by 38% of respondents ($n = 2721$), and online blogs and support communities were used by 17% ($n = 1196$). Internet searches, websites for healthcare professionals such as academic journals, TV news reports, and social media were all utilized more heavily in the beginning of the pandemic relative to later waves ($p < 0.01$ for wave one versus waves two, three, and four, individually). Online blogs and support communities were utilized significantly more in wave two (19%, $n = 426$, $p < 0.01$) than in any other wave, coinciding with peak lockdown times in the United States (April 2020).

When asked which types of information and/or support would be most helpful to respondents, the three most commonly chosen sources were up-to-date and accurate information about COVID-19 (47%, $n = 2921$), emotional and/or mental health support (25%, $n = 1581$), and financial support for other necessities/bills (22%, $n = 1356$). Most notably, the desire for emotional and/or mental support was highest in the final wave—wave four ($p = 0.01$ compared to wave two and $p < 0.01$ for wave three).

3.4. Burnout, Isolation, and Mental Health Distress

The impacts on mental and emotional health were assessed most directly in waves five and six, after roughly a year of the COVID-19 pandemic. Nearly 60% ($n = 1143$) of respondents in wave five said the pandemic had increased the level of stress and/or anxiety in their daily life, compared to 9% ($n = 183$) who did not share this perception ($p < 0.01$), and 62% ($n = 1240$) were currently worried about returning to "normal" activities ($p < 0.01$). Additionally, in wave five, when asked about family and friends, 38% ($n = 764$) reported that they are keeping in touch with loved ones less than before the pandemic, compared to only a quarter ($n = 500$) reporting an increase in connection ($p < 0.01$). Notably, about one quarter of those who sought telehealth for any reason in this group did so for a mental health counseling or therapy session (25%, $n = 371$).

In wave six, 60% ($n = 588$) said quality of life was somewhat or much worse as a result of the pandemic, while only 9% ($n = 86$) reported an improvement ($p < 0.01$). When asked directly about mental health impacts, 56% of this group felt their mental health was worse or much worse during the pandemic compared to pre-pandemic ($n = 547$, $p < 0.01$), and only 6% ($n = 63$) reported improvements in mental health.

3.5. One-Word Perceptions

Responses to the one-word, open-ended question about current feelings regarding COVID-19 were reviewed and a word cloud figure was generated (Figure 1). In the word cloud format, the most commonly cited words are represented as larger in the figure. As shown in Figure 1, earlier in the pandemic, aligned with lockdowns and CDC calls to practice social distancing, the terms "anxious", "scared", and "concerned' were the most frequently cited. In wave six, more than a year after the pandemic began and with vaccines becoming available, words like "frustrated", and "tired" are still prominent; however, new terms like "hopeful", "optimistic", and "cautious" are entering into the vocabulary again, occupying more prominent positions in the shared consciousness.

(a)   (b)

Figure 1. One-word perceptions of the pandemic in word-cloud format: (**a**) Wave four (July 2020); (**b**) Wave five (April 2021).

3.6. First-Hand Accounts of Isolation and Longing for Connection

Wave five included an additional open-ended response where respondents were asked about their biggest current struggle in relation to the COVID-19 pandemic. Upon the first review of responses, isolation, stress, mental health concerns, and longing for social connectedness were common themes. Several notable responses included:

"The biggest struggle for me is the isolation. The last time I was out to eat with friends or shopping in a store was the end of February 2020. It's a more mental/emotional struggle most days."

"Boredom due to seclusion. Normally I busy myself with helping others but being secluded in my room with little to do has begun to wear down my normally positive attitude."

"I don't have any help at home, and it's hard for me to manage. I feel incredibly isolated-which causes increased depression. I cannot even participate in communal worship because of immunosuppressant medications that increase my risk for COVID. I would like to work, as much as I can, but as a piano teacher, it is not possible, and that is more isolating and makes one feel more 'useless.' Being ill, unemployed, and having no one to have physical contact or interactions with is not normal. It is not conducive to mental health, [which is] hard enough for those with chronic pain and health conditions."

Alongside survey responses, Health Union's OHCs were continuously monitored for trends in member engagement. Notably, concurrent with the start of the survey fielding in late March 2020 (and the initial increasing mandates to stay at home due to the COVID-19 pandemic), the following unprompted response was shared Health Union's OHC dedicated to asthma:

"Warm greetings to everyone in the asthma world! As I sit here on day 22 of isolation during the COVID-19 pandemic, I am reminded of my reasons for wanting to share my experiences as an asthma patient and a lung cancer survivor, to name a couple of my health issues. So I am in isolation and I can't help but think about how grateful I am to have this forum to turn to. Not just in today's current environment but always. It is so helpful to hear from so many others who are in the same boat. Asthma often makes me feel isolated and alone. In reality, I am alone, but WE are together in our little corners of the planet doing our best to stay safe and healthy and live our lives. The current climate of the world has intensified that for us."

4. Discussion

The overall objective of this study was to determine OHC members' changing perceptions of the COVID-19 virus and its impact on mental wellbeing and need for emotional or social support, as well as to determine where people are seeking their information and support needs from. Although government websites, TV news reports, and Internet searches dominated as information sources early in the pandemic, online blogs and support communities (including OHCs) reached their peak reported usage during the time of mass lockdowns and uncertainty in the U.S. (April 2020). This difference in the time of the pandemic suggests that people living with chronic health conditions sought social connectedness, validation, and peer-to-peer information in a time when they were experiencing more distress. This notion was further strengthened by the overall increasing desire for emotional and/or mental support as the pandemic progressed from the time of waves two through four.

While TV news sites, general Internet searches, and need for financial support were always of high priority to respondents no matter the time frame, the steadily increasing interest in OHCs speaks to the desire for human connection and social support from others during an unpredictable time. By providing a safe and always available online space for these connections, OHCs can provide trusted information, validate concerns and emotions, and provide social support in order to enhance wellbeing.

General concern for COVID-19, specifically concern about the risk of being infected, decreased by wave six when a multitude of infections in the U.S. had already occurred (including potentially among respondents or their family members, thereby potentially decreasing their fear of the unknown). "COVID fatigue" was at an all-time high according to public opinion polls and news reports [32], and several vaccines were newly granted emergency use authorization by the FDA and were becoming widely available. The concerns associated with the earlier phases of the pandemic, such as fear of the unknown and concern about becoming infected, were replaced by new concerns. These included burnout, negative quality of life impacts, reduced connection to loved ones, and mental health deterioration. A large majority reported that the pandemic increased the stress and anxiety they felt in their daily lives, and that they were still apprehensive about returning to a pre-pandemic "normal".

During this same time period, although daily COVID-19 concerns may have been subdued, these were replaced with mental and emotional exhaustion. Despite these high levels of distress, survey participants indicated that they did not engage with formal mental health services. In spite of the high overall use of telehealth services in wave five (73%), only a quarter of these visits were for mental health concerns. This suggests that although mental health issues and the need for support rose throughout the pandemic, there are still barriers (whether they are social, financial, practical, or physical) to being able to access mental health services. From a socialization standpoint, OHCs may be able to facilitate trust among users of those communities and help normalize discussions around how to seek mental or emotional support, as well as provide encouragement and validation to others who may be uncertain about using telehealth for mental health services.

Based on responses to the one-word perceptions of COVID-19 and the first-hand accounts of isolation shared in response to open-ended questions, people with chronic health conditions are willing and ready to discuss these sensitive topics in OHCs. By continuing to utilize the Adaptive Engagement Model, conversations around emotional and mental health can be validated, and encouraged within a safe and supportive online environment.

OHCs are not without their own inherent risks, however, as shown in recent news accounts of the failure of leading social media companies to appropriately respond to negative impacts [33]. Personal attacks and factually incorrect medical information may arise, and depend on individual members to think critically about information presented and separate personal affiliation from safety [25,34,35]. This can be especially dangerous when it comes to sensitive discussions around emotional and mental health topics, and mentions of potential self-harm. In order to combat this, some OHCs, such as those operated by Health Union, utilized trained moderators, wrote community rules that were shared and enforced through moderation practices, and modeled appropriate responses to users of the OHC. Moderators may have a background in health or social services, and include experienced patient advocates and trained employees who monitor for safe discourse and provide conversation encouragement, resources, and validation where appropriate.

As mentioned above, there is also the risk for mentions of self-harm when creating open dialogue around sensitive and emotional issues. One aspect of OHCs is the unpredictability of people sharing comments or concerns that may not be germane to shared content or topics, and this is often the case when people mention mental health or emotional health concerns. As the COVID-19 pandemic shows in stark relief, existing OHCs, organized around health topics, may quickly become the source that people turn to when seeking support from others. Being prepared for such conversations, with both a strategy as well as an experienced and trained group of community moderators, is critical to quickly adapt to the demands of the pandemic and similar public health crises.

Limitations

There are several limitations to this study. First, in order to adapt to the changing nature of the COVID-19 pandemic while keeping surveys manageable for chronically ill respondents, several questions from the first survey were removed or tailored across subsequent waves. This made it impossible to compare responses to all variables across all six waves; however, the sample sizes for each survey wave were large enough to gain an understanding of data points of interest from smaller wave groups. Additionally, participants were recruited from people who visited one of the 28 OHCs, clicked on a survey advertisement, and proceeded to the survey consent page, or who had previously opted-in to receive email communications. This recruitment method may have led to sampling bias, specifically for those who had already participated in an OHC environment. Further, although the survey was designed to be as user-friendly as possible, it was not completed by all participants which may indicate that those who completed the survey are more comfortable with the online survey technology, and thus, OHCs.

Future research would be best served by investigating need for mental and/or emotional health support at this current juncture in the pandemic, and by utilizing similar or identical questions throughout all iterations that further explore why people choose to come to (or avoid) OHCs, what information they may be looking for, how they find support or give support to others within these spaces, and if they feel empowered to seek further help from a healthcare professional when needed.

5. Conclusions

Along with all of society, individuals with chronic illnesses fully experience the impacts of the COVID-19 pandemic, and are likely and are likely to seek out information early and avail themselves of risk reduction strategies. Changes in desired resources over time show an increasing interest from basic information about the virus, transmission risk, and ways to minimize risk, to seeking emotional support. The interest and willingness to talk about mental health impacts and the need for support speaks to the desire for human connectedness in times of societal upheaval and the resultant severe isolation.

The literature reviewed provides a backdrop for the current study and illustrates issues relevant to social connectedness for people with chronic health conditions. This includes the potential negative impact of social media on mental health, which has salience for people with chronic health conditions, particularly during the pandemic. While the increase in the availability of telehealth during the pandemic has the potential to increase the use of mental health services, it is less clear that people were able to avail themselves of these opportunities.

The first-hand accounts of people living with chronic health conditions struggling to find support and social contact are illustrated in the responses to open-ended questions and show a range of emotional impacts and coping strategies. Although mental and emotional distress are common, not all who need professional support feel comfortable or know how to access those services.

OHCs can provide an intermediate and safe space where conversations about mental and emotional wellbeing can be normalized and those in need are encouraged to seek additional assistance from healthcare professionals if warranted. However, OHCs cannot do this through passive engagement only. Active moderation of OHCs using trained and experienced moderators can provide a safe space with planned, real-time strategies to address crisis situations, including future pandemics or public health emergencies, as they arise. This research supports the thesis that OHCs, when managed and moderated appropriately, have the power to normalize mental health discussions, thus providing a unique value to those who experience mental health concerns.

Supplementary Materials: The following are available online at https://www.mdpi.com/article/10.3390/ijerph182312743/s1, File S1: Wave 5 Survey Questions.

Author Contributions: Conceptualization, B.M.G. and S.H.; survey questions, S.H. and L.B.H.; methodology, L.B.H.; formal analysis, L.B.H.; resources, A.B.; data curation, L.B.H.; writing—original draft preparation, C.A.H.; writing—review and editing, B.M.G., C.A.H., S.H., A.B. and L.B.H.; project administration, L.B.H. and A.B. All authors have read and agreed to the published version of the manuscript.

Funding: This research received no external funding.

Institutional Review Board Statement: This study was determined as Exempt under 45 CFR 46.101(b)(#2). The study was conducted according to the guidelines of the Declaration of Helsinki and other relevant laws in the U.S.

Informed Consent Statement: Informed consent was obtained from all survey participants involved in the study.

Data Availability Statement: The data presented in this study are available by request from the corresponding authors. Additional summary reports using Data from the surveys described in this study are available here: https://health-union.com/blog/covid-19-resources-pharma-marketing (accessed on 30 November 2021).

Conflicts of Interest: The authors declare no conflict of interest with the reported research results.

References

1. Clement, S.; Schauman, O.; Graham, T.; Maggioni, F.; Evans-Lacko, S.; Bezborodovs, N.; Morgan, C.; Rüsch, N.; Brown, J.S.L.; Thornicroft, G. What is the Impact of Mental Health-Related Stigma on Help-Seeking? A Systematic Review of Quantitative and Qualitative Studies. *Psychol. Med.* **2014**, *45*, 11–27. [CrossRef] [PubMed]
2. Michelle Obama Says She Is Dealing with 'Low-Grade Depression'. Available online: https://www.nytimes.com/2020/08/06/us/michelle-obama-depression.html (accessed on 15 September 2021).

3. Substantial Investment Needed to Avert Mental Health Crisis. Available online: https://www.who.int/news/item/14-05-2020-substantial-investment-needed-to-avert-mental-health-crisis (accessed on 15 September 2021).
4. Hossain, M.M.; Tasnim, S.; Sultana, A.; Faizah, F.; Mazumder, H.; Zou, L.; McKyer, E.L.J.; Ahmed, H.U.; Ma, P. Epidemiology of Mental Health Problems in COVID-19: A Review. *F1000Research* **2020**, *9*, 636. [CrossRef]
5. The Lancet Psychiatry. COVID-19 and Mental Health. *Lancet Psychiatry* **2021**, *8*, 87. [CrossRef]
6. Pierce, M.; Hope, H.; Ford, T.; Hatch, S.; Hotopf, M.; John, A.; Kontopantelis, E.; Webb, R.; Wessely, S.; McManus, S.; et al. Mental Health Before and During the COVID-19 Pandemic: A Longitudinal Probability Sample Survey of the UK Population. *Lancet Psychiatry* **2020**, *7*, 883–892. [CrossRef]
7. Mental Health and Substance Use State Fact Sheets. Available online: https://www.kff.org/statedata/mental-health-and-substance-use-state-fact-sheets/ (accessed on 15 September 2021).
8. Can Social Prescribing Support the COVID-19 Pandemic? Available online: https://www.cebm.net/covid-19/can-social-prescribing-support-the-covid-19-pandemic/ (accessed on 15 September 2021).
9. Everson-Rose, S.A.; Lewis, T.T. Psychosocial Factors And Cardiovascular Diseases. *Annu. Rev. Public Health* **2005**, *26*, 469–500. [CrossRef]
10. Uchino, B.N. Social Support and Health: A Review of Physiological Processes Potentially Underlying Links to Disease Outcomes. *J. Behav. Med.* **2006**, *29*, 377–387. [CrossRef]
11. Holt-Lunstad, J.; Smith, T.B.; Baker, M.; Harris, T.; Stephenson, D. Loneliness and Social Isolation as Risk Factors for Mortality. *Perspect. Psychol. Sci.* **2015**, *10*, 227–237. [CrossRef]
12. Daré, L.O.; Bruand, P.E.; Gérard, D. Co-morbidities of mental disorders and chronic physical diseases in developing and emerging countries: A meta-analysis. *BMC Public Health* **2019**, *19*, 304. [CrossRef] [PubMed]
13. Niall, A.; Ozakinci, G. Effectiveness of Psychological Interventions to Improve Quality of Life in People with Long-Term Conditions: Rapid Systematic Review of Randomised Controlled Trials. *BMC Psychol.* **2018**, *6*, 1. [CrossRef]
14. Wonggom, P.; Kourbelis, C.; Newman, P.; Du, H.; Clark, R.A. Effectiveness of avatar-based technology in patient education for improving chronic disease knowledge and self-care behavior: A systematic review. *JBI Database Syst. Rev. Implement. Rep.* **2019**, *17*, 1101–1129. [CrossRef] [PubMed]
15. Magharei, M.; Jaafari, S.; Mansouri, P.; Safarpour, A.; Taghavi, S.A. Effects of Self-Management Education on Self-Efficacy and Quality of Life in Patients with Ulcerative Colitis: A Randomized Controlled Clinical Trial. *Int. J. Community Based Nurs. Midwifery* **2019**, *7*, 32–42. [CrossRef]
16. Smith, T.B.; Workman, C.; Andrews, C.; Barton, B.; Cook, M.; Layton, R. Effects of psychosocial support interventions on survival in inpatient and outpatient healthcare settings: A meta-analysis of 106 randomized controlled trials. *PLoS Med.* **2021**, *18*, e1003595. [CrossRef]
17. Gonzalez, J.S.; Penedo, F.J.; Antoni, M.H.; Durán, R.E.; McPherson-Baker, S.; Ironson, G.; Isabel Fernandez, M.; Klimas, N.G.; Fletcher, M.A.; Schneiderman, N. Social Support, Positive States of Mind, and HIV Treatment Adherence in Men and Women Living With HIV/AIDS. *Health Psychol.* **2004**, *23*, 413–418. [CrossRef]
18. Mirsky, J.B.; Horn, D.M. Chronic Disease Management in the COVID-19 Era. *Am. J. Manag. Care* **2020**, *26*, 329–330. [CrossRef]
19. Liu, C.; Zhao, Y.; Okwan-Duodu, D.; Basho, R.; Cui, X. COVID-19 in Cancer Patients: Risk, Clinical Features, and Management. *Cancer Biol. Med.* **2020**, *17*, 519–527. [CrossRef] [PubMed]
20. Ejaz, H.; Alsrhani, A.; Zafar, A.; Javed, H.; Junaid, K.; Abdalla, A.E.; Abosalif, K.O.A.; Ahmed, Z.; Younas, S. COVID-19 and Comorbidities: Deleterious Impact on Infected Patients. *J. Infect. Public Health* **2020**, *13*, 1833–1839. [CrossRef]
21. Sepúlveda-Loyola, W.; Rodríguez-Sánchez, I.; Pérez-Rodríguez, P.; Ganz, F.; Torralba, R.; Oliveira, D.V.; Rodríguez-Mañas, L. Impact of Social Isolation Due to COVID-19 on Health in Older People: Mental and Physical Effects and Recommendations. *J. Nutr. Health Aging* **2020**, *24*, 938–947. [CrossRef]
22. Zhou, X.; Snoswell, C.L.; Harding, L.E.; Bambling, M.; Edirippulige, S.; Bai, X.; Smith, A.C. The Role of Telehealth in Reducing the Mental Health Burden from COVID-19. *Telemed. e-Health* **2020**, *26*, 377–379. [CrossRef] [PubMed]
23. March, S.; Day, J.; Ritchie, G.; Rowe, A.; Gough, J.; Hall, T.; Yuen, C.Y.J.; Donovan, C.L.; Ireland, M. Attitudes Toward E-Mental Health Services in a Community Sample of Adults: Online Survey. *J. Med. Internet. Res.* **2018**, *20*, e59. [CrossRef]
24. Sundar, K.R. A Patient with COVID-19 is Left Behind as Care Goes Virtual. *Health Aff.* **2020**, *39*, 1453–1455. [CrossRef] [PubMed]
25. van der Eijk, M.; Faber, M.J.; Aarts, J.W.; Kremer, J.A.; Munneke, M.; Bloem, B.R. Using Online Health Communities to Deliver Patient-Centered Care to People With Chronic Conditions. *J. Med. Internet. Res.* **2013**, *15*, e115. [CrossRef]
26. Li, Y.; Yan, X. How Could Peers in Online Health Community Help Improve Health Behavior. *Int. J. Environ. Res. Public Health* **2020**, *17*, 2995. [CrossRef] [PubMed]
27. Merolli, M.; Gray, K.; Martin-Sanchez, F. Health Outcomes and Related Effects of Using Social Media in Chronic Disease Management: A Literature Review and Analysis of Affordances. *J. Biomed. Inform.* **2013**, *46*, 957–969. [CrossRef] [PubMed]
28. Steadman, J.; Pretorius, C. The Impact of an Online Facebook Support Group for People with Multiple Sclerosis on Non-Active Users. *Afr. J. Disabil.* **2014**, *3*, 132. [CrossRef]
29. van Uden-Kraan, C.F.; Drossaert, C.H.; Taal, E.; Seydel, E.R.; van de Laar, M.A. Self-Reported Differences in Empowerment Between Lurkers and Posters in Online Patient Support Groups. *J. Med. Internet. Res.* **2008**, *10*, e18. [CrossRef]

30. Green, B.M.; Van Horn, K.T.; Gupte, K.; Evans, M.; Hayes, S.; Bhowmick, A. Assessment of Adaptive Engagement and Support Model for People With Chronic Health Conditions in Online Health Communities: Combined Content Analysis. *J. Med. Internet. Res.* **2020**, *22*, e17338. [CrossRef]
31. Trump Administration Drives Telehealth Services in Medicaid and Medicare. Available online: https://www.cms.gov/newsroom/press-releases/trump-administration-drives-telehealth-services-medicaid-and-medicare (accessed on 15 September 2021).
32. 'COVID Fatigue' Is Hitting Hard. Fighting It Is Hard, Too, Says UC Davis Health Psychologist. Available online: https://health.ucdavis.edu/health-news/newsroom/covid-fatigue-is-hitting-hard-fighting-it-is-hard-too-says-uc-davis-health-psychologist/2020/07 (accessed on 15 September 2021).
33. COVID-19 Misinformation Remains Difficult To Stop on Social Media. Available online: https://www.forbes.com/sites/petersuciu/2020/04/17/covid-19-misinformation-remains-difficult-to-stop-on-social-media/ (accessed on 15 September 2021).
34. Coulson, N.S. Receiving Social Support Online: An Analysis of a Computer-Mediated Support Group for Individuals Living with Irritable Bowel Syndrome. *Cyberpsychol. Behav.* **2005**, *8*, 580–584. [CrossRef] [PubMed]
35. Willis, E. The Making of Expert Patients: The Role of Online Health Communities in Arthritis Self-Management. *J. Health Psychol.* **2013**, *19*, 1613–1625. [CrossRef]

International Journal of
Environmental Research and Public Health

MDPI

Article

The Interactions of Media Use, Obesity, and Suboptimal Health Status: A Nationwide Time-Trend Study in China

Qinliang Liu and Xiaojing Li *

School of Media & Communication, Shanghai Jiao Tong University, Shanghai 200240, China; liuql201807@sjtu.edu.cn
* Correspondence: lixiaojing@sjtu.edu.cn

Abstract: Obesity and suboptimal health status (SHS) have been global public health concerns in recent decades. A growing number of works have explored the relationships between media use and obesity, as well as SHS. This study aimed to examine the time trend of the associations between media use (including traditional media and new media) and obesity, as well as SHS. The data were derived from three national random samples of the Chinese General Social Survey (CGSS), which was separately conducted in 2013, 2015, and 2017. In total, 34,468 respondents were included in this study, consisting of 16,624 males and 17,844 females, and the average age was 49.95 years old (SD = 16.72). It found that broadcast use and television use were positively associated with obesity and showed an increasing trend over time. Cellphone use emerged as a risk factor for obesity in 2017 and showed an increasing trend. By contrast, newspaper use, television use, and internet use were negatively associated with SHS, and television use showed a decreasing trend in the association with SHS, while internet and newspaper use showed an increasing trend. In conclusion, media use was positively associated with obesity while negatively associated with SHS. It showed a decreasing trend in the associations between traditional media use and obesity, while revealing an increasing trend in the associations between new media use and obesity, as well as SHS. The practical implications of the findings are discussed.

Keywords: media use; obesity; suboptimal health status; time trend; China

Citation: Liu, Q.; Li, X. The Interactions of Media Use, Obesity, and Suboptimal Health Status: A Nationwide Time-Ttrend Study in China. *Int. J. Environ. Res. Public Health* **2021**, *18*, 13214. https://doi.org/10.3390/ijerph182413214

Academic Editors: Irene Torres-Sanchez and Marie Carmen Valenza

Received: 26 October 2021
Accepted: 13 December 2021
Published: 15 December 2021

Publisher's Note: MDPI stays neutral with regard to jurisdictional claims in published maps and institutional affiliations.

1. Introduction

Over the recent decades, communication technology has changed dramatically. A variety of new media, including the internet, computers, cellphones, and social media have rapidly come to coexist with traditional media (e.g., television, newspaper, and broadcast media), which dominate individual's leisure activities and spare time [1]. Media use has also been integrated into daily life to communicate with peers and maintain social relationships [2], search for or share information, and have fun or entertainment [3]. Despite the ease and frequency of media use having changed radically in recent years, these changes did not always bring positive impacts on health status, but did bring negative consequences [4]. In other words, media use is a two-sided coin that may improve one's health status, but also could result in harmful health outcomes. Widespread and prolonged media use has contributed to the increasing and ongoing debate on its impacts on physical and psychological health, such as obesity or overweight, and the symptoms of suboptimal health status (SHS).

1.1. Media Use and Obesity

Obesity has been a global public health concern which has caused various health problems and social burden [5,6]. Recent work has indicated that obesity has become widespread in adolescents as well as in young and older adults [7,8], with an increasing trend in recent decades [9]. The rapid rise of obesity suggests that environmental factors along with biological disposition might be responsible for people's weight gain [10].

Among the environmental factors, ubiquitous information communication technology and pervasive media use might account for the widespread obesity [11]. One of the possible explanations lies in the fact that electronic media use or screen media use have displaced time for physical activities or outdoor sports. The more or longer one uses such media, the more likely s/he is to become obese [12,13]. A large body of studies have also examined the associations between media use and obesity or overweight [14–16]. The scope of these studies has involved almost all media types, such as television [17], video games [18], internet and computer [19], social media [20], and digital media [21]. The research outcomes have demonstrated that media use has caused a decrease in physical activity and an increase in sedentary behaviors, which has resulted in the risk of obesity [22,23].

1.2. Media Use and Suboptimal Health Status

SHS is an intermediate state between health and disease conditions, characterized by declines in vitality, physiological function, and the capacity for adaptation [24]. It includes symptoms of fatigue, non-specific pain, dizziness, anxiety, depression, and functional system disorders [25]. In recent years, more and more people are in SHS. SHS has become a significant public health concern worldwide due to its potential risk for chronic diseases across multiple populations, such as type-II diabetes mellitus, cardiovascular and stroke [26,27].

According to the "biological-psychological-social" medical model, an individual's health status is dictated by his/her biological, psychological, and social factors. Thus, SHS is usually measured or assessed by disturbance symptoms in terms of physiological, psychological, and sociological characteristics, namely physical SHS (PS), mental SHS (MS), and social SHS (SS) [28]. Given that SHS is a medically undiagnosed and functional somatic syndrome, it has been attributed to various factors. Existing studies have revealed that lifestyle and environmental factors were the critical predictors of SHS, such as sleep quality, physical exercise, unhealthy diet, smoking, alcohol drinking, adverse life events, and family living status [29]. In addition, the prevalence of SHS varies among different populations. Many demographic variables, such as age, gender, education, occupation, geographical region, and economic status, were associated with SHS [30].

As a critical predictor of SHS, screen-based media (SBM) has occupied a considerable portion of young peoples' discretionary leisure time [3]. The overuse of electronic devices might be one of the underlying causes of SHS [31]. However, the results related to SHS were mixed. According to the social causation hypothesis, unreasonable media use could induce health problems since excessive or problematic media use increased the risk of poor sleep quality, obesity, lower life satisfaction, and anxiety [32,33]. In addition, social media provided a natural platform for social comparison, which led to more mental health disorders [34]. For instance, more social media use was proved to be positively correlated with weight dissatisfaction and poor body image, which resulted in lower self-esteem and higher depressive symptoms [35]. Those active on Facebook tended to score higher in terms of poor health status than those who used other social media platforms [2].

However, others have argued that individuals could significantly benefit from media use, whether in terms of physical health or psychological health. The social compensation hypothesis interprets that media use can compensate for what individuals lack in real life, including social resources and support [36]. Primarily, individuals view online social media as an attractive way to attain social support and social engagement which they lack offline [37]. These online media are easy to access and anonymous, providing users with a sense of connectedness and reducing feelings of isolation [38]. The new media enable individuals and communities to acquire and maintain social capital. All these benefits have a positive effect on mental health [39]. Besides, electronic or digital media have become the most popular platform for accessing health information [40]. They play critical roles in changing one's attitudes toward health behaviors and persuading individuals to participate in health protection [41]. Furthermore, health-related media use was positively linked to health literacy [42], as well as eHealth literacy [43], which in turn influenced health behaviors and contributed to physical health [44].

1.3. The Current Study

Although prior studies have provided valuable knowledge to understand the relationships between media use, obesity and SHS [32,45–47], several limitations still exist in the body of scholarship. Firstly, most studies used cross-sectional data and demonstrated a concurrent relationship between media use and obesity or SHS. They could not demonstrate any causal effect over time [48]. Secondly, existing studies preferred to focus on a single type of media use to explore its associations with obesity or SHS, such as television, internet or social media, lacking a comparative perspective among different media [49–51]. Besides, most have focused on new media, ignoring traditional media which are still relevant, especially for certain age groups. Thirdly, most studies were executed in developed countries. However, developing countries like China are experiencing severe obesity and SHS problems [52] and encountering dramatic changes in media technologies in the 21st century.

Therefore, to gain a comprehensive and longitudinal view of the influence of media use, it is necessary to test the time trend of the supposed causality between media use and health status, as well as to consider the disparities between different media uses. The current study aims to reveal the time changes in the associations of media use with obesity and SHS from 2013 to 2017. Further, to compare the effects of different media use, we divided the media genres into traditional media (newspaper, magazine, broadcast, and television) and new media (internet and cellphone). Hence, the specific research questions were proposed:

- What are time trends in the associations of media use with obesity (RQ1) and SHS (RQ2) in the period 2013 to 2017?
- What are the differences in time trends between new media use and traditional media use, separately, in their associations with obesity (RQ3) and SHS (RQ4)?

2. Materials and Methods

2.1. Data Collection

The data was drawn from the Chinese General Social Survey (CGSS) in 2013, 2015, and 2017. The CGSS is a nationally presentative and continuous social survey project, implemented on the Chinese mainland by the Renmin University of China and Hong Kong University of Science and Technology annually since 2003. It has been broadly considered as the Chinese counterpart of the General Social Survey (GSS) in the United States. Using a stratified multi-stage probability proportionate to size (PPS) sampling method, samples were drawn from households in all 31 provincial units in Mainland China with three levels in each sampling frame: county or district, community, and household. In a selected county (district), four community-level units (neighborhood committees or village committees) were randomly selected. In a selected community-level unit, 25 households were selected, and in each selected household one adult was randomly selected for the survey [53,54]. CGSS is a household survey aimed at adults, and adolescents under 18 years were excluded. All the questionnaires were completed by face-to-face interview with the help of interviewers. 12,000, 10,968, and 12,582 sample cases were collected from 480 villages and urban neighborhood communities from 140 residential districts in 2013, 2015, and 2017, respectively. After excluding the missing answers, e.g., one participant who responded to the options "I don't know" or declined to answer, or in the case of an incomplete questionnaire, then the case was removed from the database. Finally, 11,263, 10,383, and 12,367 respondents were retained in the samples for 2013, 2015, and 2017, respectively. All the samples were pooled as the analysis database.

2.2. Measures

2.2.1. Media Use

Media use was measured by the following question: "In the past year, how often did you use the following media?" Media types consisted of newspaper, magazine, broadcast, television, internet, and cellphone. The answers were independently rated as "never", "seldom", "sometimes", "often", and "always".

2.2.2. Obesity

Obesity was indicated by body mass index (BMI) that was calculated as weight/height^2. According to obesity criteria, the WHO defined overweight in adults as a BMI of 25.0–29.9 kg/m^2 and obesity as a BMI of 30.0 kg/m^2 or higher. However, accumulated evidence has consistently supported the use of lower BMI cutoffs in Chinese than those in whites [55–57], and the Working Group on Obesity in Asia also recommended BMI cutoffs of 24·0 kg/m^2 to define overweight and 28·0 kg/m^2 to define obesity [58]. As proposed by Zhou (2002) [59] for Chinese adults, the standard weight was defined as a BMI < 24 kg/m^2, overweight as a BMI of 24–27.9 kg/m^2, and obesity as a BMI $\geq$ 28 kg/m^2. Consequently, each respondent was characterized as obese or not by the cut-off value of 28. The attribute of obesity was marked as "yes" or "no".

2.2.3. SHS

SHS was measured by three questions with self-defined items. According to the "biological-psychological-social" medical model, SHS was defined as physical SHS (PS), mental SHS (MS), and social SHS (SS) [28]. We selected the three most related questions responding to the three facets. The first two questions were: "In the past four weeks, what is the frequency of health problems affecting your work and other daily lives?" and "In the past four weeks, how often did you feel depressed?" The answers were rated as "never or seldom", "occasionally", "sometimes", "often", and "always". The third question was, "What is your current health status?" The answers were rated from 1 to 5, as "excellent", "good", "fair", "poor", and "very poor". All the answers were inversely coded.

According to prior studies' outcomes for Cronbach [60] and Bland & Altman [61], Cronbach's alpha was used to investigate the internal consistency of the questionnaire, and the alpha should be more than 0.7 to 0.8 for research purposes and at least be 0.90 for clinical purposes. In this study, the Cronbach's alpha of SHS was 0.774 (α = 0.774), indicating a better internal consistency for the constructed measurement items, which enabled SHS to be measured reasonably. Finally, the summed score of the three questions was marked as the measurement of SHS, and a higher score indicated a more severe SHS.

2.2.4. Covariates

Covariates included sociodemographic characteristics, such as gender (male/female), age (years old), educational level (illiterate/primary school/middle school/high school/college or bachelor's degree and above), and physical activity (inversely coded as "never", "several times/year", "several times/month", "several times/week", and "every day"). Hukou (rural/urban) was included to control for the unobserved geographic characteristics of the respondents' residential location. In China, people are mainly divided into two groups (agricultural hukou holders and non-agricultural hukou holders) based on their birthplace and lineage. Generally, people with agricultural hukou live in rural areas with disadvantages in terms of the economy, education, medical, housing, and other social resources. By contrast, non-agricultural hukou holders who live in urban areas with more public and social resources have a high quality of life. So agricultural hukou and non-agricultural hukou are also called rural hukou and urban hukou [62]. Obesity was included in the regression analysis of the association between media use and SHS, coded as "yes" or "no".

2.3. Statistical Analysis

Descriptive statistics were employed to describe the characteristics of sociodemographic indicators, media use, rate of obesity, and SHS. One-way ANOVA was used to analyze the P-trend of each variable by period. Logistic regression models were applied to estimate the associations between media use and obesity separately in 2013, 2015, and 2017. The analysis results were presented as odds ratios (ORs) with 95% confidence intervals (CIs), with all covariates being controlled. Multiple linear regression models were applied to estimate the associations between media use and SHS separately in 2013, 2015, and 2017. The results were presented as Beta and SE (standard Error) and controlled all covariates. To test whether the associations between media use and obesity as well as SHS varied over time, the cross-sections were pooled pairwise (e.g., 2013–2015, 2015–2017, and 2013–2017), and dummy-coded as an independent variable separately. For example, in the period of 2013–2015, 2013 was coded "0" and 2015 was coded "1". Interactions of media type with the pairwise year were included separately in the adjusted models to determine whether the association changed significantly over the study periods [63,64]. All statistical analyses were conducted via SPSS 25.0. *p*-values were two-sided, and a value of less than 0.05 was considered statistically significant.

3. Results

3.1. Characteristics of Demographics, Media Use, Rate of Obesity, and SHS

Table 1 shows the characteristics for demographics, media use, rate of obesity, and SHS. In total, 34,468 respondents were included in this study, comprising 16,624 males and 17,844 females, and the average age was 49.95 years old (SD = 16.72). More than half of the respondents had less educational experience (64.08%) and were rural Hukou (58.98%), and all the respondents did physical activities (Mean = 2.34, SD = 1.52) at a relatively low level. As for the frequency of media use, television was rated most among the six media types, while it showed a decline and downward trend from 2013 to 2017, as well as newspapers, magazines, and broadcast. However, internet and cellphone use had an increasing tendency during the same time period. In addition, the prevalence of obesity increased from 5.7% to 7.6%, and the score of SHS increased from 6.39 (SD = 2.61) to 6.84 (SD = 2.65) during 2013 to 2017. All changes between different years were statistically significant.

3.2. Associations between Media Use and Obesity, and Time Changes

Table 2 shows the outcomes for logistic regression analysis on the associations between media use and obesity after controlling for age, gender, education, and physical activity. The associations varied by each time point. In 2013, television use was significantly and positively associated with higher odds of obesity, while magazine use was associated with decreasing odds of obesity. Broadcast and television were the significant predictors of obesity in 2015, whereas cellphone use emerged as a new predictor of obesity in 2017. Internet use was not significantly associated with obesity at any time point. These results answered RQ1.

As for the time trend of the association of media use with obesity (see Table 3), the significant interaction term indicates a time trend in pairwise time, but whether it is a decreasing or increasing trend depended on the changes of the predictive effect of media use on obesity during that period. Therefore, newspaper use showed a decreasing trend in the association with obesity since the positively predictive effect declined from 2013 to 2017. Magazine use negatively predicted obesity in 2013, but the predictive effect declined significantly during 2013–2015 and 2015–2017, which showed decreasing association trends over those periods. Broadcast use exhibited a decreasing time trend in the association with obesity during 2015–2017, because the positively predictive effect declined in that period. The positively predictive effect of television use on obesity declined from 2013 to 2017, but these changes were not significant. Therefore, there was no time trend in the association between television use and obesity. Cellphone use positively predicted obesity in 2017 and the predictive effect increased significantly during 2015–2017, thus the association between cellphone use and obesity showed an increasing trend in that period. These results answered RQ3.

Table 1. Descriptive characteristics of demographics, media use, obesity, and SHS in 2013, 2015, and 2017.

	Total (*N* = 34,468)		2013 (*N* = 11,263)		2015 (*N* = 10,838)		2017 (*N* = 12,367)		*P* for Time Change
	M/N	*SD/%*	*M/N*	*SD/%*	*M/N*	*SD/%*	*M/N*	*SD/%*	
Male	16624	48.23%	5685	50.48%	5078	46.85%	5861	47.39%	
Female	17844	51.77%	5578	49.52%	5760	53.15%	6506	52.61%	
Age	49.95	16.72	48.56	16.37	50.32	16.88	50.90	16.80	
Hukou									
Rural	20297	58.89%	6739	59.83%	6862	63.31%	6696	54.14%	
Urban	14171	41.11%	4524	40.17%	3976	36.69%	5671	45.86%	
Education									
Illiterate	22087	64.08%	7265	64.50%	7114	65.64%	7708	62.32%	
Primary	6312	18.31%	2154	19.12%	1953	18.02%	2205	17.82%	
Junior high Middle school	2730	7.92%	915	8.12%	781	7.21%	1034	8.36%	
High school	2944	8.54%	837	7.43%	874	8.06%	1233	9.97%	
College or Bachelor and above	395	1.15%	92	0.82%	116	1.07%	187	1.51%	
Physical activity	2.35	1.52	2.07	1.38	2.46	1.53	2.50	1.60	
Newspaper	1.92	1.16	2.10	1.22	1.90	1.12	1.77	1.11	1–2, 1–3, 2–3
Magazine	1.71	0.95	1.83	1.01	1.72	0.94	1.61	0.90	1–2, 1–3, 2–3
Broadcast	1.80	1.11	1.86	1.12	1.80	1.09	1.75	1.11	1–2, 1–3, 2–3
Television	3.93	1.05	4.10	0.96	3.92	1.04	3.78	1.13	1–2, 1–3, 2–3
Internet	2.48	1.66	2.20	1.55	2.37	1.64	2.82	1.72	1–2, 1–3, 2–3
Cellphone	1.65	1.14	1.63	1.10	1.62	1.09	1.70	1.21	1–3, 2–3
Obesity (%)	6.5		5.7		6.2		7.6		1–3, 2–3
SHS	6.65	2.62	6.39	2.61	6.69	2.57	6.84	2.65	1–2, 1–3, 2–3

Note. N-Number, M-Mean, SD-Standard Deviation. Time change was tested by one-way analysis of variance (ANOVA) with Tamhane's T2 since the unequal variances. 1–2 = significant difference in mean between 2013 and 2015, 1–3 = significant difference in mean between 2013 and 2015, 2–3 = significant difference in mean between 2013 and 2017. Detailed information is presented in supplementary material, Table S1.

Table 2. Logistic regression analysis of associations between media use and obesity in 2013, 2015, and 2017.

	2013			2015			2017		
	OR	95% CI		OR	95% CI		OR	95% CI	
Gender	0.98	0.83	1.16	0.95	0.81	1.11	0.96	0.84	1.10
Age	1.00	0.99	1.01	0.99 *	0.99	1.00	1.00	1.00	1.01
Rural/Urban	1.60 ***	1.32	1.94	1.52 ***	1.26	1.83	1.30 **	1.10	1.53
Illiterate	Reference	N/A		Reference	N/A		Reference	N/A	
Primary	0.75 *	0.59	0.96	0.90	0.71	1.13	0.89	0.73	1.08
Middle school	0.73	0.51	1.04	0.62 *	0.43	0.90	0.80	0.61	1.06
High school	0.78	0.54	1.14	0.55 **	0.37	0.81	0.48 ***	0.35	0.66
College or Bachelor and above	0.58	0.20	1.62	0.41	0.15	1.15	0.31 **	0.13	0.72
Physical activity	1.02	0.96	1.09	1.04	0.98	1.10	1.02	0.98	1.07
Newspaper	1.06	0.97	1.16	0.97	0.87	1.07	0.95	0.87	1.03
Magazine	0.82 ***	0.73	0.91	0.96	0.85	1.08	1.00	0.90	1.11
Broadcast	1.07	1.00	1.16	1.09 *	1.01	1.18	0.99	0.92	1.05
Television	1.14 **	1.04	1.25	1.08 *	1.00	1.17	1.04	0.97	1.10
Internet	1.03	0.95	1.11	0.97	0.90	1.05	1.02	0.96	1.08
Cellphone	1.04	0.96	1.14	0.97	0.89	1.06	1.08 **	1.02	1.15
Constant	0.02 ***			0.05***			0.06 ***		

Note. Adjusted for gender, age, rural/urban, education, physical activity. * $p < 0.05$, ** $p < 0.01$, *** $p < 0.001$. N/A—not applicable. OR—odds ratio, 95% CI—95% confidence interval.

Table 3. Logistic regression analysis for the change in the associations between media use and obesity in 2013, 2015, and 2017.

	2013–2015 [a]			2015–2017 [b]			2013–2017 [c]		
	OR	**95% CI**		**OR**	**95% CI**		**OR**	**95% CI**	
Gender	0.96	0.86	1.08	0.96	0.86	1.06	1.08	0.97	1.20
Age	1.00	0.99	1.00	1.00	0.99	1.00	1.00	1.00	1.01
Rural/Urban	1.56 ***	1.36	1.78	1.39 ***	1.23	1.57	1.15 *	1.02	1.30
Illiterate	Reference	N/A		Reference	N/A		Reference	N/A	
Primary	0.82 *	0.69	0.97	0.89	0.76	1.03	0.94	0.80	1.10
Middle school	0.68 **	0.52	0.87	0.73 **	0.58	0.91	0.88	0.63	1.23
High school	0.66 **	0.51	0.86	0.50 ***	0.39	0.64	0.98	0.69	1.39
College or Bachelor and above	0.48 *	0.23	1.00	0.34 ***	0.18	0.65	0.74	0.26	2.06
Physical activity	1.03	0.99	1.08	1.03	0.99	1.07	1.03	0.99	1.07
Newspaper	1.06	0.97	1.16	0.97	0.88	1.07	1.09	0.99	1.19
Magazine	0.81 ***	0.73	0.91	0.97	0.86	1.09	0.80 ***	0.72	0.90
Broadcast	1.08 *	1.00	1.16	1.09*	1.01	1.17	1.09 *	1.01	1.17
Television	1.14 **	1.04	1.25	1.08	1.00	1.17	1.15 **	1.05	1.26
Internet	1.02	0.95	1.09	1.00	0.94	1.07	1.03	0.95	1.11
Cellphone	1.04	0.95	1.13	0.98	0.89	1.06	1.04	0.95	1.13
Year (Time period)	1.52	0.86	2.69	2.00 **	1.23	3.28	1.85 *	1.08	3.17
Newspaper*Year	0.90	0.79	1.03	0.98	0.87	1.12	0.84 **	0.73	0.95
Magazine*Year	1.18 *	1.00	1.39	1.02	0.87	1.20	1.34 ***	1.14	1.58
Broadcast*Year	1.00	0.90	1.11	0.91 *	0.82	1.00	0.92	0.83	1.02
Television*Year	0.95	0.84	1.08	0.96	0.87	1.07	0.93	0.83	1.04
Internet*Year	0.97	0.89	1.05	1.00	0.93	1.07	0.98	0.90	1.07
Cellphone*Year	0.94	0.83	1.06	1.11 *	1.00	1.23	1.05	0.94	1.18
Constant	0.02 ***			0.04 ***			0.02 ***		

Note. Adjusted for gender, age, rural/urban, education, physical activity, * $p < 0.05$, ** $p < 0.01$, *** $p < 0.001$. OR—odds ratio, 95% CI—95% confidence interval. [a]: The pairwise time points are dummy-coded as the independent variable "Year", 2013 was coded "0" and 2015 was coded "1". [b]: The pairwise time points are dummy-coded as the independent variable "Year", 2015 was coded "0" and 2017 was coded "1". [c]: The pairwise time points are dummy-coded as the independent variable "Year", 2013 was coded "0" and 2017 was coded "1".

3.3. Associations between Media Use and SHS and Time Changes

Table 4 shows the outcomes of linear regression analysis on the associations between media use and SHS after controlling for age, gender, education, physical activity, and obesity. Compared to obesity, the findings were somewhat different; excluding broadcast, the use of newspaper, television, and internet were negatively and significantly associated with SHS, separately in 2013, 2015, and 2017. Broadcast use was positively and significantly associated with SHS only in 2013. Magazine use was a significant predictor of SHS in 2017. Cellphone use was not a significant predictor of SHS at any time point. Thus, RQ2 was answered.

As for the time trend analysis of the association of media use and SHS (see Table 5), compared with 2013 and 2015 the negatively predictive effect of newspaper use on SHS increased significantly in 2017, thus the association of newspaper use and SHS had increasing trends both in 2015–2017 and 2013–2017. On the contrary, the positively predictive effect of broadcast use on SHS declined significantly during 2013–2015 and 2013–2017, so it presented a decreasing trend in the association and SHS, and the trend was more significant during 2013–2017. In addition, the negatively predictive effect of television use on SHS declined rapidly from 2013 to 2017, and all the changes were significant among these periods, which indicated decreasing trends in the associations between television use and SHS. By contrast, internet use negatively predicted SHS in 2015 and the predictive effect increased significantly in 2013–2015 and 2015–2017, thus the association between internet use and SHS showed an increasing trend in 2015–2017 and 2013–2017. In addition, magazine and cellphone use showed no significant time trend. These outcomes answered RQ4.

Table 4. Hierarchical multiple regression analysis of associations between media use and SHS in 2013, 2015, and 2017.

| | 2013 | | 2015 | | 2017 | |
	Model1	Model2	Model1	Model2	Model1	Model2
	Beta	Beta	Beta	Beta	Beta	Beta
Gender	0.05 ***	0.05 ***	0.07 ***	0.06 ***	0.06 ***	0.06 ***
Age	0.33 ***	0.30 ***	0.32 ***	0.27 ***	0.30 ***	0.24 ***
Rural/Urban	−0.06 ***	−0.04 ***	−0.07 ***	−0.04 ***	−0.11 ***	−0.07 ***
Illiterate	Reference	N/A	Reference	N/A	Reference	N/A
Primary	−0.05 ***	−0.03 *	−0.06 ***	−0.03 **	−0.06 ***	−0.03 ***
Middle school	−0.03 ***	−0.01	−0.06 ***	−0.03 **	−0.06 ***	−0.03 ***
High school	−0.03 **	−0.01	−0.03 ***	0.01	−0.04 ***	−0.02 *
College or Bachelor and above	−0.01	−0.01	−0.02	−0.01	−0.01	0.01
Obesity	0.01	0.01	0.03 ***	0.03 ***	0.04 ***	0.04 ***
Physical activity	−0.09 **	−0.07 ***	−0.11 ***	−0.07 ***	−0.15 ***	−0.12 ***
Newspaper		−0.06 ***		−0.06 ***		−0.07 ***
Magazine		0.01		−0.01		0.03 *
Broadcast		0.05 ***		0.01		0.01
Television		−0.12 ***		−0.10 ***		−0.06 ***
Internet		−0.08 ***		−0.13 ***		−0.16 ***
Cellphone		0.01		0.01		0.01
Adjusted R2	0.156	0.178	0.163	0.183	0.188	0.206
R2 change	0.156	0.023	0.164	0.020	0.189	0.018
F	231.569 ***	52.839 ***	235.457 ***	44.619 ***	319.696 ***	45.924 ***

Note. Adjusted for gender, age, rural/urban, education, physical activity, and obesity. * $p < 0.05$, ** $p < 0.01$, *** $p < 0.001$. N/A—not applicable.

Table 5. Hierarchical multiple regression analysis for the change in associations between media use and SHS in 2013, 2015, and 2017.

| | 2013–2015 [a] | | | 2015–2017 [b] | | | 2013–2017 [c] | | |
	Model1	Model2	Model3	Model1	Model2	Model3	Model1	Model2	Model3
	Beta	Beta	Beta	Beta	Beta	Beta	Beta	Beta	Beta
Gender	0.06 ***	0.05 ***	0.05 ***	0.06 ***	−0.06 ***	0.06 ***	0.06 ***	0.05 ***	0.05 ***
Age	0.33 ***	0.28 ***	0.28 ***	0.30 ***	0.25 ***	0.25 ***	0.32 ***	0.27 ***	0.27 ***
Rural/Urban	−0.07 ***	−0.04 ***	−0.04 ***	−0.08 ***	−0.08 ***	−0.08 ***	−0.12 ***	−0.08 ***	−0.08 ***
Illiterate	Reference	N/A		Reference	N/A		Reference	N/A	
Primary	−0.06 ***	−0.03 ***	−0.03 ***	−0.07 ***	−0.03 ***	−0.03 ***	−0.06 ***	−0.03 ***	−0.03 ***
Middle school	−0.04 ***	−0.02 **	−0.02 **	−0.07 ***	−0.03 ***	−0.03 ***	−0.05 ***	−0.02 ***	−0.02 ***
High school	−0.03 ***	−0.01	−0.01	−0.05 ***	−0.01 *	−0.01	−0.03	−0.02 *	−0.01 *
College or Bachelor and above	−0.01	−0.01	−0.01	−0.02 **	−0.01	−0.01	−0.01	0.01	0.01
Obesity	0.02 **	0.02 **	0.02 **	0.03 ***	0.03 ***	0.03 ***	0.02 ***	0.02 ***	0.02 ***
Physical activity	−0.09 ***	−0.07 ***	−0.07 ***	−0.14 ***	−0.10 ***	−0.10 ***	−0.13 ***	−0.10 ***	−0.10 ***
Newspaper		−0.06 ***	−0.06 ***		−0.06 ***	−0.04 ***		−0.07 ***	−0.05 ***
Magazine		0.01	0.01		0.01	−0.01		0.01	0.01
Broadcast		0.03 ***	0.05 ***		0.01	0.02 *		0.02 ***	0.05 ***
Television		−0.11 ***	−0.13 ***		−0.08 ***	−0.10 ***		−0.09 ***	−0.13 ***
Internet		−0.11 ***	−0.10 ***		−0.15 ***	−0.13 ***		−0.12 ***	−0.10 ***
Cellphone		0.01	0.01		0.01	0.01		0.01	0.01
Year (Time period)		0.04 ***	0.01		−0.02	−0.03		0.03 **	−0.02
Newspaper*Year			0.01			−0.04 *			−0.04 ***
Magazine*Year			0.01			0.03			0.03
Broadcast*Year			−0.04 **			−0.03			−0.07 ***
Television*Year			0.07 **			0.08 ***			0.15 ***
Internet*Year			−0.02			−0.03 *			−0.05 **
Cellphone*Year			0.01			−0.01			−0.01
Adjusted R2	0.160	0.183	0.183	0.175	0.194	0.195	0.177	0.195	0.198
R2 change	0.160	0.023	0.001	0.175	0.020	0.001	0.77	0.018	0.003
F	467.193 ***	89.975 ***	3.276 **	547.582 ***	80.405 ***	5.116 ***	565.152 ***	73.536 ***	15.976 ***

Note. Adjusted for gender, age, rural/urban, education, physical activity, and obesity. * $p < 0.05$, ** $p < 0.01$, *** $p < 0.001$. N/A—not applicable. [a]: The pairwise time points are dummy-coded as the independent variable "Year", 2013 was coded "0", and 2015 was coded "1". [b]: The pairwise time points are dummy-coded as the independent variable "Year", 2015 was coded "0", and 2017 was coded "1". [c]: The pairwise time points are dummy-coded as the independent variable "Year", 2013 was coded "0", and 2017 was coded "1".

4. Discussion

Obesity and SHS have been worldwide health issues as living environments and life patterns have changed rapidly in the past decade [29,65]. The relationships between media use and obesity as well as SHS have been well documented in previous work [35,66,67]. This study aimed to better understand the media use profiles of Chinese adults and the changes in relationships between media use and obesity as well as SHS from a time trend perspective, based on a longitudinal national sample conducted among Chinese adults in 2013, 2015, and 2017.

It revealed that the profiles of media use changed significantly in China from 2013 to 2017, and the changes varied depending on different media types. Traditional media, such as newspapers, magazines, broadcast media and television use, presented a decreasing trend, and newspapers and television declined more significantly. On the contrary, the use frequency of new media (i.e., internet and cellphone) maintained a significantly increasing trend from 2013 to 2017, and internet use increased more than cellphone use. These trends were consistent with the dramatic progress in communication technologies and the rapid growth in Chinese internet users. According to the 34–39th Statistical Report on Internet Development in China, the number of Chinese internet users, internet penetration rate, and cellphone users had increased by 22.2%, 10.9%, and 41.7%, respectively, from 2013 to 2017 [68,69]. However, the opposite development trends of traditional media use and new media use did not mean that new media (i.e., internet and cellphone) had totally displaced the role of traditional media (i.e., newspapers, magazines, broadcast, and television) in China. Conversely, as our study revealed, television was still more popular than other media types among Chinese adults during the period to 2017. Internet was ranked as the second primary media use, behind television use. The possible explanation is that most of the participants had an older age, and may be prone to use traditional media (i.e., television) more than new media (i.e., Internet or cellphone) due to their relatively low level of media literacy or digital skills [70]. In addition, more than half of them were from rural areas, and may have a lower level of education and household economy, which reduces their chances of internet use [71].

Consistent with the time trend of new media use, both the prevalence of obesity and the magnitude of SHS increased significantly from 2013 to 2017, indicating that a higher proportion of Chinese adults suffered from overweight and health problems during that period. These findings were in line with the latest research findings that the prevalence of Chinese obesity changed from 3.1% (2.5–3.7) in 2004 to 8.1% (7.6–8.7) in 2018 [72], and more than half of Chinese adults showed SHS symptoms [24]. Obesity and SHS have become serious public health issues in China, but most current studies attribute the cause to economic developments, socio-cultural norms, food systems, and environment [73]. This study provided useful evidence for the assumption that media use was associated with obesity and SHS and found a significant time trend in these associations from 2013 to 2017.

Broadcast, television, and cellphone use were positively associated with obesity at different time points. Magazine use was negatively associated with obesity in 2013 while the association disappeared in 2015 and 2017. These findings suggested that both traditional and new media use might be risk factors for obesity. There was a significant time trend in the associations between media use and obesity in the study periods, and traditional media and new media demonstrated totally opposite tendencies. Traditional media use (i.e., magazine, broadcast, and television) showed a declining tendency in association with obesity from 2013 to 2017, while new media use (i.e., cellphone) showed an increasing tendency during that period. Cellphone use was increasingly important in predicting obesity and this trend was more significant in 2017 (see Table 2). The underlying mechanism of how media use influences obesity perhaps lies in the fact that media use increases sedentary possibilities and decreases physical activities.

Cellphones are usually characterized by multiple functions and are easy to use for all age groups compared with traditional media. Previous studies also found that cellphone did indeed occupy a large amount of leisure time and even led to cellphone addiction (extensive use or overuse), which brought about obesity or other health problems [74,75]. Furthermore, cellphone use was the only significant predictor of obesity in 2017, emphasizing that the cellphone had displaced traditional media and had the leading risk factor of obesity among the six media types in 2017. Nevertheless, it is worth mentioning that television use showed a decreasing tendency in associations with obesity from 2013 to 2017, while the time trend was not significant. In other words, television use was a stable predictor of obesity among the six media types over the study period. Therefore, television may be a potential risk factor for obesity in the future years.

In the existing literature, media use was demonstrated as a beneficial factor for SHS since it provided health information resources, social engagement, and social support [76]. In this study, newspaper use, television use, and internet use were negatively associated with SHS at each time point. Meanwhile, there were significant trends in the associations of media use with SHS, and the trends differed between media types.

Television showed a rapidly decreasing trend in the association with SHS from 2013 to 2017, while internet use showed a significant increasing trend. Further, the associations between internet use and SHS were stronger than the associations between television use and SHS in 2015 and 2017, indicating that internet use has displaced television and become the strongest predictor of SHS since 2015. The possible explanation is that the internet has been the leading platform and carrier of health information and has become more popular for health information access than television in the digital age [40]. Free access to vast online health information sources has created opportunities for empowerment, information exchange, and engagement in health-promoting behaviors [77]. In addition, online activities could promote well-being by reducing loneliness and facilitating social engagement [78], which might contribute to fewer SHS symptoms. Therefore, the internet should attach great importance to public health promotion in daily life. However, we should not overlook the role of traditional media in health communication, especially for television and newspapers, which were significant predictors of SHS in our study. Traditional media are still valuable resources for providing health-related information for people of low socio-economic status [79].

5. Limitations and Future Studies

The current study is one of the few that looked at time changes in the associations of different media use with obesity and SHS. Although substantial outcomes were seen when interpreting the results, the following limitations should be considered. Firstly, the repeated cross-sectional data did not allow us to draw causal inferences about the relationships between media use and obesity or SHS. Thus, a panel study and follow-up data are recommended to assess the causal effect of media use on obesity or SHS in future. Secondly, based on second-hand data, the measures of SHS were on a self-defined scale with limited items, which failed to measure SHS comprehensively and precisely. Although we made efforts to better match the three facets of SHS (i.e., physical SHS, mental SHS, and social SHS), the measures were single item and participants may have responded to the question more subjectively in terms of self-rated physical or psychological health, which might underestimate or overestimate the accurate SHS level. Thus, our findings for SHS need to be validated in future studies and a more comprehensive measure of SHS, such as the Suboptimal Health Status Questionnaire-25 (SHSQ-25) [80], should be employed. Thirdly, as the outcomes indicated, most of the respondents were middle-aged or older adults, whose media use status differs from that of young adults or adolescents, or conversely. It was unclear whether the findings could be expanded to adolescents or young adults. Therefore, future studies on time trends of media use and health status should cover more age groups and compare adolescents and older adults.

6. Conclusions

Based on three nationally representative samples of CGSS from 2013 to 2017 we tested the time trend of media use and the associations with obesity and SHS. We considered the differences between traditional media use and new media use. The results confirmed significant time trends between media use and obesity as well as SHS from 2013 to 2017. Furthermore, our study highlighted that the trend differences between traditional media and new media use were significant. Namely, traditional media use showed decreasing trends, while new media displayed an increasing trend. These findings contribute to a better understanding of how changing media use is impacting adults' health status.

Supplementary Materials: The following are available online at https://www.mdpi.com/article/10.3390/ijerph182413214/s1, Table S1. The difference of media use, obesity and SHS among 2013, 2015, and 2017(ANOVA).

Author Contributions: X.L. and Q.L. conceptualized and designed the study together. Q.L. formally analyzed and explained the data, and prepared the original draft. X.L. reviewed and revised the manuscript, and provided funding as well as project administration. All authors have read and agreed to the published version of the manuscript.

Funding: This work was funded by Chinese National Funding of Social Sciences (grant number 18AXW005).

Institutional Review Board Statement: Ethical review and approval were waived for this study, due to the data in this article was second-handed, which didn't join in the original national survey.

Informed Consent Statement: "Not applicable" for studies not involving humans.

Data Availability Statement: Publicly available datasets were analyzed in this study. This data can be found here: http://cnsda.ruc.edu.cn/index.php?r=site/article&id=180.

Acknowledgments: We thank CGSS (Chinese General Social Survey) research team for providing access to the survey and every respondent in the study for their contributions. Also, we are grateful to the journal editors and the anonymous reviewers for their helpful comments and suggestions.

Conflicts of Interest: All authors declare that there is no conflict of interest.

References

1. Sevcikova, A. Two sides of the same coin: Communication technology, media use, and our kids' health. *Int. J. Public Health* **2015**, *60*, 129–130. [CrossRef] [PubMed]
2. Jairoun, A.; Shahwan, M. Assessment of university students' suboptimal health and social media Use: Implications for health regulatory authorities. *J. Community Health* **2021**, *46*, 653–659. [CrossRef] [PubMed]
3. Straker, L.M.; Smith, A.; Hands, B.; Olds, T.; Abbott, R. Screen-based media use clusters are related to other activity behaviours and health indicators in adolescents. *BMC Public Health* **2013**, *13*, 1174. [CrossRef] [PubMed]
4. Cleary, M.; West, S.; Visentin, D. The mental health impacts of smartphone and social media use. *Issues Ment. Health Nurs.* **2020**, *41*, 755–757. [CrossRef] [PubMed]
5. Seema, S.; Rohilla, K.K.; Kalyani, V.C.; Babbar, P. Prevalence and contributing factors for adolescent obesity in present era: Cross-sectional Study. *J. Fam. Med. Prim. Care* **2021**, *10*, 1890–1894. [CrossRef]
6. Ampofo, A.G.; Boateng, E.B. Beyond 2020: Modelling obesity and diabetes prevalence. *Diabetes Res. Clin. Pract.* **2020**, *167*. [CrossRef]
7. Gao, Q.; Mei, F.; Shang, Y.; Hu, K.; Chen, F.; Zhao, L.; Ma, B. Global prevalence of sarcopenic obesity in older adults: A systematic review and meta-analysis. *Clin. Nutr.* **2021**, *40*, 4633–4641. [CrossRef] [PubMed]
8. Suh, J.; Jeon, Y.W.; Lee, J.H.; Song, K.; Choi, H.S.; Kwon, A.; Chae, H.W.; Kim, H.C.; Kim, H.-S.; Suh, I. Annual incidence and prevalence of obesity in childhood and young adulthood based on a 30-year longitudinal population-based cohort study in Korea: The Kangwha study. *Ann. Epidemiol.* **2021**, *62*, 1–6. [CrossRef]
9. Li, Y.-J.; Xie, X.-N.; Lei, X.; Li, Y.-M.; Lei, X. Global prevalence of obesity, overweight and underweight in children, adolescents and adults with autism spectrum disorder, attention-deficit hyperactivity disorder: A systematic review and meta-analysis. *Obes. Rev.* **2020**, *21*, e13123. [CrossRef]
10. Bozkurt, H.; Ozer, S.; Sahin, S.; Sonmezgoz, E. Internet use patterns and Internet addiction in children and adolescents with obesity. *Pediatric Obes.* **2018**, *13*, 301–306. [CrossRef]
11. Marshall, S.J.; Biddle, S.J.H.; Gorely, T.; Cameron, N.; Murdey, I. Relationships between media use, body fatness and physical activity in children and youth: A meta-analysis. *Int. J. Obes.* **2004**, *28*, 1238–1246. [CrossRef] [PubMed]
12. Tsitsika, A.K.; Andrie, E.K.; Psaltopoulou, T.; Tzavara, C.K.; Sergentanis, T.N.; Ntanasis-Stathopoulos, I.; Bacopoulou, F.; Richardson, C.; Chrousos, G.P.; Tsolia, M. Association between problematic Internet use, socio-demographic variables and obesity among European adolescents. *Eur. J. Public Health* **2016**, *26*, 617–622. [CrossRef] [PubMed]

13. Spengler, S.; Mess, F.; Woll, A. Do media use and physical activity compete in adolescents? Results of the MoMo study. *PLoS ONE* **2015**, *10*, e0142544. [CrossRef] [PubMed]
14. Yore, M.M.; Fulton, J.E.; Nelson, D.E.; Kohl, H.W., III. Cigarette smoking status and the association between media use and overweight and obesity. *Am. J. Epidemiol.* **2007**, *166*, 795–802. [CrossRef] [PubMed]
15. Melkevik, O.; Haug, E.; Rasmussen, M.; Fismen, A.S.; Wold, B.; Borraccino, A.; Sigmund, E.; Balazsi, R.; Bucksch, J.; Inchley, J.; et al. Are associations between electronic media use and BMI different across levels of physical activity? *BMC Public Health* **2015**, *15*, 497. [CrossRef] [PubMed]
16. Aghasi, M.; Matinfar, A.; Golzarand, M.; Salari-Moghaddam, A.; Ebrahimpour-Koujan, S. Internet use in relation to overweight and obesity: A systematic review and meta-analysis of cross-sectional studies. *Adv. Nutr.* **2020**, *11*, 349–356. [CrossRef] [PubMed]
17. Bener, A.; Al-Mahdi, H.S.; Ali, A.I.; Al-Nufal, M.; Vachhani, P.J.; Tewfik, I. Obesity and low vision as a result of excessive Internet use and television viewing. *Int. J. Food Sci. Nutr.* **2011**, *62*, 60–62. [CrossRef] [PubMed]
18. Ballard, M.; Gray, M.; Reilly, J.; Noggle, M. Correlates of video game screen time among males: Body mass, physical activity, and other media use. *Eat. Behav.* **2009**, *10*, 161–167. [CrossRef] [PubMed]
19. Vandelanotte, C.; Sugiyama, T.; Gardiner, P.; Owen, N. Associations of leisure-time Internet and computer use with overweight and obesity, physical activity and sedentary behaviors: Cross-sectional study. *J. Med. Internet Res.* **2009**, *11*, e28. [CrossRef] [PubMed]
20. Sandercock, G.R.H.; Alibrahim, M.; Bellamy, M. Media device ownership and media use: Associations with sedentary time, physical activity and fitness in English youth. *Prev. Med. Rep.* **2016**, *4*, 162–168. [CrossRef] [PubMed]
21. Engberg, E.; Leppanen, M.H.; Sarkkola, C.; Viljakainen, H. Physical activity among preadolescents modifies the long-term association between sedentary time spent using digital media and the increased risk of being overweight. *J. Phys. Act. Health* **2021**, *18*, 1105–1112. [CrossRef] [PubMed]
22. Matusitz, J.; McCormick, J. Sedentarism: The effects of Internet use on human obesity in the United States. *Soc. Work Public Health* **2012**, *27*, 250–269. [CrossRef] [PubMed]
23. Jordan, A.B.; Kramer-Golinkoff, E.K.; Strasburger, V.C. Does adolescent media use cause obesity and eating disorders? *Adolesc. Med. State Art Rev.* **2008**, *19*, 431–449. [PubMed]
24. Xu, T.; Zhu, G.; Han, S. Prevalence of suboptimal health status and the relationships between suboptimal health status and lifestyle factors among Chinese adults using a multi-Level generalized estimating equation model. *Int. J. Environ. Res. Public Health* **2020**, *17*, 763. [CrossRef] [PubMed]
25. Sun, Q.; Xu, X.; Zhang, J.; Sun, M.; Tian, Q.; Li, Q.; Cao, W.; Zhang, X.; Wang, H.; Liu, J.; et al. Association of suboptimal health status with intestinal microbiota in Chinese youths. *J. Cell. Mol. Med.* **2020**, *24*, 1837–1847. [CrossRef] [PubMed]
26. Yan, Y.X.; Dong, J.; Liu, Y.Q.; Yang, X.H.; Li, M.; Shia, G.; Wang, W. Association of suboptimal health status and cardiovascular risk factors in urban Chinese workers. *J. Urban Health-Bull. N. Y. Acad. Med.* **2012**, *89*, 329–338. [CrossRef] [PubMed]
27. Adua, E.; Roberts, P.; Wang, W. Incorporation of suboptimal health status as a potential risk assessment for type II diabetes mellitus: A case-control study in a Ghanaian population. *Epma J.* **2017**, *8*, 345–355. [CrossRef] [PubMed]
28. Miao, J.; Liu, J.; Wang, Y.; Zhang, Y.; Yuan, H. Reliability and validity of SHMS v1.0 for suboptimal health status assessment of Tianjin residents and factors affecting sub-health a cross-sectional study. *Medicine* **2021**, *100*, e25401. [CrossRef] [PubMed]
29. Xue, Y.; Huang, Z.; Liu, G.; Zhang, Z.; Feng, Y.; Xu, M.; Jiang, L.; Li, W.; Xu, J. Associations of environment and lifestyle factors with suboptimal health status: A population-based cross-sectional study in urban China. *Glob. Health* **2021**, *17*. [CrossRef] [PubMed]
30. Mahara, G.; Liang, J.; Zhang, Z.; Ge, Q.; Zhang, J. Associated factors of suboptimal health status among adolescents in China: A cross-sectional study. *J. Multidiscip. Healthc.* **2021**, *14*, 1063–1071. [CrossRef] [PubMed]
31. Ma, C.; Xu, W.; Zhou, L.; Ma, S.; Wang, Y. Association between lifestyle factors and suboptimal health status among Chinese college freshmen: A cross-sectional study. *Bmc Public Health* **2018**, *18*, 105. [CrossRef]
32. Alonzo, R.; Hussain, J.; Stranges, S.; Anderson, K.K. Interplay between social media use, sleep quality, and mental health in youth: A systematic review. *Sleep Med. Rev.* **2021**, *56*, 101414. [CrossRef] [PubMed]
33. Buda, G.; Lukoseviciute, J.; Salciunaite, L.; Smigelskas, K. Possible effects of social media use on adolescent health behaviors and perceptions. *Psychol Rep* **2021**, *124*, 1031–1048. [CrossRef]
34. Dibb, B. Social media use and perceptions of physical health. *Heliyon* **2019**, *5*, e00989. [CrossRef] [PubMed]
35. Kelly, Y.; Zilanawala, A.; Booker, C.; Sacker, A. social media use and adolescent mental health: Findings from the UK millennium cohort study. *Eclinicalmedicine* **2018**, *6*, 59–68. [CrossRef] [PubMed]
36. Ruggieri, S.; Santoro, G.; Pace, U.; Passanisi, A.; Schimmenti, A. Problematic Facebook and anxiety concerning use of social media in mothers and their offspring: An actor-partner interdependence model. *Addict. Behav. Rep.* **2020**, *11*, 100256. [CrossRef] [PubMed]
37. Sujarwoto, S.; Tampubolon, G.; Pierewan, A.C. A Tool to help or harm? Online social media use and adult mental health in indonesia. *Int. J. Ment. Health Addict.* **2019**, *17*, 1076–1093. [CrossRef]
38. Brown, G.; Rathbone, A.L.; Prescott, J. Social media use for supporting mental health (SMILE). *Ment. Health Rev. J.* **2021**, *26*, 279–297. [CrossRef]
39. Karim, F.; Oyewande, A.; Abdalla, L.F.; Ehsanullah, R.C.; Khan, S. Social media use and its connection to mental health: A systematic review. *Cureus* **2020**, *12*. [CrossRef] [PubMed]

40. Haluza, D.; Boehm, I. Mobile and online health information: Exploring digital media use among austrian parents. *Int. J. Environ. Res. Public Health* **2020**, *17*, 6053. [CrossRef] [PubMed]

41. Tang, L.; Wang, J. Effects of new media use on health behaviors: A case study in china. *Iran. J. Public Health* **2021**, *50*, 949–958. [CrossRef]

42. Rosenbaum, J.E.; Johnson, B.K.; Deane, A.E. Health literacy and digital media use: Assessing the Health Literacy Skills Instrument—Short Form and its correlates among African American college students. *Digit. Health* **2018**, *4*, 1–8. [CrossRef]

43. Li, X.; Liu, Q. Social media use, eHealth literacy, disease knowledge, and preventive behaviors in the COVID-19 pandemic: Cross-sectional study on Chinese netizens. *J. Med Internet Res.* **2020**, *22*. [CrossRef] [PubMed]

44. Niu, Z.; Willoughby, J.; Zhou, R. Associations of health literacy, social media use, and self-efficacy with health information-seeking intentions among social media users in China: Cross-sectional survey. *J. Med. Internet Res.* **2021**, *23*. [CrossRef]

45. Hingle, M.; Kunkel, D. Childhood obesity and the media. *Pediatric Clin. North Am.* **2012**, *59*, 677–692. [CrossRef] [PubMed]

46. Malaeb, D.; Salameh, P.; Barbar, S.; Awad, E.; Haddad, C.; Hallit, R.; Sacre, H.; Akel, M.; Obeid, S.; Hallit, S. Problematic social media use and mental health (depression, anxiety, and insomnia) among Lebanese adults: Any mediating effect of stress? *Perspect. Psychiatr. Care* **2021**, *57*, 539–549. [CrossRef]

47. Robinson, T.N.; Banda, J.A.; Hale, L.; Lu, A.S.; Fleming-Milici, F.; Calvert, S.L.; Wartella, E. Screen media exposure and obesity in children and adolescents. *Pediatrics* **2017**, *140*, S97–S101. [CrossRef]

48. Szabo, A.; Allen, J.; Stephens, C.; Alpass, F. Longitudinal analysis of the relationship between purposes of Internet use and well-being among older adults. *Gerontologist* **2019**, *59*, 58–68. [CrossRef]

49. Mendoza, J.A.; Zimmerman, F.J.; Christakis, D.A. Television viewing, computer use, obesity, and adiposity in US preschool children. *Int. J. Behav. Nutr. Phys. Act.* **2007**, *4*, 44. [CrossRef]

50. Thomee, S. Mobile Phone Use and Mental Health. A review of the research that takes a psychological perspective on exposure. *Int. J. Environ. Res. Public Health* **2018**, *15*, 2692. [CrossRef] [PubMed]

51. Vandewater, E.A.; Denis, L.M. Media, Social networking, and pediatric obesity. *Pediatric Clin. North Am.* **2011**, *58*, 1509–1519. [CrossRef] [PubMed]

52. Hou, H.; Feng, X.; Li, Y.; Meng, Z.; Guo, D.; Wang, F.; Guo, Z.; Zheng, Y.; Peng, Z.; Zhang, W.; et al. Suboptimal health status and psychological symptoms among Chinese college students: A perspective of predictive, preventive and personalised health. *Epma J.* **2018**, *9*, 367–377. [CrossRef] [PubMed]

53. Niu, G.; Zhao, G. Survey data on political attitudes of China's urban residents compiled from the Chinese General Social Survey (CGSS). *Data Brief* **2018**, *20*, 591–595. [CrossRef]

54. Bian, Y.; Li, L. The Chinese general social survey (2003-8) sample designs and data evaluation. *Chin. Sociol. Rev.* **2012**, *45*, 70–97. [CrossRef]

55. Bei-Fan, Z.; Cooperative Meta Anal Grp, W. Predictive values of body mass index and waist circumference for risk factors of certain related diseases in Chinese adults: Study on optimal cut-off points of body mass index and waist circumference in Chinese adults. *Asia Pac. J. Clin. Nutr.* **2002**, *11*, S685–S693. [CrossRef]

56. Chen, Y.M.; Ho, S.C.; Lam, S.S.H.; Chan, S.S.G. Validity of body mass index and waist circumference in the classification of obesity as compared to percent body fat in Chinese middle-aged women. *Int. J. Obes.* **2006**, *30*, 918–925. [CrossRef] [PubMed]

57. He, W.; Li, Q.; Yang, M.; Jiao, J.; Ma, X.; Zhou, Y.; Song, A.; Heymsfield, S.B.; Zhang, S.; Zhu, S. Lower BMI cutoffs to define overweight and obesity in China. *Obesity* **2015**, *23*, 684–691. [CrossRef]

58. WHO. Appropriate body-mass index in Asian populations and its implications for policy and intervention strategies. *Lancet* **2004**, *363*, 157–163, Erratum in **2004**, *363*, 902.

59. Zhou, B.F. Effect of body mass index on all-cause mortality and incidence of cardiovascular diseases—Report for meta-analysis of prospective studies on optimal cut-off points of body mass index in Chinese adults. *Biomed. Environ. Sci.* **2002**, *15*, 245–252. [PubMed]

60. Cronbach, L.J. Coefficient alpha and the internal structure of tests. *Psychometrika* **1951**, *16*, 297–334. [CrossRef]

61. Bland, J.M.; Altman, D.G. Cronbach's alpha. *Br. Med. J.* **1997**, *314*, 572. [CrossRef] [PubMed]

62. Long, C.; Han, J.; Yi, C. Does the effect of Internet use on Chinese citizens' psychological well-being differ based on their Hukou category? *Int. J. Environ. Res. Public Health* **2020**, *17*, 6680. [CrossRef] [PubMed]

63. Pedersen, W.; von Soest, T. Adolescent alcohol use and binge drinking: An 18-year trend study of prevalence and correlates. *Alcohol Alcohol.* **2015**, *50*, 219–225. [CrossRef]

64. Kachi, Y.; Abe, A.; Ando, E.; Kawada, T. Socioeconomic disparities in psychological distress in a nationally representative sample of Japanese adolescents: A time trend study. *Aust. N. Z. J. Psychiatry* **2017**, *51*, 278–286. [CrossRef]

65. Chen, J.; Xiang, H.; Jiang, P.; Yu, L.; Jing, Y.; Li, F.; Wu, S.; Fu, X.; Liu, Y.; Kwan, H.; et al. The role of healthy lifestyle in the implementation of regressing suboptimal health status among college students in China: A nested case-control study. *Int. J. Environ. Res. Public Health* **2017**, *14*, 240. [CrossRef]

66. Huang, H.-M.; Chien, L.-Y.; Yeh, T.-C.; Lee, P.-H.; Chang, P.-C. Relationship between media viewing and obesity in school-aged children in Taipei, Taiwan. *J. Nurs. Res.* **2013**, *21*, 195–202. [CrossRef] [PubMed]

67. Schneider, M.; Dunton, G.F.; Cooper, D.M. Media use and obesity in adolescent females. *Obesity* **2007**, *15*, 2328–2335. [CrossRef] [PubMed]

68. CNNIC. The 41th China Statistical Report on Internet Development. Available online: http://www.cnnic.net.cn/hlwfzyj/hlwxzbg/hlwtjbg/201803/P020180305409870339136.pdf (accessed on 29 September 2021).
69. CNNIC. The 34th China Statistical Report on Internet Development. Available online: http://www.cnnic.net.cn/hlwfzyj/hlwxzbg/hlwtjbg/201407/P020140721507223212132.pdf (accessed on 28 September 2021).
70. Thanasrivanitchai, J.; Moschis, G.P.; Shannon, R. Explaining older consumers' low use of the Internet. *Int. J. Internet Mark. Advert.* **2017**, *11*, 355–375. [CrossRef]
71. Seifert, A.; Schelling, H.R. Old and offline? Findings on the use of the Internet by people aged 65 years and older in Switzerland. *Z. Fur Gerontol. Und Geriatr.* **2016**, *49*, 619–625. [CrossRef] [PubMed]
72. Pan, X.-F.; Wang, L.; Pan, A. Epidemiology and determinants of obesity in China. *Lancet Diabetes Endocrinol.* **2021**, *9*, 373–392. [CrossRef]
73. Wang, L.; Zhou, B.; Zhao, Z.; Yang, L.; Zhang, M.; Jiang, Y.; Li, Y.; Zhou, M.; Wang, L.; Huang, Z.; et al. Body-mass index and obesity in urban and rural China: Findings from consecutive nationally representative surveys during 2004-18. *Lancet* **2021**, *398*, 53–63. [CrossRef]
74. Zou, Y.; Xia, N.; Zou, Y.; Chen, Z.; Wen, Y. Smartphone addiction may be associated with adolescent hypertension: A cross-sectional study among junior school students in China. *BMC Pediatrics* **2019**, *19*, 310. [CrossRef] [PubMed]
75. Kenney, E.L.; Gortmaker, S.L. United States adolescents' television, computer, videogame, smartphone, and tablet use: Associations with sugary drinks, sleep, physical activity, and obesity. *J. Pediatrics* **2017**, *182*, 144–149. [CrossRef]
76. Bekalu, M.A.; McCloud, R.F.; Viswanath, K. Association of social media use with social well-being, positive mental health, and self-rated health: Disentangling routine use from emotional connection to use. *Health Educ. Behav.* **2019**, *46*, 69–80. [CrossRef] [PubMed]
77. Alhusseini, N.; Banta, J.E.; Oh, J.; Montgomery, S.B. Social media use for health purposes by chronic disease patients in the United States. *Saudi J. Med. Med Sci.* **2021**, *9*, 51–58. [CrossRef] [PubMed]
78. Schehl, B. Outdoor activity among older adults: Exploring the role of informational Internet use. *Educ. Gerontol.* **2020**, *46*, 36–45. [CrossRef]
79. Ko, D.; Myung, E.; Moon, T.-J.; Shah, D.V. Physical activity in persons with diabetes: Its relationship with media use for health information, socioeconomic status and age. *Health Educ. Res.* **2019**, *34*, 257–267. [CrossRef] [PubMed]
80. Adua, E.; Afrifa-Yamoah, E.; Frimpong, K.; Adama, E.; Karthigesu, S.P.; Anto, E.O.; Aboagye, E.; Yan, Y.; Wang, Y.; Tan, X.; et al. Construct validity of the Suboptimal Health Status Questionnaire-25 in a Ghanaian population. *Health Qual. Life Outcomes* **2021**, *19*, 180. [CrossRef]

International Journal of
***Environmental Research
and Public Health***

Review

Chronic Obstructive Pulmonary Disease Patients' Acceptance in E-Health Clinical Trials

Saeed M. Alghamdi [1,2,*]**, Ahmed M. Al Rajah** [3]**, Yousef S. Aldabayan** [3]**, Abdulelah M. Aldhahir** [4]**,
Jaber S. Alqahtani** [5,6] **and Abdulaziz A. Alzahrani** [1,7]

[1] Department of Respiratory Care, College of Applied Health Science, Umm Al Qura University,
Makkah 21955, Saudi Arabia; aaa717@student.bham.ac.uk
[2] National Heart and Lung Institute, Imperial College London, London SW3 6NP, UK
[3] Respiratory Care Department, College of Applied Medical Sciences, King Faisal University,
Al-Ahsa 31982, Saudi Arabia; amalrajeh@kfu.edu.sa (A.M.A.R.); yaldabayan@kfu.edu.sa (Y.S.A.)
[4] Respiratory Care Department, Faculty of Applied Medical Sciences, Jazan University,
Jazan 45142, Saudi Arabia; abdulelah.aldhahir.17@ucl.ac.uk
[5] UCL Respiratory, University College London, London WC1E 6BT, UK; jaber.alqahtani.18@ucl.ac.uk
[6] Department of Respiratory Care, Prince Sultan Military College of Health Sciences,
Dammam 34313, Saudi Arabia
[7] Institute of Clinical Sciences, University of Birmingham, Birmingham B15 2TT, UK
[*] Correspondence: s.alghamdi18@imperial.ac.uk; Tel.: +44-330-128-4058

Citation: Alghamdi, S.M.;
Rajah, A.M.A.; Aldabayan, Y.S.;
Aldhahir, A.M.; Alqahtani, J.S.;
Alzahrani, A.A. Chronic Obstructive
Pulmonary Disease Patients'
Acceptance in E-Health Clinical Trials.
Int. J. Environ. Res. Public Health **2021**,
18, 5230. https://doi.org/10.3390/
ijerph18105230

Academic Editors:
Irene Torres-Sanchez, Marie
Carmen Valenza and Paul
B. Tchounwou

Received: 29 March 2021
Accepted: 10 May 2021
Published: 14 May 2021

Publisher's Note: MDPI stays neutral
with regard to jurisdictional claims in
published maps and institutional affiliations.

Abstract: Introduction: Telehealth (TH) interventions with Chronic Obstructive Pulmonary Disease
(COPD) management were introduced in the literature more than 20 years ago with different labeling,
but there was no summary for the overall acceptance and dropout rates as well as associated variables.
Objective: This review aims to summarize the acceptance and dropout rates used in TH interventions
and identify to what extent clinical settings, sociodemographic factors, and intervention factors might
impact the overall acceptance and completion rates of TH interventions. Methods: We conducted
a systematic search up to April 2021 on CINAHL, PubMed, MEDLINE (Ovid), Cochrane, Web of
Sciences, and Embase to retrieve randomized and non-randomized control trials that provide TH
interventions alone or accompanied with other interventions to individuals with COPD. Results:
Twenty-seven studies met the inclusion criteria. Overall, the unweighted average of acceptance and
dropout rates for all included studies were 80% and 19%, respectively. A meta-analysis on the pooled
difference between the acceptance rates and dropout rates (weighted by the sample size) revealed a
significant difference in acceptance and dropout rates among all TH interventions 51% (95% CI 49%
to 52; $p < 0.001$) and 63% (95% CI 60% to 67; $p < 0.001$), respectively. Analysis revealed that acceptance
and dropout rates can be impacted by trial-related, sociodemographic, and intervention-related
variables. The most common reasons for dropouts were technical difficulties (33%), followed by
complicated system (31%). Conclusions: Current TH COPD interventions have a pooled acceptance
rate of 51%, but this is accompanied by a high dropout rate of 63%. Acceptance and dropout levels
in TH clinical trials can be affected by sociodemographic and intervention-related factors. This
knowledge enlightens designs for well-accepted future TH clinical trials. PROSPERO registration
number CRD4201707854.

Keywords: systematic review; meta-analysis; telehealth; chronic obstructive pulmonary disease; COPD

1. Introduction

More than 10% of the population worldwide aged 40 years or older are affected
by Chronic Obstructive Pulmonary Disease (COPD) [1]. In general, COPD is caused
by smoking cigarettes, which may lead to death or disability [2]. The prevalence of
COPD has increased dramatically over the past 30 years and is predicted to be the third-
leading cause of death by 2030 [1–3]. According to the Global Initiative for Chronic
Obstructive Lung Disease (GOLD), COPD is a common, preventable, and treatable disease

characterized by persistent airflow limitation that is usually progressive and associated with enhanced chronic inflammatory responses to noxious gases in the airways and the lungs [4]. Airflow limitations lead to worsening respiratory symptoms—exacerbation—and, often, hospitalization [5,6]. COPD patients require an appropriate management strategy that aims to minimize the frequency of hospitalizations [7].

Recently, research has focused on delivering COPD care via telehealth (TH) to offer prompt access to healthcare and to increase the capacity of COPD care [8]. Published systematic reviews have found that delivering COPD care using TH may provide a mechanism to encourage collaboration between patients and healthcare providers and may enhance patient knowledge and skills in learning how to deal with their conditions [8,9]. Moreover, it facilitates regular monitoring of patients' clinical data, such as vital signs, to allow the healthcare team to detect any disease deterioration at an early stage before it worsens and provide the necessary care to minimize hospital admissions due to respiratory exacerbation [8,9]. Additionally, TH plays an important role in supporting public health precautions and in mitigating the spread of infections such as COVID-19 [10]. Taking the above into account, TH technologies have different applications to provide health care for COPD patients [8,11].

Using TH in COPD management has been found feasible, valuable, and accessible, but recent evidence shows variation regarding the completion of such interventions [11–13]. Previous clinical trials show that individuals with COPD have positive attitudes toward participating in TH interventions [14–19]. However, evidence about the impact of TH on health service outcomes or patient-related outcomes is still inconclusive [8,11]. This lack of knowledge about the effectiveness of TH for COPD management might lead to poor acceptance of TH interventions and/or a high dropout rate and withdrawal of participants from TH studies [20]. While TH use is promising in COPD management, it is unclear which factors are most associated with acceptance and dropout rates and whether these factors are trial-related, sociodemographic, or intervention-related. Therefore, our review aims to (1) assess the overall acceptance and dropout rates in TH clinical trials, (2) summarize the reasons for dropouts from TH interventions, and (3) explore factors that have an impact on overall acceptance and dropout rates.

2. Methods

The current systematic review was reported according to the Preferred Reporting Items for Systematic Review and Meta-analysis (PRISMA) guidelines [21].

2.1. The Inclusion Criteria

1. Controlled clinical trials with or without randomization that examined TH interventions;
2. Studies that include patients diagnosed with COPD (defined as forced expiratory volume in 1 s (FEV_1)/forced vital capacity (FVC) ratio < 70%, FEV_1 < 80% predicted);
3. The intervention included in this review is telehealth. As telehealth interventions have different labels (e.g., telemonitoring, telerehabilitation) in the literature, no restrictions have been applied on intervention labeling. TH interventions with different labels which use internet or electronic health information and communication technologies to support distance health care and/or exchange information between patients and healthcare providers were included.

2.2. The Exclusion Criteria

1. Studies that targeted non-COPD individuals and/or a general population;
2. Trials published in a language other than English;
3. Studies that did not describe TH, including the content of the intervention, delivery method, mode of administration, and frequency of data transmission;
4. Studies that did not report the number of COPD individuals who were approached, consented, and dropped out.

2.3. Search Strategy

An electronic search of the following databases up to April 2021 was undertaken to retrieve relevant articles: CINAHL, PubMed, MEDLINE (Ovid), Cochrane Library, and Embase. Medical subject headings, subject headings, and/or their combinations used in all databases were as follows: telehealth, telecare, telehomecare, telemonitoring, telerehabilitation, telemedicine, home monitoring, digital monitoring, web-based interventions, internet-based monitoring, e-health, chronic obstructive pulmonary disease, chronic obstructive lung disease, and COPD. The search was conducted in collaboration with the health sciences librarian to ensure that our search included the appropriate and necessary keywords in the review. Keywords and subject terms were customized in each database. A full search strategy from all databases is provided in Supplementary materials Files S1–S10.

2.4. Search Procedures

The search was performed by the main reviewer (SA), after which all articles were imported to EndNote version X9.3 (Clarivate, Philadelphia, PA, USA), and duplicates were removed. All titles and abstracts were screened by two reviewers (SA and AA). A third reviewer (YA) was available to resolve any persisting disagreements. A manual search of the reference lists of relevant studies was undertaken to identify any potentially relevant articles that were missed by the database search but that might be suitable for inclusion in the review. A full-text review of all suitable articles was undertaken, and any study that did not meet the inclusion criteria was excluded.

2.5. Data Extraction

A standardized Excel (Microsoft Corporation, Redmond, WA, USA) spreadsheet was created for data extraction. The spreadsheet included information on acceptance and dropout rates, as well as trial-related, sociodemographic, and intervention-related factors. Trial-related factors include study place, study design, and recruitment location. The sociodemographic factors include age, status at recruitment, and smoking history. The intervention-related factors include the components of the intervention, methods of delivery, display, frequency, and duration of the intervention. For any missing data, the authors attempted to contact the corresponding publishers of the included studies and completed the data extraction form.

The quality of the studies was defined based on the Cochrane risk-of-bias tool [22]. Data extraction and quality assessment were performed by two independent reviewers (SA and AA). Any disagreement between reviewers (SA and AA) was resolved by consensus.

2.6. Data Analysis

The synthesized results in this review focused on the key outcomes of interest, including the acceptance and dropout rates, reasons for dropout, and the possible factors that might impact acceptance and dropout rates. The overall acceptance rate in this paper refers to the number of participants who consented to participate divided by the number of participants who were approached to participate in the trial [23,24], and dropout rate refers to the total number of participants in each treatment arm who dropped out from the clinical trial divided by the total number of the participants who consented to participate in the clinical trial [23].

Possible explanatory trial-related factors, sociodemographic factors, and intervention-related factors that might be associated with acceptance and dropout rates were identified from the literature [25–30]. All factors were converted to binary data. Trial-related factors are categorized as the place of the study (Europe vs. non-Europe), design of the study (Randomized Control Trial vs. non-Randomized Control Trial), and the recruitment location (one recruitment location vs. more than one). The sociodemographic factors are categorized as age (<69 vs. $\geq$69), the status of COPD at recruitment (stable vs. non-stable), and smoking history (yes vs. no). The intervention-related factors include components of the intervention (one component vs. more than one), methods of delivery (internet-based vs.

other), interactive display (interactive vs. not interactive), frequency (daily vs. weekly), and the duration of the intervention ($\leq$20 weeks vs. >20 weeks). Overall, acceptance and dropout rates were computed for each category to compare factors. Continuous variables are expressed as mean and standard deviations. Categorical variables are expressed as frequency and percentages.

We performed an additional meta-analysis to estimate the pooled difference of acceptance and dropout rates between treatment arms. The estimation of rates weighted by the sample size in each clinical trial, and data were pooled using random-effects models. We expressed rates as proportions and 95% confidence intervals (CIs). Heterogeneity among included studies was assessed using the I-square (I^2) value. All statistical analyses in this study were performed using STATA software (StataCorp, College Station, TX, USA). More information about the statistical methods can be found in Supplementary File S1.

3. Results

3.1. Study Selection

The search identified a total of 1463 relevant articles, 885 after duplicates had been excluded, with a total of 112 articles maintained for full-text review following the title and abstract screening. The remaining articles were read in full, and 27 articles were considered for the review, as shown in the PRISMA flow diagram (Figure 1).

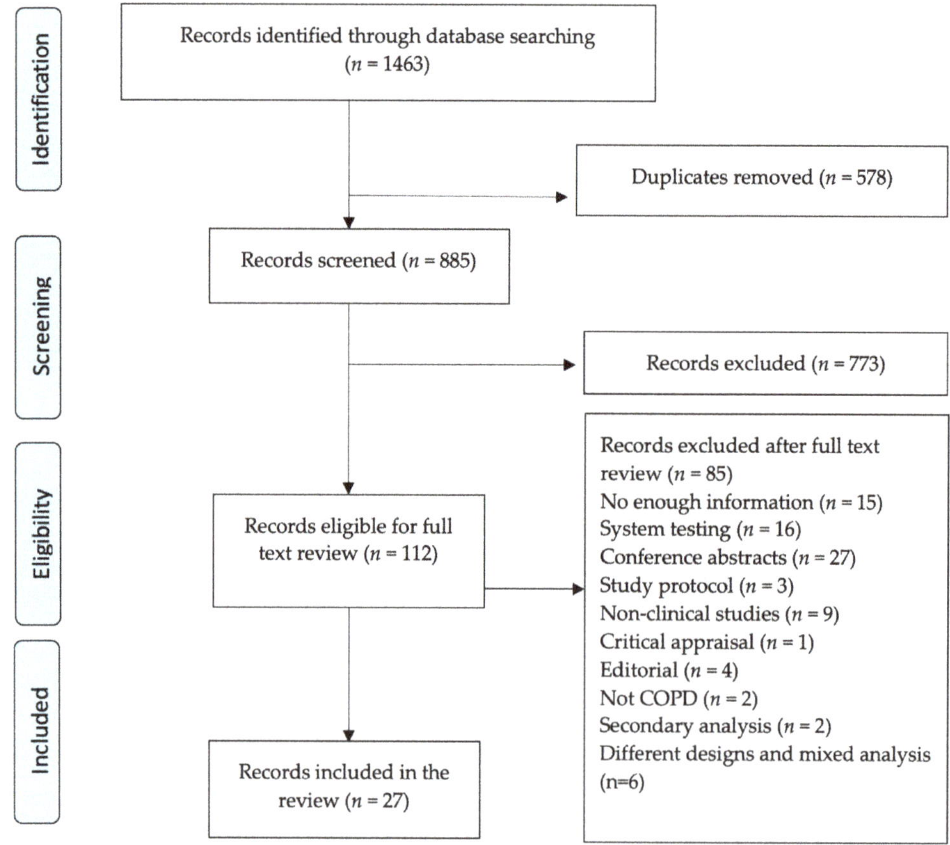

Figure 1. PRISMA flow diagram showing studies related to the TH (telehealth) interventions in COPD (with Chronic Obstructive Pulmonary Disease).

3.2. Study Characteristics

The majority of the included studies were published in Europe (*n* = 18). Most of the studies were RCT (*n* = 24). Twenty-one out of twenty-seven clinical trials recruited patients from different clinical settings.

3.3. Patient Characteristics

The included trials [18,31–56] collectively comprised a total of 4157 COPD patients. Participation age (mean ± standard deviation) was 65 ± 7.1 years. In the included studies, 18 studies out of 27 (65%) reported smoking history. COPD GOLD III was the most common GOLD grade among included participants: 3118/4157 (75%). More descriptive details about the studies and sociodemographic characteristics can be found in Supplementary File S3.

3.4. TH Intervention Characteristics

A variety of TH interventions were provided in the included studies, either self-management programs via telemonitoring or self-management programs combined with other interventions (e.g., exercise, pulmonary rehabilitation, home care). Twenty of the studies out of twenty-seven had more than one component. Fourteen of the studies out of twenty-seven delivered interventions using web-based video or telephone calls. Eighteen of the studies out of twenty-seven delivered the intervention with daily frequencies, while nine studies out of twenty-seven delivered the intervention in weekly frequencies. The duration of the intervention ranged from three weeks to forty-eight weeks. Sixteen of the studies out of twenty-seven had an intervention of >20 weeks. The control groups, for example, were provided with the usual care, written self-management, and written exercise guidelines. More descriptive details about TH and control intervention characteristics can be found in Supplementary Files S4 and S5.

3.5. Acceptance and Dropout Rates

The total number of individuals with COPD approached to participate in the included clinical trial was 8085. Of these, 3928 patients were ineligible and excluded. A total of 4157 consented and enrolled in the clinical trials. Overall, the unweighted average of acceptance and dropout rates for all included studies were 80% and 19%, respectively. The acceptance and dropout rates in all included studies were stratified by factors, as provided in Table 1.

Table 1. Overall acceptance and dropout rates across included studies stratified by trial-related, sociodemographic, and intervention-related factors (*n* = 27) [a].

Factors	Number of Studies (%)	TH Acceptance Rate (Mean ± SD)	TH Dropout Rate (Mean ± SD)
Trial-related factors			
Place of the study			
Europe	18 (65%)	82% ± 14%	19% ± 14%
Non-Europe	9 (35%)	76% ± 14%	19% ± 16%
Study design			
Randomized clinical trial	24 (88%)	81% ± 14%	18% ± 14%
Non-Randomized clinical trial	3 (12%)	77% ± 18%	27% ± 14%
Recruitment location			
One location	21 (78%)	79% ± 14%	20% ± 14%
More than one	6 (22%)	83% ± 19%	16% ± 19%
Sociodemographic factors			
COPD status at recruitment			
Stable	18 (65%)	80% ± 16%	19% ± 16%
Non-stable	9 (35%)	81% ± 12%	18% ± 12%

Table 1. *Cont.*

Factors	Number of Studies (%)	TH Acceptance Rate (Mean ± SD)	TH Dropout Rate (Mean ± SD)
Smoking history			
Yes	18 (65%)	82% ±14%	17% ± 14%
No	9 (35%)	76% ±16%	23% ± 15%
Age			
<69	5 (18%)	83% ± 10%	16% ± 10%
≥69	22(81%)	80% ± 15%	20% ± 15%
Intervention-related factors			
Telehealth component (s)			
One component	7 (25%)	78% ± 18%	21% ± 18%
More than one	20 (75%)	81% ± 13%	18% ± 13%
Methods of delivery			
Web-based	14 (51%)	78% ± 17%	21% ± 17%
Other	13 (49%)	82% ± 13%	17% ± 13%
Interactive display			
Interactive	14 (51%)	78% ± 17%	21% ± 17%
Not interactive	13 (49%)	83% ± 11%	16% ± 11%
Frequency of TH			
Daily	18 (65%)	81% ± 16%	18% ± 16%
Weekly	9 (35%)	78% ± 12%	21% ± 12%
Duration of TH			
20 weeks or less	11 (40%)	86% ± 12%	14% ± 12%
More than 20 weeks	16 (60%)	77% ± 12%	23% ± 12%

[a] Data presented as frequency and percentages or means and standard deviations.

3.6. Meta-Analysis

An additional meta-analysis of acceptance and dropout rates in all included studies (weighted by the sample size) demonstrated significant differences in the acceptance and dropout rates between TH and control groups. As shown in Table 2, the pooled difference of the acceptance and dropout rates of TH interventions with corresponding 95% CIs was 51% (49% to 52%); $p < 0.001$ and 63% (60% to 67%); $p < 0.001$, respectively. More information about the pooled acceptance and dropout rates in each treatment arm can be found in Supplementary Files S6–S9.

Table 2. Overall weighted acceptance and dropout rates of all included studies ($n = 27$) [a].

Overall Rates	Weighted (Estimation)	S.E.	*p*-Value	95% CIs
Acceptance rate in TH	51%	0.2	<0.001	49% to 52%
Acceptance rate in control	49%	0.3	<0.001	48% to 51%
Dropout rate in TH	63%	0.2	<0.001	60% to 67%
Dropout rate in control	37%	0.3	<0.001	33% to 40%

[a] S.E: standard error.

3.7. Reasons for Dropout

After randomization or allocation to treatment, 1152 participants completed their TH interventions, and 946 withdrew before the end of the study. Of these, only 513 participants provided the reasons for study withdrawal. Reasons for dropout classified as TH-related reasons and individual-related reasons are shown in Table 3.

Table 3. Most common reasons for dropout from TH interventions ($n = 513$) [a].

Dropout Reasons	Number of Patients (%)
TH-related reasons	
Technical difficulties	122 (24%)
Complicated system	117 (23%)
Time constraints	9 (2%)
Individual-related reasons	
Hospital admission	138 (27%)
Deceased	68 (13%)
Not interested in continuing	45 (9%)
Moved from the study location	14 (3%)

[a] Data presented as frequency and percentage.

3.8. Quality Assessment

Quality assessment was performed using the Cochrane risk-of-bias assessment tool [22]. The studies included showed variation in their risk of bias, but most were limited by a lack of blinding. A summary of our judgments on the potential risk of bias can be found in Supplementary File S10.

4. Discussion

In the context of COPD, exploring the current acceptance and dropout rates of a different form of TH intervention and its associated factors is a crucial objective in terms of whether TH intervention is feasible or not; furthermore, these data have not been previously pooled in forest plots to summarize the overall acceptance and dropout rates. Our findings revealed that the overall average of acceptance and dropout rates was reasonable for patients with COPD, considering the trial-related, sociodemographic, and intervention-related factors. The findings indicated that 20% of the eligible participants were expected to refuse the TH interventions, and once participants were allocated to TH interventions, around 19% were expected to drop out and not complete the clinical trial. The weighted acceptance rate was 51% in TH groups, suggesting that TH interventions were accepted among people with COPD. Moreover, the dropout rate was higher in the TH intervention groups versus control groups (63% vs. 37%). This finding was expected because most of the dropout reasons from TH studies were TH intervention-related reasons. However, we identified certain factors that have the potential to improve acceptance and reduce the dropout rate of TH interventions with COPD.

In the context of clinical research on COPD, the acceptance rate was higher in multicenter short-term (<20 weeks) trials that were done in European countries. A possible explanation is that advanced countries have good adoption and success rates of TH implementations compared to low- and middle-class countries [57–59], but it could also be due to other financial, organizational, and economic factors since large multicenter clinical trials need considerable research funding, collaboration between clinical settings, qualified researchers, and a large number of patients [57,60].

Moreover, a high acceptance rate in the studies that remotely monitored people with COPD after hospital discharge has been observed. This is due to the benefits of TH in detecting COPD deterioration in early stages and minimizing hospital readmission due to exacerbations. Moreover, some evidence found that offering TH interventions to people with COPD seems to be an opportunity to provide quality healthcare and to minimize unnecessary hospitalization [61,62]. A patient's smoking history was identified as an additional explanatory factor to increase TH acceptability. Similar findings were found by Watson et al. that recruiting smokers in TH interventions for less than 20 weeks significantly increased the acceptance rate [63]. This result could be because these patients have been chronic smokers for years, and smoking cessation interventions are not working well for them. The increased participation of smokers with COPD in TH interventions could indicate that those patients are searching for more accessible smoking cessation interventions [64].

Furthermore, the high acceptance rate of TH that provides several components (e.g., telemonitoring, tele-coaching, and self-management education) has been inspected. A possible explanation is that these TH interventions provide more individual-based interventions that meet the patient's needs. There were similar findings in the literature about using more approaches to address the patient's needs (e.g., uptake pulmonary rehabilitation), resulting in greater patient engagement and satisfaction [65]. Moreover, TH interventions that provide daily monitoring through convenient methods (e.g., telephone calls and text messages) had a high acceptance rate because they used a simple method of communication [32].

Refusal to complete TH interventions is primarily attributed to the interventions themselves [66,67]. It was noted that TH interventions with multiple components were fraught with complexities and technical difficulties that have resulted in decreased treatment sessions or even termination [18,32]. This could lead to participant dissatisfaction and, ultimately, dropping out of the study. Essential steps at the designing and implementation levels are needed to plan and enable more technical support during the delivery of TH interventions [68]. Another possible explanation for dropout rates could be the discrepancy between the patients' expectations and their abilities to operate and use TH [69]. Most of the studies did not provide training sessions for the participants; specifically, there were no measurements of participants' competency in using the technology or cognitive abilities to use TH. Thus, attention must be paid to accommodating these factors in the design of future TH interventions [70].

Telehealth application is important, and COPD management guidelines have supported its implementation [71–73]. Current systematic reviews provide a comprehensive description and summary of the acceptance and dropout rates of different forms of TH interventions for patients with COPD. Generally, TH interventions are acceptable among the COPD population [11,20,23,24,67,68,74,75]. Nevertheless, methodological queries remain regarding the design of a more acceptable and feasible TH intervention, the best strategy to provide TH interventions (individual vs. community), components of TH interventions (solo vs. joint other treatments), technological aspects (classic vs. advanced), and COPD phenotypes (stable vs. non-stable) that will obtain more benefits from TH interventions. This suggests that we need a more in-depth understanding of the acceptability and feasibility of TH interventions developed for COPD patients, as well as an in-depth understanding of the sociodemographic characteristics of the COPD population and their cognitive abilities when we propose TH solutions [75]. Here, the high dropout rates in TH interventions happened due to both intervention-related reasons and patient-related reasons. An additional contribution of this review is to inform future clinical trials regarding the acceptance and dropout rates of existing TH trials in the context of COPD. Moreover, this review will help researchers to identify the reasons that prevent individuals with COPD from completing TH interventions, and it provides evidence to guide future researchers in designing prospective TH clinical trials accordingly.

5. Limitations

Some limitations must be considered when attempting to interpret the results of the current review. There was a lack of information and incomplete data across all included studies, which impeded the possibility of exploring more variables that might influence or be influenced by the acceptance and dropout rate. Moreover, there was considerable variation in the quality of the included studies, which could limit the ability to generalize the results of this review.

6. Conclusions

Current TH interventions with COPD have a pooled acceptance rate of 51%, but this is accompanied by a high dropout rate of 63%. Acceptance and dropout levels in TH clinical trials may be affected by sociodemographic and intervention-related factors. Taken together, continued efforts are needed to improve patients' acceptance by understanding and mitigating the issues contributing to the acceptance and dropout rates that would support the preparation and establishment of effective and well-accepted TH interventions.

Supplementary Materials: The following are available online at https://www.mdpi.com/article/10.3390/ijerph18105230/s1, File S1: Search Strategies of Databases, File S2: More Information about Statistical Methods for Acceptance and Dropout Rates, File S3: Study and Sociodemographic-Related Characteristics among Included Studies, File S4: A Detailed Description of TH Interventions and Parallel Groups, File S5: TH Intervention Characteristics of Included Studies, File S6: Forest Plot of Acceptance Rates in TH Interventions, File S7: Forest Plot of Dropout Rates in TH Interventions, File S8: Forest Plot of Acceptance Rates in Controls, File S9: Forest Plot of Dropout Rates in Controls, File S10: Assessment Risk of Bias for Included Studies.

Author Contributions: S.M.A., A.M.A.R., Y.S.A., A.M.A., J.S.A. and A.A.A. developed the idea and designed the study protocol. S.M.A., J.S.A. and A.M.A. designed and wrote the search strategy and the first protocol draft. S.M.A., J.S.A. and A.M.A. planned data extraction and statistical analysis. A.M.A.R., Y.S.A. and A.A.A. provided critical insights. All authors have read and agreed to the published version of the manuscript.

Funding: This research received no external funding.

Institutional Review Board Statement: Not applicable.

Informed Consent Statement: Not applicable.

Data Availability Statement: Not applicable.

Conflicts of Interest: The authors of this paper declare no conflict of interest.

Abbreviations

TH	Telehealth
TM	Telemonitoring
SF	Self-management
COPD	Chronic Obstructive Pulmonary Disease
RCT	Randomized control trials
NRCT	Non-randomized control trials

References

1. Mathers, C.D.; Loncar, D. Projections of global mortality and the burden of disease from 2002 to 2030. *PLoS Med.* **2006**, *3*, e442. [CrossRef]
2. Buist, A.S.; McBurnie, M.A.; Vollmer, W.M.; Gillespie, S.; Burney, P.; Mannino, D.M.; Menezes, A.M.; Sullivan, S.D.; A Lee, T.; Weiss, K.B.; et al. International variation in the prevalence of COPD (The BOLD Study): A population-based prevalence study. *Lancet* **2007**, *370*, 741–750. [CrossRef]
3. World Health Organization Fact Sheet for Chronic Obstructive Lung Disease. World Health Organization 2017. Available online: https://www.who.int/news-room/fact-sheets/detail/chronic-obstructive-pulmonary-disease-(COPD) (accessed on 22 October 2020).
4. Global Initiative for Chronic Obstructive Lung Disease (GOLD) Global Strategy for the Diagnosis, Management, and Prevention of Chronic Obstructive Pulmonary Disease (2020 Report). 2020. Available online: https://goldcopd.org/ (accessed on 23 October 2020).
5. Fletcher, M.J.; Upton, J.; Taylor-Fishwick, J.; Buist, S.A.; Jenkins, C.; Hutton, J.; Barnes, N.; Van Der Molen, T.; Walsh, J.W.; Jones, P.; et al. COPD uncovered: An international survey on the impact of chronic obstructive pulmonary disease [COPD] on a working age population. *BMC Public Health* **2011**, *11*, 612. [CrossRef]
6. Halbert, R.J.; Natoli, J.L.; Gano, A.; Badamgarav, E.; Buist, A.S.; Mannino, D.M. Global burden of COPD: Systematic review and meta-analysis. *Eur. Respir. J.* **2006**, *28*, 523–532. [CrossRef]
7. Yang, F.; Xiong, Z.-F.; Yang, C.; Li, L.; Qiao, G.; Wang, Y.; Zheng, T.; He, H.; Hu, H. Continuity of care to prevent readmissions for patients with chronic obstructive pulmonary disease: A systematic review and meta-analysis. *COPD J. Chronic Obstr. Pulm. Dis.* **2017**, *14*, 251–261. [CrossRef]

8. Selzler, A.-M.; Wald, J.; Sedeno, M.; Jourdain, T.; Janaudis-Ferreira, T.; Goldstein, R.; Bourbeau, J.; Stickland, M.K. Telehealth pulmonary rehabilitation: A review of the literature and an example of a nationwide initiative to improve the accessibility of pulmonary rehabilitation. *Chronic Respir. Dis.* **2017**, *15*, 41–47. [CrossRef]

9. Zwerink, M.; Brusse-Keizer, M.; Van Der Valk, P.D.; Zielhuis, G.A.; Monninkhof, E.M.; Van Der Palen, J.A.M.; Frith, P.A.; Effing, T. Self management for patients with chronic obstructive pulmonary disease. *Cochrane Database Syst. Rev.* **2014**, *2014*, CD002990. [CrossRef]

10. Alghamdi, S.M.; Alqahtani, J.S.; Aldhahir, A.M. Current status of telehealth in Saudi Arabia during COVID-19. *J. Fam. Community Med.* **2020**, *27*, 208–211. [CrossRef]

11. Alghamdi, S.; Alqahtani, J.; Aldhahir, A.; Alrajeh, A.; Aldabayan, Y. Effectiveness of telehealth-based interventions with chronic obstructive pulmonary disease: A systematic review and meta-analysis. *Am. Thorac. Soc.* **2020**. [CrossRef]

12. Alwashmi, M.; Hawboldt, J.; Davis, E.; Marra, C.; Gamble, J.-M.; Abu Ashour, W. The effect of smartphone interventions on patients with chronic obstructive pulmonary disease exacerbations: A systematic review and meta-analysis. *JMIR mHealth uHealth* **2016**, *4*, e105. [CrossRef]

13. Alghamdi, S.M.; Aldhahir, A.M.; Alqahtani, J.S.; Aldabayan, Y.S.; Al Rajeh, A.M. Telemonitoring parameters used in home-based rehabilitation with COPD: A systematic review. *Respir. Care* **2020**, *65*, 344226.

14. Almojaibel, A.A. Delivering pulmonary rehabilitation for patients with chronic obstructive pulmonary disease at home using telehealth: A review of the literature. *Saudi J. Med. Med Sci.* **2016**, *4*, 164. [CrossRef]

15. Burkow, T.M.; Vognild, L.K.; Østengen, G.; Johnsen, E.; Risberg, M.J.; Bratvold, A.; Hagen, T.; Brattvoll, M.; Krogstad, T.; Hjalmarsen, A. Internet-enabled pulmonary rehabilitation and diabetes education in group settings at home: A preliminary study of patient acceptability. *BMC Med. Inform. Decis. Mak.* **2013**, *13*, 33. [CrossRef]

16. Holland, A.E.; Hill, C.J.; Rochford, P.; Fiore, J.; Berlowitz, D.J.; McDonald, C.F. Telerehabilitation for people with chronic obstructive pulmonary disease: Feasibility of a simple, real time model of supervised exercise training. *J. Telemed. Telecare* **2013**, *19*, 222–226. [CrossRef]

17. Paneroni, M.; Colombo, F.; Papalia, A.; Colitta, A.; Borghi, G.; Saleri, M.; Cabiaglia, A.; Azzalini, E.; Vitacca, M. Is telerehabilitation a safe and viable option for patients with COPD? A feasibility study. *COPD J. Chronic Obstr. Pulm. Dis.* **2014**, *12*, 217–225. [CrossRef]

18. Tabak, M.; Van Der Valk, P.; Hermens, H.; Vollenbroek-Hutten, M.; Brusse-Keizer, M. A telehealth program for self-management of COPD exacerbations and promotion of an active lifestyle: A pilot randomized controlled trial. *Int. J. Chronic Obstr. Pulm. Dis.* **2014**, *9*, 935. [CrossRef]

19. Zanaboni, P.; Lien, L.A.; Hjalmarsen, A.; Wootton, R. Long-term telerehabilitation of COPD patients in their homes: Interim results from a pilot study in Northern Norway. *J. Telemed. Telecare* **2013**, *19*, 425–429. [CrossRef]

20. Cruz, J.; Brooks, D.; Marques, A. Home telemonitoring in COPD: A systematic review of methodologies and patients' adherence. *Int. J. Med. Inform.* **2014**, *83*, 249–263. [CrossRef]

21. Moher, D.; Liberati, A.; Tetzlaff, J.; Altman, D.G.; Group, P. Preferred reporting items for systematic reviews and meta-analyses: The PRISMA statement. *PLoS Med.* **2009**, *6*, e1000097. [CrossRef]

22. Higgins, J.P.T.; Altman, D.G.; Gøtzsche, P.C.; Jüni, P.; Moher, D.; Oxman, A.D.; Savović, J.; Schulz, K.F.; Weeks, L.; Sterne, J.A.C.; et al. The Cochrane Collaboration's tool for assessing risk of bias in randomised trials. *BMJ* **2011**, *343*, d5928. [CrossRef]

23. Arafah, A.M.; Bouchard, V.; Mayo, N.E. Enrolling and keeping participants in multiple sclerosis self-management interventions: A systematic review and meta-analysis. *Clin. Rehabil.* **2017**, *31*, 809–823. [CrossRef]

24. Alghamdi, S.M.; Barker, R.E.; Alsulayyim, A.S.S.; Alasmari, A.M.; Banya, W.A.S.; I Polkey, M.; Birring, S.S.; Hopkinson, N.S. Use of oscillatory positive expiratory pressure (OPEP) devices to augment sputum clearance in COPD: A systematic review and meta-analysis. *Thorax* **2020**, *75*, 855–863. [CrossRef]

25. Heerema-Poelman, A.; Stuive, I.; Wempe, J.B. Adherence to a maintenance exercise program one year after pulmonary rehabilitation: What are the predictors of dropout? *J. Cardiopulm. Rehabil. Prev.* **2013**, *33*, 419–426. [CrossRef] [PubMed]

26. Brown, A.T.; Hitchcock, J.; Schumann, C.; Wells, J.M.; Dransfield, M.T.; Bhatt, S.P. Determinants of successful completion of pulmonary rehabilitation in COPD. *Int. J. Chronic Obstr. Pulm. Dis.* **2016**, *11*, 391.

27. Selzler, A.-M.; Simmonds, L.; Rodgers, W.M.; Wong, E.Y.; Stickland, M.K. Pulmonary rehabilitation in chronic obstructive pulmonary disease: Predictors of program completion and success. *COPD J. Chronic Obstr. Pulm. Dis.* **2012**, *9*, 538–545. [CrossRef] [PubMed]

28. Juretic, M.; Hill, R.; Hicken, B.; Luptak, M.; Rupper, R.; Bair, B. Predictors of attrition in older users of a home-based monitoring and health information delivery system. *Telemed. e-Health* **2012**, *18*, 709–712. [CrossRef] [PubMed]

29. Turner, J.; Wright, E.; Mendella, L.; Anthonisen, N. Predictors of patient adherence to long-term home nebulizer therapy for COPD. *Chest* **1995**, *108*, 394–400. [CrossRef]

30. Milner, S.C.; Bourbeau, J.; Ahmed, S.; Janaudis-Ferreira, T. Improving acceptance and uptake of pulmonary rehabilitation after acute exacerbation of COPD: Acceptability, feasibility, and safety of a PR "taster" session delivered before hospital discharge. *Chronic Respir. Dis.* **2019**, *16*, 1479973119872517. [CrossRef]

31. Antoniades, N.C.; Rochford, P.D.; Pretto, J.J.; Pierce, R.J.; Gogler, J.; Steinkrug, J.; Sharpe, K.; McDonald, C.F. Pilot study of remote telemonitoring in COPD. *Telemed. e-Health* **2012**, *18*, 634–640. [CrossRef] [PubMed]

32. Berkhof, F.F.; Berg, J.W.V.D.; Uil, S.M.; Kerstjens, H.A. Telemedicine, the effect of nurse-initiated telephone follow up, on health status and health-care utilization in COPD patients: A randomized trial. *Respirology* **2015**, *20*, 279–285. [CrossRef]

33. Calvo, G.S.; Gómez-Suárez, C.; Soriano, J.; Zamora, E.; Gónzalez-Gamarra, A.; González-Béjar, M.; Jordán, A.; Tadeo, E.; Sebastián, A.; Fernández, G.; et al. A home telehealth program for patients with severe COPD: The PROMETE study. *Respir. Med.* **2014**, *108*, 453–462. [CrossRef] [PubMed]

34. Cameron-Tucker, H.L.; Wood-Baker, R.; Joseph, L.; Walters, J.A.; Schüz, N.; Walters, E.H. A randomized controlled trial of telephone-mentoring with home-based walking preceding rehabilitation in COPD. *Int. J. Chronic Obstr. Pulm. Dis.* **2016**, *11*, 1991. [CrossRef] [PubMed]

35. Chau, J.P.-C.; Lee, D.T.-F.; Yu, D.S.-F.; Chow, A.Y.-M.; Yu, W.-C.; Chair, S.-Y.; Lai, A.S.; Chick, Y.-L. A feasibility study to investigate the acceptability and potential effectiveness of a telecare service for older people with chronic obstructive pulmonary disease. *Int. J. Med. Inform.* **2012**, *81*, 674–682. [CrossRef]

36. De San Miguel, K.; Smith, J.; Lewin, G. Telehealth remote monitoring for community-dwelling older adults with chronic obstructive pulmonary disease. *Telemed. e-Health* **2013**, *19*, 652–657. [CrossRef]

37. Dinesen, B.; Haesum, L.K.; Soerensen, N.; Nielsen, C.; Grann, O.; Hejlesen, O.; Toft, E.; Ehlers, L. Using preventive home monitoring to reduce hospital admission rates and reduce costs: A case study of telehealth among chronic obstructive pulmonary disease patients. *J. Telemed. Telecare* **2012**, *18*, 221–225. [CrossRef] [PubMed]

38. Farmer, A.; Williams, V.; Velardo, C.; Shah, S.A.; Yu, L.-M.; Rutter, H.; Jones, L.; Williams, N.; Heneghan, C.; Price, J.; et al. Self-management support using a digital health system compared with usual care for chronic obstructive pulmonary disease: Randomized controlled trial. *J. Med. Internet Res.* **2017**, *19*, e144. [CrossRef]

39. Halpin, D.M.G.; Laing-Morton, T.; Spedding, S.; Levy, M.L.; Coyle, P.; Lewis, J.; Newbold, P.; Marno, P. A randomised controlled trial of the effect of automated interactive calling combined with a health risk forecast on frequency and severity of exacerbations of COPD assessed clinically and using EXACT PRO. *Prim. Care Respir. J.* **2011**, *20*, 324–331. [CrossRef]

40. Ho, T.-W.; Huang, C.-T.; Chiu, H.-C.; Ruan, S.-Y.; Tsai, Y.-J.; Yu, C.-J.; Lai, F. Effectiveness of telemonitoring in patients with chronic obstructive pulmonary disease in Taiwan—A randomized controlled trial. *Sci. Rep.* **2016**, *6*, 23797. [CrossRef]

41. Jakobsen, A.S.; Laursen, L.C.; Rydahl-Hansen, S.; Østergaard, B.; Gerds, T.A.; Emme, C.; Schou, L.H.; Phanareth, K. Home-Based telehealth hospitalization for exacerbation of chronic obstructive pulmonary disease: Findings from "The Virtual Hospital" trial. *Telemed. e-Health* **2015**, *21*, 364–373. [CrossRef]

42. Koff, P.; Jones, R.H.; Cashman, J.M.; Voelkel, N.F.; Vandivier, R. Proactive integrated care improves quality of life in patients with COPD. *Eur. Respir. J.* **2009**, *33*, 1031–1038. [CrossRef]

43. Lewis, K.E.; Annandale, J.A.; Warm, D.L.; Rees, S.E.; Hurlin, C.; Blyth, H.; Syed, Y.; Lewis, L. Does home telemonitoring after pulmonary rehabilitation reduce healthcare use in optimized COPD?? A pilot randomized trial. *COPD J. Chronic Obstr. Pulm. Dis.* **2011**, *7*, 44–50. [CrossRef] [PubMed]

44. Lilholt, P.H.; Udsen, F.W.; Ehlers, L.; Hejlesen, O.K. Telehealthcare for patients suffering from chronic obstructive pulmonary disease: Effects on health-related quality of life: Results from the Danish 'TeleCare North' cluster-randomised trial. *BMJ Open* **2017**, *7*, e014587. [CrossRef] [PubMed]

45. McDowell, J.E.; McClean, S.; FitzGibbon, F.; Tate, S. A randomised clinical trial of the effectiveness of home-based health care with telemonitoring in patients with COPD. *J. Telemed. Telecare* **2015**, *21*, 80–87. [CrossRef] [PubMed]

46. Nield, M.; Hoo, G.W.S. Real-time telehealth for COPD self-management using Skype™. *COPD J. Chronic Obstr. Pulm. Dis.* **2012**, *9*, 611–619. [CrossRef]

47. Pedone, C.; Chiurco, D.; Scarlata, S.; Incalzi, R.A. Efficacy of multiparametric telemonitoring on respiratory outcomes in elderly people with COPD: A randomized controlled trial. *BMC Heal Serv. Res.* **2013**, *13*, 82. [CrossRef]

48. Pinnock, H.; Hanley, J.; McCloughan, L.; Todd, A.; Krishan, A.; Lewis, S.; Stoddart, A.; Van Der Pol, M.; MacNee, W.; Sheikh, A.; et al. Effectiveness of telemonitoring integrated into existing clinical services on hospital admission for exacerbation of chronic obstructive pulmonary disease: Researcher blind, multicentre, randomised controlled trial. *BMJ* **2013**, *347*, f6070. [CrossRef]

49. Ringbæk, T.; Green, A.; Laursen, L.C.; Frausing, E.; Brøndum, E.; Ulrik, C.S. Effect of telehealth care on exacerbations and hospital admissions in patients with chronic obstructive pulmonary disease: A randomized clinical trial. *Int. J. Chronic Obstr. Pulm. Dis.* **2015**, *10*, 1801.

50. Ringbaek, T.J.; Lavesen, M.; Lange, P. Tablet computers to support outpatient pulmonary rehabilitation in patients with COPD. *Eur. Clin. Respir. J.* **2016**, *3*, 31016. [CrossRef]

51. Schou, L.; Østergaard, B.; Rydahl-Hansen, S.; Rasmussen, L.S.; Emme, C.; Jakobsen, A.S.; Phanareth, K. A randomised trial of telemedicine-based treatment versus conventional hospitalisation in patients with severe copd and exacerbation–effect on self-reported outcome. *J. Telemed. Telecare* **2013**, *19*, 160–165. [CrossRef]

52. Shany, T.; Hession, M.; Pryce, D.; Roberts, M.; Basilakis, J.; Redmond, S.; Lovell, N.; Schreier, G. A small-scale randomised controlled trial of home telemonitoring in patients with severe chronic obstructive pulmonary disease. *J. Telemed. Telecare* **2017**, *23*, 650–656. [CrossRef]

53. Stickland, M.K.; Jourdain, T.; Wong, E.Y.; Rodgers, W.M.; Jendzjowsky, N.G.; Macdonald, G.F. Using Telehealth technology to deliver pulmonary rehabilitation to patients with chronic obstructive pulmonary disease. *Can Respir. J.* **2011**, *18*, 216–220. [CrossRef]

54. Trappenburg, J.C.; Niesink, A.; Oene, G.H.D.W.-V.; Van Der Zeijden, H.; Van Snippenburg, R.; Peters, A.; Lammers, J.-W.J.; Schrijvers, A.J. Effects of telemonitoring in patients with chronic obstructive pulmonary disease. *Telemed. e-Health* **2008**, *14*, 138–146. [CrossRef]

55. Tsai, L.L.Y.; McNamara, R.J.; Moddel, C.; Alison, J.A.; McKenzie, D.K.; McKeough, Z.J. Home-based telerehabilitation via real-time videoconferencing improves endurance exercise capacity in patients with COPD: The randomized controlled TeleR Study. *Respirology* **2017**, *22*, 699–707. [CrossRef]

56. Vianello, A.; Fusello, M.; Gubian, L.; Rinaldo, C.; Dario, C.; Concas, A.; Saccavini, C.; Battistella, L.; Pellizzon, G.; Zanardi, G.; et al. Home telemonitoring for patients with acute exacerbation of chronic obstructive pulmonary disease: A randomized controlled trial. *BMC Pulm. Med.* **2016**, *16*. [CrossRef]

57. Al-Samarraie, H.; Ghazal, S.; Alzahrani, A.I.; Moody, L. Telemedicine in middle eastern countries: Progress, barriers, and policy recommendations. *Int. J. Med. Inform.* **2020**, *141*, 104232. [CrossRef]

58. Poenaru, C.; Poenaru, E.; Vinereanu, D. Current perception of Telemedicine in an EU Country. *Maedica* **2014**, *9*, 367–374.

59. Huang, B.; De Vore, D.; Chirinos, C.; Wolf, J.; Low, D.; Willard-Grace, R.; Tsao, S.; Garvey, C.; Donesky, D.; Su, G.; et al. Strategies for recruitment and retention of underrepresented populations with chronic obstructive pulmonary disease for a clinical trial. *BMC Med. Res. Methodol.* **2019**, *19*, 39. [CrossRef] [PubMed]

60. Irving, S.Y.; Curley, M.A. Challenges to conducting multicenter clinical research: Ten points to consider. *AACN Adv. Crit. Care* **2008**, *19*, 164–169. [CrossRef] [PubMed]

61. Barbosa, M.T.; Sousa, C.S.; Morais-Almeida, M.; Simões, M.J.; Mendes, P. Telemedicine in COPD: An overview by topics. *COPD J. Chronic Obstr. Pulm. Dis.* **2020**, *17*, 601–617. [CrossRef]

62. Fan, K.G.; Mandel, J.; Agnihotri, P.; Tai-Seale, M. Remote patient monitoring technologies for predicting chronic obstructive pulmonary disease exacerbations: Review and comparison. *JMIR mHealth uHealth* **2020**, *8*, e16147. [CrossRef] [PubMed]

63. Watson, N.L.; Heffner, J.L.; Mull, K.E.; McClure, J.B.; Bricker, J.B. Comparing treatment acceptability and 12-month cessation rates in response to web-based smoking interventions among smokers who do and do not screen positive for affective disorders: Secondary analysis. *J. Med. Internet Res.* **2019**, *21*, e13500. [CrossRef] [PubMed]

64. Naughton, F.; Jamison, J.; Boase, S.; Sloan, M.; Gilbert, H.; Prevost, A.T.; Mason, D.; Smith, S.; Brimicombe, J.; Evans, R.; et al. Randomized controlled trial to assess the short-term effectiveness of tailored web-and text-based facilitation of smoking cessation in primary care (i Quit in Practice). *Addiction* **2014**, *109*, 1184–1193. [CrossRef]

65. Early, F.; Wellwood, I.; Kuhn, I.; Deaton, C.; Fuld, J. Interventions to increase referral and uptake to pulmonary rehabilitation in people with COPD: A systematic review. *Int. J. Chronic Obstr. Pulm. Dis.* **2018**, *13*, 3571–3586. [CrossRef]

66. Dixon, L.J.; Linardon, J. A systematic review and meta-analysis of dropout rates from dialectical behaviour therapy in randomized controlled trials. *Cogn. Behav. Ther.* **2020**, *49*, 181–196. [CrossRef] [PubMed]

67. Alwashmi, M.F.; Fitzpatrick, B.; Farrell, J.; Gamble, J.-M.; Davis, E.; Van Nguyen, H.; Farrell, G.; Hawboldt, J. Perceptions of patients regarding mobile health interventions for the management of chronic obstructive pulmonary disease: Mixed methods study. *JMIR mHealth uHealth* **2020**, *8*, e17409. [CrossRef]

68. Alrajeh, A.M.; Aldabayan, Y.S.; Aldhahir, A.; Pickett, E.; Quaderi, S.A.; Alqahtani, J.S.; Lipman, M.; Hurst, J.R. Global use, utility, and methods of tele-health in COPD: A health care provider survey. *Int. J. Chronic Obstr. Pulm. Dis.* **2019**, *14*, 1713–1719. [CrossRef]

69. Al Rajeh, A.; Steiner, M.C.; Aldabayan, Y.; Aldhahir, A.; Pickett, E.; Quaderi, S.; Hurst, J.R. Use, utility and methods of telehealth for patients with COPD in England and Wales: A healthcare provider survey. *BMJ Open Respir. Res.* **2019**, *6*. [CrossRef]

70. Gaveikaite, V.; Grundstrom, C.; Winter, S.; Chouvarda, I.; Maglaveras, N.; Priori, R. A systematic map and in-depth review of European telehealth interventions efficacy for chronic obstructive pulmonary disease. *Respir. Med.* **2019**, *158*, 78–88. [CrossRef]

71. Wedzicha, J.A.; Miravitlles, M.; Hurst, J.R.; Calverley, P.M.; Albert, R.K.; Anzueto, A.; Criner, G.J.; Papi, A.; Rabe, K.F.; Rigau, D.; et al. Management of COPD exacerbations: A European respiratory society/American thoracic society guideline. *Eur. Respir. J.* **2017**, *49*, 1600791. [CrossRef] [PubMed]

72. Hopkinson, N.S.; Molyneux, A.; Pink, J.; Harrisingh, M.C. Chronic obstructive pulmonary disease: Diagnosis and management: Summary of updated NICE guidance. *BMJ* **2019**, *366*, l4486. [CrossRef] [PubMed]

73. National Institute for Health and Care Excellence (NICE) Guideline. *Chronic Obstructive Pulmonary Disease in Over 16s: Diagnosis and Management*. 2018. Available online: nice.org.uk/guidance/ng115 (accessed on 19 October 2020).

74. Dunphy, E.; Gardner, E.C. Telerehabilitation to Address the rehabilitation gap in anterior cruciate ligament care: Survey of patients. *JMIR Form. Res.* **2020**, *4*, e19296. [CrossRef] [PubMed]

75. Loeckx, M.; A Rabinovich, R.; Demeyer, H.; Louvaris, Z.; Tanner, R.; Rubio, N.; Frei, A.; De Jong, C.; Gimeno-Santos, E.; Rodrigues, F.M.; et al. Smartphone-based physical activity telecoaching in chronic obstructive pulmonary disease: Mixed-methods study on patient experiences and lessons for implementation. *JMIR mHealth uHealth* **2018**, *6*, e200. [CrossRef] [PubMed]

International Journal of
*Environmental Research
and Public Health*

Review

Nurse-Coordinated Blood Pressure Telemonitoring for Urban Hypertensive Patients: A Systematic Review and Meta-Analysis

Woo-Seok Choi [1,2,*], Nam-Suk Kim [3], Ah-Young Kim [1] and Hyung-Soo Woo [1]

[1] Moon Soul Graduate School of Future Strategy, Korea Advanced Institute of Science and Technology (KAIST), Daejeon 34141, Korea; journalist@kaist.ac.kr (A.-Y.K.); h.woo@kaist.ac.kr (H.-S.W.)
[2] Keyu Internal Medicine Clinic, Daejeon 35250, Korea
[3] Public Health and Welfare Bureau, Daejeon City Hall, Daejeon 35242, Korea; heewasu@korea.kr
* Correspondence: keyuim@kaist.ac.kr; Tel.: +82-42-483-7554; Fax: +82-42-485-7554

Abstract: Coronavirus disease 2019 (COVID-19) has put hypertensive patients in densely populated cities at increased risk. Nurse-coordinated home blood pressure telemonitoring (NC-HBPT) may help address this. We screened studies published in English on three databases, from their inception to 30 November 2020. The effects of NC-HBPT were compared with in-person treatment. Outcomes included changes in blood pressure (BP) following the intervention and rate of BP target achievements before and during COVID-19. Of the 1916 articles identified, 27 comparisons were included in this review. In the intervention group, reductions of 5.731 mmHg (95% confidence interval: 4.120–7.341; $p < 0.001$) in systolic blood pressure (SBP) and 2.342 mmHg (1.482–3.202; $p < 0.001$) in diastolic blood pressure (DBP) were identified. The rate of target BP achievement was significant in the intervention group (risk ratio, RR = 1.261, 1.154–1.378; $p < 0.001$). The effects of intervention over time showed an SBP reduction of 3.000 mmHg (-5.999–11.999) before 2000 and 8.755 mmHg (5.177–12.334) in 2020. DBP reduced by 2.000 mmHg (-2.724–6.724) before 2000 and by 3.529 mmHg (1.221–5.838) in 2020. Analysis of the target BP ratio before 2010 (RR = 1.101, 1.013–1.198) and in 2020 (RR = 1.906, 1.462–2.487) suggested improved BP control during the pandemic. NC-HBPT more significantly improves office blood pressure than UC among urban hypertensive patients.

Keywords: blood pressure; hypertension; COVID-19; nurse; coordination; telemonitoring; urban

Citation: Choi, W.-S.; Kim, N.-S.; Kim, A.-Y.; Woo, H.-S. Nurse-Coordinated Blood Pressure Telemonitoring for Urban Hypertensive Patients: A Systematic Review and Meta-Analysis. *Int. J. Environ. Res. Public Health* **2021**, *18*, 6892. https://doi.org/10.3390/ijerph18136892

Academic Editors: Irene Torres-Sánchez and Marie Carmen Valenza

Received: 29 May 2021
Accepted: 24 June 2021
Published: 27 June 2021

Publisher's Note: MDPI stays neutral with regard to jurisdictional claims in published maps and institutional affiliations.

1. Introduction

Essential hypertension is the primary modifiable risk factor for cardiovascular disease, which is a leading cause of death according to the World Health Organization (WHO) [1]. Approximately 45% of American adults had elevated blood pressure in the 2017–2018 time period [2,3], but only 34% were being managed to bring the blood pressure down to the recommended treatment level [4]. It has been reported that managing blood pressure to remain within the recommended level reduced the incidence of stroke by 35–40%, heart attack by 20–25%, and heart failure by more than 50% [5]. Moreover, the American Heart Association/American College of Cardiology in 2017 and the European Society of Hypertension/European Society of Cardiology in 2018 issued guidelines recommending that the target blood pressure in hypertensive patients should be regulated more strictly than recommended by previous guidelines [6,7]. Additionally, recent studies have reported that essential hypertension is one of the most common comorbidities that cause lung damage and mortality due to the coronavirus disease 2019 (COVID-19) [8,9]. Thus, strict blood pressure management is crucial to reduce the incidence of cardiovascular disease (CVD) in patients with risk factors.

Despite preventive measures such as hand washing, self-isolation, mask wearing, and social distancing, COVID-19, which first emerged in Wuhan, China, spread rapidly in high-income countries such as the United States, Spain, Germany, Italy, and Korea in the early days of the pandemic [10]. Although the number of newly confirmed patients

183

has been decreasing with the administration of the vaccine since December 2020 [11], the COVID-19 pandemic is still ongoing. According to the WHO, as of April 2021, there were approximately 1.6 billion cases of severe acute respiratory syndrome coronavirus 2 (SARS-CoV-2)—the virus that causes COVID-19—infections worldwide, including 3 million deaths [12]. However, with the exception of some countries, the global vaccination rate is still too low (below 40%) [12]. In particular, in most countries in Africa, the Middle East, and Southeast Asia, as of the end of April 2021, only approximately 10% of the population had received at least one dose of a vaccine [12]. The drastic increase in demand for medical care during the pandemic has exposed the limitations of traditional medical systems and is adversely affecting the existing medical systems for non-communicable diseases (NCDs) such as essential hypertension [13].

During infectious disease pandemics, a high frequency of direct contact between individuals in densely populated cities can increase the infection risk even in well-established in-person medical infrastructures [14]. Thus, the need for treatment through non-face-to-face interaction between doctors and patients is being emphasized more than ever before in cities [15]. Home blood pressure telemonitoring (HBPT) is widely used as an alternative measure in the management of hypertensive patients and is known to be effective in overcoming clinical inertia that may occur in face-to-face care settings [3,16–18]. In addition, remote monitoring of home blood pressure is helpful in managing patients by finding white-coat effects and masked hypertension that may be overlooked by doctors in medical institutions [19,20]. Studies on the effectiveness of nurse-coordinated HBPT (NC-HBPT) to prevent CVD attacks in hypertensive patients existed prior to the COVID-19 pandemic [21–25]. However, evidence on whether or not HBPT statistically enhances blood pressure control by improving communication between patients and healthcare providers has been mixed. Additionally, there is a paucity of literature indicating a solid basis for the effectiveness of nurse-led remote monitoring in the avoidance of CVD.

A coordinator plays a vital role in remote monitoring in chronic diseases [21,26], and nurses are core coordinators in the telemonitoring teams of medical institutions, working either independently or as members of a team [27]. Nurses can provide more regular follow-up, high-quality care, favorable health outcomes, and higher patient satisfaction, all equivalent to that achieved by physicians [28]. A 2005 review recommended nurse-led monitoring as an additional measure to support face-to-face therapy [29]. The literature classifies ancillary interventions to help control blood pressure into one of six categories: care led by health professionals, patient monitoring, education of medical professionals, education of patients, organizational interventions, and appointment reminder systems. Particularly, for the step-by-step care of hypertension that is not controlled by drug treatment, nurse-led regular monitoring was emphasized. However, there is uncertainty about the effectiveness of nurse-led remote monitoring in cities with established medical infrastructure and abundant medical resources. Although many studies have reported the nurses' role in HBPT intervention [3,20,28–31], no study has systematically reviewed the effectiveness of NC-HBPT in urban areas and, particularly, none have presented quantitative results according to the temporal progress. Therefore, comprehensive comparative analysis of outcome data of NC-HBPT performed in urban areas before and during the COVID-19 pandemic is important to overcome the challenges of NCD and communicable disease (CD) management and to determine the future directions for remote monitoring policies.

This study hypothesizes that NC-HBPT in urban adult hypertensive patients is not more effective than usual care (UC) in preventing CVD and that its effect over time would be identical to that of UC. To derive robust results, we conducted a systematic review and meta-analysis of randomized controlled trials (RCTs), which are placed at the top in the hierarchy of evidence-based research. Previous meta-analyses investigated the effect of HBPT in hypertensive patients [32,33], but none have integrated the results of NC-HBPT for efficient implementation in cities yet. Thus, this study aimed to examine and compare the following: mean changes in systolic blood pressure (SBP) and diastolic blood pressure

(DBP); rates of achieving the target blood pressure after NC-HBPT; and effects of NC-HBPT over time in an urban area.

2. Materials and Methods

2.1. Literature Search and Identification of Eligible Studies

We conducted a review in accordance with the guidelines summarized in the Cochrane Handbook for Systematic Reviews of Interventions [34]. A protocol for this study has been published in the PROSPERO [CRD42020222789], which is an international prospective register of systematic reviews operated by the Center for Reviews and Dissemination at the University of York [35]. Studies evaluating HBPT that were published between the date of inception of the utilized databases and 30 November 2020, were identified. The electronic databases we used included PubMed, EBSCOhost, and the Cochrane library (CENTRAL), and the search was limited to studies published in peer-reviewed journals in English. The related search keywords included "urban", "hypertension", "remote monitoring", "telemonitoring", "telemedicine", and "randomized controlled trials." The adopted search formula was constructed by combining free terms of relevant keywords and Medical Subject Headings (MeSH) terms via truncation, Boolean operators, and phrasing. We first searched in PubMed using this structured formula and sequentially performed additional searches according to the syntax of each database (Appendix A).

All retrieved data were exported to EndNote X8.2 (Thomson Reuters, Philadelphia, PA, USA). Titles and abstracts of each study were screened, and the main text was reviewed as required. We searched for all meta-analyses conducted previously on the topic and reviewed all primary studies and relevant references in those meta-analyses. To find grey literature, we referred to related websites in the United States and Europe (e.g., OpenGrey: http://opengrey.eu/ (accessed on 30 November 2020); Grey Literature Report: https://www.greylit.org/ (accessed on 10 December 2020). To ensure objectivity and transparency of the eligibility assessment, two of the authors (WSC and AYK) independently conducted the literature search and arrived at a mutually-agreed selection of studies.

2.2. Inclusion and Exclusion Criteria

An intervention group was defined as one in which patients measured their blood pressure on their own at home, reported the measurements to their doctors, and regularly visited the medical institute for follow-up, and in which a coordinator—including a nurse with or without other health care professionals—was involved in the process. As for nurses, only situations where the registered nurse monitored the patient's home blood pressure using an automatic sphygmomanometer without face-to-face contact were included. All the following actions were also included: consultation with the patient using a telephone line, mobile phone, computer, or letter; education on the disease or intervention process; execution of accompanying interventions such as a text message reminder service; and sending information regarding the patient's health status.

A control group was defined as one in which patients received routine in-person examinations at the doctor's office. Since NC-HBPT is not yet a standardized treatment itself, none of the included studies provided an equivalent of NC-HBPT, and so no active control groups were included in our analyses. Their ethnicity, level of income, and severity of hypertension were not considered separately.

We included studies (1) involving patients with essential hypertension (SBP $\geq$ 130 mmHg, DBP $\geq$ 80 mmHg) regardless of hypertension onset or history of pharmacotherapy; (2) including patients who received treatment at an urban medical institution; (3) having an intervention that was provided for $\geq$2 months; (4) utilizing a RCT design; and (5) involving adults aged $\geq$17 years. The diagnostic criteria for hypertension in Europe are SBP $\geq$ 140 or DBP $\geq$ 90, which is higher than that in the US [7]. Thus, the diagnostic guideline for hypertension published by the 2018 ACC/AHA associations [6], which has a wider diagnostic range, was adopted in searching databases.

However, we excluded studies (1) conducted in several regions with unclear study locations; (2) having a research location that was a mix of urban and rural areas; (3) involving patients with acute CVD or stroke with a drastic change in blood pressure; (4) involving women in the peripartum period; (5) having an intervention provided to patients in nursing homes or care facilities; (6) having an intervention provided as part of another intervention program for a different disease; (7) using a different intervention in the UC group; and (8) utilizing a cluster- or cross-over RCT design. In this study, an "urban area" was defined according to the administrative functions and population size.

2.3. Study Selection

A further search was conducted by reviewing the full-text manuscripts of studies identified during the first round of screening and their reference lists. All articles related to "coordinator" and "nurse" were additionally identified. Among the blinded RCTs that regularly verified the effects of these two groups, those that reported changes in blood pressure measurements before and after interventions were selected for data synthesis.

Two of the authors (NSK and HW) independently classified and excluded duplicate studies and those that did not meet the inclusion criteria. The main body of potentially valid studies was reviewed, and any disagreement between the two authors was resolved by a senior author (WSC).

2.4. Outcomes

The primary outcome was the weighted mean difference (WMD) of office SBP and DBP between the baseline and follow-up in the NC-HBPT and UC groups. The secondary outcome was the rate of target blood pressure achievement.

2.5. Data Extraction and Coding

Two researchers (NSK and AYK) independently extracted data from the selected primary studies. The data were coded in an electronic sheet using Comprehensive Meta-Analysis Software Version 2.2.064 (CMA, Biostat, Englewood, NJ, USA). Patients' age and sex, duration of remote monitoring (months), accompanying interventions, intervention pathway, the coordinator's profession, and outcome data were extracted. When a single article had different follow-up periods for the intervention or different sample sizes, or compared two or more interventions [21,22,26,27,36–40], the results of each intervention were classified as independent comparative studies. For missing or inaccurate data in the primary research materials [27], we referred to journal websites or public trial registries (e.g., the US National Institutes of Health ongoing trials register) or directly contacted the authors. In cases where the standard deviation (SD) values for the mean changes were not presented [38], the values were imputed by calculating the mean of individual studies included in the review, and a sensitivity test was conducted to check for bias [40,41]. For the one study that did not provide baseline data [36], the data from the first assessment and the last follow-up were compared. In case of any disagreement between the researchers, a third researcher (WSC) adjudicated.

2.6. Quality Assessment

Two researchers (NSK and HW) independently assessed the risk of bias using Review Manager (RevMan, Version 5.4, Copenhagen: The Nordic Cochrane Center) by the Cochrane Collaboration [42], evaluating the RCTs in terms of selection, performance, detection, attrition, reporting biases, and other bias domains. A senior researcher (WCS) resolved any disagreement between the two researchers.

2.7. Quantitative Data Synthesis

Two researchers (WSC and NSK) analyzed the coded data. A random effects model (REM) was adopted because most of the primary studies were performed in different research institutions or by different researchers.

For computational options for data synthesis, WMD was set as a continuous variable and relative risk (RR) as a dichotomous variable; the changes in the mean office SBP and DBP of the NC-HBPT and UC groups before and after the intervention were extracted first. Hedges' g (g) was additionally calculated to determine the appropriateness of the effect size. If the *g*-value converted from the WMD was at least 0.5, the effect size was deemed appropriate for analysis [43]. RR was used to determine the rate of achieving the target blood pressure and was calculated using the number of samples that reached the target blood pressure during each follow-up period. All results are reported with 95% confidence intervals (CIs), and CMA software was used for statistical analysis.

Overall statistics, including weighted values, were analyzed by combining independent data. A χ^2 test was used to assess differences between studies, and a *p*-value < 0.10 was deemed statistically significant. Clinical heterogeneity was assessed using the Q statistic, tau-square (τ^2), and I^2 statistic. Q statistic and τ^2 interpreted the numerical values, and I^2 was considered significant at $\geq$50% [43,44]. To determine the bias in each study, examine its effect on the distribution of summary effect size, and detect outliers, a sensitivity analysis using the "one study removed" method was conducted [45]. In addition, a cumulative meta-analysis was performed to identify the chronological patterns of the effect size of each study [41] (Appendix B).

After determining the effect size of each of the primary studies, their temporal changes were analyzed, and the results before and during the COVID-19 pandemic were compared.

3. Results

3.1. Study Characteristics

In total, 1916 potentially relevant articles were initially identified from databases, reference lists of each retrieved paper, and other electronic sources (Figure 1). Six datasets were additionally acquired directly from the authors, and 423 duplicates were excluded. The titles and abstracts of 1499 references were screened, and 1101 references were excluded because they were deemed inappropriate for analysis. The full copies of the remaining 398 potentially eligible articles were scrutinized. Of these, 382 studies that did not meet the aforementioned inclusion criteria—that is, studies with participants less than 17 years of age (n = 19), no obtainable data (n = 183), no nurse-coordinated intervention (n = 86), cluster or cross-over RCTs (n = 12), patients with life-threatening CVDs or stroke (n = 25), studies not performed in an urban setting (n = 43), and studies that were part of a research program for another disease (n = 14)—were excluded. A total of 27 individual comparisons met the inclusion criteria and were selected as the final material for data synthesis.

In the meta-analysis, the mean length of NC-HBPT was 7.26 months, and the mean age of participants was 61.35 years in the intervention group and 61.62 years in the control group. Of the 27 comparisons, 20 were nurse alone-led cases [21–23,26,27,39–41,46–49], and seven cases additionally involved experts from other fields [24,25,27,36,38]. The characteristics of these studies are summarized in Table 1.

The medium of administering the intervention was mobile phones for 2 cases [39], mobile-web for 5 cases [38,40,47], web-based for 5 cases [21,25,37], telephone for 14 cases [22–24,26,27,36,46,49], and telephone-linked computers for 1 case [48]. A city was considered metropolitan if its population was at least one million; 11 cases were in metropolitan cities [21–23,36,38,39,47,48], and 16 were in smaller cities.

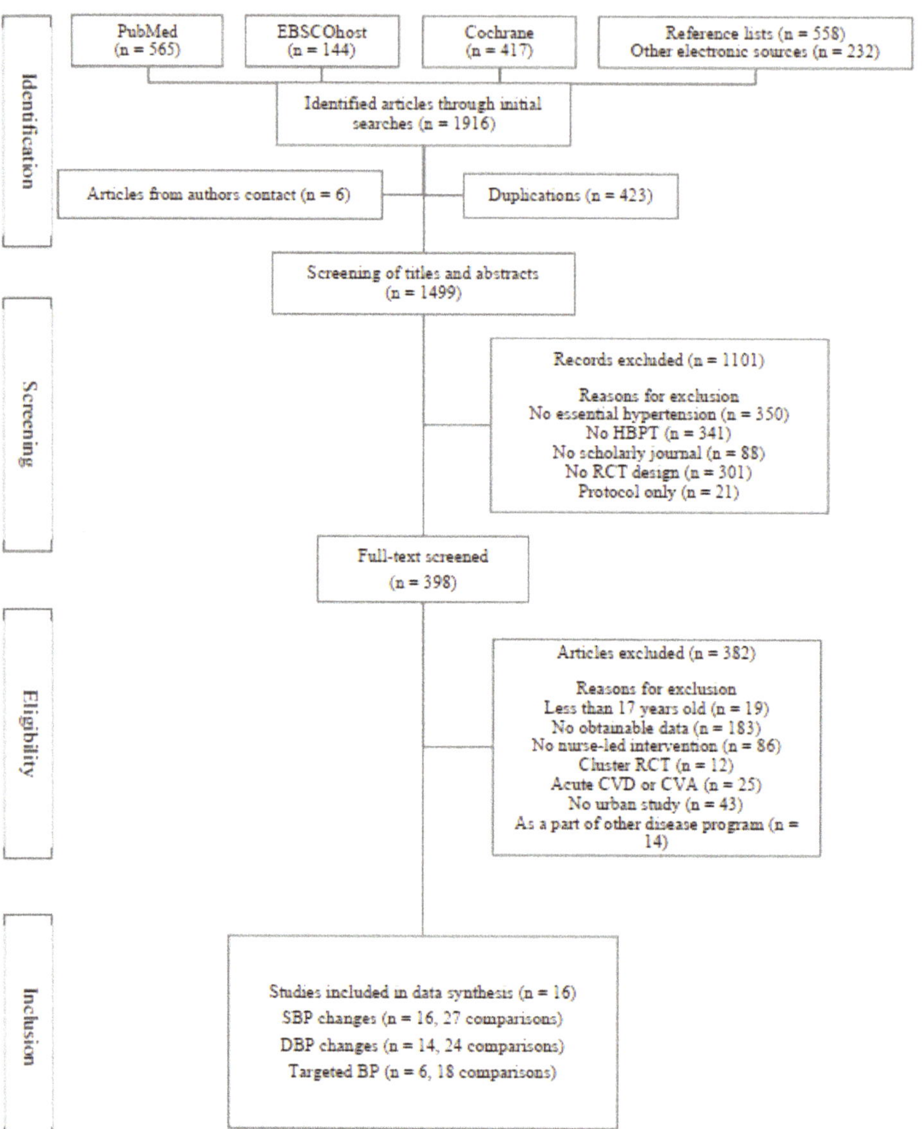

Figure 1. PRISMA flow diagram of the selection of studies included in the systematic review. Note. HBPT, home blood pressure telemonitoring; RCT, randomized controlled trial; CVD, cardiovascular disease; CVA, cerebrovascular accident; SBP, systolic blood pressure; DBP, diastolic blood pressure; BP, blood pressure.

Table 1. Characteristics of included individual studies.

Study	Number of Subjects (UC vs. HBPT Group)	Inclusion Criteria	The Profession of Coordinator	Name of City (Country)	Size of Urban Population (Over 1,000,000)	Description of Intervention Pathway	Additional Support	Main Outcome Measures
Artinian 2007A [26]	193/194	African-American hypertensive patients	Trained registered nurse	Detroit (USA)	No (916,952 in 2007)	Telephonic transmission	Telecounseling call and patient education	Changes in office BP
Artinian 2007B [26]	157/164	African-American hypertensive patients	Trained registered nurse	Detroit (USA)	No (916,952 in 2007)	Telephonic transmission	Telecounseling call and patient education	Changes in office BP
Artinian 2007C [26]	169/167	African-American hypertensive patients	Trained registered nurse	Detroit (USA)	No (916,952 in 2007)	Telephonic transmission	Telecounseling call and patient education	Office BP changes
Bosworth 2007A [27]	150/150	Hypertensive patients who are using a BP-lowering medication with poor (inadequate) BP control	Intervention nurses	Durham (USA)	No (217,847 in 2007)	Home BP monitoring + Behavioral intervention	Tailored behavioral intervention	Improved rates of BP control
Bosworth 2007B [27]	150/150	Hypertensive patients who are using a BP-lowering medication with poor (inadequate) BP control	Intervention nurses supervised with a physician	Durham (USA)	No (217,847 in 2007)	Home BP monitoring + Medication management	Medication management with a validated decision support system (DSS)	Improved rates of BP control
Bosworth 2007C [27]	150/150	Hypertensive patients who are using a BP-lowering medication with poor (inadequate) BP control	Intervention nurses	Durham (USA)	No (217,847 in 2007)	Home BP monitoring + Combined intervention	Combined behavioral and medication management	Improved rates of BP control
Bosworth 2011 [46]	137/127	Treated hypertensive patients	Registered nurse	Durham (USA)	No (228,354 in 2010)	Telephonic transmission	Behavioral management	Change in BP control, SBP, and DBP

Table 1. *Cont.*

Study	Number of Subjects (UC vs. HBPT Group)	Inclusion Criteria	The Profession of Coordinator	Name of City (Country)	Size of Urban Population (Over 1,000,000)	Description of Intervention Pathway	Additional Support	Main Outcome Measures
Cicolini 2013A [21]	98/100	Treated and untreated hypertensive patients	Registered nurse	Chieti (Italy)	No (51,484 in 2011)	Web-based	Email reminder and phone call	1. Changes in BP2. BMI, alcohol consumption, cigarette smoking, adherence to therapy
Cicolini 2013B [21]	98/100	Treated and untreated hypertensive patients	Registered nurse	Chieti (Italy)	No (51,484 in 2011)	Web-based	Email reminder and phone call	1. Changes in BP2. BMI, alcohol consumption, cigarette smoking, adherence to therapy
Hebert 2011A [23]	83/85	Uncontrolled hypertensive patients	Registered nurse	New York (USA)	Yes (8,174,959 in 2010)	Telephonic transmission	Information on the use of home BP monitor	BP reduction
Hill 1999 [24]	77/78	Black or African-American hypertensive young male resident within the Johns-Hopkins Hospital catchment area	Nurse-community health worker team (registered nurse and health worker team)	Baltimore (USA)	No (763.014 in 1990)	Telephonic transmission	Individualized counseling, monthly telephone call, and a home visit (educational-behavioral intervention)	Changes in office BP
Kerry 2012A [36]	169/168	Hypertensive patients with history of stroke or transient ischemic attack	Nurse	London (UK)	Yes (8,135,667 in 2011)	Telephonic transmission	Nurse-led telephone support	Reduction of SBP
Kerry 2012B [36]	169/168	Hypertensive patients with history of stroke or transient ischemic attack	Nurse	London (UK)	Yes (8,135,667 in 2011)	Telephonic transmission	Nurse-led telephone support	Reduction of SBP

Table 1. *Cont.*

Study	Number of Subjects (UC vs. HBPT Group)	Inclusion Criteria	The Profession of Coordinator	Name of City (Country)	Size of Urban Population (Over 1,000,000)	Description of Intervention Pathway	Additional Support	Main Outcome Measures
Kim KB 2014A [37]	192/191	Uncontrolled Korean-American hypertensive seniors	Bilingual RNs and nutritionist	Ellicott City (USA)	No (68,507 in 2014)	Web-based	2 h weekly education and training for 6 weeks and monthly telephone counseling	Changes in SBP and DBP
Kim KB 2014B [37]	187/185	Uncontrolled Korean-American hypertensive seniors	Bilingual RNs and nutritionist	Ellicott City (USA)	No (68,507 in 2014)	Web-based	2 h weekly education and training for 6 weeks and monthly telephone counseling	Changes in SBP and DBP
McMahon 2005 [25]	35/37	Poorly controlled diabetics and hypertensive patients	Advanced practice nurse and certified diabetes educator	Boston (USA)	No (559,034 in 2005)	Web-based	Telephone to encourage website usage	Changes in HbA_{1c}, BP, lipid profiles
Mohsen 2019A [22]	50/50	Hypertensive patients who are on antihypertensive medication	Staff nurse	Shibin El Kom (Egypt)	No (249,611 in 2018)	Telenursing	Follow-up phone calls	Reducing arterial blood pressure and patients' anthropometric measurement
Mohsen 2019B [22]	50/50	Hypertensive patients who are on antihypertensive medication	Staff nurse	Shibin El Kom (Egypt)	No (249,611 in 2018)	Telenursing	Follow-up phone calls	Reducing arterial blood pressure and patients' anthropometric measurement

Table 1. *Cont.*

Study	Number of Subjects (UC vs. HBPT Group)	Inclusion Criteria	The Profession of Coordinator	Name of City (Country)	Size of Urban Population (Over 1,000,000)	Description of Intervention Pathway	Additional Support	Main Outcome Measures
Pan 2018A [38]	55/52	Hypertensive patients with uncontrolled BP	GP, a hypertensionspecialist, a general nurse, an information manager	Beijing (China)	Yes (19,612,368 in 2010)	Smartphone application	Follow-up phone calls	The reduction in systolic blood pressure
Pan 2018B [38]	55/52	Hypertensive patients with uncontrolled BP	GP, a hypertensionspecialist, a general nurse, an information manager	Beijing (China)	Yes (19,612,368 in 2010)	Smartphone application	Follow-up phone calls	The reduction in systolic blood pressure
Park MJ 2009 [47]	21/28	Obese hypertensive patients	Registered nurse	Seoul (S. Korea)	Yes (10,208,302 in 2009)	Mobile or internet transmission	Short message service by cellular phone and internet recommendation	Changes in BP, body weight, waist circumference, and serum lipid profile
Rahmani Pour 2019A [39]	21/21	Hypertensive patients, use of antihypertensive medications	Instructors of a faculty of nursing and midwifery and several cardiologists	Tehran (Iran)	Yes (8,693,706 in 2016)	Short message service	Communicate with the first author	Improved treatment adherence, no significant differences among the groups with respect to the baseline SBP and DBP
Rahmani Pour 2019A [39]	21/21	Hypertensive patients, use of antihypertensive medications	Instructors of a faculty of nursing and midwifery and several cardiologists	Tehran (Iran)	Yes (8,693,706 in 2016)	Non-ISMS	Follow-up	No significant differences among the groups with respect to the baseline SBP and DBP
Shea 2006 [48]	347/333	Diabetic hypertensive patients	Nurse case manager	New York (USA)	Yes (8,143,197 in 2005)	Telephone-linked web-enabled computer system	Web-based messaging	Changes in HbA_{1c}, BP, and cholesterol level

Table 1. *Cont.*

Study	Number of Subjects (UC vs. HBPT Group)	Inclusion Criteria	The Profession of Coordinator	Name of City (Country)	Size of Urban Population (Over 1,000,000)	Description of Intervention Pathway	Additional Support	Main Outcome Measures
Wakefield 2011 [49]	97/83	Type 2 diabetics and hypertensive patients	Registered nurse	Iowa City (USA)	No (68,036 in 2010)	Telephonic transmission	Telephonic transmission and nurse care management	Changes in HbA_{1c} and SBP
Zha 2020A [40]	13/12	Underserved and vulnerable urban population on HTN medication with diagnosed uncontrolled hypertension	Two nurses in the community health center	Newark (USA)	No (281,764 in 2016)	Mobile Health (Smart phone or tablet)	Three training sessions	No significant change in systolic BP, the potential to facilitate hypertension management
Zha 2020B [40]	13/12	Underserved and vulnerable urban population on HTN medication with diagnosed uncontrolled hypertension	Two nurses in the community health center	Newark (USA)	No (281,764 in 2016)	Mobile Health (Smart phone or tablet)	Three training sessions	No significant change in systolic BP, the potential to facilitate hypertension management

Notes. UC, usual care; HBPT, home blood pressure telemonitoring; GP, general practitioner; BP, blood pressure; SBP, systolic blood pressure; DBP, diastolic blood pressure; HTN, hypertension; Non-ISMS, non-interactive short message service; BMI, body mass index; HbA1c, glycated hemoglobin.

3.2. Assessment of Risk of Bias

The selection and performance processes for most trials were appropriate. In the case of attrition or reporting bias domain, there were studies with an unclear or high risk of bias. A primary study at a low risk of bias in at least four domains was deemed to be of high quality [50], and 23 comparisons were identified as having a high risk of bias in fewer than four domains, suggesting that the overall quality of the materials was relatively high.

Publication bias was assessed using the trim-and-fill method [51]. The point estimates of SBP based on a funnel plot were as follows: WMD = 5.327 mmHg and g = 0.723 (0.445–1.002, p < 0.001); there were no trimmed studies (Figure 2). For DBP, g = 0.362 (0.222–0.503, p < 0.001), which represented a meaningful effect size (Figure 3). The funnel plots of both SBP and DBP showed good visual symmetry, and there were no imputed studies for SBP and DBP. Thus, it was concluded that the potential publication biases did not affect the primary outcomes of the materials included in this study.

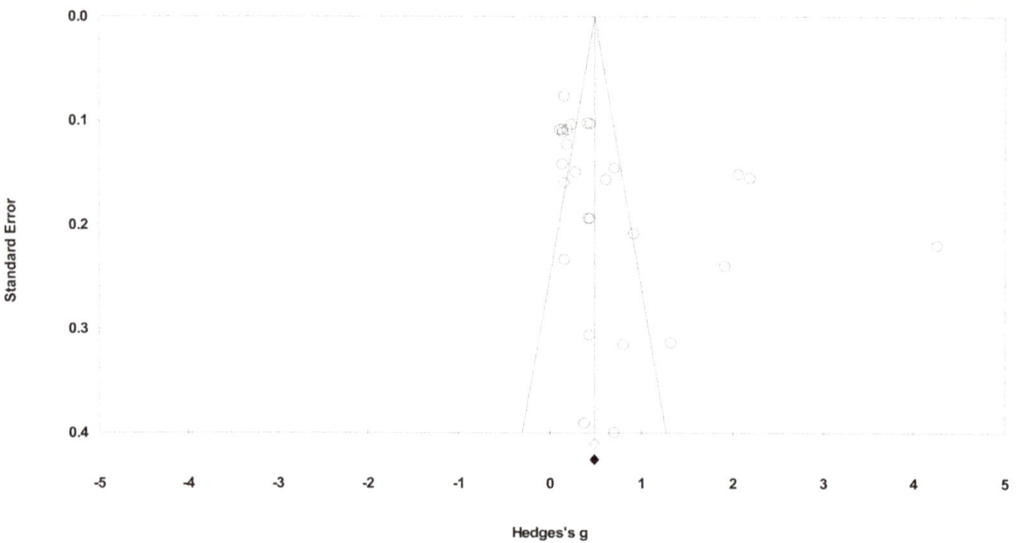

Figure 2. Funnel plot of systolic blood pressure by Hedges' g (plot observed and imputed), random effects model. Notes. Summary effect size (◇); Summary effect size of imputed studies (♦); Individual studies (○).

Publication bias was also assessed for the rate of reaching the target office blood pressure as a secondary outcome. The funnel plot showed good visual symmetry (Figure 4). One study was imputed, but the difference in point estimates was not significant (observed RR, 1.261, vs. adjusted RR, 1.240). Thus, the publication bias by the potentially unpublished study did not affect the RR effect size.

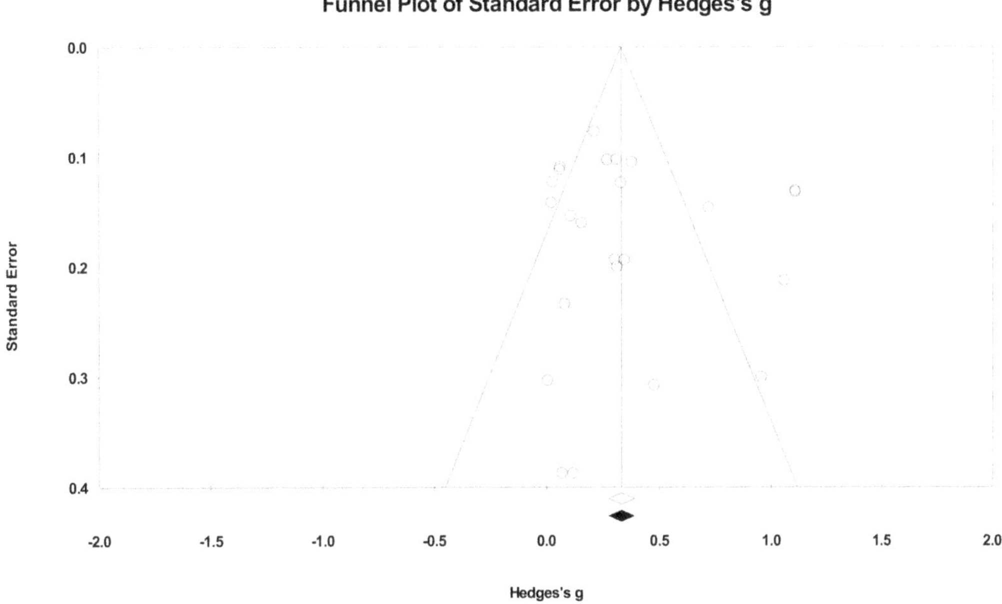

Figure 3. Funnel plot of diastolic blood pressure by Hedges' g (plot observed and imputed), random effects model. Notes. Summary effect size (◇); Summary effect size of imputed studies (♦); Individual studies (○).

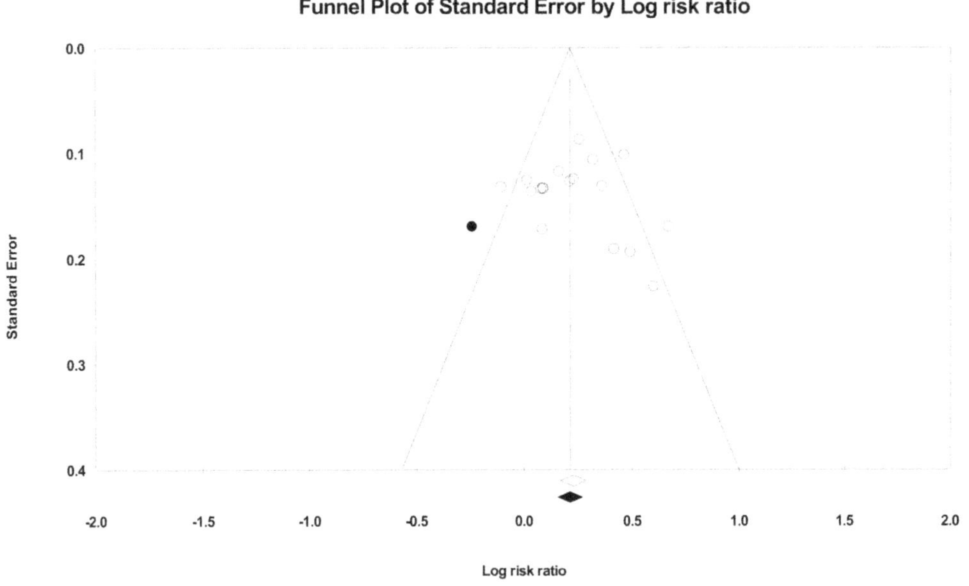

Figure 4. Funnel plot of rate of target blood pressure by log risk ratio (plot observed and imputed), random effects model. Notes. Summary effect size (◇); Imputed study (●); Summary effect size of imputed studies (♦); Individual studies (○).

3.3. Primary Outcomes

3.3.1. SBP Changes by Nurse-Coordinated HBPT

Data were pooled from the 27 comparisons (16 studies) that included 2860 patients in the NC-HBPT group and 2918 patients in the control group, comprising a total of 5778 patients in both groups [21–27,36–40,46–49]. A significant reduction in blood pressure was observed in the nurse-led intervention group compared with the UC group (5.731 mmHg; 4.120–7.341; $p < 0.001$; Figure 5), and heterogeneity was significant among the studies ($I^2 = 71.717\%$; $p < 0.001$). In the sensitivity test [45], the individual studies did not affect the summary point estimates (WMD: 3.822–7.578 mmHg).

Author (year) (ref.)	Outcome	Difference in means	Lower limit	Upper limit	p-Value
Artinian,2007A	SBP	8.100	4.190	12.010	0.000
Artinian,2007B	SBP	2.900	-1.547	7.347	0.201
Artinian,2007C	SBP	4.000	-0.413	8.413	0.076
Bosworth,2007A	SBP	3.100	-1.074	7.274	0.146
Bosworth,2007B	SBP	6.400	2.009	10.791	0.004
Bosworth,2007C	SBP	3.300	-1.125	7.725	0.144
Bosworth,2011	SBP	3.100	-1.158	7.358	0.154
Cicolini,2013A	SBP	1.000	-1.827	3.827	0.488
Cicolini,2013B	SBP	5.000	2.180	7.820	0.001
Hebert,2011A	SBP	13.100	7.394	18.806	0.000
Hill,1999	SBP	3.000	-2.889	8.889	0.318
Kerry,2013A	SBP	2.100	-2.284	6.484	0.348
Kerry,2013B	SBP	2.700	-1.737	7.137	0.233
Kim KB,2014A	SBP	8.000	4.251	11.749	0.000
Kim KB,2014B	SBP	4.000	0.262	7.738	0.036
McMahon,2005	SBP	3.000	-6.200	12.200	0.523
Mohsen,2020A	SBP	9.400	5.184	13.616	0.000
Mohsen,2020B	SBP	20.000	15.723	24.277	0.000
Pan,2018A	SBP	6.900	1.745	12.055	0.009
Pan,2018B	SBP	6.600	1.445	11.755	0.012
Park MJ,2009	SBP	11.900	6.839	16.961	0.000
Rahmani Pour,2020A	SBP	6.670	-3.070	16.410	0.180
Rahmani Pour,2020B	SBP	11.720	2.203	21.237	0.016
Shea,2006	SBP	3.190	-0.213	6.593	0.066
Wakefield,2011	SBP	4.770	-0.243	9.783	0.062
Zha,2020A	SBP	2.170	-1.862	6.202	0.292
Zha,2020B	SBP	3.600	-0.115	7.315	0.058
		5.731	4.120	7.341	0.000

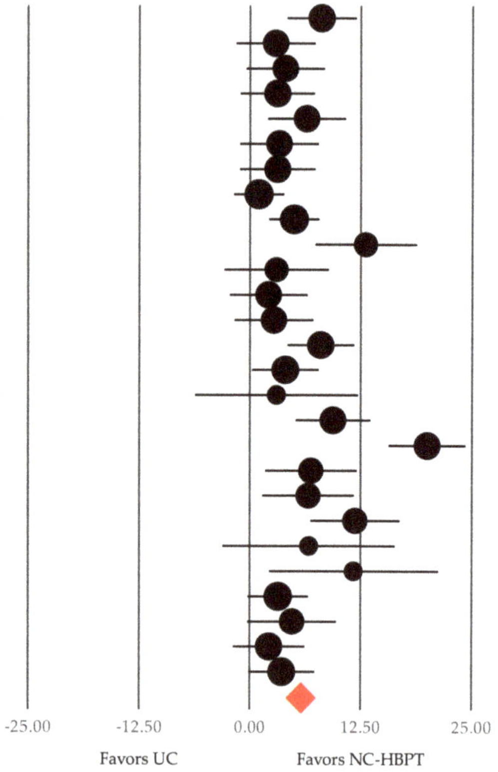

Figure 5. Difference in means of office systolic blood pressure changes by nurse-coordinated intervention. Notes. Point estimate of individual study (●); Summary effect size (◆); SBP, systolic blood pressure; UC, usual care; NC-HBPT, nurse-coordinated home blood pressure telemonitoring.

Changes in WMD over time were examined across four time frames. The SBP reduction during time frame I (inception to 2000) was 3.000 mmHg (−5.999 to 11.999; $p = 0.514$) [24]. SBP reductions of 5.150 mmHg (2.383–7.898; $p < 0.001$) and 4.990 mmHg (2.565–7.415; $p < 0.001$) were observed in time frames II (2001–2010) [25–27,47,48] and III (2011–2019) [21,23,33–35,43,46], respectively (Figure 6). A significant SBP reduction of 8.755 mmHg (5.177–12.334; $p < 0.001$) was observed in time frame IV (2020) [22,39,40], the year in which the COVID-19 pandemic began.

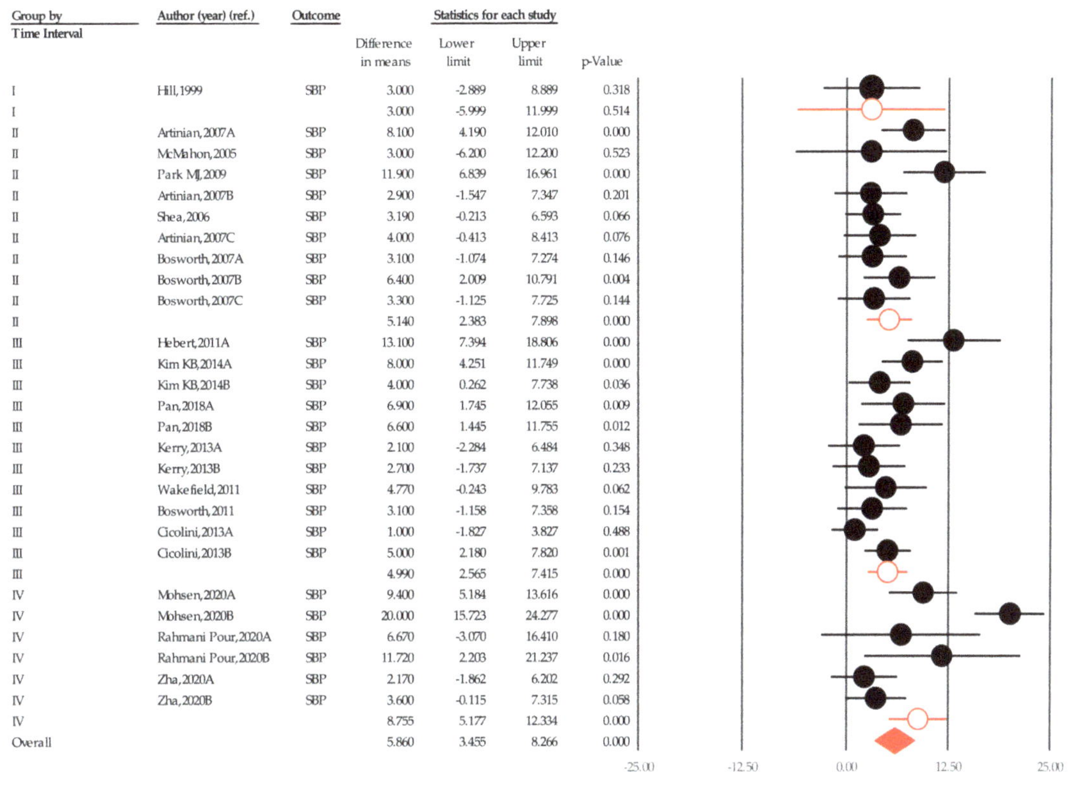

Figure 6. Difference in means of office systolic blood pressure changes over time. Notes. Point estimate of individual study (•); Subgroup (○); Summary effect size (◆); DBP, diastolic blood pressure; UC, usual care; NC-HBPT, nurse-coordinated home blood pressure telemonitoring.

3.3.2. DBP Changes by Nurse-Led Coordination

The WMD analysis of DBP was possible using data pooled from 24 comparisons (14 studies) [21–27,37–40,46–48], and data from 4928 patients (2445 from the NC-HBPT group and 2483 from the UC group) were analyzed. There was a significant decrease in blood pressure in the intervention group compared with the control (2.342 mmHg, 1.482–3.202; $p < 0.001$; Figure 7). Significant heterogeneity was observed between the comparisons ($I^2 = 51.380\%$; $p = 0.002$). In the sensitivity test, individual studies did not significantly alter the summary effect size (WMD: 1.343–3.359 mmHg).

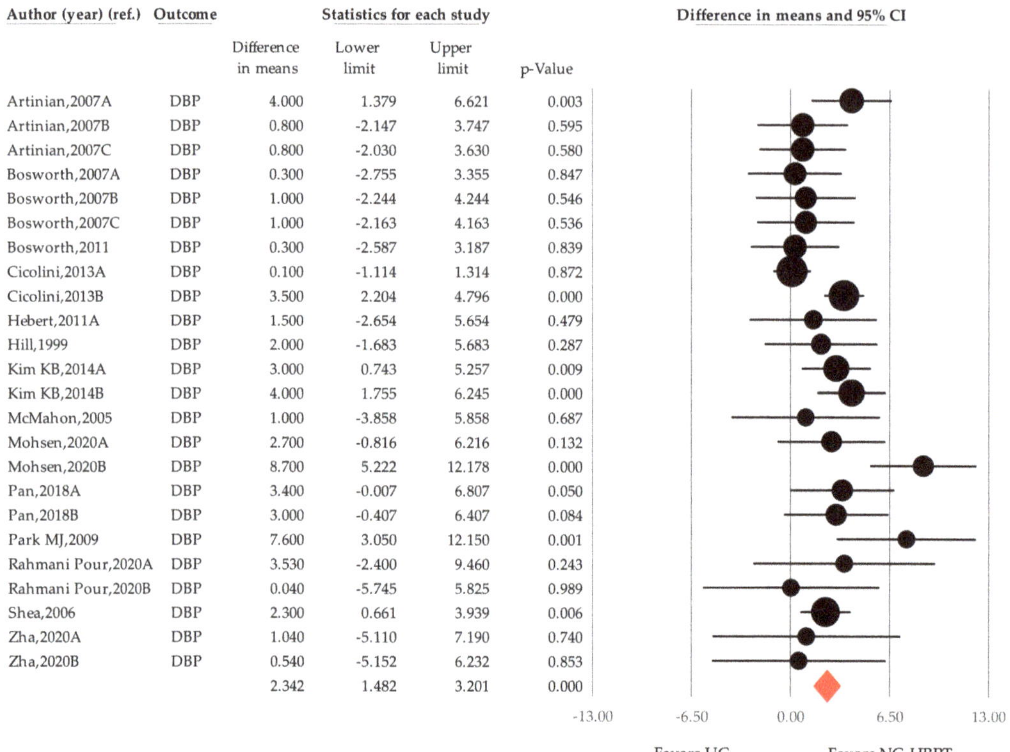

Author (year) (ref.)	Outcome	Difference in means	Lower limit	Upper limit	p-Value
Artinian,2007A	DBP	4.000	1.379	6.621	0.003
Artinian,2007B	DBP	0.800	-2.147	3.747	0.595
Artinian,2007C	DBP	0.800	-2.030	3.630	0.580
Bosworth,2007A	DBP	0.300	-2.755	3.355	0.847
Bosworth,2007B	DBP	1.000	-2.244	4.244	0.546
Bosworth,2007C	DBP	1.000	-2.163	4.163	0.536
Bosworth,2011	DBP	0.300	-2.587	3.187	0.839
Cicolini,2013A	DBP	0.100	-1.114	1.314	0.872
Cicolini,2013B	DBP	3.500	2.204	4.796	0.000
Hebert,2011A	DBP	1.500	-2.654	5.654	0.479
Hill,1999	DBP	2.000	-1.683	5.683	0.287
Kim KB,2014A	DBP	3.000	0.743	5.257	0.009
Kim KB,2014B	DBP	4.000	1.755	6.245	0.000
McMahon,2005	DBP	1.000	-3.858	5.858	0.687
Mohsen,2020A	DBP	2.700	-0.816	6.216	0.132
Mohsen,2020B	DBP	8.700	5.222	12.178	0.000
Pan,2018A	DBP	3.400	-0.007	6.807	0.050
Pan,2018B	DBP	3.000	-0.407	6.407	0.084
Park MJ,2009	DBP	7.600	3.050	12.150	0.001
Rahmani Pour,2020A	DBP	3.530	-2.400	9.460	0.243
Rahmani Pour,2020B	DBP	0.040	-5.745	5.825	0.989
Shea,2006	DBP	2.300	0.661	3.939	0.006
Zha,2020A	DBP	1.040	-5.110	7.190	0.740
Zha,2020B	DBP	0.540	-5.152	6.232	0.853
		2.342	1.482	3.201	0.000

Figure 7. Difference in means of office diastolic blood pressure changes by nurse-coordinated intervention. Notes. Point estimate of individual study (•); Summary effect size (◆); DBP, diastolic blood pressure; UC, usual care; NC-HBPT, nurse-coordinated home blood pressure telemonitoring.

The WMD of DBP in the NC-HBPT group was 2.000 mmHg (-2.724 to 6.724; $p = 0.407$) in time frame I [24], 1.947 mmHg (0.524–3.370; $p = 0.007$) in time frame II [25–27,47,48], 2.327 mmHg (0.958–3.695; $p < 0.001$) in time frame III [21,23,37,38,46], and 3.529 mmHg (1.221–5.838; $p = 0.003$) in time frame IV [22,39,40], suggesting that the effect of the intervention was statistically greater closer to the COVID-19 outbreak period (Figure 8).

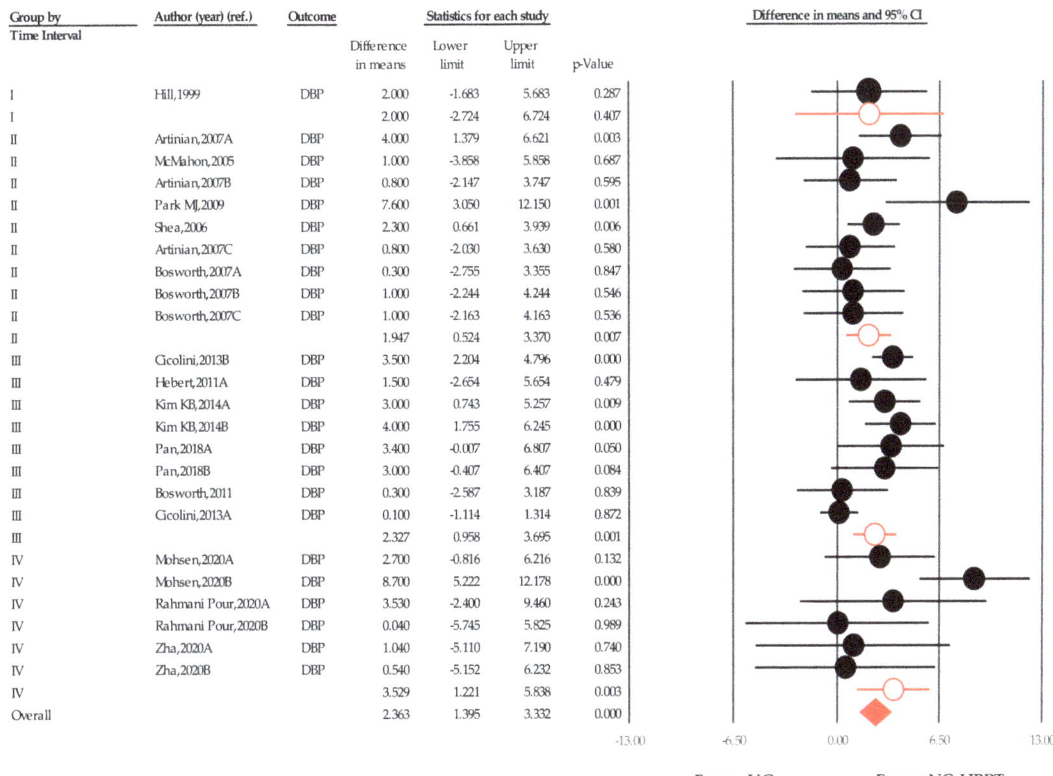

Figure 8. Difference in means of office diastolic blood pressure changes over time. Notes. Summary effect size (•); Point estimate of individual study (•); Subgroup (○); UC, usual care; NC-HBPT, nurse-coordinated home blood pressure telemonitoring.

3.4. Secondary Outcomes

Rate of Reaching the Target Office Blood Pressure

Using 18 available comparisons (6 studies) [21–23,27,37,38], we calculated the rate of reaching the target office blood pressure in the NC-HBPT group. Data from 4078 patients (2021 from the intervention group and 2057 from the UC group) showed that the rate of reaching the target office blood pressure was significantly higher in the intervention group than in the UC group (RR 1.261, 1.154–1.378; $p < 0.001$). Heterogeneity between the studies was substantial ($I^2 = 51.6\%$). In the sensitivity test, study removal did not have a significant effect on the summary effect size ($p < 0.001$; range: 1.133–1.399).

Changes in the rate of reaching the target blood pressure over time were analyzed by combining the data obtained after the year 2000. RR values of 1.101 (1.013–1.198; $p = 0.024$), 1.400 (1.279–1.534; $p < 0.001$), and 1.906 (1.462–2.487; $p < 0.001$) were found for time frames II [27], III [21,23,37,38], and IV [22], respectively, showing a clear improvement rate.

3.5. Subgroup Analysis
3.5.1. Size of City

Cities were classified based on their population size, where small- to medium-sized cities had fewer than 1 million residents, and large cities had more than 1 million residents. In small to medium cities (n = 16; participants – 3713) [18,21,24–27,37,40,49], the decrease in SBP by NC-HBPT was 4.134 mmHg (2.275–5.992, $p < 0.001$; $I^2 = 7.934$), while SBP decreased by 8.355 mmHg (5.937–10.773, $p < 0.001$; $I^2 = 82.533$) in large cities (n = 11; participants = 2069), compared with the UC group [22,23,36,38,39,47,48] (Appendix C).

3.5.2. Setting

When groups were classified according to the setting in which the study was conducted, the reduction in SBP due to NC-HBPT in primary care clinics (n = 5; participants = 1741) was 3.793 mmHg (0.450–7.136, $p = 0.026$; $I^2 = 0.000$) [27,46,48], 4.353 mmHg (1.877–6.829, $p = 0.001$; $I^2 = 34.921$) in community health centers (n = 9; participants = 1704) [21,26,38,40], and 7.781 mmHg of SBP (1.375–5.483, $p < 0.001$; $I^2 = 79.627$) in hospitals (n = 13; participants = 2337) compared with the UC group [22–25,36,37,39,47,49].

3.5.3. Duration

For the full duration of NC-HBPT, a total of 5278 people (NC-HBPT group = 2612 vs. UC = 2666) were analyzed. The WMD of SBP was consistently decreased throughout the intervention, by 6.694 mmHg after 3 months (3.644–9.744, $p < 0.001$; $I^2 = 72.511$) [21,22,26,38–40,47], 6.608 mmHg after 6 months (3.777–9.444, $p < 0.001$; $I^2 = 85.761$) [21,22,26,37–39,41,50], and 3.573 mmHg at 12 months of intervention (0.796–6.349, $p = 0.012$; $I^2 = 12.877$) compared with the UC group [24,25,27,36,37,46,48].

3.5.4. Coordinator's Profession

A total of 4398 patients were included in the analysis to derive the effectiveness of the nurse-only interventions (n = 20). The decrease in WMD of SBP by nurse alone was 6.132 mmHg (4.262–8.002, $p < 0.001$; $I^2 = 78.074$) [21,22,26,27,37,39,40,46–49]. Cases where nurses collaborated with other professionals (n = 7) included a total of 1384 people and showed a decrease in SBP of 4.465 mmHg (1.122–7.807, $p = 0.009$; $I^2 = 75.373$) [24,25,27,36,38]. When classified according to the professions nurses collaborated with, SBP decreased by 2.399 mmHg (−3571–8.369, $I^2 = 0.000$) for nutritionists, 3.000 mmHg (−8.682–14.682, $I^2 = 0.000$) for lifestyle educators, and 3.000 mmHg (−6.300–12.300, $I^2 = 34.534$) for community health workers compared with UC. When the collaboration was with a doctor, SBP was reduced by 6.626 mmHg (1.599–11.652, $I^2 = 67.129$) compared with the control group.

3.5.5. Medically Underserved Area

The effect in medically underserved areas reported in each primary study (participants = 5278) was analyzed. In underserved areas (n = 10; participants = 2852), the decrease in SBP was 5.100 mmHg (2.484–7.717, $p < 0.001$; $I^2 = 49.707$) [23,24,26,37,40,48], whereas SBP was reduced by 6.150 mmHg (4.041–8.259, $p < 0.001$) in non-marginalized areas (n = 17; participants = 2930) [21,22,25,27,36,38,39,46,47,49]. In the latter areas, the heterogeneity was substantial ($I^2 = 78.192$).

4. Discussion

This study assessed the effectiveness of NC-HBPT in patients with hypertension, a common NCD, in an urban setting. If remote medical services are defined as delivering patients' biological information and managing diseases using information and communications technology (ICT) [52], NC-HBPT can be considered a safe, effective, and timely intervention in the current pandemic situation, which has drastically increased the demand for medical resources, in addition to social measures to prevent the spread of COVID-19 [53].

In this meta-analysis, NC-HBPT achieved an SBP reduction of 5.731 mmHg (4.120–7.341, $p < 0.001$) at an average of 7.26 months. A large-scale meta-analysis using individual patient data showed that an SBP reduction of 4 mmHg can reduce the CVD incidence to 10 events/1000 cases in hypertensive patients with a moderate 5-year CVD risk (11–15%) [54]. A previous meta-analysis examining the effects of HBPT in the same setting without consideration of coordinators reported an SBP reduction of 3.482 mmHg

(2.459–4.505, $p < 0.001$) [55]. Thus, the increased effect of NC-HPBT in preventing CVD can be considered clinically significant.

When examining the effect of NC-HBPT over time, SBP was relatively low at the beginning of the intervention but increased by 3.000 mmHg (-5.999–11.999, $p = 0.514$), and the maximum reduction of 8.755 mmHg (5.177–12.334, $p < 0.001$) was achieved in 2020, when the pandemic began. Although the number of studies included in time frame I is insufficient and lacks significance, the data cannot be completely ignored, as the aforementioned trend undeniably exists based on the comparison of time frame II values (5.140 mmHg, 2.383–7.898, $p < 0.001$) with those in time frames III and IV. Greater SBP reduction was achieved by NC-HBTP over time. Considering the preference for non-face-to-face contact in 2020 and 2021 due to the COVID-19 pandemic and the increased demand for remote medical services, it can be extrapolated from our results that an increase in the potential effectiveness of remote medical services with nurse coordination is possible.

Based on evidence from previous literature suggesting that NC-HBPT is effective [3,20], we examined the effect of a nurse-alone intervention through 20 comparisons [21–23,26, 27,37,39,40,46–49]. HBPT coordinated by a nurse alone achieved an SBP reduction of 6.132 mmHg (4.262–8.002, $p < 0.001$), which was not inferior to the mean reduction of interventions by all coordinators (5.731 mmHg). Moreover, compared with NC-HBPT additionally coordinated by an expert from another field, HBPT was even more effective when a physician was involved in the intervention (n = 3; 6.626 mmHg, 1.599–11.652) [27,38]. HBPT showed limited effectiveness when coordinated by community health workers (3.000 mmHg, -6.300–12.300) [24], nutritionists (2.399 mmHg, -3.571–8.369) [36], or lifestyle educators (3.000 mmHg, -8.682–14.682) [25].

We also found that NC-HBPT was associated with a larger reduction in DBP (2.342 mmHg, 1.482–3.201, $p < 0.001$) than HBPT in the same setting but without consideration of co-ordinators (1.638 mmHg, 1.084–2.192, $p < 0.001$) [55]. Similar patterns were observed in DBP and SBP changes over time. A higher reduction in DBP of 3.529 mmHg (1.221–5.838, $p = 0.003$) was observed in time frame IV [22,39,40] than in time frame II (1.947 mmHg, 0.524–3.370; $p = 0.007$) [25–27,47,48].

HBPT coordinated by a nurse alone achieved a DBP reduction of 2.389 mmHg (1.393–3.384, $p < 0.001$) [21–24,26,27,37,39,40,47–49], and additional coordination by a doctor achieved a DBP reduction of 2.440 mmHg (-0.166–5.047, $p = 0.066$) [26,38], showing a significantly greater DBP reduction when HBPT is additionally coordinated by experts from other specific fields than in NC-HBPT [24–26,38]. Although the reason for the differences in blood pressure reductions according to the profession of the additional coordinator is unclear, it may have to do with social and organizational factors, the coordinator's level of medical knowledge and experience, and similarity in the education received by the coordinators [56,57]. However, since the number of cases in which NC-HBPT was coordinated by experts from other fields was small and the results were not statistically significant, further research is needed to more accurately determine the validity of the effect of the intervention.

Palmas et al. (2006) reported that a lack of awareness of the benefits of remote medical technologies and the burden of using these technologies have contributed to the low expectations for remote medical services in urban areas [58]. However, the high percentage of hypertensive patients who are highly susceptible to COVID-19 and the environmental factors that are found in densely populated cities contribute to excessive medical demands that cannot be handled by traditional medical systems [59]. Therefore, the importance of remote monitoring technology as a means to efficiently provide medical services with limited resources is being increasingly emphasized. To the best of our knowledge, this is the first meta-analysis to examine the effect of NC-HBPT on urban hypertensive patients over time. We have derived meaningful results regarding the benefits of nurse-coordinated interventions.

In this study, the heterogeneity of the summary effect sizes was found to be substantial. Specifically, the I^2 for SBP and DBP were 71.717% ($p < 0.001$) and 51.380% ($p = 0.002$), respectively. Thus, the authors applied a random effects model to the analysis, which did not completely remove the heterogeneity. Therefore, subgroup analysis was performed to assess the causes of the heterogeneity, and several moderators were explored that revealed clinical implications, along with evaluation of heterogeneity.

Despite selecting well-founded blinded RCTs through a transparent and systematic process and deriving solid and integrated results for the primary outcomes without publication biases, our study has some limitations. First, an extensive literature search was conducted on reliable and relevant databases using a structured formula, but the number of studies included was low. By building a more precise search formula, the reliability of our results could be improved. Second, since each time frame did not contain an equal number of comparisons, the analysis results for different periods were not based on the same number of studies. Including as many studies as possible, with an equal number of studies per time frame, can overcome this limitation. Third, while the researchers reasonably set the duration of each time frame to 9 and 10 years according to the technological changes and historical events to explore temporal patterns of outcomes of NC-HBPT, the distinction between each time frame may not have been clear. Thus, it may be necessary to set the time frames based on the turning points of ICT development to increase the precision of the findings. Lastly, for studies that compared interventions with different coordinators or lengths of follow-up, each comparison was counted as an independent primary study, but there were cases where multiple comparisons were included in one study. Although there was substantial heterogeneity between each comparison, and no statistical errors were observed, there may be a lack of optimal scientific robustness due to the methodological limitations of a random effects model. This limitation can also be overcome by updating the results based on a larger number of studies.

5. Conclusions

This study revealed that NC-HBPT for urban hypertensive patients can deliver clinically and statistically good BP reduction in terms of avoiding CVD outbreak when compared with UC. Our findings may have meaningful implications for policymakers in urban areas that are planning to introduce remote monitoring systems or in areas with inefficient telemedicine systems. However, some included studies in this analysis lack quality. Thus, although the evidence for the benefit of NC-HBPT was found in this review, further research is necessary on the nurses' roles as coordinators. Additionally, future work must consider the effect of multiple variables on NC-HBPT for more efficient implementation of the intervention system in urban areas.

Author Contributions: Conceptualization, N.-S.K. and W.-S.C.; methodology, N.-S.K., A.-Y.K., H.-S.W. and W.-S.C.; software, A.-Y.K. and W.-S.C.; validation, N.-S.K., A.-Y.K., H.-S.W. and W.-S.C.; formal analysis, N.-S.K., A.-Y.K. and W.-S.C.; investigation, H.-S.W. and W.-S.C.; writing—original draft preparation, N.-S.K. and W.-S.C.; writing—review and editing, H.-S.W. and W.-S.C. All authors have read and agreed to the published version of the manuscript.

Funding: This research received no external funding.

Institutional Review Board Statement: Not applicable.

Informed Consent Statement: Not applicable.

Conflicts of Interest: The authors declare no conflict of interest.

Appendix A. Searching Formula

((((((("Hypertension"[Mesh]) OR ((hypertensi*) OR high blood pressure))) AND (((((("Urban Population"[Mesh]) OR "Urban Health"[Mesh]) OR "Urban Health Services" [Mesh]) OR "Cities"[Mesh])) OR ((((urban*) OR city) OR cities) OR central cit*)))) AND ((((("Telemedicine"[Mesh]) OR "Telemetry"[Mesh]) OR "Blood Pressure Monitoring, Ambulatory"[Mesh])) OR ((((((((((telemedicine) OR telemetry) OR telenurs*) OR telemonitor*) OR eHealth) OR telehealth) OR remote monitor*) OR technology) OR telephone) OR smartphone) OR internet)))) AND ((((((((randomised controlled trial) OR randomized controlled) OR controlled clinical trial)) OR ((((((((randomised[Title/Abstract]) OR randomized[Title/Abstract]) OR placebo[Title/Abstract]) OR drug therapy[Title/Abstract]) OR groups[Title/Abstract]) OR clinical trials as topic[Title/Abstract]) OR randomly[Title/Abstract]) OR trial[Title/Abstract]))) NOT cluster randomized controlled trials)) NOT cross over study)

Appendix B. Cumulative Meta-Analysis Presenting the Summative Effect at Each Time Point

Author (year) (ref.)	Outcome	Point	p-Value
Hill,1999	SBP	3.000	0.318
McMahon,2005	SBP	3.000	0.236
Shea,2006	SBP	3.129	0.029
Artinian,2007A	SBP	4.779	0.001
Artinian,2007B	SBP	4.397	0.000
Artinian,2007C	SBP	4.346	0.000
Bosworth,2007A	SBP	4.143	0.000
Bosworth,2007B	SBP	4.433	0.000
Bosworth,2007C	SBP	4.306	0.000
Park MJ,2009	SBP	4.950	0.000
Bosworth,2011	SBP	4.761	0.000
Hebert,2011A	SBP	5.340	0.000
Wakefield,2011	SBP	5.287	0.000
Zha,2020A	SBP	5.026	0.000
Cicolini,2013A	SBP	4.658	0.000
Cicolini,2013B	SBP	4.663	0.000
Kerry,2013A	SBP	4.504	0.000
Kerry,2013B	SBP	4.395	0.000
Kim KB,2014A	SBP	4.624	0.000
Kim KB,2014B	SBP	4.578	0.000
Pan,2018A	SBP	4.664	0.000
Pan,2018B	SBP	4.731	0.000
Mohsen,2020A	SBP	4.959	0.000
Mohsen,2020B	SBP	5.692	0.000
Rahmani Pour,2020A	SBP	5.710	0.000
Rahmani Pour,2020B	SBP	5.831	0.000
Zha,2020B	SBP	5.731	0.000
		5.731	0.000

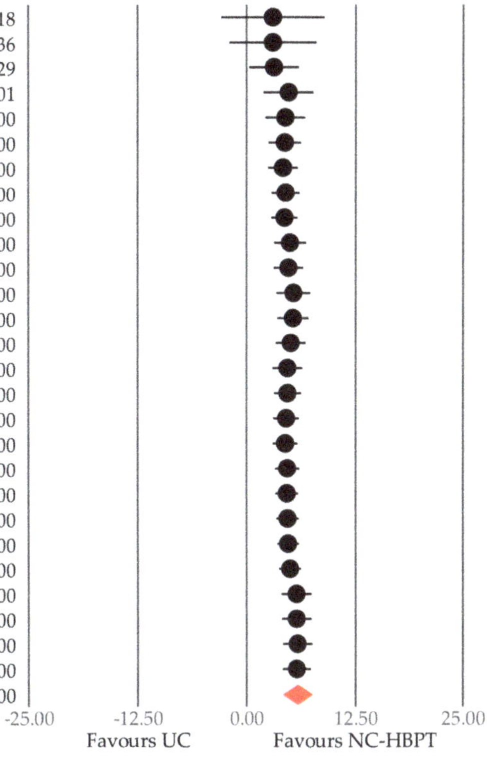

Figure A1. Cumulative meta-analysis presenting the summative effect at each time point. Note. SBP, systolic blood pressure; UC, usual care; NC-HBPT, nurse-coordinated home blood pressure telemonitoring; Summary effect size (♦); Point estimate of individual study (●).

Appendix C. Subgroup Analysis

Table A1. Subgroup analysis.

Category	Number of Studies	Summary WMD of SBP, mmHg (95% CI)	Heterogeneity, I^2 (%) by FEM	*p*-Value of I^2
Overall	27	5731	71.717	$p < 0.001$
City size (population)				
Under 1 million	16	4.134 (2.275–5.992)	7.934	$p = 0.063$
Over 1 million	11	8.355 (5.937–10.773)	82.533	$p < 0.001$
Setting				
Primary care clinic	5	3.793 (0.450–7.136)	0.000	$p = 0.086$
Community health center	9	4.353 (1.877–6.829)	34.921	$p = 0.039$
Hospital	13	7.781 (5.483–10.079)	79.627	$p < 0.001$
Duration (month)				
3 or less	8	6.694 (3.644–9.744)	72.511	$p = 0.001$
6	8	6.608 (3.771–9.444)	85.761	$p < 0.001$
12	9	3.573 (0.796–6.349)	12.877	$P = 0.042$
Medically underserved area				
Not underserved	17	6.150 (4.041–8.259)	78.192	$p < 0.001$
Underserved	10	4.424 (2.484–7.717)	49.707	$p = 0.036$
Collaboration with other professionals				
No (Nurse-alone)	20	6.132	78.074	$p < 0.001$
Yes	7	4.465	75.373	$p = 0.633$

Note. WMD, weighted mean difference; SBP, systolic blood pressure; CI, confidence interval; FEM, fixed effect model.

References

1. WHO. Hypertension. Available online: https://www.who.int/news-room/fact-sheets/detail/hypertension (accessed on 27 May 2021).
2. Ostchega, Y.; Fryar, C.D.; Nwankwo, T.; Nguyen, D.T. *Hypertension Prevalence Among Adults Aged 18 and Over: United States, 2017–2018. NCHS Data Brief, No 364*; National Center for Health Statistics: Hyattsville, MD, USA, 2020.
3. Clark, C.E.; Smith, L.F.P.; Taylor, R.S.; Campbell, J.L. Nurse-led interventions to improve control of blood pressure in people with hypertension: Systematic review and meta-analysis. *BMJ* **2010**, *341*. [CrossRef]
4. Egan, B.M.; Li, J.; Hutchison, F.N.; Ferdinand, K.C. Hypertension in the United States, 1999 to 2012: Progress toward Healthy People 2020 goals. *Circulation* **2014**, *130*, 1692–1699. [CrossRef]
5. Chobanian, A.V.; Bakris, G.L.; Black, H.R.; Cushman, W.C.; Green, L.A.; Izzo, J.L., Jr.; Jones, D.W.; Materson, B.J.; Oparil, S.; Wright, J.T., Jr.; et al. Seventh report of the Joint National Committee on Prevention, Detection, Evaluation, and Treatment of High Blood Pressure. *Hypertension* **2003**, *42*, 1206–1252. [CrossRef] [PubMed]
6. Whelton, P.K.; Carey, R.M.; Aronow, W.S.; Casey, D.E.; Collins, K.J.; Dennison Himmelfarb, C.; DePalma, S.M.; Gidding, S.; Jamerson, K.A.; Jones, D.W.; et al. 2017 ACC/AHA/AAPA/ABC/ACPM/AGS/APhA/ASH/ASPC/NMA/PCNA Guideline for the Prevention, Detection, Evaluation, and Management of High Blood Pressure in Adults: Executive Summary: A Report of the American College of Cardiology/American Heart Association Task, F. *Hypertension* **2018**, *71*, 1269–1324. [CrossRef] [PubMed]
7. Williams, B.; Mancia, G.; Spiering, W.; Rosei, E.A.; Azizi, M.; Burnier, M.; Clement, D.L.; Coca, A.; de Simone, G.; Dominiczak, A.; et al. 2018 ESC/ESH Guidelines for the management of arterial hypertension. *Eur. Heart J.* **2018**, *39*, 3021–3104. [CrossRef] [PubMed]
8. Zhou, F.; Yu, T.; Du, R.; Fan, G.; Liu, Y.; Liu, Z.; Xiang, J.; Wang, Y.; Song, B.; Gu, X.; et al. Clinical course and risk factors for mortality of adult inpatients with COVID-19 in Wuhan, China: A retrospective cohort study. *Lancet* **2020**, *395*, 1054–1062. [CrossRef]

9.   Omboni, S.; McManus, R.J.; Bosworth, H.B.; Chappell, L.C.; Green, B.B.; Kario, K.; Logan, A.G.; Magid, D.J.; Mckinstry, B.; Margolis, K.L.; et al. Evidence and Recommendations on the Use of Telemedicine for the Management of Arterial Hypertension. *Hypertension* **2020**, *76*, 1368–1383. [CrossRef]

10.   Wilkins, A. Local response in health emergencies: Key considerations for addressing the COVID-19 pandemic in informal urban settlements. *Env. Urban.* **2020**, *32*, 503–522. [CrossRef]

11.   Sharma, A.; Ahmed, S.; Kaur, J.; Chawla, R.; Rejeeth, C. Exploring status of emergency drugs and vaccine development in Covid-19 pandemic: An update. *VirusDisease* **2021**, 1–13. [CrossRef]

12.   World Health Organization. WHO Coronavirus Disease (COVID-19) Dashboard. 2020. Available online: https://covid19.who.int/ (accessed on 27 May 2021).

13.   Bello, B.; Usehv, U. COVID-19: Are Non-Communicable Diseases Risk Factors for Its Severity? *Am. J. Health Promot.* **2021**, *35*, 720–729. [CrossRef]

14.   Miner, H.; Fatehi, A.; Ring, D.; Reichenberg, J.S. Clinician Telemedicine Perceptions During the COVID-19 Pandemic. *Telemed. e-Health* **2021**, *27*. [CrossRef] [PubMed]

15.   Ing, E.B.; Xu, Q.; Salimi, A.; Torun, N. Physician deaths from corona virus (COVID-19) disease. *Occup. Med.* **2020**, *70*, 370–374. [CrossRef]

16.   Phillips, L.S.; Branch, W.T.; Cook, C.B.; Doyle, J.P.; El Kebbi, I.M.; Gallina, D.L.; Miller, C.D.; Zeimer, D.C.; Barnes, C.S. Clinical inertia. *Ann. Intern. Med.* **2001**, *135*, 825–834. [CrossRef]

17.   Monahan, M.; Jowett, S.; Nickless, A.; Frassen, M.; Grant, S.; Greenfield, S.; Hobbs, F.D.R.; Hodgkinson, J.; Mant, J.; McManus, R. Cost-Effectiveness of Telemonitoring and Self-Monitoring of Blood Pressure for Antihypertensive Titration in Primary Care (TASMINH4). *Hypertension* **2019**, *73*, 1231–1239. [CrossRef] [PubMed]

18.   Omboni, S.; Guarda, A. Impact of Home Blood Pressure Telemonitoring and Blood Pressure Control: A Meta-Analysis of Randomized Controlled Studies. *Am. J. Hypertens.* **2011**, *24*, 989–998. [CrossRef] [PubMed]

19.   Filho, S.P.; Paffer, M.T.; Paffer, P.T.; Figeuiredo, M.C.; Veras, C.O.; Fonsêca, F.B. Home blood pressure monitoring: Report of a database of 1474 patients. *J. Hypertens.* **2021**, *39*, e121–e122. [CrossRef]

20.   Clark, C.E.; Horvath, I.A.; Taylor, R.S.; Campbell, J.L. Doctors record higher blood pressures than nurses: Systematic review and meta-analysis. *Br. J. Gen. Pract.* **2014**, *64*, e223–e232. [CrossRef]

21.   Cicolini, G.; Simonetti, V.; Comparcini, D.; Celiberti, I.; Nicola, M.D.; Capasso, L.M.; Flacco, M.E.; Bucci, M.; Mezzetti, A.; Manzoli, L. Efficacy of a nurse-led email reminder program for cardiovascular prevention risk reduction in hypertensive patients: A randomized controlled trial. *Int J. Nurs. Stud.* **2014**, *51*, 833–843. [CrossRef]

22.   Mohsen, M.M.; Riad, N.A.; Badawy, A.E.; Abd El Gafar, S.E.; Abd El-Hammed, B.M.; Eltomy, E.M. Tele-nursing versus Routine Outpatient Teaching for Improving Arterial Blood Pressure and Body Mass Index for Hypertensive Patients. *Am. J. Nurs. Res.* **2020**, *8*, 18–26.

23.   Hebert, P.; Sisk, J.E.; Tuzzio, L.; Casabianca, J.M.; Pogue, V.A.; Wang, J.J.; Chen, Y.; Cowles, C.; McLaughlin, M.A. Nurse-led disease management for hypertension control in a diverse urban community: A randomized trial. *J. Gen. Intern. Med.* **2012**, *27*, 630–639. [CrossRef] [PubMed]

24.   Hill, M.N.; Bone, L.R.; Hilton, S.C.; Roary, M.C.; Kelen, G.D.; Levine, D.M. A clinical trial to improve high blood pressure care in young urban black men: Recruitment, follow-up, and outcomes. *Am. J. Hypertens.* **1999**, *12*, 548–554. [CrossRef]

25.   McMahon, G.T.; Gomes, H.E.; Hickson Hohne, S.; Hu, T.M.; Levine, B.A.; Conlin, P.R. Web-based care management in patients with poorly controlled diabetes. *Diabetes Care* **2005**, *28*, 1624–1629. [CrossRef] [PubMed]

26.   Artinian, N.T.; Flack, J.M.; Nordstrom, C.K.; Hockman, E.M.; Washington, O.G.M.; Jen, K.C.; Fathy, M. Effects of nurse-managed telemonitoring on blood pressure at 12-month follow-up among urban African Americans. *Nurs. Res.* **2007**, *56*, 312–322. [CrossRef]

27.   Bosworth, H.B.; Olsen, M.K.; McCant, F.; Harrelson, M.; Gentry, P.; Rose, C.; Goldstein, M.K.; Hoffman, B.B.; Powers, B.; Oddone, E.Z. Hypertension Intervention Nurse Telemedicine Study (HINTS): Testing a multifactorial tailored behavioral/educational and a medication management intervention for blood pressure control. *Am. Heart J.* **2007**, *153*, 918–924.

28.   Laurant, M.; Reeves, D.; Hermens, R.; Braspenning, J.; Grol, R.; Sibbald, B. Substitution of doctors by nurses in primary care. *Cochrane Database Syst. Rev.* **2005**, CD001271. [CrossRef]

29.   Fahey, T.; Schroeder, K.; Ebrahim, S.; Glynn, L. Interventions used to improve control of blood pressure in patients with hypertension. *Cochrane Database Syst. Rev.* **2010**, CD005182. [CrossRef]

30.   McManus, R.; Mant, J.; Franssen, M.; Nickless, A.; Schwartz, C.; Hodgkinson, J.; Bradburn, P.; Farmer, A.; Grant, S.; Greenfield, S.; et al. Telemonitoring and/or self-monitoring of blood pressure in hypertension (TASMINH4): A randomised controlled trial. *J. Hypertens.* **2018**, *36*, e5. [CrossRef]

31.   Glynn, L.G.; Murphy, A.W.; Smith, S.M.; Schroeder, K.; Fahey, T. Self-monitoring and other non-pharmacological interventions to improve the management of hypertension in primary care: A systematic review. *Br. J. Gen. Pract.* **2010**, *60*, e476–e488. [CrossRef] [PubMed]

32.   Omboni, S.; Gazzola, T.; Carabelli, G.; Parati, G. Clinical usefulness and cost effectiveness of home blood pressure telemonitoring: Meta-analysis of randomized controlled studies. *J. Hypertens.* **2013**, *31*, 455–468. [CrossRef]

33.   Duan, Y.; Xie, Z.; Dong, F.; Wu, Z.; Lin, Z.; Sun, N.; Xu, J. Effectiveness of home blood pressure telemonitoring: A systematic review and meta-analysis of randomised controlled studies. *J. Hum. Hypertens.* **2017**, *31*, 427–437. [CrossRef]

34. Cumpston, M.; Li, T.; Page, M.J.; Chandler, J.; Welch, V.A.; Higgins, J.P.; Thomas, J. Updated guidance for trusted systematic reviews: A new edition of the Cochrane Handbook for Systematic Reviews of Interventions. *Cochrane Database Syst. Rev.* **2019**, *10*, ED000142. [CrossRef]

35. Choi, W.S.; Kim, N.S.; Kim, A.Y.; Woo, H. A Systematic Review and Meta-Analysis of Nurse-Coordinated Blood Pressure Telemonitoring for Urban Hypertensive Patients During the Covid-19 Era. *PROSPERO.* 2020, p. CRD42020222789. Available online: https://www.crd.york.ac.uk/PROSPERO/display_record.php?ID=CRD42020222789 (accessed on 30 April 2020).

36. Kerry, S.M.; Markus, H.S.; Khong, T.K.; Cloud, G.C.; Tulloch, J.; Coster, D.; Ibison, J.; Oakeshott, P. Home blood pressure monitoring with nurse-led telephone support among patients with hypertension and a history of stroke: A community-based randomized controlled trial. *CMAJ* **2013**, *185*, 23–31. [CrossRef]

37. Kim, K.B.; Han, H.R.; Huh, B.; Nguyen, T.; Lee, H.; Kim, M.T. The effect of a community-based self-help multimodal behavioral intervention in Korean American seniors with high blood pressure. *Am. J. Hypertens.* **2014**, *27*, 1199–1208. [CrossRef] [PubMed]

38. Pan, F.; Wu, H.; Liu, C.; Zhang, X.; Peng, W.; Wei, X.; Gao, W. Effects of home telemonitoring on the control of high blood pressure: A randomised control trial in the Fangzhuang Community Health Center, Beijing. *Aust. J. Prim. Health* **2018**, *24*, 398–403. [CrossRef] [PubMed]

39. Pour, R.E.; Aliyari, S.; Farsi, Z.; Ghelich, Y. Comparing the effects of interactive and noninteractive education using short message service on treatment adherence and blood pressure among patients with hypertension. *Nurs. Midwifery Stud.* **2020**, *9*, 68–76.

40. Zha, P.; Qureshi, R.; Porter, S.; Chao, Y.Y.; Pacquiao, D.; Chase, S.; O'Brien-Richardson, P. Utilizing a Mobile Health Intervention to Manage Hypertension in an Underserved Community. *West. J. Nurs. Res.* **2020**, *42*, 201–209. [CrossRef]

41. Borenstein, M.; Hedges, L.V.; Higgins, J.P.; Rothstein, H.R. *Introduction to Meta-Analysis*; John Wiley & Sons: Hoboken, NJ, USA, 2011.

42. Review Manager (RevMan) [Computer program]. Version 5.4. The Cochrane Collaboration, 2020.

43. Higgins, J.P.; Thompson, S.G.; Deeks, J.J.; Altman, D.G. Measuring inconsistency in meta-analyses. *BMJ* **2003**, *327*, 557–560. [CrossRef] [PubMed]

44. Cohen, J. *Statistical Power Analysis for the Behavioral Sciences*, 2nd ed.; Lawrence Erlbaum Associates: Mahwah, NJ, USA, 1988.

45. Borenstein, M.; Hedges, L.; Higgins, J.; Rothstein, H. *Comprehensive Meta-Analysis, Version 2; 104*; Biostat: Englewood, NJ, USA, 2005.

46. Bosworth, H.B.; Powers, B.J.; Olsen, M.K.; McCant, F.; Grubber, J.; Smith, V.; Gentry, P.W.; Rose, C.; Van Houtven, C.; Wang, V.; et al. Home blood pressure management and improved blood pressure control: Results from a randomized controlled trial. *Arch. Intern. Med.* **2011**, *171*, 1173–1180. [CrossRef] [PubMed]

47. Park, M.; Kim, H.; Kim, K. Cellular phone and Internet-based individual intervention on blood pressure and obesity in obese patients with hypertension. *Int. J. Med. Inform.* **2009**, *78*, 704–710. [CrossRef] [PubMed]

48. Shea, S.; Weinstock, R.S.; Starren, J.; Teresi, J.; Palmas, W.; Field, L.; Morin, P.; Goland, R.; Izquierdo, R.E.; Wolff, T.; et al. A randomized trial comparing telemedicine case management with usual care in older, ethnically diverse, medically underserved patients with diabetes mellitus. *J. Am. Med. Inform. Assoc.* **2006**, *13*, 40–51. [CrossRef]

49. Wakefield, B.J.; Holman, J.E.; Ray, A.; Scherubel, M.; Adams, M.R.; Hillis, S.L.; Rosenthal, G.E. Effectiveness of home telehealth in comorbid diabetes and hypertension: A randomized, controlled trial. *Telemed. e-Health* **2011**, *17*, 254–261. [CrossRef]

50. Higgins, J.P.T.; Altman, D.G.; Sterne, J.A.C. Assessing risk of bias in included studies. In *Cochrane Handbook for Systematic Reviews of Interventions. Version 5.1.0 (Updated March 2011)*; Higgins, J.P.T., Green, S., Eds.; The Cochrane Collaboration, John Wiley & Sons: Chichester, UK, 2011; Available online: http//www.handbook.cochrane.org (accessed on 31 July 2019).

51. Duval, S.; Tweedie, R. Trim and fill: A simple funnel-plot–based method of testing and adjusting for publication bias in meta-analysis. *Biometrics* **2000**, *56*, 455–463. [CrossRef] [PubMed]

52. World Health Organization. *A Health Telematics Policy in Support of WHO's Health-For-All Strategy for Global Health Development. Report of the WHO Group Consultation on Health Telematics, December 11–16, 1997*; World Health Organization: Geneva, Switzerland, 1998.

53. Chu, D.K.; Akl, E.A.; Duda, S.; Solo, K.; Yaacoub, S.; Schünemann, H.J. Physical distancing, face masks, and eye protection to prevent person-to-person transmission of SARS-CoV-2 and COVID-19: A systematic review and meta-analysis. *Lancet* **2020**, *395*, 1973–1987. [CrossRef]

54. Blood Pressure Lowering Treatment Trialists' Collaboration. Blood pressure-lowering treatment based on cardiovascular risk: A meta-analysis of individual patient data. *Lancet* **2014**, *384*, 591–598. [CrossRef]

55. Choi, W.S.; Choi, J.H.; Oh, J.; Shin, I.; Yang, J. Effects of Remote Monitoring of Blood Pressure in Management of Urban Hypertensive Patients: A Systematic Review and Meta-Analysis. *Telemed. e-Health* **2020**, *26*, 744–759. [CrossRef] [PubMed]

56. Aas, I.H.M. Telemedical work and cooperation. *J. Telemed. Telecare* **2001**, *7*, 212–218. [CrossRef]

57. Jacob, C.; Sanchez-Vazquez, A.; Ivory, C. Social, Organizational, and Technological Factors Impacting Clinicians' Adoption of Mobile Health Tools: Systematic Literature Review. *JMIR mHealth uHealth* **2020**, *8*, e15935. Available online: https://preprints.jmir.org/preprint/15935 (accessed on 10 April 2020). [CrossRef]

58. Palmas, W.; Teresi, J.; Morin, P.; Wolff, L.T.; Field, L.; Eimicke, J.P.; Capps, L.; Prigollini, A.; Orbe, I.; Weinstock, R.S.; et al. Recruitment and enrollment of rural and urban medically underserved elderly into a randomized trial of telemedicine case management for diabetes care. *Telemed. e-Health* **2006**, *12*, 601–607. [CrossRef] [PubMed]

59. Smith, A.C.; Thomas, E.; Snoswell, C.L.; Haydon, H.; Mehrotra, A.; Clemensen, J.; Caffery, L.J. Telehealth for global emergencies: Implications for coronavirus disease 2019 (COVID-19). *J. Telemed. Telecare* **2020**, *26*, 309–313. [CrossRef]

International Journal of
Environmental Research and Public Health

Systematic Review

eHealth Interventions to Treat Substance Use in Pregnancy: A Systematic Review and Meta-Analysis

Katherine Silang [1], Hangsel Sanguino [1], Pooja R. Sohal [1], Charlie Rioux [2], Hyoun S. Kim [3] and Lianne M. Tomfohr-Madsen [1,4,5,*]

1   Department of Psychology, University of Calgary, Calgary, AB T2N 1N4, Canada;
    katherine.silang@ucalgary.ca (K.S.); hangsel.sanguino@ucalgary.ca (H.S.); pooja.sohal@ucalgary.ca (P.R.S.)
2   Department of Educational Psychology and Leadership, Texas Tech University, Lubbock, TX 79409, USA;
    charlie.rioux@ttu.edu
3   Department of Psychology, Ryerson University, Toronto, ON M5B 2K3, Canada; andrewhs.kim@ryerson.ca
4   Alberta Children's Hospital Research Institute (ACHRI), Calgary, AB T3B 6A8, Canada
5   Department of Pediatrics, University of Calgary, Calgary, AB T2N 1N4, Canada
*   Correspondence: ltomfohr@ucalgary.ca

Abstract: Substance use during pregnancy is associated with adverse pregnancy and neonatal outcomes; eHealth interventions offer a potential accessible treatment option. The objective of this systematic review and meta-analysis was to evaluate the effectiveness of eHealth interventions for the treatment of substance use during pregnancy. A comprehensive search of PsycINFO, Medline, CINAHL, Cochrane and Embase databases was conducted from May 2020 to April 2021. The protocol for this study was registered with Prospero (CRD42020205186) through the University of York Centre for Reviews and Dissemination. Two independent reviewers completed screening, data extraction, and quality assessment. RCTs were included if they reported: (a) administration of an eHealth intervention for (b) substance use outcomes, among (c) pregnant individuals. Comprehensive Meta-Analysis Software (CMA) was used to calculate pooled effect sizes (Odds Ratio) to determine the effect of eHealth interventions on substance use outcomes. Six studies were identified with substance use outcomes that included: smoking ($n = 3$), alcohol ($n = 2$), and other ($n = 1$). eHealth interventions were delivered through the internet ($n = 1$), computer ($n = 3$), telephone ($n = 1$), and text ($n = 1$). Results suggested that eHealth interventions significantly reduced substance use in pregnant individuals compared to controls (OR = 1.33, 95% CI = 1.06 to 1.65, $p = 0.013$). eHealth interventions offer a promising and accessible treatment option to reduce substance use during pregnancy. This work was supported by the generous donors of the Alberta Children's Hospital Foundation, the Canadian Child Health Clinician Scientist Program (CCHCSP), the Canadian Institute of Health Research and the Fonds de Recherche du Québec—Santé.

Keywords: pregnancy; substance-related disorders; randomized controlled trials; smoking; alcohol; cannabis; drug use; internet intervention; telemedicine; digital intervention

Citation: Silang, K.; Sanguino, H.; Sohal, P.R.; Rioux, C.; Kim, H.S.; Tomfohr-Madsen, L.M. eHealth Interventions to Treat Substance Use in Pregnancy: A Systematic Review and Meta-Analysis. *Int. J. Environ. Res. Public Health* **2021**, *18*, 9952. https://doi.org/10.3390/ijerph18199952

Academic Editors: Irene Torres-Sanchez and Marie Carmen Valenza

Received: 15 August 2021
Accepted: 17 September 2021
Published: 22 September 2021

1. Introduction

Heavy substance use is associated with serious physical and psychological consequences [1]. The risk of developing a substance use disorder is heightened during reproductive years [2] and substance use is prevalent during pregnancy. 11–15% of pregnant individuals reporting use of alcohol, tobacco, cannabis and/or illicit substances [3–5]. The actual prevalence may be higher, as stigma and judgement may cause some pregnant people to be hesitant to report substance use [6].

Heavy substance use in pregnancy has serious short and long-term consequences, including elevated risk of miscarriage [7], low birthweight [8], infant mortality [9], and sudden infant death syndrome [10]. Long term outcomes for children exposed to substances in-utero vary [11]. For example, prenatal cannabis use has been linked to reduced

207

attention and executive functioning skills, poor academic achievement, and increased behavioural problems [12]. Prenatal drinking has been linked to multiple long-term effects including cognitive and behavioural issues [13], executive functioning deficits [14], and poor psychosocial outcomes [15].

Given the high prevalence of substance use in pregnancy and its serious associated harms, it is imperative that pregnant individuals receive access to evidence-based supports. Despite results which have shown the effectiveness of psychological interventions in treating substance abuse, the literature has consistently found that in the general population, people often do not seek addiction and mental health services [16]. Potential obstacles to treatment include limited resources, time conflicts, and stigma [17–19]. Pregnant individuals, especially individuals belonging to marginalized ethnic and socioeconomic groups, are also more likely to be experience arrest, prosecution, conviction and/or child removal related to substance use disclosure, contributing to increased hesitancy to seek help [20]. Concerns about separation from family, as well as a lack of childcare are also known treatment barriers for pregnant individuals who use substances [21].

eHealth is an emerging field that is attracting attention for a variety of mental health conditions. eHealth focuses on the delivery of health services and information through web-based programs, remote monitoring, teleconsultation, and mobile device-supported care. eHealth is a potential avenue to address substance use treatment barriers in pregnancy [22], particularly during COVID-19, which has disrupted a number of face-to-face psychotherapy services. Beyond COVID-19, eHealth initiatives have the potential for broad scale health promotion for substance use [23]. eHealth is accessible, which may make it more appealing to those in remote locations. Additionally, it is cost-effective, and can be flexibly incorporated into one's schedule [24]. Given the accessible nature of eHealth interventions, some pregnant individuals may prefer to use eHealth interventions as opposed to traditional face to face treatment. Treatment preference is important to consider because matching patients to their treatment preferences has been shown to result in greater reduction of substance use behaviours [25]. Moreover, patient centered care (PCC) is one of the techniques that has been recommended to improve the quality of substance use disorder treatment—and a key aspect of patient centered care is shared decision making [26].

A number of meta-analyses of eHealth interventions for treatment of substance use disorders in the general population have been conducted, with promising results [27,28]; however, the literature in for eHealth interventions treating substance use in pregnancy has yet to be integrated as a review.

Accordingly, the primary objective of this systematic review and meta-analysis was to evaluate the effectiveness of randomized controlled trials (RCTs) on eHealth interventions delivered during pregnancy with the goal of reducing substance use, where substance use was defined broadly to include any kind of reported alcohol, tobacco, or other drugs. This definition, which includes a variety of substances at varying levels of use was justified by guidelines suggesting that all substance use should be avoided during pregnancy [29]. Substance use was measured by self-reported and objective reports of abstinence.

2. Materials and Methods

Methods outlined by the Cochrane Collaboration's Handbook [30] and the standards set by Preferred Reporting Items for Systematic review and Meta-Analysis (PRISMA) were used [31,32]. The protocol was registered with Prospero (CRD42020205186) through the University of York Centre for Reviews and Dissemination.

2.1. Search Strategy

A preliminary search found that the majority of papers investigating substance use in pregnancy were published in psychology and nursing journals. Thus, we searched the five databases that were most likely to capture the literature within these disciplines. Articles published between 1 January 2000and 19 April 2021were identified from Medline®, PsychINFO, EMBASE, CINAHL, and Cochrane CENTRAL. The most recent search took place on 18 April 2021. English-language restrictions were applied. The search terms included database specific controlled vocabulary, field codes, operators, relevant keywords, and subject headings to identify the participant population (pregnant individuals), the exposure (eHealth interventions), and the outcome (substance use) [33,34]. Key terms used to conduct the search were related to telepsychology, randomized control trials, substance use and pregnancy (see Appendix A). Duplicate articles were removed. The remaining articles were divided and were screened independently by two reviewers from a eight member research team. Pairs of reviewers screened the titles and abstracts to determine eligibility for inclusion in the full-text review, and the first author reviewer (KS) supervised and reviewed ~100 records to ensure >85% consistency. Out of 2560 abstracts, 159 conflicts were identified. The types of conflicts included whether the study was targeting the right population (pregnant people), whether the intervention fit our definition for eHealth, or whether the study included extractable outcomes. These conflicts were resolved by the first author (KS).

2.2. Inclusion and Exclusion Criteria

Studies were eligible for inclusion if they included: (a) a RCT; (b) an empirical journal article; (c) an eHealth intervention (e.g., video therapy sessions, telephone, SMS, recorded therapy sessions); (d) the goal of the intervention was to reduce substance use; (e) the sample consisted of pregnant individuals; (f) extractable outcomes with respect to substance use; and (g) the intervention took place during pregnancy. If more than one article reported results from the same intervention in the same sample, the more recent study was included in the study. Studies were excluded on the basis that they did not meet inclusion criteria or were irretrievable/unavailable in English.

2.3. Data Extraction

Two team members independently extracted data into a Microsoft Excel file and conflicts were resolved by consensus with the coders and the first and second authors. Extracted data included authors' names, publication years, country, sample demographics, pregnancy characteristics, substance use parameters, intervention characteristics and administration, and mental health assessments for all groups. Sample characteristics that were extracted when provided included sample size, age, gestational age, ethnicity, race, and gender breakdown. Study characteristics that were extracted when provided included the name of intervention, description, method of administration, degree of interactivity (i.e., completely online or some in-person component), degree of guidance, and participant time spent on the intervention. The outcomes extracted were odds ratios (OR) measuring substance use outcomes post-intervention. Corresponding authors of included articles were contacted if studies had missing or incomplete data.

2.4. Data Analysis

A random effects meta-analysis was conducted using Comprehensive Meta-Analysis Software (Biostat Inc, Englewood, NJ, USA) [35]. Most studies reported ORs, and these were used to calculate meta-estimates of substance use post-intervention in the intervention groups compared to the control groups. Ref. [36] was the only study to report chi-squares, which were transformed to ORs through the CMA software. Some studies had several post-tests (e.g., immediately post-intervention, later follow-up) and outcomes (e.g., smoking, alcohol use, general substance use). To meet the assumption of independence, effect sizes from the same study were aggregated in CMA and the single effect size estimate for each study was used to calculate pooled ORs. A forest plot was also created to display the ORs for each individual study as well as the pooled OR from all the studies. To test for publication bias, the Begg and Mazumdar rank correlation test [37] as well as the Egger's regression test [38] were performed to assess bias by regressing standardized effect size to the studies precision. A significant test indicates publication bias, or significant funnel plot asymmetry [36,38,39]. Meta-regression analyses were originally planned to explore significant moderators and explore secondary outcomes; however, not enough studies were included to complete these analyses. Sensitivity analyses were also completed to assess the robustness of the synthesized results.

2.5. Quality Assessment

To assess the quality of the RCT studies, the Cochrane Risk of Bias Tool for randomized trials was used [30]. This tool assesses literature based on seven potential sources of bias within the general categories of selection bias (allocation concealment), performance and detection bias (blinding), attrition bias (incomplete data) and selective reporting bias [39]. Bias was judged individually by a team member and then cross-referenced by the judgment of another team member to complete a 100% check. Total scores range from 0 (unlikely to alter the results), to 7 (greatly weakens confidence in the results). Higher scores indicate lower study quality and a higher risk of biased results. The study informally defined scores from 0–2 as low risk, scores from 3–5 as moderate risk, and scores between 6–7 as high risk.

2.6. Primary Outcome

The current review aimed to determine whether eHealth interventions delivered during pregnancy reduced substance use when compared to a control group. Substance use was measured using self-reports of frequency and quantity of substances taken, as well as self-report measures of abstinence and objective measures of abstinence. Objective forms of abstinence were defined as a biochemical measure of substance use. For example, in certain studies where smoking was the outcome, carbon monoxide (CO) readings and/or saliva samples were tested for a certain amount of cotinine.

3. Results
3.1. Study Selection

This search was originally conducted with an associated study [40], which reviewed eHealth interventions in pregnancy for treatment of depression, anxiety, and insomnia. A wider search was conducted to include substance use for the purposes of this paper. The search identified 5505 relevant articles, with 2945 duplicates removed. In total, 2367 of the articles were excluded after title and abstract review and 193 articles were reviewed at the full-text level. A total of 6 articles met inclusion criteria for this review. See Figure 1 for the PRISMA diagram [32].

3.2. Characteristics of Included Studies

Table 1 provides characteristics of the included studies. Participant baseline age ranged from 18–37 years old. Gestational age ranged from 4–23 weeks. Of the four studies which reported ethnicity, three studies had a total sample where >85% of participants were of European descent [36,41,42].

With respect to the type of eHealth interventions, most of the interventions were created in a way that communication of services took place through the use of technology (i.e., telephone/text), rather than the use of a specific app to reduce substance use behaviours. Four of the eHealth interventions were delivered via computer or the internet [42–45], one was delivered through text message (SMS) [41] and one was delivered via telephone [36]. The types of interventions that were delivered included: motivational interviewing in one study [43], the use of general health advice (presented educational information regarding substance use without a psychological component) in three studies [42,45,46], and psychoeducation in two studies [36,41]. Three studies assigned control participants to receive treatment as usual from their healthcare providers [41,43,46], one study provided control group participants access to a website with standard advice, [42] and two studies used a time-matched placebo condition [43,44].

Interventions varied with respect to whether the eHealth intervention was guided or unguided, which was defined by whether a therapist/healthcare professional facilitated treatment. Most of the included studies were guided (*n* = 4) [36,42–44]. Two studies were unguided [41,45]. Length of follow-up varied across studies with follow-up occurring at 4 weeks after baseline [41], 8 weeks after baseline [42], 12 weeks after baseline [43], 16 weeks after baseline [44], up to 22 weeks after baseline [36].

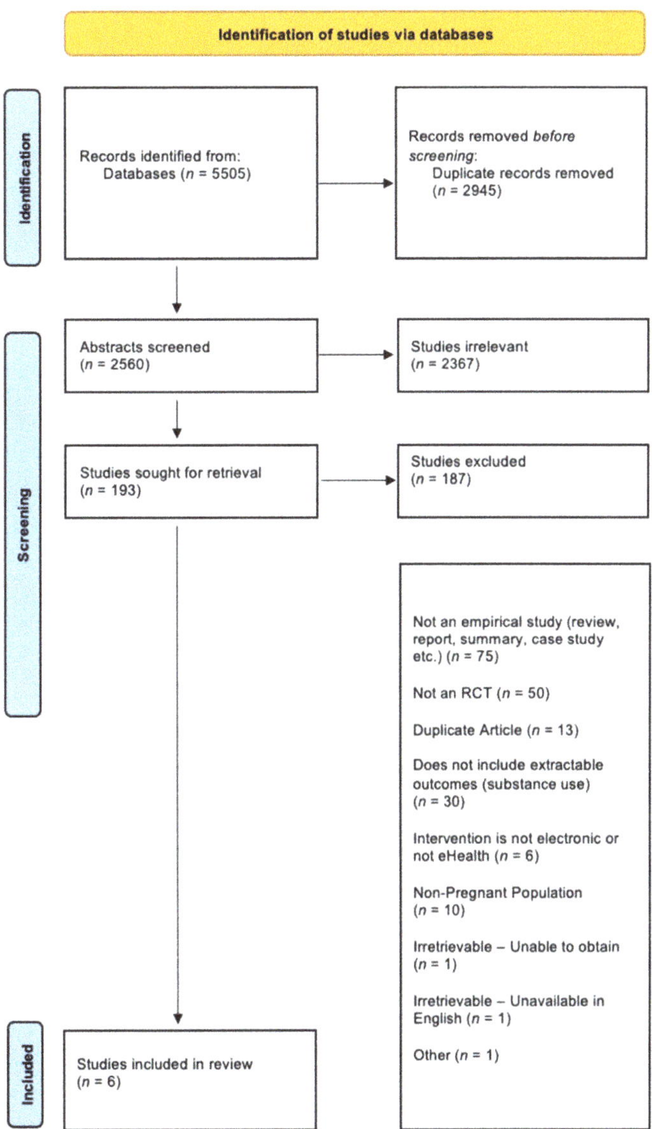

Figure 1. PRISMA diagram detailing the database searches, the number of abstracts screened, and the full texts retrieved.

Table 1. Study Characteristics.

References	Title	Year	Country	Intervention (N); Control (N)	Intervention, Age of Sample (M, SD), Range (y)	Control, Age of Sample (M, SD), Range (y)	Intervention, Gestational Age (M, SD), Range (w) Control, Gestational Age (M, SD), Range (w)	Type of eHealth Intervention	Description of Intervention	Time of Follow Up
Bullock, L., et al. [36]	Baby BEEP: A Randomized Controlled Trial of Nurses' Individualized Social Support for Poor Rural Pregnant Smokers	2009	United States	Telephone Social Support Group and Booklet Group (117), Control (119)	Telephone Social Support Group and Booklet Group (23.1, 4.3) Telephone Social Support Only Group (24.0, 4.7) Booklet Only Group (23.6, 4.8)	Control (23.9, 4.8)	13.5 weeks 13.5 weeks	Telephone Intervention	Baby BEEP intervention consisted of a scheduled weekly telephone call and 24-h access to the nurse for any additional social support needed	T2 (28–32 weeks gestation; T3: 6 weeks postpar-tum)
Naughton, F. [41]	Large multi-centre pilot randomized controlled trialtesting a low-cost, tailored, self-help smoking cessationtext message intervention for pregnant smokers (MiQuit)	2017	England	203 204	26.6 (5.7), 16.9–40.0	26.4 (5.7), 16.6–41.3	14.6 weeks (4.2), 4–23 14.7 weeks (4.5), 3–24	Text Message Based Intervention	MiQuit is an automated text support service that delivers information or motivational messages	4 weeks post-randomization up until 36 weeks gestation
Herbec, A., et al. [42]	Pilot randomized controlled trial of an internet-based smoking cessation intervention for pregnant smokers ('MumsQuit')	2014	England	99 101	27.6 (6.0)	26.1 (5.8)	NR	Internet-Based Intervention	MumsQuit is a personalized, interactive quitting plan that mimics advisory support from a smoking cessation expert	8 weeks post baseline
Ondersma, S. J., et al. [43]	Computer-Delivered Screening and Brief Intervention for Alcohol Use in Pregnancy: A Pilot Randomized Trail	2015	United States	24 24	18–25 (50.0%) 26–33 (33.3%) 34–37 (16.7%)	18–25 (58.3%) 26–33 (33.3%) 34–37 (8.3%)	12.5 weeks (5.6) 12.0 weeks (5.3)	Computer-delivered screening and brief intervention (eSBI)	A brief 20-min video was delivered via tablet while waiting for a prenatal care appointment and three separated tailored mailings followed	3 month follow up
Tzilos, Wernette. G., et al. [44]	A Pilot Randomized Controlled Trial of a Computer-Delivered Brief Intervention for Substance Use and Risky Sex During Pregnancy	2018	United States	31 19	25.1 (5.79)	23.2 (4.21)	12.9 (4.76) 13.9 (4.21)	Computer Delivered Intervention	A single motivational session and a booster session	4 month follow up

Table 1. *Cont.*

References	Title	Year	Country	Intervention (N); Control (N)	Intervention, Age of Sample (M, SD), Range (y)	Control, Age of Sample (M, SD), Range (y)	Intervention, Gestational Age (M, SD), Range (w) Control, Gestational Age (M, SD), Range (w)	Type of eHealth Intervention	Description of Intervention	Time of Follow Up
van de Wulp, N., et al. [45]	Reducing Alcohol Use During Pregnancy Via Health Counseling by Midwives and Internet-Based Computer-Tailored Feedback: A Cluster Randomized Trial	2014	Netherlands	Computer tailoring (111); Usual care (124)	32.31 (4.22)	33.53; (3.85)	7.73 weeks (2.06) 7.92 weeks (1.99)	Computer delivered intervention	Respondents in the computer-tailoring group received usual care from their midwife and computer-tailored feedback via the Internet. This feedback was tailored to the participant's alcohol use, knowledge, risk perception, attitude, social influence, self-efficacy, intention, and action and coping plans	6 months after baseline

With respect to the type of substances studied, three of the studies assessed smoking using time sensitive self-reported abstinence, as well as dose and dependence tests of drugs [36,41,42]. Two studies assessed alcohol use with time sensitive self-reported abstinence [43,45]. One study measured general substance use using time sensitive self-reported abstinence [44]. Five studies used self-reports of either abstinence or daily substance use behaviours as the outcome [41–45]. Two studies used validated tests of dose and dependence [36,41] carbon monoxide (CO) readings and/or saliva samples were tested for a certain amount of cotinine (<30 mg/mL). One study provided both self-reports of substance use and reports of validated tests of dose and dependence [41]. Only one of the included studies showed a statistically significant benefit of eHealth interventions over the control group [45].

3.3. Risk of Bias in Included Studies RCTs

The results of the quality assessments are shown in Figure 2. Overall, risk of bias was rated as low for 5 of the 6 studies that were included, where low was defined as a risk of bias score between 0–2. The most common risk of bias was due to attrition (missing data). Other common risks of bias within the current review were detection bias as well as selection bias.

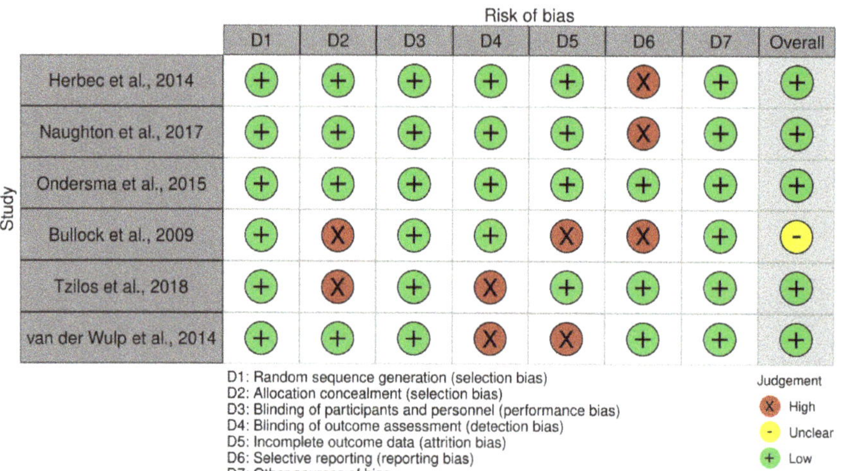

Figure 2. Visual plot demonstrating results from quality assessment. Recall that the study informally defined scores from 0–2 as low risk, scores from 3–5 as moderate risk, and scores between 6–7 as high risk.

3.4. Efficacy of eHealth Interventions on Substance Use

Using a random-effects model, the efficacy of the eHealth interventions was tested by calculating a pooled OR across 1176 participants and comparing the intervention group (n = 585) to control group (n = 591). Results showed that pregnant participants who received an eHealth intervention for the treatment of substance use had 1.3 times greater likelihood of reduced substance use compared to those who were assigned to a control group (OR = 1.325, 95% CI = 1.062–1.654, Z = 2.490, p = 0.013; see Figure 3 and Table 2).

Table 2. Primary Outcomes.

References	Title	Year	Type of Substance	Measurement used for Assessment	Intervention Effect on Substance Use	Odds Ratio	Confidence Interval	p-Value	Quality Assessment Rating
Bullock, L., et al. [36]	Baby BEEP: A Randomized Controlled Trial of Nurses' Individualized Social Support for Poor Rural Pregnant Smokers	2009	Smoking	Readiness to Stop Smoking, The Fagerstrom Test for Nicotine Dependence and Dosage (cotinine < 30ng/mL)	The nurse-delivered social support telephone intervention was not more effective than booklets alone or usual care in reducing smoking behaviour.	1.18	0.90–1.54	0.239	3
Naughton, F. [41]	Large multi-centre pilot randomized controlled trial testing a low-cost, tailored, self-help smoking cessation text message intervention for pregnant smokers (MiQuit)	2017	Smoking	Validated 4-week continuous abstinence (CO readings < 9ppm), Self-reported 4-week continuous abstinence, 7-day point prevalence for 4-weeks continuous abstinence	No statistical significance was found between the MiQuit intervention group and the usual care control group.	1.32	0.68–2.56	0.409	1
Herbec, A., et al. [42]	Pilot randomized controlled trial of an internet-based smoking cessation intervention for pregnant smokers ('MumsQuit')	2014	Smoking	Self-reported 4-week continuous abstinence	The analysis determined no significant difference when measuring continuous abstinence rates between the MumsQuit intervention and control group.	1.50	0.79–2.86	0.217	1
Ondersma, S. J., et al. [43]	Computer-Delivered Screening and Brief Intervention for Alcohol Use in Pregnancy: A Pilot Randomized Trail	2015	Alcohol Use	Self-reported 90-day abstinence period	No statistical significance was found between the intervention group and the control group when comparing abstinence.	3.20	0.52–19.78	0.211	0
Tzilos, Wernette. G , et al. [44]	A Pilot Randomized Controlled Trial of a Computer-Delivered Brief Intervention for Substance Use and Risky Sex During Pregnancy	2018	General substance use	Self-reported substance use behaviours using a calendar and multiple prompts	The final analysis determined no significant reduction of substance use in the intervention group compared to the control group.	2.06	0.59–7.31	0.249	2
van de Wulp, N., et al. [45]	Reducing Alcohol Use During Pregnancy Via Health Counseling by Midwives and Internet-Based Computer-Tailored Feedback: A Cluster Randomized Trial	2014	Alcohol	Average alcohol consumption during pregnancy was assessed with the 5-item Dutch Quantity-Frequency-Variability (QFV) Questionnaire	The final analysis showed that computer-tailoring respondents used alcohol significantly less often when compared to usual care respondents.	2.77	1.05–7.32	0.040	2

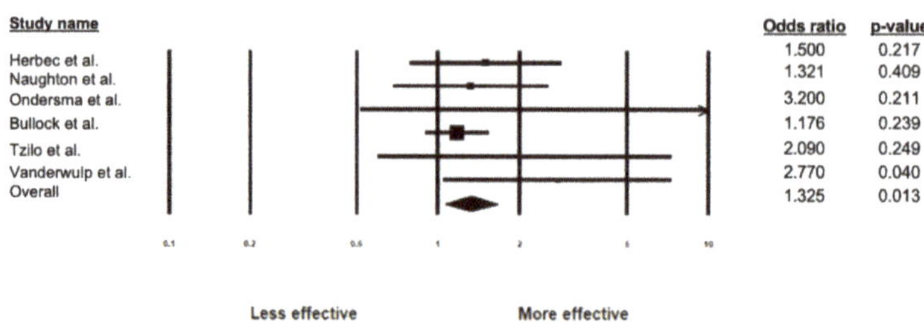

Figure 3. Forest Plot displaying individual and overall effect sizes examining the effectiveness of eHealth interventions.

Significant heterogeneity was observed across studies (Q = 4.505, p = 0.479, I^2 = 0.000). Egger's regression test (B = 1.39, $t(4)$ = 4.43, SE = 0.324, p = 0.012) and the Begg and Mazumdar rank correlation test (*Tau* = 0.600, Z = 1.69, p = 0.090) showed mixed findings regarding the presence of publication bias, potentially due to the rank correlation test having lower power [46]. The funnel plot showed clear asymmetry on the positive side which suggests that the overall effect size may be smaller than estimated (See Figure 4).

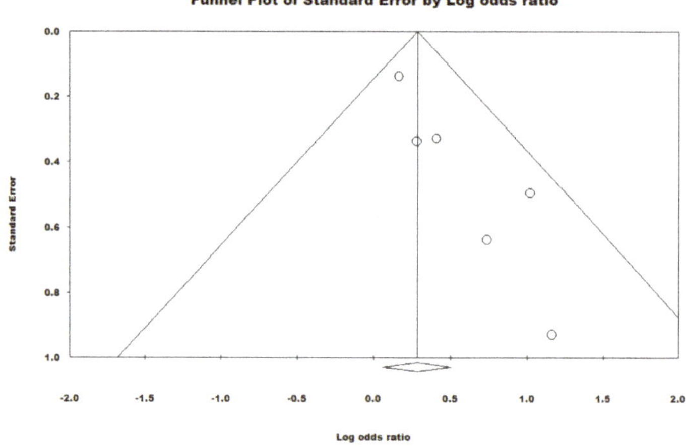

Figure 4. Funnel Plot demonstrating the presence of publication bias.

A subgroup analysis was also conducted on the studies which measured abstinence (n = 4), which also revealed a small size effect size where the odds of increased abstinence was 1.25 times greater for pregnant individuals in the intervention group than for individuals within the control group (OR = 1.251, 95% CI = 0.993–1.577). Although secondary aims of the current study were to evaluate potential moderators of treatment for substance use, not enough studies were identified to conduct a moderator analysis.

3.5. Sensitivity Analysis

After systematically removing one study at a time, it was observed that two studies affected the meta-estimate of the effect size of eHealth intervention during pregnancy by more than 5% [36,41]. Specifically, the two studies affected the meta-estimate such that it made the estimate larger [36,41]. When Bullock et al., (2011) was removed, the pooled effect size increased to 1.69 [36]. When Naughton et al., was removed, the pooled effect size increased to 1.41 [41].

4. Discussion

4.1. Primary Findings

The current systematic review and meta-analysis synthesized existing evidence from 6 RCTs on the efficacy of eHealth interventions for substance use among a pregnant population by comparing participants using an eHealth intervention (N = 585) to participants in a control group (N = 591). Participant ages ranged from 18–37 years old and gestational ages ranged from 7.73–14.7 weeks. All the studies took place in an economically advantaged country which speaks to the potential difference in the accessibility for eHealth interventions for developing countries [47]. Most of the included RCTs measured smoking and drinking outcomes, whereas studies on eHealth interventions for harder drug use were more uncommon. The lack of studies measuring harder drug use among this population may be due to the stigma associated with these substances—particularly during pregnancy [48]. The majority of the eHealth interventions were delivered via computer/internet which is consistent with other reviews on eHealth where most eHealth interventions were internet-based. This may be attributed to the rise in technology use in recent years and that most of the included studies recently took place between the years of 2014–2018. With respect to the types of interventions used, most of the included interventions were brief in nature and had minimal clinician guidance, which could have reduced the current effect sizes. There was also variability in the time of follow-up among studies, with some studies having a shorter follow-up ranging from 4 weeks to 6 months. All of the included studies were assessed to have a low risk of bias which provides support for the high quality of the studies in the current review. The most common type of bias noted among studies was attrition bias, though this is common for eHealth interventions [49].

Results of the meta-analysis showed that eHealth interventions delivered in pregnancy reduced substance use when compared to control conditions (OR = 1.33, p = 0.013). This effect size was calculated using a predominantly intent-to-treat sample (ITT). With the exception of one study [43], each of the included studies reported an OR which were included based on an ITT analysis. Though it should be noted that the study which did not use an ITT sample completed a sensitivity analysis and found that the completers sample as opposed to the use of an ITT sample did not significantly differ [43]. Results suggest that eHealth interventions are significantly more effective than control conditions in reducing substance use in pregnancy. Findings from the sensitivity analyses found that two studies when removed, made the effect size larger [36,41]. Of note, both of these studies had smoking as an outcome which suggests that smoking may be less amenable to treatment in comparison to other substances. Furthermore, both the [36] study and the [41] study used a telephone to deliver their intervention. This is in comparison to the other studies which largely relied on the computer/internet. This may suggest that telephone and or text interventions may not be as effective in comparison to other modes of eHealth delivery.

4.2. Consistency with Existing Literature

The finding of a small yet significant effect of eHealth interventions is consistent with current literature of eHealth interventions for substance use within non-pregnant populations [27,50]. For example, in a meta-analysis examining the effectiveness of internet interventions for adults with substance abuse issues, results showed that internet interventions significantly decreased substance use with a small to moderate effect size (*Hedge's g* = 0.36) [27]. Results are also consistent with a meta-analysis examining the effectiveness of internet interventions for adult alcohol misuse, which also found a small but significant effect size in favour of the internet interventions (*Hedge's g* = 0.20) [50]. With respect to the effectiveness of eHealth interventions in a pregnant population, no review to date has synthesized this information. However, one review has observed the effectiveness of technology-based interventions for substance use among participants who were of a child-bearing age (ages 18–45) [51]. Results from this meta-analysis found that technology-based interventions were efficacious in comparison to control groups in preventing and reducing substance use for individuals at a child-bearing age, though the effect size was small ($d = 0.19$) [51] which is in line with the current review.

The effect sizes of eHealth interventions are generally consistent with those of face-to-face interventions for substance use within the general population. For example, a meta-analysis of psychosocial interventions for substance use in the general population found a significant yet small to moderate effect size for treatment (*Cohen's d = 0.45*) [28]. In another meta-analysis evaluating the effectiveness of motivational interviewing on substance use behaviours in adolescence, interventions were effective in improving substance use outcomes, but the effect sizes were again small (*Cohen's d = 0.17*) [52].

4.3. Obstacles to eHealth Interventions

In the studies included in this review, there were high attrition rates and varied engagement levels [53,54]. The observed attrition rates varied widely from 2% [44] to 33.5% [42] which is consistent with other eHealth interventions which typically range from 19% to 50% [55,56]. Individuals with substance use problems are more likely to terminate treatment than individuals with other psychosocial problems, with substance use treatment programs reporting the highest dropout rates when compared to individuals with other psychosocial concerns [57]. Taken together, these findings suggest that individuals taking part in a substance use eHealth intervention may require additional guidance and human monitoring to decrease levels of dropout. In the current review, only one study involved human monitoring, though this did not appear to increase effect sizes over and above other studies [36].

Furthermore, there were relatively high levels of participant attrition within the reviewed studies. This is a common feature among substance use eHealth studies and may in part be due to the lack of human monitoring. Across all eHealth programs, retaining pregnant participants to enhance positive outcomes continues to be a challenge. This speaks to the need to evaluate different ways to reduce attrition by improving participant engagement in treatment (e.g., gamification) [49,58] and implementing more rigorous designs which could include human monitoring and follow-ups by clinicians to reduce the high participant dropout rates. For example, the one study which did implement human monitoring had lower attrition rates relative to most of the other included studies [36]. In doing this, future eHealth interventions for substance use would maximize its clinical effectiveness [54,59].

4.4. Strengths, Limitations and Future Directions

Prior systematic reviews and meta-analyses were limited by the population studied, which to date, have only included the effects of eHealth on substance use within a general population. While some reviews exist on the effect of eHealth interventions on participants of "child-bearing age," these participants are not recruited during pregnancy and do not complete the eHealth intervention during pregnancy. The current study is the first, to our knowledge, to have synthesized the findings from the literature within a specifically pregnant population. This is incredibly important because as mentioned previously, many mental health and substance use behaviours in pregnancy persist into the postpartum period. Moreover, substance use has unique negative effects during pregnancy and pregnant populations may require support that is tailored to their needs.

Despite these strengths, the findings from this meta-analysis should be interpreted in the context of several limitations. Notably, the current review included a small number of studies ($n = 6$). Due to the limited number of studies included in the current analysis, there was not enough statistical power to conduct moderator analyses. Future research should investigate if demographic variables and/or study characteristics moderate the relationship between eHealth interventions and substance use during pregnancy. A larger research base is needed to better understand specifically what types of eHealth interventions and methods of delivery are most effective in pregnancy, and for whom. Substance use outcomes also varied—where most of the included studies assessed smoking cessation.

Consequently, future reviews may benefit from sub-group analyses to investigate if the effectiveness of eHealth interventions vary as a function of the type of substance use being treated and/or differences in using self-report and objective measures to assess treatment success. There was also wide variability in the types of eHealth interventions that were used, with delivery occurring through internet, phone, and text messages, which likely have also influenced the heterogeneity of the reported outcomes in this review. It should also be noted that since all reviewed studies took place in economically advantaged countries (i.e., United States, England, Netherlands), the lack of studies from other countries may limit the generalizability of our findings. Additionally, publication bias was observed in the outcomes Egger's Regression test ($p = 0.012$), which may suggest that effect sizes were over-estimated within the current review [38].

Future interventions should consider the high comorbidity that substance use has with various mental health concerns, including anxiety [60], depression [60], bipolar disorder, [61], attention-deficit hyperactivity disorder [62], and suicidality [63,64]. Moreover, individuals in the general population who were treated in programs providing specific treatment to target concurrent disorders had higher rates of using mental health services, which predicted improvements in both mental health and substance use following treatment [65]. Lastly, future studies should also examine the impact of eHealth treatments during pregnancy and in the postpartum period and beyond, with the hope that eHealth interventions will be able to create long-lasting treatment effects which persist beyond the intervention period.

4.5. Clinical Implications

According to the World Health Organization, all pregnant people should have access to affordable treatment options that respects their autonomy [66]. eHealth interventions may be best used as a first line intervention in stepped care models, as they may increase accessibility for some clients, and be less costly than more intensive in-person approaches [67,68]. The privacy and anonymity afforded by eHealth interventions may increase the likelihood of seeking support for substance use in pregnancy. eHealth interventions are also in line with international guidelines for the treatment of substance use during pregnancy [66,69]. eHealth interventions have the potential to be tailored to track substance use and treatment progress, as well as provide information on where pregnant people can receive more intensive substance treatment, which is in line with the National Institute for Health and Care Excellence guidelines in the treatment of substance use during pregnancy [69].

However, it is important to recognize that not everyone has access to reliable internet. Indeed, according to a report from the United Nations, over half of the world population does not have access to reliable internet, and there are important sociodemographic disparities in internet access within countries [70]. For example, only 24% of First Nation reserves in Canada have access to reliable internet [71]. Moreover, internet use may be limited be lack of devices (i.e., only one computer for one family), and compromised internet speed due to multiple devices. In another study which looked at telemedicine use in Peru, noted that almost 60% of the population in Peru belong to the lowest socioeconomic strata, preventing them from owning devices such as a computer or smartphone with internet access [72]. Furthermore, due to the lack of owning these devices, these individuals may lack the technological skills to know how to access and utilize telemedicine services, even if a device is provided to them [72]. Additionally, in a study of the disparities in digital access among Medicare beneficiaries, results found that individuals who lacked digital access were higher among those with low socioeconomic status, those 85 years or older, and in certain ethnic communities [73].

Accordingly, improved internet coverage and digital access has been highlighted as an important step into making eHealth more accessible. The future of eHealth should include determining how these interventions can be properly incorporated into the current healthcare system to increase patient accessibility to mental health services.

5. Conclusions

Taken together, the review found that eHealth interventions are effective in treating substance use during pregnancy. Furthermore, eHealth interventions are a promising healthcare intervention that increase access to care. In order to increase effect sizes, adaptations should be considered with development occurring, in conjunction with patient and provider partners. For example, future eHealth interventions could be more integrated such that treatment is completed in conjunction with therapist or peer support and additional guidance could be provided by having increased interactions with clinicians as opposed to pre-set modules which are to be completed at the patient's pace. This suggestion is supported by the finding that guided eHealth interventions were significantly superior to unguided interventions [74].

Author Contributions: Conceptualization, L.M.T.-M. and K.S.; methodology, H.S.; software, P.R.S.; validation, K.S., H.S. and P.R.S.; formal analysis, P.R.S.; investigation, K.S.; resources, L.M.T.-M.; data curation, H.S.; writing—original draft preparation, K.S.; writing—review and editing, K.S.; C.R.; H.S.K.; L.M.T.-M.; visualization, K.S.; supervision, L.M.T.-M.; project administration, L.M.T.-M.; funding acquisition, L.M.T.-M. All authors have read and agreed to the published version of the manuscript.

Funding: This work was supported by the generous donors of the Alberta Children's Hospital Foundation, the Canadian Child Health Clinician Scientist Program (CCHCSP), the Canadian Institute of Health Research and the Fonds de Recherche du Québec—Santé.

Institutional Review Board Statement: Not applicable.

Informed Consent Statement: Not applicable.

Data Availability Statement: Not applicable.

Acknowledgments: We thank the team of research assistants who contributed to data collection and extraction, including Roshni Sohail, Jasleen Kaur, Beatrice Valmana, Songyang (Mark) Jin, and Makayla Freeman.

Conflicts of Interest: The authors report no conflict of interest.

Appendix A

Table A1. Terms used in Search Strategy.

eHealth/Telepsychology	And	Study Design	And	Mental Health and Substance Use	And	Pregnancy
eHealth/e-Health		RCT		Substance Use		pregnant
internet		efficacy		externalizing		perinatal/peri-natal
online/on-line		random allocation		substance-related disorder/substance related disorder		prepartum/pre-partum
app/apps		effectiveness		substance abuse/substance-abuse		antenatal/ante-natal
web-based/web based		randomized controlled trial		substance dependence/substance-dependence		birth
smart-phone/smartphone/smart phone		trial		addiction		childbirth/child-birth
mobile phone/mobile-phone		controlled clinical trial		drug abuse/drug-abuse		labor
mobile health		clinical trial		drug dependence/drug-dependence		labour
mHealth				alcohol abuse/alcohol-abuse		gestation

Table A1. *Cont.*

eHealth/Telepsychology	And	Study Design	And	Mental Health and Substance Use	And	Pregnancy
app-based				alcohol dependence/ alcohol-dependence		
computer systems				alcoholism/alcoholic		
computers						
cell phone/cell-phone/ cellphone						
website						
computer						
social media						
web-based/web based						
SMS						
mobile						
text-based/text based						
digital						
self-directed/self directed						
technology-assisted/ technology assisted						
self-help/self help						
self-guided/self guided						
telecommunications/ telecommunication						

References

1. Whiteford, H.A.; Ferrari, A.J.; Degenhardt, L.; Feigin, V.; Vos, T. The global burden of mental, neurological and substance use disorders: An analysis from the Global Burden of Disease Study 2010. *PLoS ONE* **2015**, *10*, e0116820. [CrossRef]
2. Compton, W.M.; Thomas, Y.F.; Stinson, F.S.; Grant, B.F. Prevalence, correlates, disability, and comorbidity of DSM-IV drug abuse and dependence in the United States: Results from the national epidemiologic survey on alcohol and related conditions. *Arch. Gen. Psychiatry* **2007**, *64*, 566–576. [CrossRef]
3. Results from the 2010 National Survey on Drug Use and Health: Summary of National Finding. Available online: www.samhsa.gov/data/sites/default/files/NSDUHresults2010/NSDUHresults2010.htm (accessed on 10 September 2021).
4. Chasnoff, I.J.; McGourty, R.F.; Bailey, G.W.; Hutchins, E.; Lightfoot, S.O.; Pawson, L.L.; Fahey, C.; May, B.; Brodie, P.; McCulley, L. The 4P's Plus© screen for substance use in pregnancy: Clinical application and outcomes. *J. Perinatol.* **2005**, *25*, 368–374. [CrossRef]
5. Havens, J.R.; Simmons, L.A.; Shannon, L.M.; Hansen, W.F. Factors associated with substance use during pregnancy: Results from a national sample. *Drug Alcohol Depend.* **2009**, *99*, 89–95. [CrossRef] [PubMed]
6. Garg, M.; Garrison, L.; Leeman, L.; Hamidovic, A.; Borrego, M.; Rayburn, W.F.; Bakhireva, L. Validity of self-reported drug use information among pregnant women. *Matern. Child Health J.* **2016**, *20*, 41–47. [CrossRef] [PubMed]
7. Avalos, L.A.; Roberts, S.C.; Kaskutas, L.A.; Block, G.; Li, D.-K. Volume and type of alcohol during early pregnancy and the risk of miscarriage. *Subst. Use Misuse* **2014**, *49*, 1437–1445. [CrossRef] [PubMed]
8. Pinto, S.; Dodd, S.; Walkinshaw, S.; Siney, C.; Kakkar, P.; Mousa, H. Substance abuse during pregnancy: Effect on pregnancy outcomes. *Eur. J. Obstet. Gynecol. Reprod. Biol.* **2010**, *150*, 137–141. [CrossRef]
9. O'Leary, C.M.; Jacoby, P.J.; Bartu, A.; D'Antoine, H.; Bower, C. Maternal alcohol use and sudden infant death syndrome and infant mortality excluding SIDS. *Pediatrics* **2013**, *131*, e770–e778. [CrossRef] [PubMed]
10. Ali, K.; Greenough, A. Sudden infant death syndrome (SIDS), substance misuse, and smoking in pregnancy. *Res. Rep. Neonatol.* **2012**, *2*, 95–101.
11. Forray, A. Substance use during pregnancy. *F1000Research* **2016**, *5*, 887. [CrossRef] [PubMed]
12. Warner, T.D.; Roussos-Ross, D.; Behnke, M. It's not your mother's marijuana: Effects on maternal-fetal health and the developing child. *Clin. Perinatol.* **2014**, *41*, 877–894. [CrossRef] [PubMed]
13. Green, C.R.; Roane, J.; Hewitt, A.; Muhajarine, N.; Mushquash, C.; Sourander, A.; Lingley-Pottie, P.; McGrath, P.; Reynolds, J.N. Frequent behavioural challenges in children with fetal alcohol spectrum disorder: A needs-based assessment reported by caregivers and clinicians. *J. Popul. Ther. Clin. Pharmacol.* **2014**, *21*, 21.
14. Fuglestad, A.J.; Whitley, M.L.; Carlson, S.M.; Boys, C.J.; Eckerle, J.K.; Fink, B.A.; Wozniak, J.R. Executive functioning deficits in preschool children with Fetal Alcohol Spectrum Disorders. *Child Neuropsychol.* **2015**, *21*, 716–731. [CrossRef] [PubMed]
15. Rangmar, J.; Hjern, A.; Vinnerljung, B.; Strömland, K.; Aronson, M.; Fahlke, C. Psychosocial outcomes of fetal alcohol syndrome in adulthood. *Pediatrics* **2015**, *135*, e52–e58. [CrossRef] [PubMed]

16. Saloner, B.; Karthikeyan, S. Changes in substance abuse treatment use among individuals with opioid use disorders in the United States, 2004–2013. *JAMA* **2015**, *314*, 1515–1517. [CrossRef] [PubMed]
17. Substance Abuse and Mental Health Services Administration. *2017 National Survey on Drug Use and Health: Detailed Tables*; Substance Abuse and Mental Health Services Administration Rockville: Rockville, MD, USA, 2018.
18. Xu, J.; Rapp, R.C.; Wang, J.; Carlson, R.G. The multidimensional structure of external barriers to substance abuse treatment and its invariance across gender, ethnicity, and age. *Subst. Abus.* **2008**, *29*, 43–54. [CrossRef] [PubMed]
19. Substance Abuse and Mental Health Services Administration. *National Survey on Drug Use and Health*; Substance Abuse and Mental Health Services Administration: Rockville, MD, USA, 2014.
20. Stone, R. Pregnant women and substance use: Fear, stigma, and barriers to care. *Health Justice* **2015**, *3*, 1–15. [CrossRef]
21. Frazer, Z.; McConnell, L.M. Treatment for substance use disorders in pregnant women: Motivators and barriers. *Drug Alcohol Depend.* **2019**, *205*, 107652. [CrossRef]
22. World Health Organization. *mHealth: New Horizons for Health through Mobile Technologies*; World Health Organization: Geneva, Switzerland, 2011.
23. Ondersma, S.J.; Walters, S.T. Clinician's Guide to Evaluating and Developing eHealth Interventions for Mental Health. *Psychiatr. Res. Clin. Pract.* **2020**, *2*, 26–33. [CrossRef]
24. Smit, F.; Lokkerbol, J.; Riper, H.; Majo, C.; Boon, B.; Blankers, M. Modeling the cost-effectiveness of health care systems for alcohol use disorders: How implementation of eHealth interventions improves cost-effectiveness. *J. Med. Internet Res.* **2011**, *13*, e56. [CrossRef]
25. Friedrichs, A.; Spies, M.; Härter, M.; Buchholz, A. Patient preferences and shared decision making in the treatment of substance use disorders: A systematic review of the literature. *PLoS ONE* **2016**, *11*, e0145817. [CrossRef] [PubMed]
26. Marchand, K.; Beaumont, S.; Westfall, J.; MacDonald, S.; Harrison, S.; Marsh, D.C.; Schechter, M.T.; Oviedo-Joekes, E. Conceptualizing patient-centered care for substance use disorder treatment: Findings from a systematic scoping review. *Subst. Abus. Treat. Prev. Policy* **2019**, *14*, 37. [CrossRef]
27. Boumparis, N.; Karyotaki, E.; Schaub, M.P.; Cuijpers, P.; Riper, H. Internet interventions for adult illicit substance users: A meta-analysis. *Addiction* **2017**, *112*, 1521–1532. [CrossRef]
28. Dutra, L.; Stathopoulou, G.; Basden, S.L.; Leyro, T.M.; Powers, M.B.; Otto, M.W. A meta-analytic review of psychosocial interventions for substance use disorders. *Am. J. Psychiatry* **2008**, *165*, 179–187. [CrossRef] [PubMed]
29. Public Health Agency of Canada. Your Guide to a Healthy Pregnancy. Available online: https://www.canada.ca/en/public-health/services/health-promotion/healthy-pregnancy/healthy-pregnancy-guide.html (accessed on 2 June 2021).
30. Higgins, J.P.; Altman, D.G.; Gøtzsche, P.C.; Jüni, P.; Moher, D.; Oxman, A.D.; Savović, J.; Schulz, K.F.; Weeks, L.; Sterne, J.A. The Cochrane Collaboration's tool for assessing risk of bias in randomised trials. *BMJ* **2011**, *343*, d5928. [CrossRef]
31. Moher, D.; Shamseer, L.; Clarke, M.; Ghersi, D.; Liberati, A.; Petticrew, M.; Shekelle, P.; Stewart, L.A. Preferred reporting items for systematic review and meta-analysis protocols (PRISMA-P) 2015 statement. *Syst. Rev.* **2015**, *4*, 1. [CrossRef] [PubMed]
32. Page, M.J.; Moher, D.; Bossuyt, P.M.; Boutron, I.; Hoffmann, T.C.; Mulrow, C.D.; Shamseer, L.; Tetzlaff, J.M.; Akl, E.A.; Brennan, S.E. PRISMA 2020 explanation and elaboration: Updated guidance and exemplars for reporting systematic reviews. *BMJ* **2021**, *372*, n160. [CrossRef]
33. Covidence Systematic Review Software. Available online: www.covidence.org (accessed on 15 July 2021).
34. Veritas Health Innovation. *Covidence Systematic Review Software*; Veritas Health Innovation: Melbourne, Australia, 2017.
35. Borenstein, M.; Hedges, L.; Higgins, J.; Rothstein, H. *Comprehensive Meta-Analysis*; Version 3 Biostat: Englewood, NJ, USA, 2013.
36. Bullock, L.; Everett, K.D.; Mullen, P.D.; Geden, E.; Longo, D.R.; Madsen, R. Baby BEEP: A randomized controlled trial of nurses' individualized social support for poor rural pregnant smokers. *Matern. Child Health J.* **2009**, *13*, 395–406. [CrossRef]
37. Begg, C.B.; Mazumdar, M. Operating characteristics of a rank correlation test for publication bias. *Biometrics* **1994**, *50*, 1088–1101. [CrossRef]
38. Egger, M.; Smith, G.D.; Schneider, M.; Minder, C. Bias in meta-analysis detected by a simple, graphical test. *BMJ* **1997**, *315*, 629–634. [CrossRef]
39. Sterne, J.A.; Savović, J.; Page, M.J.; Elbers, R.G.; Blencowe, N.S.; Boutron, I.; Cates, C.J.; Cheng, H.-Y.; Corbett, M.S.; Eldridge, S.M. RoB 2: A revised tool for assessing risk of bias in randomised trials. *BMJ* **2019**, *366*, l4898. [CrossRef]
40. Silang, K.A.; Sohal, P.R.; Bright, K.; Leason, J.; Roos, L.; Lebel, G.G.; Tomfohr-Madsen, L.M. eHealth Interventions for Treatment and Prevention of Depression, Anxiety and Insomnia during Pregnancy: A Systematic Review and Meta-Analysis. *J. Med. Internet Res.*. submitted for publication.
41. Naughton, F.; Cooper, S.; Foster, K.; Emery, J.; Leonardi-Bee, J.; Sutton, S.; Jones, M.; Ussher, M.; Whitemore, R.; Leighton, M. Large multi-centre pilot randomized controlled trial testing a low-cost, tailored, self-help smoking cessation text message intervention for pregnant smokers (MiQuit). *Addiction* **2017**, *112*, 1238–1249. [CrossRef] [PubMed]
42. Herbec, A.; Brown, J.; Tombor, I.; Michie, S.; West, R. Pilot randomized controlled trial of an internet-based smoking cessation intervention for pregnant smokers ('MumsQuit'). *Drug Alcohol Depend.* **2014**, *140*, 130–136. [CrossRef]
43. Ondersma, S.J.; Beatty, J.R.; Svikis, D.S.; Strickler, R.C.; Tzilos, G.K.; Chang, G.; Divine, G.W.; Taylor, A.R.; Sokol, R.J. Computer-delivered screening and brief intervention for alcohol use in pregnancy: A pilot randomized trial. *Alcohol. Clin. Exp. Res.* **2015**, *39*, 1219–1226. [CrossRef]

44. Wernette, G.T.; Plegue, M.; Kahler, C.W.; Sen, A.; Zlotnick, C. A pilot randomized controlled trial of a computer-delivered brief intervention for substance use and risky sex during pregnancy. *J. Women's Health* **2018**, *27*, 83–92. [CrossRef]
45. van der Wulp, N.Y.; Hoving, C.; Eijmael, K.; Candel, M.J.; van Dalen, W.; De Vries, H. Reducing alcohol use during pregnancy via health counseling by midwives and internet-based computer-tailored feedback: A cluster randomized trial. *J. Med. Internet Res.* **2014**, *16*, e274. [CrossRef]
46. Sterne, J.A.; Egger, M. Regression methods to detect publication and other bias in meta-analysis. *Publ. Bias Meta-Anal. Prev. Assess. Adjust.* **2005**, *99*, 110.
47. Qureshi, Q.A.; Shah, B.; Kundi, G.M.; Nawaz, A.; Miankhel, A.K.; Chishti, K.A.; Qureshi, N.A. Infrastructural barriers to e-health implementation in developing countries. *Eur. J. Sustain. Dev.* **2013**, *2*, 163.
48. Eggertson, L. Stigma a major barrier to treatment for pregnant women with addictions. *Can. Med. Assoc.* **2013**, *185*, 1562. [CrossRef] [PubMed]
49. Geraghty, A.W.; Torres, L.D.; Leykin, Y.; Pérez-Stable, E.J.; Muñoz, R.F. Understanding attrition from international internet health interventions: A step towards global eHealth. *Health Promot. Int.* **2013**, *28*, 442–452. [CrossRef]
50. Riper, H.; Spek, V.; Boon, B.; Conijn, B.; Kramer, J.; Martin-Abello, K.; Smit, F. Effectiveness of E-self-help interventions for curbing adult problem drinking: A meta-analysis. *J. Med. Internet Res.* **2011**, *13*, e42. [CrossRef] [PubMed]
51. Hai, A.H.; Hammock, K.; Velasquez, M.M. The efficacy of technology-based interventions for alcohol and illicit drug use among women of childbearing age: A systematic review and meta-analysis. *Alcohol. Clin. Exp. Res.* **2019**, *43*, 2464–2479. [CrossRef]
52. Jensen, C.D.; Cushing, C.C.; Aylward, B.S.; Craig, J.T.; Sorell, D.M.; Steele, R.G. Effectiveness of motivational interviewing interventions for adolescent substance use behavior change: A meta-analytic review. *J. Consult. Clin. Psychol.* **2011**, *79*, 433. [CrossRef] [PubMed]
53. Licciardone, J.C.; Pandya, V. Feasibility Trial of an eHealth Intervention for Health-Related Quality of Life: Implications for Managing Patients with Chronic Pain during the COVID-19 Pandemic. *Healthcare* **2020**, *8*, 381. [CrossRef] [PubMed]
54. Milward, J.; Drummond, C.; Fincham-Campbell, S.; Deluca, P. What makes online substance-use interventions engaging? A systematic review and narrative synthesis. *Digit. Health* **2018**, *4*, 2055207617743354. [CrossRef]
55. Taylor, G.M.; Dalili, M.N.; Semwal, M.; Civljak, M.; Sheikh, A.; Car, J. Internet-based interventions for smoking cessation. *Cochrane Database Syst. Rev.* **2017**, *2017*, CD007078. [CrossRef]
56. Magill, M.; Ray, L.A. Cognitive-behavioral treatment with adult alcohol and illicit drug users: A meta-analysis of randomized controlled trials. *J. Stud. Alcohol Drugs* **2009**, *70*, 516–527. [CrossRef]
57. Cournoyer, L.G.; Brochu, S.; Landry, M.; Bergeron, J. Therapeutic alliance, patient behaviour and dropout in a drug rehabilitation programme: The moderating effect of clinical subpopulations. *Addiction* **2007**, *102*, 1960–1970. [CrossRef]
58. de Paiva Azevedo, J.; Delaney, H.; Epperson, M.; Jbeili, C.; Jensen, S.; McGrail, C.; Weaver, H.; Baglione, A.; Barnes, L.E. Gamification of ehealth interventions to increase user engagement and reduce attrition. In Proceedings of the 2019 Systems and Information Engineering Design Symposium (SIEDS), Charlottesville, VA, USA, 26 April 2019; pp. 1–5.
59. Eysenbach, G.; Group, C.-E. CONSORT-EHEALTH: Improving and standardizing evaluation reports of Web-based and mobile health interventions. *J. Med. Internet Res.* **2011**, *13*, e126. [CrossRef]
60. Lai, H.M.X.; Cleary, M.; Sitharthan, T.; Hunt, G.E. Prevalence of comorbid substance use, anxiety and mood disorders in epidemiological surveys, 1990–2014: A systematic review and meta-analysis. *Drug Alcohol Depend.* **2015**, *154*, 1–13. [CrossRef]
61. Cerullo, M.A.; Strakowski, S.M. The prevalence and significance of substance use disorders in bipolar type I and II disorder. *Subst. Abus. Treat. Prev. Policy* **2007**, *2*, 29. [CrossRef] [PubMed]
62. van Emmerik-van Oortmerssen, K.; van de Glind, G.; van den Brink, W.; Smit, F.; Crunelle, C.L.; Swets, M.; Schoevers, R.A. Prevalence of attention-deficit hyperactivity disorder in substance use disorder patients: A meta-analysis and meta-regression analysis. *Drug Alcohol Depend.* **2012**, *122*, 11–19. [CrossRef]
63. Wu, L.-T.; Zhu, H.; Ghitza, U.E. Multicomorbidity of chronic diseases and substance use disorders and their association with hospitalization: Results from electronic health records data. *Drug Alcohol Depend.* **2018**, *192*, 316–323. [CrossRef] [PubMed]
64. Rioux, C.; Huet, A.-S.; Castellanos-Ryan, N.; Fortier, L.; Le Blanc, M.; Hamaoui, S.; Geoffroy, M.-C.; Renaud, J.; Séguin, J.R. Substance use disorders and suicidality in youth: A systematic review and meta-analysis with a focus on the direction of the association. *PLoS ONE* **2021**, *16*, e0255799. [CrossRef] [PubMed]
65. Grella, C.E.; Stein, J.A. Impact of program services on treatment outcomes of patients with comorbid mental and substance use disorders. *Psychiatr. Serv.* **2006**, *57*, 1007–1015. [CrossRef] [PubMed]
66. World Health Organization. *Guidelines for Identification and Management of Substance Use and Substance Use Disorders in Pregnancy*; World Health Organization: Geneva, Switzerland, 2014.
67. Shidhaye, R.; Lund, C.; Chisholm, D. Closing the treatment gap for mental, neurological and substance use disorders by strengthening existing health care platforms: Strategies for delivery and integration of evidence-based interventions. *Int. J. Ment. Health Syst.* **2015**, *9*, 40. [CrossRef] [PubMed]
68. Scogin, F.R.; Hanson, A.; Welsh, D. Self-administered treatment in stepped-care models of depression treatment. *J. Clin. Psychol.* **2003**, *59*, 341–349. [CrossRef]
69. National Institute for Health and Clinical Excellence. Descriptions of Services for Pregnant Women with Complex Social Factors. Available online: https://www.nice.org.uk/guidance/cg110/resources/service-descriptions-pdf-136153837 (accessed on 1 September 2021).

70. Downs, R. UN: Majority of World's Population Lacks Internet Access. Available online: https://www.upi.com/Top_News/World-News/2017/09/18/UN-Majority-of-worlds-population-lacks-internet-access/6571505782626/ (accessed on 3 June 2021).
71. Canadian Radio-television and Telecommunications Commission. *Communication Monitoring Report*; Canadian Radio-television and Telecommunications Commission: Ottawa, Canada, 2018; pp. 138–143.
72. Alvarez-Risco, A.; Del-Aguila-Arcentales, S.; Yáñez, J.A. Telemedicine in Peru as a result of the COVID-19 pandemic: Perspective from a country with limited internet access. *Am. J. Trop. Med. Hyg.* **2021**, *105*, 6.
73. Roberts, E.T.; Mehrotra, A. Assessment of disparities in digital access among Medicare beneficiaries and implications for telemedicine. *JAMA Intern. Med.* **2020**, *180*, 1386–1389. [CrossRef]
74. Baumeister, H.; Reichler, L.; Munzinger, M.; Lin, J. The impact of guidance on Internet-based mental health interventions—A systematic review. *Internet Interv.* **2014**, *1*, 205–215. [CrossRef]

International Journal of
***Environmental Research
and Public Health***

MDPI

Systematic Review

Adherence to Telemonitoring by Electronic Patient-Reported Outcome Measures in Patients with Chronic Diseases: A Systematic Review

Jim Wiegel [1,2,*], Bart Seppen [1,2], Marike van der Leeden [1,3], Martin van der Esch [1,4], Ralph de Vries [5] and Wouter Bos [1]

[1]   Amsterdam Rheumatology and Immunology Center, Reade, 1056 AA Amsterdam, The Netherlands; b.seppen@reade.nl (B.S.); m.vd.leeden@reade.nl (M.v.d.L.); m.vd.esch@reade.nl (M.v.d.E.); w.bos@reade.nl (W.B.)
[2]   VU Medical Center, Department of Rheumatology, Amsterdam UMC, 1081 HV Amsterdam, The Netherlands
[3]   VU Medical Center, Department of Rehabilitation Medicine, 1081 HV Amsterdam, The Netherlands
[4]   CoE Urban Vitality, Faculty Health, Amsterdam University of Applied Sciences, 1081 HV Amsterdam, The Netherlands
[5]   Medical Library, Vrije Universiteit Amsterdam, 1081 HV Amsterdam, The Netherlands; r2.de.vries@vu.nl
*   Correspondence: j.wiegel@reade.nl; Tel.: +31-20-2421805

Citation: Wiegel, J.; Seppen, B.; van der Leeden, M.; van der Esch, M.; de Vries, R.; Bos, W. Adherence to Telemonitoring by Electronic Patient-Reported Outcome Measures in Patients with Chronic Diseases: A Systematic Review. *Int. J. Environ. Res. Public Health* **2021**, *18*, 10161. https://doi.org/10.3390/ijerph181910161

Academic Editors: Marie Carmen Valenza and Irene Torres-Sanchez

Received: 9 August 2021
Accepted: 17 September 2021
Published: 27 September 2021

Publisher's Note: MDPI stays neutral with regard to jurisdictional claims in published maps and institutional affiliations.

Abstract: *Background:* Effective telemonitoring is possible through repetitive collection of electronic patient-reported outcome measures (ePROMs) in patients with chronic diseases. Low adherence to telemonitoring may have a negative impact on the effectiveness, but it is unknown which factors are associated with adherence to telemonitoring by ePROMs. The objective was to identify factors associated with adherence to telemonitoring by ePROMs in patients with chronic diseases. *Methods:* A systematic literature search was conducted in PubMed, Embase, PsycINFO and the Cochrane Library up to 8 June 2021. Eligibility criteria were: (1) interventional and cohort studies, (2) patients with a chronic disease, (3) repetitive ePROMs being used for telemonitoring, and (4) the study quantitatively investigating factors associated with adherence to telemonitoring by ePROMs. The Cochrane risk of bias tool and the risk of bias in nonrandomized studies of interventions were used to assess the risk of bias. An evidence synthesis was performed assigning to the results a strong, moderate, weak, inconclusive or an inconsistent level of evidence. *Results:* Five studies were included, one randomized controlled trial, two prospective uncontrolled studies and two retrospective cohort studies. A total of 15 factors potentially associated with adherence to telemonitoring by ePROMs were identified in the predominate studies of low quality. We found moderate-level evidence that sex is not associated with adherence. Some studies showed associations of the remaining factors with adherence, but the overall results were inconsistent or inconclusive. *Conclusions:* None of the 15 studied factors had conclusive evidence to be associated with adherence. Sex was, with moderate strength, not associated with adherence. The results were conflicting or indecisive, mainly due to the low number and low quality of studies. To optimize adherence to telemonitoring with ePROMs, mixed-method studies are needed.

Keywords: adherence; patient reported outcomes; patient reported outcome measures; telemonitoring

1. Introduction

The increasing incidence of chronic diseases and the proportional increase in health care costs require efficacy in healthcare [1–3]. To improve efficacy, new ways of delivering healthcare were evaluated, such as the use of telemonitoring for chronic diseases. Telemonitoring, or remote patient monitoring, is defined as the use of technology to monitor patients at a distance [4]. One way of telemonitoring is collecting repetitive electronic patient reported outcome measures (ePROMs) at home, which allows for a quick and easy way of a frequent capture of important disease-specific outcomes and is already

225

applied in a variety of chronic diseases such as chronic obstructive pulmonary disease, chronic kidney disease, rheumatoid arthritis, inflammatory bowel disease or congestive heart failure [5–10]. The benefits of telemonitoring by repetitive ePROMs are (i) a reduced number of outpatient visits, (ii) a reduced workload, and (iii) lower healthcare costs [11–13]. Patients that use repetitive ePROMs for telemonitoring reported (iv) improved satisfaction, (v) improved communication with the provider, (vi) more insight into their disease activity, and (vii) more control over their disease [14–16]. However, one of the main problems of telemonitoring by ePROMs is a lack of adherence, with adherence rates down to 20% [17,18]. Although adherence toward health technology is increasingly investigated, there is no comprehensive review regarding what quantitative factors are associated with adherence to telemonitoring by ePROMs specifically.

The low rate of adherence may affect the effectiveness of ePROMs. It is important to identify which factors affect adherence to telemonitoring by repetitive ePROMs, since this allows for a targeted approach in order to improve adherence and therefore its effectiveness [19]. A recent systematic review qualitatively investigated patients' experience of telemonitoring (including but not limited to telemonitoring with ePROMs) and suggested that older patients and patients with less experience with technology were concerned about their ability to use telemonitoring [20]. Furthermore, the coexistence of comorbidity, social support, self-discipline and the use of persuasive design principles in the telemonitoring tool (i.e., use of reminders) were all qualitatively identified as possible factors affecting adherence [21–24]. However, it is not yet known if they influence adherence to telemonitoring by ePROMs specifically and if the quantitative evidence supports and extends these observations.

A systematic review was performed with the aim to identify factors quantitatively associated with adherence to telemonitoring by repetitive ePROMs in patients with chronic diseases in all studies.

2. Materials and Methods

2.1. Search Strategy

A literature search was performed from inception to 6 June 2020 and was updated on 8 June 2021, based on the preferred reporting items for systematic reviews and meta-analyses (PRISMA) statement (www.prisma-statement.org, accessed on 5 March 2020). At the initiation of this review, the new PRIMSA guidelines were not yet available; therefore, this review was conducted following the 2009 PRISMA guidelines [25]. To identify all relevant publications, we conducted systematic searches in the bibliographic databases PubMed, Embase, PsycINFO (Ebsco) and the Cochrane Library (Wiley), in collaboration with a medical information specialist. The following terms were used (including synonyms and closely related words) as index terms or free-text words: "electronic patient reported outcome", "e-Health", "Telemonitoring", "Remote patient monitoring", "Mobile applications", "Chronic disease", "Adherence", "Usage", "Dropout", Engagement" and "Compliance". The references of the identified articles were searched for relevant publications. Duplicate articles were excluded. The full search strategies for all databases can be found in Appendix A (Table A1).

2.2. Selection Process

Two reviewers (JW and BS) independently screened all potentially relevant titles and abstracts for eligibility. We selected articles where (P) patients with chronic diseases (I) telemonitored their symptoms with repetitive ePROMs, (C) no comparison groups were necessary, and (O) where factors potentially positively or negatively associated with adherence were investigated, (S) in trial or cohort studies. Differences in judgement were resolved through consensus. Studies were included if they met the following criteria:

1. Type of study is a randomized controlled trial (RCT), randomized controlled crossover trial, clinical trial, prospective uncontrolled studies and cohort studies.
2. Population consists of patients with a chronic disease

3. Repetitive ePROMs are used for telemonitoring
4. Adherence to reporting repetitive ePROMs is described
5. The study quantitatively analyzes at least one factor potentially associated with adherence to telemonitoring by repetitive ePROMs
6. Written in Dutch or English

Exclusion criteria were:

The described chronic disease is a mental disorder according to the *Diagnostic and Statistical Manual of Mental Disorders-5* [26].

If necessary, the full text article was checked for the presence of all inclusion criteria.

2.3. Data Collection Process and Data Items

Data were extracted with the aid of a standardized data form by one reviewer (JW) [27]. The second reviewer (BS) sampled 50% of the articles for accuracy. The extracted information of each study consisted of: author, year of publication, study design, region, participant characteristics (sample size, diagnosis, age, sex, education level, work status and clinical characteristics), telemonitoring characteristics (design, frequency of intended usage, definition of adherence and automated reminders) and outcome measures (adherence rates and results for each investigated factor potentially affecting adherence).

2.4. Methodological Quality of Individual Studies

Both researchers (JW and BS) assessed the risk of bias of the included studies independently. Disagreements were solved by discussion, and if necessary, a third party was consulted (WB). For RCTs the Cochrane risk of bias tool was used and for cohort studies the risk of bias in nonrandomized studies of interventions (ROBINS-1) [28,29]. The Cochrane risk of bias tool aims to disclose relevant information to the risk of bias within a fixed set of domains through signaling questions in randomized trials. The domains of bias investigated were (1) random sequence generation, (2) allocation concealment, (3) selective reporting, (4) other sources of bias, (5) blinding of participants and personnel, (6) blinding of outcome assessment, and (7) incomplete outcome data and can be judged as "high risk", "low risk" or "unclear risk", if the data are insufficient for judgement. The ROBINS-1 tool evaluates the risk of bias in nonrandomized studies of interventions such as cohort studies and is based on the Cochrane risk of bias tool. The domains of bias that were assessed were (1) confounding, (2) selection of participants, (3) classification of intervention, (4) deviation from intended intervention, (5) missing data, (6) measurement of outcomes, and (7) selection of the reported results. Through signaling questions, each domain was classified as "low risk", "moderate risk", "serious risk", "critical risk" or "no information". The overall risk of bias classification for the article was equal to the classification of the domain with the highest risk of bias. Articles describing subgroup analyses of previously performed RCTs were assessed as an RCT, and the original publication was retrieved to assess the methodological quality properly.

2.5. Data Synthesis

The identified factors were categorized according to the WHO five dimensions associated with adherence, (1) social/economic factors (i.e., age and level of education), (2) health system/health care team-related factors (i.e., quality of consultations), (3) therapy related factors (i.e., treatment duration), (4) condition-related factors (i.e., symptom severity), and (5) patient related factors (i.e., self-efficacy) [30]. Although this categorization was intended for medication adherence or adherence to health therapy, we used the five dimensions for aggregating the results solely since such a classification does not exist for ePROM/technology adherence. If the factors were investigated in the included studies but were not described by the WHO, we categorized the factor in the most suitable dimension. To assess whether a meta-analysis was possible, we explored the heterogeneity of the included studies based on intervention (length of follow-up, app or web based intervention, intended frequency of ePROMs and used description of adherence), context (diagnosis) and target participants (demographics, severity of symptoms and comorbidity) following the guidelines of Pigott et al. [31]. An evidence synthesis was performed when quantitative analysis was not possible due to high heterogeneity. In the evidence synthesis, the results were assigned to one of five levels of evidence: strong, moderate, weak, inconclusive or inconsistent following the criteria adapted from Ariëns et al.; see Table 1 [32]. When available, the results of univariable analysis were used for synthesis; otherwise, the results of multivariable analysis were used.

Table 1. Strength of evidence criteria [32].

Strength of Evidence	Criteria
Strong	At least 2 high-quality studies with consistent findings
Moderate	1 high-quality study and at least 2 low-quality studies with consistent findings
Weak	At least 2 low-quality studies with consistent findings
Inconclusive	Insufficient or conflicting studies
Inconsistent	Agreement of findings <75% of studies

3. Results

The literature search generated a total of studies: 3943 in PubMed, 4339 in Embase, 714 in PsycINFO and 2485 in the Cochrane Library. After removing the duplicates of references that were selected from more than one database, 7275 studies remained. All the remaining studies were screened based on title and abstract. A total of 51 full-text studies were assessed for eligibility, and finally five studies were included in the synthesis. The flow chart of the search and selection process is shown in Figure 1.

The included studies consisted of one article that analyzed the subgroups of RCTs, two prospective uncontrolled studies and two retrospective cohort studies [33–37]. Study populations comprised of patients with rheumatoid arthritis, heart failure, chronic pain and systemic auto-immune diseases. Publication dates varied between 2013 and 2020; see Table 1. A detailed description of the study characteristics is presented in Table 2. Table 3 shows the tool and intervention characteristics.

The RCT was judged at high risk of bias on one domain, blinding of participants and personnel [33]. Both prospective uncontrolled studies were judged at serious risk due to risk of confounding [34,35]. Both retrospective cohort studies were judged at moderate risk [36,37]. The full quality assessment is presented in Figure 2 for RCTs and Figure 3 for prospective uncontrolled and retrospect cohort studies. The heterogeneity of the studies was considered too high to perform a meta-analysis; therefore, an evidence synthesis was performed.

3.1. Adherence

Overall adherence to repetitive ePROMs ranged between 61% and 96% in the studies. The definition of adherence varied between studies. In three studies, the percentage of completed questionnaires was used as a proxy for adherence and reported as a continuous variable, and two studies had a predefined number as cut-off point for high/low.

3.2. Factors Affecting Adherence Classified by the Five WHO Dimensions

A total of five studies investigated 15 unique factors. Of the factors investigated, eight belonged to the social/economic dimension (1), two factors to condition related (2), three to the patient related dimension (3), one to the therapy related dimension (4), and one to the healthcare team related dimension (5); see Table 4.

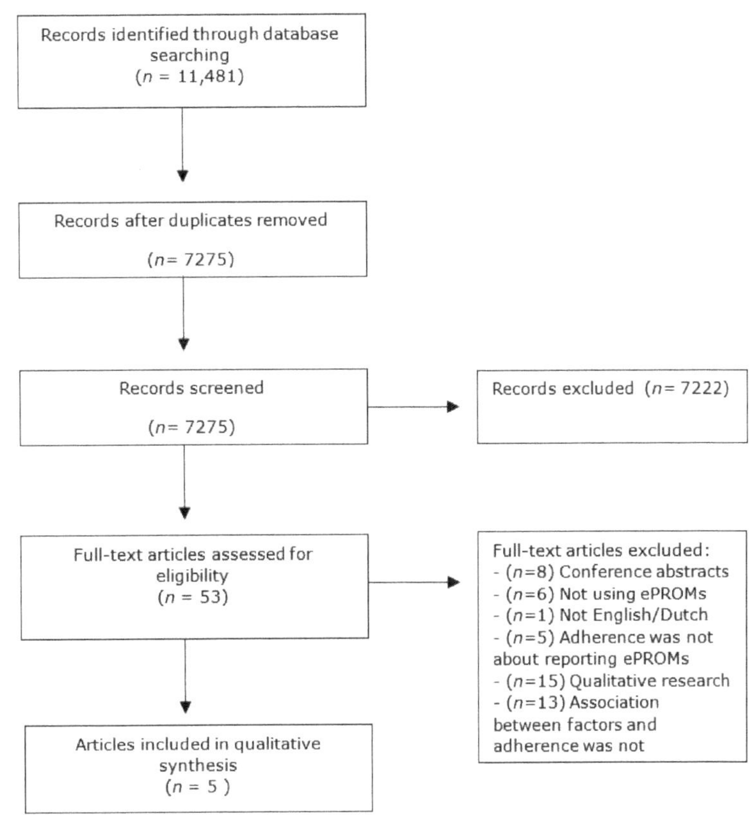

Figure 1. Flowchart of the search and selection process.

Table 2. Study characteristics of the included studies.

Author	Study Design	Year	Disease	Study Duration	Sample Size	Age, Mean (SD)	Men (%)
Colls J. et al.	RCT	2020	Rheumatoid Arthritis	6 months	78	55 (10.7)	15 (19%)
Jamilloux Y. et al.	Prospective uncontrolled study	2015	SLE, primary Sjögren and IBD	6 months	128	42 (median)	27 (21%)
Rosen D. et al.	Prospective uncontrolled study	2017	Heart failure	4 months	50	61 (-)	14 (29%)
Guzman, J.R.S. et al.	Retrospective cohort study	2013	Heart failure	3 months	248	76 (7.1)	240 (97%)
Ross, E.L. et al.	Retrospective cohort study	2020	Chronic pain	3 months	253	51 (14)	71 (28%)

* SLE = Systemic Lupus Erythematosus, IBD = Inflammatory Bowel Disease, COPD = Chronic Obstructive Pulmonary Disease.

Table 3. Tool and adherence characteristics.

Authors	Tool Medium	Intended Frequency of Reporting	Definition of Adherence	Overall Adherence
Colls J. et al.	App or tablet	Daily	% completed	79%
Jamilloux Y. et al.	Web-based	Monthly	Good: 5 or 6 reported ePROMs Bad: <5	82%
Rosen D. et al.	Tablet	Daily	% completed	96%
Guzman, J.R.S. et al.	Tablet	Daily	<80% low >80% high adherence	61%
Ross. E.L. et al.	App	Daily	% of completed	69%

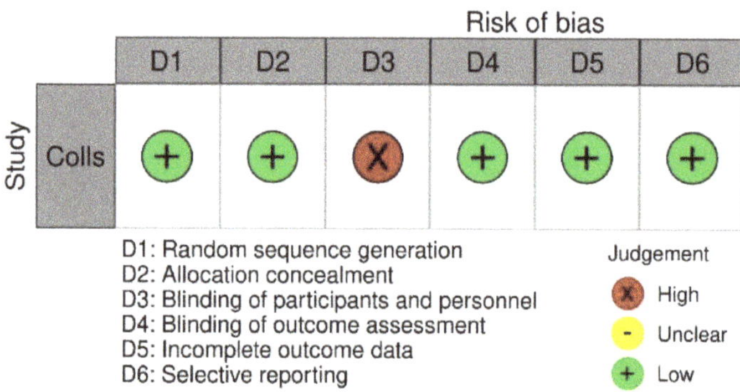

Figure 2. Risk of bias assessment following the Cochrane risk of bias tool for RCTs.

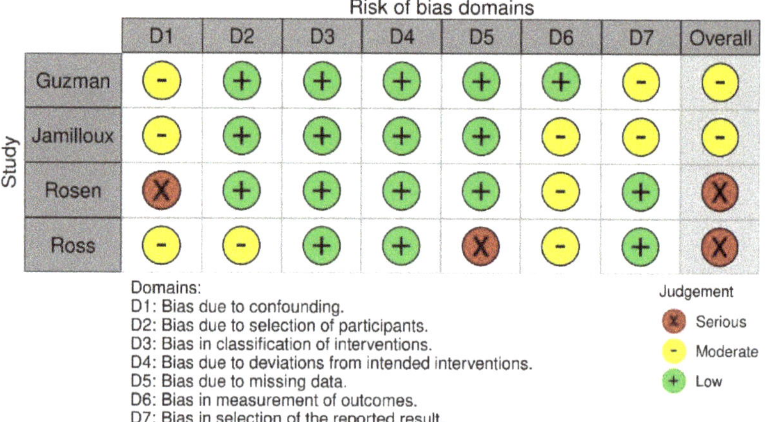

Figure 3. Risk of bias assessment following the ROBINS-1 for nonrandomized interventions.

Table 4. Overview of the investigated factors and their association with adherence.

		Colls J. et al.	Jamilloux Y. et al.	Rosen D. et al.	Guzman J.R.S. et al.	Ross E.L. et al.	Strength of Evidence
Social/economic dimension	Sex	[↔]	[↔]	[↔]	[↔]	[↔]	Moderate
	Increasing age	↑	[↔]	[↔]	[↔]	[↔]	Inconsistent
	Higher education	[↔]					Inconsistent
	Married		↑	[↔]	[↔]		Inconsistent
	Annual income				[↔]		Inconclusive
	Number of children at home		↑				Inconclusive
	Area of residence		[↔]				Inconclusive
	Employment status		[↔]				Inconclusive
Condition related dimension	Sever symptoms	↓				↑	Inconsistent
	Longer symptom duration					[↔]	Inconclusive
Patient related dimension	Depression			[↔]			weak
	Comorbidity				↓		Inconsistent
	Existing experience in online surveys		↑				Inconclusive
Therapy/intervention related dimension	Satisfaction with the app/web-based intervention	[↔]				↑	Inconsistent
Healthcare team related dimension	Primary care compared to specialized care				↑		Inconclusive

↑ = significant positive association with adherence, ↓ = significant negative association with adherence, [↔] = no significant association with adherence.

3.3. Social/Economic Dimension

There was moderate-level evidence that sex is not associated with adherence, as none of the five studies found sex was associated with adherence. The evidence for the association of age and marital status with adherence was inconsistent, while education level, annual income, number of children, area of residence and employment status were inconclusive. Colls et al. showed that patients over 65 years old had higher adherence compared to patients <45 years old, 77% vs. 62% $p = 0.02$ (32). By contrast, the four other studies did not find any difference regarding age and adherence [34–37]. Jamilloux et al. found an association between higher response rate and being married/having a partner (OR = 2.89, $p = 0.02$), in contrast to two studies who did not find any difference in adherence rates regarding marital status [34–36]. Higher education, annual income, area of residence (urban or not), employment status and number of children at home were all investigated by a single study, of which only a greater number of children at home was associated with higher adherence (average of 1.1 children at nonadherent group vs. 1.3 children at adherent group $p = 0.004$) [34].

3.4. Condition-Related Dimension

There was inconsistent and inconclusive evidence that symptom severity and disease duration, respectively, affected adherence. Colls et al. found that lower baseline disease activity, or lower symptom severity, was associated with higher adherence in comparison with higher disease activity (76% vs. 58%, $p = 0.02$), in contrast to Ross et al. who found that a higher symptom severity was associated with more daily ePROM reports [33,37].

The association between symptom duration and adherence was only investigated by Ross et al., who did not find an association.

3.5. Patient-Related Dimension

The coexistence of comorbidity (including depression) and previous participation in online surveys was investigated in only one study [36]. Guzman et al. showed that a higher score on the Charlson comorbidity index was correlated with lower adherence (beta = $-0.13, p < 0.05$) (35). Jamilloux et al. used previous participation in online surveys as surrogate for eHealth literacy and found it to be associated with higher adherence (61% in the adherent group vs. 38% in the nonadherent group OR = 2.56 95% CI 1.02–6.72 $p = 0.04$) [34]. According to Rosen et al., depression was not associated with adherence [35].

3.6. Therapy/Intervention-Related Dimension

Evidence for the association of satisfaction with the app/web-based intervention and adherence was inconsistent. Ross et al. showed that the number of daily pain entries was positively associated with satisfaction of the app, while Colls et al. did not find an association between global treatment satisfaction and adherence [33,37].

3.7. Healthcare-Team-Related Dimension

Guzman et al. found a positive association between higher adherence and patients treated at primary care, compared to patients treated at specialized care (OR 2.51 95%CI 1.22–5.19 $p < 0.03$) [36]. Since this was investigated by only one study, the result was classified as inconclusive.

4. Discussion

One of the main problems of telemonitoring by ePROMs is a lack of adherence and the lack of knowledge of the factors that influence adherence to telemonitoring. This review aimed to identify which factors are associated with adherence to telemonitoring by ePROMs in patients with chronic diseases. The definition of adherence varied between studies, with reported adherence ranging from 61% to 96%. Symptom severity, comorbidity, marital status, eHealth literacy and satisfaction with the intervention may be associated with higher adherence, but the evidence was not conclusive for any of the identified factors. Moderate evidence was found that sex is not associated to adherence.

We quantitatively investigated adherence to telemonitoring with ePROMs, in addition to previously performed qualitative research. Recent qualitative research toward telemonitoring (including but not limited to telemonitoring by ePROMs) found that coexistence of comorbidity, older age and eHealth illiteracy may negatively affect adherence [20,22]. In addition, increased social support, higher self-discipline and the usage of persuasive design in the telemonitoring tool were all qualitatively identified as possible factors positively affecting adherence [21,23,24]. Three of these factors were investigated quantitatively in our included studies: older age, eHealth literacy and comorbidity.

Colls et al. reported higher adherence in older patients, which is contrary to the earlier mentioned qualitative research and the widely accepted unified theory of acceptance and use of technology (UTAUT2) framework [38]. The UTAUT2 describes that older people may have more difficulty in adapting new technologies and hypothesized that older people are less frequently eHealth literate. However, a recent study in patients with chronic diseases showed that only eHealth literacy, and not age, was predictive for eHealth adherence [39]. The fact that smartphone usage in 55+ years old increased from 40% in 2014 to 90% in 2017 in the Netherlands suggests that the eHealth illiterate group might be decreasing in patients of older age [40]. Therefore, age seems to be of diminishing importance in eHealth adherence, while eHealth illiteracy remains a potential barrier. Guzman et al. identified the presence of comorbidity as a factor negatively affecting adherence, in accordance with qualitative research [22]. Multimorbidity is frequently present in patients with chronic diseases, for example in RA with an estimated prevalence of two-thirds of the patients [41].

Described rationales for this negative impact on adherence are that (1) comorbidity may physically limit the use of mHealth, and (2) it may shift the patients' priority away from the primary disease [23]. Therefore, patients with multimorbidity are potentially in extra need for attention or help in order to achieve higher adherence. Measures such as the index of coexistent diseases are best-suited to study the impact of comorbidity on adherence, since they include both the disease severity and physical impairment of the comorbidities [42,43].

Symptom severity was not identified by qualitative research as a barrier for telemonitoring adherence but showed remarkable contradictive results in the included studies, acting both as a facilitator and barrier [33,37]. In studies investigating medicine adherence, it has mainly been described as a potential barrier. Patients considered high symptom severity as a barrier (i) when they are becoming skeptical toward the efficacy if previous treatment had failed, (ii) when establishing routines for disease management are difficult when symptoms are severe, (iii) and when patients experience a reduced quality of interaction with their healthcare provider when symptoms are severe [44–46]. However, high symptom severity may act as a facilitator if the patients experience greater benefits from the intervention when the symptoms are severe [37,47]. Therefore, the role of symptom severity on adherence may depend on the patients' perceived benefits of telemonitoring and how easy the tool is to use even when symptom severity is high.

Only one factor was investigated in both the intervention-related and healthcare-related dimension. However, investigating intervention and healthcare factors may be valuable. For example, a high level of assistance of the healthcare practitioner may have a positive influence on adherence [48,49]. Furthermore, the frequency at which patients need to report ePROMs influences adherence is unknown. A higher frequency may lead to reporting fatigue and therefore lower adherence, but a higher frequency may also facilitate adherence in acquiring the habit more easily. Future studies investigating ePROM adherence should focus more on healthcare- and intervention-related dimensions. Not many factors are investigated both qualitatively and quantitatively, mainly due to the low number of studies investigating adherence of telemonitoring with ePROMs quantitatively. Quantitative research may help to identify which factors lead toward a decrease or increase in adherence on a group level. Furthermore, it gives the possibility of monitoring the effect of adjustments for improving adherence on group level. Therefore, it is of importance to combine qualitative and quantitative studies in prospective mixed-method studies in order to investigate thoroughly the adherence of telemonitoring by ePROMs.

Given the presented results, where the required skills and ease of use of the telemonitoring tool were mentioned as possible reasons why factors such as eHealth illiteracy, higher disease state and comorbidity may negatively impact adherence, it seems logical to design and use telemonitoring tools with a highly perceived ease of use as a possible first step to improve adherence. It is already known that this is an important factor in the successful implementation of eHealth; however, it also seems to be of importance for the sustained usage over time for telemonitoring [50]. This emphasizes the importance of a user-centered design [51].

Limitations

This review has several limitations which should be considered when interpreting the results. First, comparing results of the studies is difficult because of the large heterogeneity in diagnosis, intervention and outcome measurements between the included studies. This is partly due to the inclusion of all chronic diseases which was necessary since studies investigating ePROM adherence quantitatively are scarce, despite the increasing interest in telemonitoring by ePROMs especially during the COVID-19 pandemic. During the pandemic, patients with chronic diseases were often instantly telemonitored since outpatient clinic visits limited their access to high-priority care only. Therefore, there must be many currently unpublished data, which can help us understand which patients need extra assistance or are better suitable for telemonitoring. We hope that by underlining this knowledge gap, we can motivate others to analyze and publish their adherence data.

Secondly, the methodological quality of the studies was low, leading to weak, inconsistent or inconclusive evidence. This underlines the need for large, well-designed prospective cohort studies with the primary focus on investigating adherence. Thirdly, there is a persisting general lack of knowledge about how often ePROMs should be reported in order for telemonitoring to be effective [52]. This makes determining adherence mostly arbitrary.

5. Conclusions

The factor sex is not associated with adherence to repetitive ePROMs. Although several studies reported various associations between factors and adherence, the results show to be inconsistent or inconclusive. A limited number of telemonitoring studies by ePROMs report in-depth adherence data, making it difficult to draw conclusions from the studied factors. Future ePROM-guided telemonitoring studies should report adherence data in a standardized way. Mixed-method analyses are preferred for studies investigating telemonitoring adherence with ePROMs.

Author Contributions: J.W. and R.d.V. performed the search. J.W. and B.S. independently screened all potentially relevant titles and abstracts for eligibility. J.W. performed the data extraction which B.S. sampled. J.W, B.S., W.B., M.v.d.L. and M.v.d.E. had substantial contribution to the methodology and writing of this manuscript. All authors have read and agreed to the published version of the manuscript.

Funding: This research received no external funding.

Institutional Review Board Statement: Not applicable.

Informed Consent Statement: Not applicable.

Conflicts of Interest: The authors declare no conflict of interest.

Appendix A

Table A1. Pubmed search results.

Search	Query	Items Found
#13	#4 AND #11 AND #12	3943
#12	"Treatment Adherence and Compliance"[Mesh:NoExp] OR "Patient Compliance"[Mesh:NoExp] OR "Patient Dropouts"[Mesh] OR "Patient Participation"[Mesh] OR adherenc*[tiab] OR nonadherenc*[tiab] OR attrition*[tiab] OR dropout*[tiab] OR "drop-out*"[tiab] OR complian*[tiab] OR noncomplian*[tiab] OR engagement[tiab] OR disengagement[tiab] OR persist*[tiab] OR nonusage[tiab] OR "non-usage"[tiab] OR perseveran*[tiab]	916,734
#11	#5 OR #6 OR #7 OR #8 OR #9 OR #10	2,682,001
#10	"Inflammatory Bowel Diseases"[Mesh] OR "Inflammatory Bowel Disease*"[tiab] OR "Crohn Disease"[tiab] OR "Crohn's Disease"[tiab] OR "Crohns Disease"[tiab] OR "Granulomatous Colitis"[tiab] OR "Colitis Ulcerosa"[tiab] OR "Ulcerative Colitis"[tiab]	118,928
#9	"Heart Failure"[Mesh:NoExp] OR "Heart Failure, Diastolic"[Mesh] OR "Heart Failure, Systolic"[Mesh] OR "heart failure"[tiab] OR "cardiac failure"[tiab]	227,473
#8	"Diabetes Mellitus"[Mesh] OR diabetes[tiab] OR diabetic*[tiab] OR dm2[tiab] OR niddm[tiab] OR "dm 2"[tiab] OR t2 dm*[tiab] OR "t2 dm*"[tiab] OR "dm type 2"[tiab] OR "dm type II"[tiab] OR dm1[tiab] OR iddm[tiab] OR "dm 1"[tiab] OR t1 dm*[tiab] OR "t1 dm*"[tiab] OR "dm type 1"[tiab] OR "dm type I"[tiab]	739,709
#7	"Pulmonary Disease, Chronic Obstructive"[Mesh] OR "Asthma"[Mesh] OR COPD[tiab] OR emphysema*[tiab] OR asthma*[tiab]	265,144

Table A1. *Cont.*

Search	Query	Items Found
#6	"Arthritis, Rheumatoid"[Mesh:NoExp] OR "Spondylarthritis"[Mesh:NoExp] OR "Arthritis, Psoriatic"[Mesh] OR "Spondylitis, Ankylosing"[Mesh] OR arthriti*[tiab] OR osteoarthriti*[tiab] OR "spondyloarthritis ankylopoietica*"[tiab] OR "ankylosing spondyl*"[tiab] OR bechterew*[tiab] OR "rheumatoid spondyliti*"[tiab] OR "spondylitis ankylopoietica*"[tiab] OR spondylarthr*[tiab] OR "psoriatic arthr*"[tiab]	290,238
#5	"Chronic Disease"[Mesh] OR chronic[tiab] OR chronically[tiab]	1,341,059
#4	#1 OR #2 OR #3	218,203
#3	"Mobile Applications"[Mesh] OR "Cell Phone"[Mesh] OR "Cell Phone Use"[Mesh] OR "Smartphone"[Mesh] OR "m-health*"[tiab] OR mhealth*[tiab] OR "mobile health*"[tiab] OR "health application*"[tiab] OR android*[tiab] OR iphone*[tiab] OR app[tiab] OR apps[tiab] OR "mobile application*"[tiab] OR "mobile phone*"[tiab] OR "mobile technolog*"[tiab] OR "phone application*"[tiab] OR "smart device*"[tiab] OR smartphone*[tiab] OR "smart phone*"[tiab] OR "telephone application*"[tiab]	69,540
#2	"Internet"[Mesh] OR "world wide web*"[tiab] OR "web-based"[tiab] OR webbased[tiab] OR "web-delivered"[tiab] OR "internet delivered"[tiab] OR "internet supported"[tiab] OR "internet mediat*"[tiab] OR "internet-based"[tiab] OR "web page*"[tiab] OR "web application*"[tiab]	116,047
#1	"Telemedicine"[Mesh:NoExp] OR "Remote Consultation"[Mesh] OR "Public health informatics"[Mesh] OR telemedic*[tiab] OR telehealth*[tiab] OR "tele-medic*"[tiab] OR "tele-health*"[tiab] OR telemonitor*[tiab] OR "tele-monitor*"[tiab] OR "digital health"[tiab] OR ehealth*[tiab] OR "e-health*"[tiab] OR "e-therap*"[tiab] OR etherap*[tiab] OR "electronic pro"[tiab] OR "electronic prom"[tiab] OR epro[tiab] OR eprom[tiab] OR "e-pro"[tiab] OR "e-prom"[tiab] OR "electronic patient reported outcome*"[tiab] OR "remote monitor*"[tiab] OR "remotely monitor*"[tiab]	53,590

References

1. Safiri, S.; Kolahi, A.A.; Hoy, D.; Smith, E.; Bettampadi, D.; Mansournia, M.A.; Almasi-Hashiani, A.; Ashrafi-Asgarabad, A.; Moradi-Lakeh, M.; Qorbani, M.; et al. Global, regional and national burden of rheumatoid arthritis 1990–2017: A systematic analysis of the Global Burden of Disease study 2017. *Ann. Rheum. Dis.* **2019**, *78*, 1463–1471. [CrossRef]
2. Hootman, J.M.; Helmick, C.G.; Barbour, K.; Theis, K.A.; Boring, M.A. Updated Projected Prevalence of Self-Reported Doctor-Diagnosed Arthritis and Arthritis-Attributable Activity Limitation Among US Adults, 2015–2040. *Arthritis Rheumatol.* **2016**, *68*, 1582–1587. [CrossRef] [PubMed]
3. Risk Factors Collaborators. Global, regional, and national comparative risk assessment of 79 behavioural, environmental and occupational, and metabolic risks or clusters of risks, 1990–2015: A systematic analysis for the Global Burden of Disease Study 2015. *Lancet* **2016**, *388*, 1659–1724. [CrossRef]
4. Meystre, S. The Current State of Telemonitoring: A Comment on the Literature. *Telemed. e-Health* **2005**, *11*, 63–69. [CrossRef] [PubMed]
5. Dumais, K.M.; Dias, N.; Khurana, L.; Gary, S.T.; Witherspoon, B.; Evans, C.J.; Dallabrida, S.M. Preferences for Use and Design of Electronic Patient-Reported Outcomes in Patients with Chronic Obstructive Pulmonary Disease. *Patient Patient-Cent. Outcomes Res.* **2019**, *12*, 621–629. [CrossRef] [PubMed]
6. Vigneau, C.; Choukroun, G.; Isnard-Bagnis, C.; Pau, D.; Sinnasse-Raymond, G.; Pibre, S.; Moranne, O. Doctor, can I have less frequent injection with highly efficient treatment? A patient centered study using an electronic choice-based conjoint analysis (ePRO) to assess real world preferences regarding erythropoiesis stimulating agent to treat anaemia in chronic kidney disease (PERCEPOLIS study). *Nephrol. Ther.* **2019**, *15*, 152–161. [CrossRef] [PubMed]
7. Solomon, D.; Lu, F.; Xu, C.; Colls, J.; Suh, D.H.; Murray, M.; Corrigan, C.; Lee, Y. THU0155 Patient Adherence with a Smartphone App for Disease Monitoring in Rheumatoid Arthritis. *BMJ J.* **2019**, *78*, 350–351. [CrossRef]
8. Pincus, T. Electronic multidimensional health assessment questionnaire (eMDHAQ): Past, present and future of a proposed single data management system for clinical care, research, quality improvement, and monitoring of long-term outcomes. *Clin. Exp. Rheumatol.* **2016**, *34*, S17–S33. [PubMed]
9. de Jong, M.J.; Jong, A.E.V.D.M.-D.; Romberg-Camps, M.J.; Becx, M.C.; Maljaars, J.; Cilissen, M.; van Bodegraven, A.A.; Mahmmod, N.; Markus, T.; Hameeteman, W.M.; et al. Telemedicine for management of inflammatory bowel disease (my-IBDcoach): A pragmatic, multicentre, randomised controlled trial. *Lancet* **2017**, *390*, 959–968. [CrossRef]
10. Stut, W.; Deighan, C.; Cleland, J.; Jaarsma, T. Adherence to self-care in patients with heart failure in the HeartCycle study. *Patient Prefer. Adherence* **2015**, *9*, 1195–1206. [CrossRef] [PubMed]

11. Elbert, N.J.; van Os-Medendorp, H.; van Renselaar, W.; Ekeland, A.G.; Hakkaart-van Roijen, L.; Raat, H.; Nijsten, T.E.; Pasmans, S.G. Effectiveness and cost-effectiveness of ehealth interventions in somatic diseases: A systematic review of systematic reviews and me-ta-analyses. *J. Med. Internet Res.* **2014**, *16*, e110. [CrossRef]

12. Seppen, B.A.-O.; L'Ami, M.J.A.-O.; Duarte Dos Santos Rico, S.A.-O.; Ter Wee, M.A.-O.; Turkstra, F.A.-O.; Roorda, L.A.-O.; Catarinella, F.A.-O.; van Schaardenburg, D.A.-O.; Nurmohamed, M.A.-O.; Boers, M.A.-O.X.; et al. A Smartphone App for Self-Monitoring of Rheumatoid Arthritis Disease Activity to Assist Patient-Initiated Care: Protocol for a Randomized Controlled Trial. *JMIR Res. Protoc.* **2020**, *9*, e15105. [CrossRef]

13. Müskens, W.D.; Dartel, S.A.A.R.-V.; Vogel, C.; Huis, A.; Adang, E.M.M.; van Riel, P.L.C.M. Telemedicine in the management of rheumatoid arthritis: Maintaining disease control with less health-care utilization. *Rheumatol. Adv. Pract.* **2021**, *5*, rkaa079. [CrossRef] [PubMed]

14. Syngle, A.; Garg, N.; Rana, R.; Sherawat, S. Innovators AB0281 Utilization of Smart Phone Application to Assess the Disease Outcomes in Rheumatoid Arthritis: Smart-RA Study. *BMJ J.* **2019**, *78*, 1598. [CrossRef]

15. Khusial, R.J.; Honkoop, P.J.; Usmani, O.; Soares, M.; Simpson, A.; Biddiscombe, M.; Meah, S.; Bonini, M.; Lalas, A.; Polychronidou, E.; et al. Effectiveness of myAirCoach: A mHealth Self-Management System in Asthma. *J. Allergy Clin. Immunol. Pract.* **2020**, *8*, 1972–1979.e8. [CrossRef]

16. Chen, J.; Ou, L.; Hollis, S.J. A systematic review of the impact of routine collection of patient reported outcome measures on patients, providers and health organisations in an oncologic setting. *BMC Health Serv. Res.* **2013**, *13*, 211. [CrossRef]

17. Huber, S.; Priebe, J.A.; Baumann, K.M.; Plidschun, A.; Schiessl, C.; Tölle, T.R. Treatment of Low Back Pain with a Digital Multidisci-plinary Pain Treatment App: Short-Term Results. *JMIR Rehabil. Assist. Technol.* **2017**, *4*, e11. [CrossRef] [PubMed]

18. Eysenbach, G. The law of attrition. *J. Med. Internet Res.* **2005**, *7*, e11. [CrossRef]

19. Rathbone, A.L.; Clarry, L.; Prescott, J. Assessing the Efficacy of Mobile Health Apps Using the Basic Principles of Cognitive Behavioral Therapy: Systematic Review. *J. Med. Internet Res.* **2017**, *19*, e399. [CrossRef]

20. Walker, R.C.; Tong, A.; Howard, K.; Palmer, S.C. Patient expectations and experiences of remote monitoring for chronic diseases: Systematic review and thematic synthesis of qualitative studies. *Int. J. Med. Inform.* **2019**, *124*, 78–85. [CrossRef]

21. Ware, P.; Dorai, M.; Ross, H.J.; Cafazzo, J.A.; Laporte, A.; Boodoo, C.; Seto, E. Patient Adherence to a Mobile Phone–Based Heart Failure Telemonitoring Program: A Longitudinal Mixed-Methods Study. *JMIR mHealth uHealth* **2019**, *7*, e13259. [CrossRef]

22. Bossen, D.; Buskermolen, M.; Veenhof, C.; De Bakker, D.; Dekker, J. Adherence to a Web-Based Physical Activity Intervention for Patients with Knee and/or Hip Osteoarthritis: A Mixed Method Study. *J. Med. Internet Res.* **2013**, *15*, e223. [CrossRef]

23. de Vries, H.J.; Kloek, C.J.; de Bakker, D.H.; Dekker, J.; Bossen, D.; Veenhof, C. Determinants of Adherence to the Online Component of a Blended Intervention for Patients with Hip and/or Knee Osteoarthritis: A Mixed Methods Study Embedded in the e-Exercise Trial. *Telemed. e-Health* **2017**, *23*, 1002–1010. [CrossRef]

24. Kelders, S.M.; Kok, R.; Ossebaard, H.C.; Van Gemert-Pijnen, J.E. Persuasive System Design Does Matter: A Systematic Review of Adherence to Web-based Interventions. *J. Med. Internet Res.* **2012**, *14*, e152. [CrossRef]

25. Moher, D.; Liberati, A.; Tetzlaff, J.; Altman, D.G. The PRISMA Group Preferred Reporting Items for Systematic Reviews and Meta-Analyses: The PRISMA Statement. *PLoS Med.* **2009**, *6*, e1000097. [CrossRef]

26. Association, AP. *Diagnostic and Statistical Manual of Mental Disorders*, 5th ed.; American Psychiatric Publishing: Arlington, VA, USA, 2013.

27. Li, T.; Vedula, S.S.; Hadar, N.; Parkin, C.; Lau, J.; Dickersin, K. Innovations in Data Collection, Management, and Archiving for Systematic Reviews. *Ann. Intern. Med.* **2015**, *162*, 287–294. [CrossRef]

28. Higgins, J.P.; Altman, D.G.; Gøtzsche, P.C.; Jüni, P.; Moher, D.; Oxman, A.D.; Savovic, J.; Schulz, K.F.; Weeks, L.; Sterne, J.A. The Cochrane Collaboration's tool for assessing risk of bias in randomised trials. *BMJ* **2011**, *343*, d5928. [CrossRef]

29. Sterne, J.; Hernán, M.; Reeves, B.C.; Savović, J.; Berkman, N.D.; Viswanathan, M.; Henry, D.; Altman, D.G.; Ansari, M.T.; Boutron, I.; et al. ROBINS-I: A tool for assessing risk of bias in non-randomised studies of interventions. *BMJ* **2016**, *355*, i4919. [CrossRef]

30. World Health Organization. Adherence to Long-Term Therapie-Evidence for Action. Available online: https://wwwwhoint/chp/knowledge/publications/adherence_full_reportpdf?ua=1.2003 (accessed on 9 September 2021).

31. Pigott, T.; Shepperd, S. Identifying, documenting, and examining heterogeneity in systematic reviews of complex interventions. *J. Clin. Epidemiol.* **2013**, *66*, 1244–1250. [CrossRef]

32. Ariëns, G.A.; Van Mechelen, W.; Bongers, P.M.; Bouter, L.M.; Van Der Wal, G. Physical risk factors for neck pain. *Scand. J. Work. Environ. Health* **2000**, *26*, 7–19. [CrossRef]

33. Colls, J.; Lee, Y.C.; Xu, C.; Corrigan, C.; Lu, F.; Marquez-Grap, G.; Murray, M.; Suh, D.H.; Solomon, D.H. Patient adherence with a smartphone app for patient-reported outcomes in rheumatoid arthritis. *Rheumatology* **2020**, *60*, 108–112. [CrossRef]

34. Jamilloux, Y.; Sarabi, M.; Kérever, S.; Boussely, N.; Le Sidaner, A.; Valgueblasse, V.; Carrier, P.; Loustaud-Ratti, V.; Sautereau, D.; Fauchais, A.L.; et al. Adherence to online monitoring of patient-reported outcomes by patients with chronic inflammatory diseases: A feasibility study. *Lupus* **2015**, *24*, 1429–1436. [CrossRef]

35. Rosen, D.; McCall, J.D.; Primack, B.A. Telehealth Protocol to Prevent Readmission Among High-Risk Patients with Congestive Heart Failure. *Am. J. Med.* **2017**, *130*, 1326–1330. [CrossRef]

36. Guzman-Clark, J.R.; van Servellen, G.; Chang, B.; Mentes, J.; Hahn, T.J. Predictors and outcomes of early adherence to the use of a home telehealth device by older veterans with heart failure. *Telemed. J. E Health* **2013**, *19*, 217–223. [CrossRef]

37. Janmohamed, T.; Whitehead, L.; Seaton, P.; Ross, E.L.; Jamison, R.N.; Nicholls, L.; Perry, B.M.; Nolen, K.D. Clinical Integration of a Smartphone App for Patients with Chronic Pain: Retrospective Analysis of Predictors of Benefits and Patient Engagement Between Clinic Visits. *J. Med. Internet Res.* **2020**, *22*, e16939.
38. Venkatesh, V.; Thong, J.Y.; Xu, X. Consumer acceptance and use of information technology: Extending the unified theory of acceptance and use of technology. *MIS Q.* **2012**, *36*, 157–178. [CrossRef]
39. Richtering, S.S.; Hyun, K.; Neubeck, L.; Coorey, G.; Chalmers, J.; Usherwood, T.; Peiris, D.; Chow, C.K.; Redfern, J. eHealth Literacy: Predictors in a Population with Moderate-to-High Cardiovascular Risk. *JMIR Hum. Factors* **2017**, *4*, e4. [CrossRef]
40. Mobile usage Netherlands. Available online: https://www.consultancy.nl/nieuws/15292/smartphonebezit-gegroeid-naar-93-van-nederlanders-veelvuldig-gebruik-storend#:~{}:text=Maar%20liefst%2093%25%20van%20de,het%20gebruik%20van%20mobiele%20apparaten.viewed01-09-2020 (accessed on 9 September 2021).
41. Radner, H. Multimorbidity in rheumatic conditions. *Wien. Klin. Wochenschr.* **2016**, *128*, 786–790. [CrossRef]
42. Miskulin, D.C.; Athienites, N.V.; Yan, G.; Martin, A.A.; Ornt, D.B.; Kusek, J.W.; Meyer, K.; Levey, A.S. for the Hemodialysis (HEMO) Study Group Comorbidity assessment using the Index of Coexistent Diseases in a multicenter clinical trial. *Kidney Int.* **2001**, *60*, 1498–1510. [CrossRef]
43. De Groot, V.; Beckerman, H.; Lankhorst, G.J.; Bouter, L. How to measure comorbidity: A critical review of available methods. *J. Clin. Epidemiol.* **2003**, *56*, 221–229. [CrossRef]
44. Horne, R.; Weinman, J. Patients' beliefs about prescribed medicines and their role in adherence to treatment in chronic physical illness. *J. Psychosom. Res.* **1999**, *47*, 555–567. [CrossRef]
45. Wagner, G.J.; Ryan, G.W. Relationship Between Routinization of Daily Behaviors and Medication Adherence in HIV-Positive Drug Users. *AIDS Patient Care STDs* **2004**, *18*, 385–393. [CrossRef]
46. Hall, J.A.; Roter, D.L.; Milburn, M.A.; Daltroy, L.H. Patients' Health as a Predictor of Physician and Patient Behavior in Medical Visits. *Med. Care* **1996**, *34*, 1205–1218. [CrossRef]
47. Seppen, B.F.; Wiegel, J.; L'Ami, M.J.; Duarte Dos Santos Rico, S.; Catarinella, F.S.; Turkstra, F.; Boers, M.; Bos, W.H. Feasibility of Self-Monitoring Rheumatoid Arthritis with a Smartphone App: Results of Two Mixed-Methods Pilot Studies. *JMIR Form. Res.* **2020**, *4*, e20165. [CrossRef]
48. Talboom-Kamp, E.P.; Verdijk, N.A.; Kasteleyn, M.J.; Harmans, L.M.; Talboom, I.J.; Numans, M.E.; Chavannes, N.H.; Liu, N.; Farmer, A.; Stellefson, M.; et al. High Level of Integration in Integrated Disease Management Leads to Higher Usage in the e-Vita Study: Self-Management of Chronic Obstructive Pulmonary Disease with Web-Based Platforms in a Parallel Cohort Design. *J. Med. Internet Res.* **2017**, *19*, e185. [CrossRef]
49. van Zelst, C.M.; Kasteleyn, M.J.; van Noort, E.M.J.; Rutten-van Molken, M.; Braunstahl, G.J.; Chavannes, N.H.; In't Veen, J. The impact of the involvement of a healthcare professional on the usage of an eHealth platform: A retrospective observational COPD study. *Respir. Res.* **2021**, *22*, 88. [CrossRef]
50. Ross, J.; Stevenson, F.; Lau, R.; Murray, E. Factors that influence the implementation of e-health: A systematic review of systematic reviews (an update). *Implement. Sci.* **2016**, *11*, 1–12. [CrossRef]
51. Triberti, S.; Brivio, E. User-Centered Design *Approaches and Methods for P5 eHealth*. In *P5 eHealth: An Agenda for the Health Technologies of the Future*; Pravettoni, G., Triberti, S., Eds.; Springer International Publishing: Cham, Switzerland, 2020; pp. 155–171.
52. Sieverink, F.; Kelders, S.M.; Van Gemert-Pijnen, J.E. Clarifying the Concept of Adherence to eHealth Technology: Systematic Review on When Usage Becomes Adherence. *J. Med. Internet Res.* **2017**, *19*, e402. [CrossRef]

International Journal of
*Environmental Research
and Public Health*

Review

"Help in a Heartbeat?": A Systematic Evaluation of Mobile Health Applications (Apps) for Coronary Heart Disease

Chiara Mack [1], Yannik Terhorst [2], Mirjam Stephan [1], Harald Baumeister [2], Michael Stach [3], Eva-Maria Messner [2], Jürgen Bengel [1] and Lasse B. Sander [1,*]

[1] Department of Rehabilitation Psychology and Psychotherapy, Institute of Psychology, Albert-Ludwigs-University of Freiburg, 79085 Freiburg, Germany; chiara.mack.96@gmail.com (C.M.); miri.stephan@mail.de (M.S.); bengel@psychologie.uni-freiburg.de (J.B.)

[2] Department of Clinical Psychology and Psychotherapy, Institute of Psychology and Education, Ulm University, 89040 Ulm, Germany; yannik.terhorst@uni-ulm.de (Y.T.); harald.baumeister@uni-ulm.de (H.B.); eva-maria.messner@uni-ulm.de (E.-M.M.)

[3] Institute of Databases and Information Systems, Ulm University, 89040 Ulm, Germany; michael.stach@uni-ulm.de

* Correspondence: lasse.sander@psychologie.uni-freiburg.de

Citation: Mack, C.; Terhorst, Y.; Stephan, M.; Baumeister, H.; Stach, M.; Messner, E.-M.; Bengel, J.; Sander, L.B. "Help in a Heartbeat?": A Systematic Evaluation of Mobile Health Applications (Apps) for Coronary Heart Disease. *Int. J. Environ. Res. Public Health* **2021**, *18*, 10323. https://doi.org/10.3390/ijerph181910323

Academic Editors: Irene Torres-Sanchez and Marie Carmen Valenza

Received: 25 August 2021
Accepted: 20 September 2021
Published: 30 September 2021

Publisher's Note: MDPI stays neutral with regard to jurisdictional claims in published maps and institutional affiliations.

Abstract: For patients with coronary heart disease (CHD) lifestyle changes and disease management are key aspects of treatment that could be facilitated by mobile health applications (MHA). However, the quality and functions of MHA for CHD are largely unknown, since reviews are missing. Therefore, this study assessed the general characteristics, quality, and functions of MHA for CHD. Hereby, the Google Play and Apple App stores were systematically searched using a web crawler. The general characteristics and quality of MHA were rated with the Mobile Application Rating Scale (MARS) by two independent raters. From 3078 identified MHA, 38 met the pre-defined criteria and were included in the assessment. Most MHA were affiliated with commercial companies (52.63%) and lacked an evidence-base. An overall average quality of MHA ($M = 3.38$, $SD = 0.36$) was found with deficiencies in information quality and engagement. The most common functions were provision of information and CHD risk score calculators. Further functions included reminders (e.g., for medication or exercises), feedback, and health management support. Most MHA (81.58%) had one or two functions and MHA with more features had mostly higher MARS ratings. In summary, this review demonstrated that a number of potentially helpful MHA for patients with CHD are commercially available. However, there is a lack of scientific evidence documenting their usability and clinical potential. Since it is difficult for patients and healthcare providers to find suitable and high-quality MHA, databases with professionally reviewed MHA are required.

Keywords: coronary heart disease (CHD); apps; mobile health; eHealth; systematic evaluation

1. Introduction

Cardiovascular diseases and especially coronary heart diseases (CHD) are one of the leading causes of death worldwide [1,2]. According to the global burden of disease study 17.8 million people died from cardiovascular diseases in 2017 [1]. According to the heart disease and stroke statistics the prevalence of CHD in the US ranges from 5.3% for female adults to 7.4% for male adults [3].

CHD and common complications like arrhythmia, myocardial infarction, and heart failure have a significant negative impact on the affected person's health, leading to high mortality and healthcare costs [4–6]. In addition, certain mental and physical conditions and factors are associated with CHD, including depression, cigarette smoking, hypertension, and obesity [7–10].

Disease management and behavior change including lifestyle changes are key aspects of CHD care but often not adequately and enduringly considered in care settings [11]. The large number of risk and lifestyle factors render the prevention and self-management of CHD extensive and complex for patients [11,12]. Therefore, means of promoting disease management and lifestyle changes as well as information are necessary to improve prevention and conventional treatment of CHD [11,13,14]. Mobile health applications (MHA) are discussed to contribute in overcoming this gap in treatment by fostering CHD management [13,15]. First, MHA may support daily monitoring of activities and symptoms [16]. Second, adherence to treatment and lifestyle changes can be increased by self-tracking, feedback, and reminder functions of MHA [16,17]. Third, MHA are accessible at all times and at relatively little costs [18] making MHA a scalable solution to provide general information about CHD, symptoms, and specific lifestyle modifications [19,20]. Fourth, MHA can increase patients' perception to play an active role in their own healthcare and hereby foster self-sufficiency, disease management, and patient autonomy [16,18,21].

However, high-quality applications with suitable content are required, while the quality of MHA is largely unknown due to an intransparent MHA market and a lack of methodologically sound quality assessments [16,22]. Previous studies focused on other cardiological conditions examining the quality of MHA for heart failure [23,24], atrial fibrillation [25], and blood pressure [26]. Hereby the quality of MHA was reported as mostly acceptable [24] or mainly poor [23,26]. This is particularly alarming because MHA can also be harmful [19,22]. Risks and constraints regarding MHA concern data security, privacy, and confidentiality, since missing privacy policies and information transfer to third parties have been observed [22,27]. Furthermore, possible misinformation poses potential risks to users and the sheer number of MHA may lead to consumer confusion [11,19,28–30]. No evaluation of MHA specifically for CHD was found [16].

Therefore, in this study we systematically searched for and conducted a standardized evaluation of MHA for CHD which are available in commercial app stores. Hereby we addressed the following research questions:

1. What is the quality of CHD applications in European commercial app stores in regard to engagement, functionality, aesthetics, and information quality in general?
2. What functions are employed in CHD applications?

2. Materials and Methods

2.1. App Search Strategy

With an automated search engine (web crawler) of the 'Mobile Health Application Database' (MHAD) [31] the Google Play store and Apple App store were systematically searched for MHA. Search terms to identify CHD applications included 'coronary heart disease', 'coronary artery disease', 'ischemic heart disease', and 'heart disease' in English and German. A list of all search terms is included in Appendix A. The searches were conducted between December 2020 and February 2021. Duplicates were automatically removed. For the assessment, MHA from the Google Play store were installed on an Honor 6X (BLL-L22) and apps from the Apple App store on an iPad Pro A1652.

2.2. Inclusion Criteria and Process

The identified MHA were examined for eligibility in a two-step procedure. In the first step, the title and app description were screened and the inclusion criteria for the download of MHA were applied. Apps were downloaded if (a) in the app title or description the subject of coronary heart disease was stated, (b) the app was developed for patients with CHD, persons at risk, or otherwise affected individuals, (c) the MHA was available in German or English language, and (d) download was possible.

In a second step, the identified apps were downloaded and the criteria for inclusion in the evaluation were examined within the app. MHA were included if (a) CHD was focused, a CHD-specific section was included, or the app description stated its use for CHD, (b) no other specific information (such as login/ access data) was required for usage of the app, (c) the application was functional, and d) there were no further technical reasons to eliminate the MHA. Technical malfunctions were tested on two devices.

2.3. Data Extraction, Evaluation Criteria, and Instruments

Two independent raters (master's degree students in psychology C.M. and M.S., under supervision of a licensed psychotherapist L.B.S.) conducted the acquisition and rating of all included MHA by applying the Mobile Application Rating Scale (MARS) in the German version [32–34]. For all sections of the MARS a good to excellent internal consistency (Omega = 0.793 to 0.904), an overall excellent internal consistency (Omega = 0.929) and a good intra-class correlation (ICC = 0.816, 95% CI: 0.810 to 0.822) were shown [34]. Therefore, with the MARS the quality of MHA can be assessed reliably [34]. The MARS contains a section for classification and for quality rating as well as three additional subscales.

To prepare for the app rating with the MARS, a free online tutorial provided by the developers of the German MARS version was viewed. For the rating, each app was tested by trying out all features. To check the agreement between the raters, the interrater reliability (IRR) was calculated. Here, the intra-class correlation (ICC) needs to be ≥ 0.75 to indicate a sufficient agreement [35]. In case of an ICC below 0.75 the supervisor (L.B.S.) was consulted.

2.4. General Characteristics of MHA

For this study the MARS classification section was adapted to include the following general characteristics: (1) app name, (2) platform (Android, iOS), (3) affiliation, (4) price, (5) embedment in therapy, (6) user star rating, (7) number of user ratings, (8) app store category, (9) methods, (10) technical aspects, and (11) security and privacy.

2.5. Quality Rating

For the quality rating with the MARS 19 items are rated on a five-point scale ranging from 1 (inadequate) to 5 (excellent). These items constitute the four dimensions: (A) engagement (five items: entertainment, interest, individual adaptability, interactivity, target group), (B) functionality (four items: performance, usability, navigation, gestural design), (C) aesthetics (three items: layout, graphics, visual appeal), and (D) information quality (seven items: accuracy of app description, goals, quality of information, quantity of information, quality of visual information, credibility, evidence base). To assess the evidence-base, for each MHA Google Scholar was searched for published studies.

2.6. Statistical Analysis

For each of these four dimensions, the mean score (M) and standard deviation (SD) were computed as well as a total mean quality score across all four objective dimensions [33]. The scores of both raters were averaged. Additionally, the three subjective subscales of the MARS: (E) therapeutic gain, (F) subjective quality, and (G) perceived impact were evaluated without effect on the overall mean score. Correlation analyses between the available user star ratings (one star to five stars) and the MARS total mean score as well as the objective dimensions were conducted if at least three ratings were available.

2.7. Assessment of Functions

Subsequently, the employed functions of the included MHA were assessed with a classification from the 'Chances and Risks of Mobile Health Apps' (CHARISMHA) study [36]. The classification is divided into six categories with one to five subcategories each. These are: provision of information (news, reference, learning material, player/viewer, broker), data acquisition, processing, and evaluation (decision support, calculator, meter, monitor, surveillance/tracker), administrative use (administration), calendar and appointment-related apps (diary, reminder, calendar), support (utility/aid, coach, health manager) and other (actuator, communicator, game, store, other). Hereby, for each MHA it was examined which functions are employed. Additionally, a correlation between the number of functions and the MARS total score was calculated.

3. Results

3.1. Search

In Figure 1 the screening and inclusion process is illustrated. A total of 3078 apps were found through the web crawler. From 1217 apps without duplicates, 38 MHA (3.12%) were included in the evaluation. Of those, 30 apps (78.95%) were available on android, seven apps (18.42%) on iOS, and one app (2.63%) for both.

3.2. General Characteristics

The characteristics of included MHA are depicted in Table 1. The apps were affiliated with commercial companies ($n = 20$, 52.63%), non-governmental organizations (NGO; $n = 2$, 5.26%), universities ($n = 2$, 5.26%), and governments ($n = 1$, 2.63%). For 13 apps (34.21%) the affiliation was unknown. The basic version was free of cost for most apps ($n = 34$, 89.47%) and required payment for four apps (10.53%) with prices ranging from EUR 1.09 to EUR 3.69 ($M = 2.57$, $SD = 1.08$). In three apps (7.89%) an upgraded or extended pro version was available or in-app purchases were possible. No app was embedded in a treatment concept or had a certification to comply for example with the medical device regulation.

For 12 apps (31.58%) a user rating was available in the Google Play store and for one app (2.63%) in the Apple App store. The median user star rating in the Google Play store was 4.4 ($M = 4.26$, $SD = 0.47$) with five to 1276 ratings ($M = 220.42$, $SD = 403.29$) and the user star rating in the Apple App store was 1.0 with one rating (user ratings last updated on 4 April 2021). MHA were classified in eight app store categories: 'Health & Fitness' ($n = 18$, 47.37%), 'Medical' ($n = 10$, 26.32%), 'Education', 'Books & Reference' ($n = 3$, 7.89% each), 'Lifestyle', 'Food & Drink', 'Entertainment', and 'Social Networking' ($n = 1$, 2.63% each). In 19 apps (50.00%) internet was required for some or all functions and one app (2.63%) had an app community. Most common methods were information and education ($n = 34$, 89.47%), tips and advice ($n = 25$, 65.79%), and feedback ($n = 16$, 42.11%). For most apps a privacy policy ($n = 26$, 68.42%) and contact information ($n = 33$, 86.84%) was provided and in seven apps (18.42%) active consent was required. Login was necessary in six apps (15.79%) and a password protection in three apps (7.89%).

For one MHA ('The Heart App') the accuracy to detect acute coronary syndromes was examined in a diagnostic accuracy study [37]. Otherwise, no study or randomized controlled trial (RCT) was found.

3.3. Quality Rating of MHA

The MARS rating results for each included MHA are presented in Table 2. The total quality of included MHA was average, with $M = 3.38$ ($SD = 0.36$) and ranged from $M = 2.50$ to $M = 4.22$. Of the four objective dimensions, the highest-rated was functionality ($M = 4.06$, $SD = 0.31$), thereafter aesthetics ($M = 3.62$, $SD = 0.47$), followed by information quality ($M = 3.18$, $SD = 0.43$), and engagement ($M = 2.64$, $SD = 0.55$). For the additional subjective subscales, the means were lower, with $M = 2.49$ ($SD = 0.35$) for therapeutic gain, $M = 2.45$ ($SD = 0.52$) for subjective quality, and $M = 1.90$ ($SD = 0.35$) for perceived impact. The IRR for the rating of all MHA was excellent (2-way mixed ICC = 0.944, 95%-CI 0.935 to 0.952) and for no single app an ICC below 0.75 was evident. No significant correlations between user star ratings and MARS total mean score (r (10) = -0.52, $p = 0.080$) or the objective subscales (r (10) = 0.001–0.52, $p > 0.05$) were found. In addition, none of the apps that required payment for the basic version were among the ten highest-rated apps.

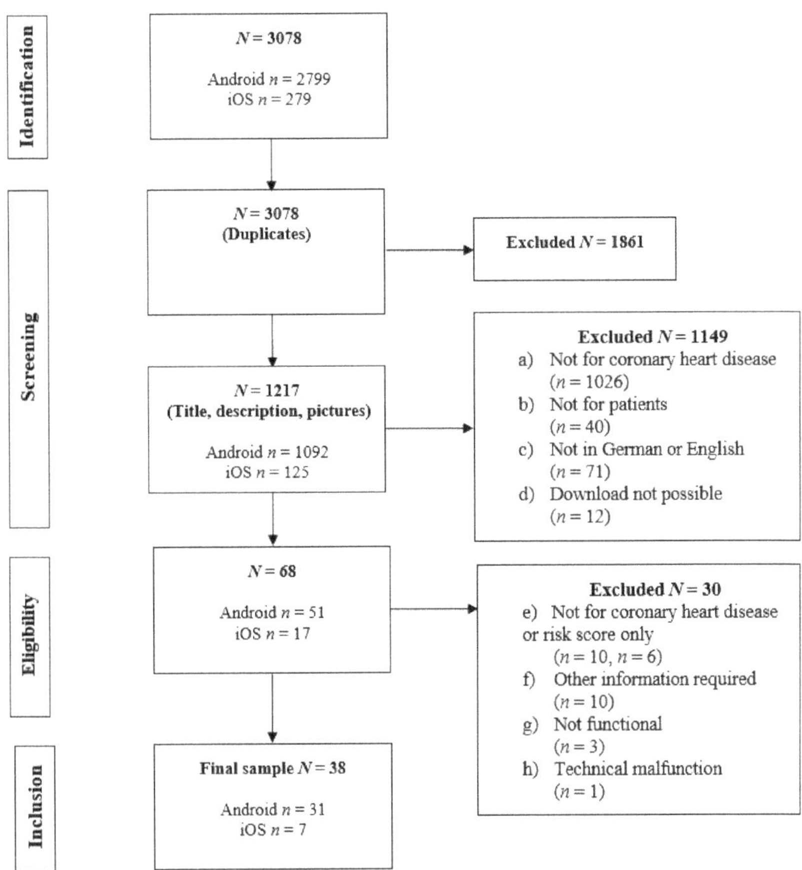

Figure 1. Flowchart of the app screening and inclusion process.

Table 1. General characteristics of included MHA for coronary heart disease.

	n (%)	*M (SD)*
Platform		
Android	30 (78.95%)	
iOS	7 (18.42%)	
Both	1 (2.63%)	
Affiliation		
Commercial company	20 (52.63%)	
NGO	2 (5.26%)	
University	2 (5.26%)	
Government	1 (2.63%)	
Unknown	13 (34.21%)	
Obligatory payment		
Google Play store	2 (5.26%)	2.84 (0.85)
Apple App store	2 (5.26%)	2.29 (1.20)

Table 1. *Cont.*

	n (%)	*M (SD)*
User ratings		
Google Play store	12 (31.58%)	4.26 (0.47)
Apple App store	1 (2.63%)	1.0 (0.00)
Technical aspects		
Internet required	19 (50.0%)	
App community	1 (2.63%)	
Methods		
Information and education	34 (89.47%)	
Tips and advice	25 (65.79%)	
Feedback	16 (42.11%)	
Alternative medicine	3 (7.89%)	
Bodily exercises	2 (5.26%)	
Security & privacy		
Privacy policy	26 (68.42%)	
Contact information	33 (86.84%)	
Informed consent	7 (18.42%)	
Login	6 (15.79%)	
Password	3 (7.89%)	

Note. *n* = number of apps; *M* = mean; *SD* = standard deviation.

3.4. Functions of MHA

The number of functions of all MHA is presented in Appendix B. The most common function was provision of information, specifically reference (information texts about e.g., CHD, risk factors, or treatment) in 37 apps (97.37%) and a player/viewer (e.g., video or audio files) in six apps (15.79%). Hereof 28 MHA (73.68%) were primarily assigned to the category provision of information. Further functions were decision support (e.g., concerning foods or goals; $n = 4$, 10.53%), calculators such as CHD risk score or body mass index (BMI) calculators ($n = 13$, 34.21%) and monitor of activity ($n = 1$, 2.63%). Hereof six MHA (15.79%) were primarily categorized under data acquisition, processing, and evaluation. Other functions were diary ($n = 1$, 2.63%), reminder (e.g., for medication or workouts) and calendar functions ($n = 3$, 7.89% each), with no MHA primarily being a calendar and appointment-related app. Three MHA (7.89%) functioned as health managers, targeting goals regarding exercise, weight, and nutrition, and were categorized as support apps. Two MHA (5.26%) were communicators and of those, one app (2.63%) was primarily a social network.

The functions of the ten highest-rated apps are shown in Table 3 and a full table depicting all employed functions per MHA is included in Appendix C. In general, many MHA had one ($n = 17$, 44.74%) or two functions ($n = 14$, 36.84%) and in seven apps (18.42%) three or more functions were employed. Of those MHA with three or more functions, five apps (71.43%) were among the ten highest-rated apps. A significant positive correlation with a large effect size was found between the MARS total score and the number of employed functions (r (36) = 0.66, $p < 0.001$).

Table 2. Means of the MARS rating from highest to lowest total score.

Name	Rated on	Total Score	Quality Rating				Subjective Subscales		
			Engagement	Functionality	Aesthetics	Information Quality	E	F	G
CardiaCare	GP	4.22	4.00	4.50	4.67	3.70	2.83	3.25	2.58
Love My Heart for Women	AA	4.00	3.80	4.25	4.17	3.78	2.50	3.00	2.42
CardioVisual: Heart Health Built by Cardiologists	GP	3.84	3.30	3.88	4.17	4.00	2.67	3.00	2.08
Heart Disease Yoga & Diet–Cardiovascular disease	GP	3.83	3.70	4.13	4.00	3.50	2.83	3.13	2.50
My Heart Age	GP	3.83	3.60	4.00	4.00	3.70	2.67	3.50	2.83
ASCVD Risk Estimator Plus	GP	3.79	2.90	4.25	4.00	4.00	3.00	2.88	2.25
Texas Heart Institute	AA	3.75	2.80	4.25	4.17	3.78	2.67	2.88	2.00
The Heart App ©	GP	3.74	3.20	4.25	4.17	3.33	3.83	3.25	2.17
Angina	GP	3.59	2.90	4.13	3.83	3.50	2.67	2.63	2.08
Heart Disease 101 Audio Book	GP	3.56	2.50	4.38	4.00	3.38	2.67	3.13	2.00
Heart Disease Support	AA	3.56	3.50	4.13	3.50	3.13	2.50	2.75	1.58
Heart Diseases & Treatment	GP	3.55	2.60	4.25	3.83	3.50	2.50	2.50	1.92
MESA CHD Risk Score	GP	3.53	2.90	3.88	3.83	3.50	2.50	2.63	1.75
Healthy Heart Guides	GP	3.51	2.90	4.13	3.83	3.20	2.83	2.88	2.33
Heart Care Health & Diet Tips	GP	3.51	2.80	4.00	4.00	3.25	2.50	2.63	1.92
Basic Cardiology	GP	3.49	2.20	4.25	4.00	3.50	2.67	2.63	1.83
CardioRisk Calc	AA	3.43	2.80	4.13	3.67	3.11	2.17	2.00	1.58
Heart Disease Guide	GP	3.40	2.60	4.13	3.67	3.20	2.50	2.38	1.83
Cardiovascular Diseases	GP	3.39	2.20	4.38	3.50	3.50	2.67	2.50	2.00
Heart Disease B	GP	3.36	2.20	4.00	3.83	3.40	2.50	2.63	2.08
Cardiovascular Care Guide	GP	3.34	2.30	4.13	3.83	3.10	2.50	2.63	1.83
Heart Health Tips	GP	3.34	2.30	4.38	3.67	3.00	2.50	2.13	1.50
Atherosclerosis	GP	3.28	2.20	4.25	3.67	3.00	2.50	2.38	2.00
Heart Disease A	GP *	3.27	3.00	3.88	3.00	3.20	2.33	2.63	2.00
Heart Disease C	GP	3.26	2.50	4.13	3.33	3.10	2.67	2.63	1.92
Heart Disease Diet-Have a Fit & Healthy Heart with Best Nutrition!	AA *	3.25	2.20	4.13	3.67	3.00	2.33	1.88	1.92
Home Remedies For Chest Pain (Angina)	GP	3.24	2.00	4.25	3.83	2.88	2.00	1.75	1.67
Cardiology consultation	GP	3.21	2.80	3.88	3.17	3.00	2.33	2.13	1.83
Natural Remedies For Chest Pain (Angina)	GP	3.16	2.00	4.25	3.50	2.88	2.17	1.88	1.67
Cardiology-Expert Consult 4 Diagnosis & Treatment	GP	3.09	2.30	4.00	3.17	2.90	2.33	2.50	2.08
Angina Pectoris Disease	GP	3.08	2.20	3.75	3.50	2.88	2.67	2.50	1.75
Cardiovascular Disease Information	GP	3.08	1.90	4.25	3.17	3.00	2.50	2.00	1.75
Herz und koronarer Herzkrankhe	AA *	3.00	2.30	4.00	3.00	2.70	2.17	1.88	1.50
Arteriosclerosis Disease	GP	2.88	2.00	3.63	3.50	2.38	2.17	1.63	1.42
Heart Disease Risk Prediction and Prevention	GP *	2.87	2.60	4.00	2.50	2.38	2.17	1.75	1.42
How To Cure Heart Disease	GP	2.84	2.30	4.13	2.67	2.25	2.00	1.63	1.33
CORONARY HEART DISEASE RISK	GP	2.76	2.40	2.75	3.00	2.88	2.00	1.63	1.42
Universal Healing Programme	AA	2.50	1.80	3.25	2.67	2.30	1.67	1.63	1.33
Total mean	-	**3.38**	**2.64**	**4.06**	**3.62**	**3.18**	**2.49**	**2.45**	**1.90**

Note. * fee required. GP = Googly Play, AA = Apple App, E = Therapeutic Gain, F = Subjective Quality, G = Perceived Impact.

Table 3. Employed functions per included MHA for the ten highest-rated apps.

Name	Provision of Information				Data Acquisition, Processing and Evaluation						Calendar and Appointment-Related			Support		Other	
	News	Reference	Learning Material	Player/ Viewer	Broker	Decision Support	Calculator	Meter	Monitor	Surveillance/ Tracker	Diary	Reminder	Calendar	Utility/Aid	Coach	Health Manager	Communicator/ Social Network
CardiaCare	-	✓	-	-	-	-	✓	-	-	-	-	✓	-	-	-	✔	-
Love My Heart for Women	-	✓	-	-	-	✓	✓	-	-	-	-	✓	✓	-	-	✔	-
CardioVisual: Heart Health Built by Cardiologists	-	✓	-	✔	-	-	✓	-	-	-	-	-	-	-	-	-	-
Heart Disease Yoga & Diet–Cardiovascular disease	-	✓	-	-	-	-	✓	-	-	-	-	✓	✓	-	-	✔	-
My Heart Age	-	✓	-	-	-	✓	✔	-	✓	-	-	-	-	-	-	-	-
ASCVD Risk Estimator Plus	-	✓	-	-	-	-	✔	-	-	-	-	-	-	-	-	-	-
Texas Heart Institute	-	✔	-	-	-	-	✓	-	-	-	-	-	-	-	-	-	-
The Heart App ©	-	✔	-	-	-	-	✓	-	-	-	-	-	-	-	-	-	-
Angina	-	✔	-	-	-	-	-	-	-	-	-	-	-	-	-	-	✓
Heart Disease 101 Audio Book	-	✔	-	✓	-	-	-	-	-	-	-	-	-	-	-	-	-

Note. ✔ primary function of the MHA, ✓ function is employed in the MHA.

4. Discussion

This study is the first to systematically review MHA for CHD by assessing the general characteristics, quality, and functions of MHA in European app stores. The overall quality of apps for CHD, as assessed with the MARS, was average ($M = 3.38$, $SD = 0.36$). Here, the functionality and aesthetics of included apps were generally high while deficits in information quality and engagement were shown. This is in line with previous studies which reported a varying, but largely acceptable or poor quality of apps for cardiological conditions such as heart failure, atrial fibrillation, or hypertension [23,24,26]. Since many MHA primarily provide information, the deficits in the average information quality are alarming [11,13]. A total of $N = 3078$ apps were identified by the web crawler and only 38 apps (3.12%) were CHD-specific and met the inclusion criteria. Above that, none of the objective subscales nor the MARS total mean scores were significantly correlated with the user star ratings which may increase the challenge of patients with CHD to identify a reliable MHA restricting the clinical use of MHA. Since none of the paid apps were among the ten highest-rated MHA, requiring payment is also not an adequate indicator of app quality. Furthermore, some MHA with very little CHD-specific information, an overwhelming amount, or questionable content were found. This included MHA implying to cure CHD only by certain yoga practices or solely by specific natural remedies. From a clinical perspective, this misleading information can have harmful effects on affected users such as not seeking professional medical advice or treatment.

Considering this, the lack of evidence regarding the usefulness and effectiveness of MHA is concerning. Only one study [37] investigating the diagnostic accuracy of one MHA could be found, which corresponds to previous app reviews, demonstrating little evidence-base for commercially available MHA [38–41]. This is in line with a validation study that included MHA for several health conditions and reported no evidence-base for 94.8% of the 1299 included MHA [34]. As many of the included MHA only consist of information texts or calculators, efficacy studies are rather inadequate, since symptom reduction due to those MHA alone is unlikely. However, studies examining for example the usability, feasibility, or user acceptance could be of importance and increase the scientific discourse on MHA. This necessity is further increased by the fact, that only few MHA (13.16%) were developed by credible sources such as universities, NGOs, or governmental organizations.

Identified functions included information, CHD risk score calculators, reminders for exercise or medication, feedback to data entries and health managers with goal setting regarding exercise, weight, and nutrition. With regard to the overall conceptualization of MHA we found that the majority comprised solely one or two functions while in only seven apps (18.42%) three or more functions were employed. The number of employed functions was positively correlated with the MARS total score, therefore those MHA with several functions were mostly among the highest-rated apps. Most MHA were limited to information and CHD risk score calculators. Thus, the majority of apps fall short of their potential to foster behavior change in patients using reminders, notifications, achievements, and encouragement and regarding important lifestyle changes like quitting smoking, being more active, or eating healthier [42]. Nevertheless, the embedment of MHA in current treatment models could most likely be valuable for patients as well as for health care providers being able to quickly access patients' data. Independent expert rating platforms or databases such as www.mhad.science (accessed on 19 September 2021) or https://mindapps.org (accessed on 19 September 2021) are necessary to support patients and providers in finding and choosing reliable MHA of high quality [43].

This study has some limitations. First, the web crawler is limited to 200 apps per search term and the European market. This might deform the results by omitting some MHA, even though apps specifically developed for the US market were also found. Second, some apps might be locally restricted and published for specific countries only. Additionally, apps that required specific login/ access information were excluded, since they are not instantly available for most users, reducing the number of MHA. Third, only MHA in

German or English language and for the chosen search terms were covered. Therefore, the number of included MHA is potentially not extensive and future studies could examine the search terms that patients use when looking for CHD apps. Fourth, the development of apps is very rapid [44], resulting in one app ('Atherosclerosis') no longer being available for rating by the second reviewer and some apps no longer being detectable between the first and second screening or for inclusion. Fifth, according to the standard procedure of the MARS two reviewers rated the apps, even though more raters would lead to more accurate estimates.

Sixth, with the MARS the quality of MHA is evaluated in regard to engagement, functionality, aesthetics, and information quality. In addition to those dimensions, other aspects might be relevant for app users as well and a high MARS score does not imply a high effectiveness of the app. Hence, in future studies this systematic review of MHA could be replicated with other instruments like ENLIGHT [45] or different suitable scales.

5. Conclusions

This first systematic evaluation of MHA for CHD demonstrated an average overall quality of MHA ($M = 3.38$, $SD = 0.36$). The most common functions were information texts and risk score calculators. Only few MHA provide a set of multiple functions and incorporate behavior change techniques limiting the potential for lifestyle changes and support in disease management of users. Most MHA were not developed by a credible source and there is a considerable lack of scientific evidence for the usefulness and efficacy of the included MHA. Nevertheless, some potentially helpful MHA were identified. The results of this study will be made publicly available to users and healthcare providers at www.mhad.science (accessed on 19 September 2021).

Author Contributions: Conceptualization, C.M., L.B.S., Y.T., and J.B.; methodology, Y.T., L.B.S., and E.-M.M.; web crawler and tooling, M.S. (Mirjam Stephan), H.B., and E.-M.M.; validation, C.M., L.B.S., and M.S. (Mirjam Stephan); formal analysis, C.M.; investigation, C.M..; data curation, C.M.; writing—original draft preparation, C.M.; writing—review and editing, L.B.S., Y.T., C.M., J.B., E.-M.M., H.B., and M.S. (Michael Stach); supervision, L.B.S.; project administration, L.B.S.; All authors have read and agreed to the published version of the manuscript.

Funding: This research received no external funding. The article processing charge was funded by the Baden-Wuerttemberg Ministry of Science, Research and Art and the University of Freiburg in the funding program Open Access Publishing.

Institutional Review Board Statement: Not applicable.

Informed Consent Statement: Not applicable.

Data Availability Statement: The raw data supporting the conclusions of this article will be made available by the corresponding author on request.

Conflicts of Interest: The authors declare no conflict of interest. The funders had no role in the design of the study; in the collection, analyses, or interpretation of data; in the writing of the manuscript, or in the decision to publish the results.

Appendix A

Table A1. List of all search terms for coronary heart disease.

English	German
Coronary heart disease	Koronare Herzkrankheit
Coronary artery disease	Koronare Herzerkrankung
Ischemic heart disease	Ischämische Herzkrankheit
Heart disease	Ischämische Herzerkrankung
	Herzerkrankung Herzkrankheit

Appendix B

Table A2. Functions of the included MHA.

	n (%) Android	*n* (%) iOS
Provision of information		
News	-	-
Reference	31 (100%)	6 (85.71%)
Learning material	-	-
Player/Viewer	5 (16.13%)	1 (14.29%)
Broker	-	-
Data acquisition, processing and evaluation		
Decision support	3 (9.68%)	1 (14.29%)
Calculator	10 (32.26%)	3 (42.86%)
Meter	-	-
Monitor	1 (3.23%)	-
Surveillance/Tracker	-	-
Administrative use		
Administration	-	-
Calendar and appointment-related		
Diary	1 (3.23%)	-
Reminder	2 (6.45%)	1 (14.29%)
Calendar	2 (6.45%)	1 (14.29%)
Support		
Utility/Aid	-	-
Coach	-	-
Health manager	2 (6.45%)	1 (14.29%)
Other		
Actuator	-	-
Communicator/Social network	1 (3.23%)	1 (14.29%)
Game	-	-
Store	-	-

Appendix C

Table A3. Employed functions per included MHA.

Name	Provision of Information					Data Acquisition, Processing and Evaluation					Calendar and Appointment-Related			Support			Other
	News	Reference	Learning Material	Player/Viewer	Broker	Decision Support	Calculator	Meter	Monitor	Surveillance/Tracker	Diary	Reminder	Calendar	Utility/Aid	Coach	Health Manager	Communicator/Social Network
CardiaCare	-	✓	-	-	-	-	✓	-	-	-	-	✓	-	-	-	✓	-
Love My Heart for Women	-	✓	-	-	-	✓	✓	-	-	-	-	✓	✓	-	-	✓	-
CardioVisual: Heart Health Built by Cardiologists	-	✓	-	✓	-	-	✓	-	-	-	-	-	-	-	-	-	-
Heart Disease Yoga & Diet– Cardiovascular disease	-	✓	-	-	-	-	✓	-	-	-	-	✓	✓	-	-	✓	-
My Heart Age	-	✓	-	-	-	✓	✓	-	✓	-	-	-	-	-	-	-	-
ASCVD Risk Estimator Plus	-	✓	-	-	-	-	✓	-	-	-	-	-	-	-	-	-	-
Texas Heart Institute	-	✓	-	-	-	-	✓	-	-	-	-	-	-	-	-	-	-
The Heart App ©	-	✓	-	-	-	-	✓	-	-	-	-	-	-	-	-	-	-
Angina	-	✓	-	-	-	-	-	-	-	-	-	-	-	-	-	-	✓
Heart Disease 101 Audio Book	-	✓	-	✓	-	-	-	-	-	-	-	-	-	-	-	-	-
Heart Disease Support	-	✓	-	-	-	-	-	-	-	-	-	-	-	-	-	-	✓
Heart Diseases & Treatment	-	✓	-	-	-	-	-	-	-	-	-	-	-	-	-	-	-
MESA CHD Risk Score	-	✓	-	-	-	-	✓	-	-	-	-	-	-	-	-	-	-
Healthy Heart Guides	-	✓	-	-	-	-	✓	-	-	-	-	-	✓	-	-	-	-
Heart Care Health & Diet Tips	-	✓	-	-	-	-	✓	-	-	-	-	-	-	-	-	-	-
Basic Cardiology	-	✓	-	-	-	-	-	-	-	-	-	-	-	-	-	-	-
CardioRisk Calc	-	✓	-	-	-	-	✓	-	-	-	-	-	-	-	-	-	-
Heart Disease Guide	-	✓	-	✓	-	-	-	-	-	-	-	-	-	-	-	-	-
Cardiovascular Diseases	-	✓	-	-	-	-	-	-	-	-	-	-	-	-	-	-	-
Heart Disease B	-	✓	-	-	-	-	-	-	-	-	-	-	-	-	-	-	-

Table A3. *Cont.*

Name		Provision of Information					Data Acquisition, Processing and Evaluation						Calendar and Appointment-Related			Support			Other
	News	Reference	Learning Material	Player/ Viewer	Broker	Decision Support	Calculator	Meter	Monitor	Surveillance/ Tracker	Diary	Reminder	Calendar	Utility/Aid	Coach	Health Manager	Communicator/ Social Network
Cardiovascular Care Guide	-	✔	-	✓	-	-	-	-	-	-	-	-	-	-	-	-	-
Heart Health Tips	-	✔	-	-	-	-	-	-	-	-	-	-	-	-	-	-	-
Atherosclerosis	-	✔	-	-	-	-	-	-	-	-	-	-	-	-	-	-	-
Heart Disease A	-	✔	-	-	-	✓	-	-	-	-	-	-	-	-	-	-	-
Heart Disease C	-	✔	-	✓	-	-	-	-	-	-	-	-	-	-	-	-	-
Heart Disease Diet-Have a Fit & Healthy Heart with Best Nutrition!	-	✔	-	-	-	-	-	-	-	-	-	-	-	-	-	-	-
Home Remedies For Chest Pain (Angina)	-	✔	-	-	-	-	-	-	-	-	-	-	-	-	-	-	-
Cardiology consultation	-	✓	-	-	-	✔	-	-	-	-	-	-	-	-	-	-	-
Natural Remedies For Chest Pain (Angina)	-	✔	-	-	-	-	-	-	-	-	-	-	-	-	-	-	-
Cardiology -Expert Consult 4 Diagnosis & Treatment	-	✔	-	-	-	-	-	-	-	-	-	-	-	-	-	-	-
Angina Pectoris Disease	-	✔	-	-	-	-	-	-	-	-	-	-	-	-	-	-	-
Cardiovascular Disease Information	-	✔	-	-	-	-	-	-	-	-	-	-	-	-	-	-	-
Herz und koronarer Herzkrankhe	-	✔	-	-	-	-	-	-	-	-	-	-	-	-	-	-	-
Arteriosclerosis Disease	-	✔	-	-	-	-	-	-	-	-	-	-	-	-	-	-	-
Heart Disease Risk Prediction and Prevention	-	✓	-	-	-	-	✔	-	-	-	✓	-	-	-	-	-	-
How To Cure Heart Disease	-	✔	-	-	-	-	-	-	-	-	-	-	-	-	-	-	-
CORONARY HEART DISEASE RISK	-	✔	-	-	-	-	-	-	-	-	-	-	-	-	-	-	-
Universal Healing Programme	-	-	-	✔	-	-	-	-	-	-	-	-	-	-	-	-	-

Note. ✔ primary function of the MHA, ✓ function is employed in the MHA.

References

1. Roth, G.A.; Abate, D.; Abate, K.H.; Abay, S.M.; Abbafati, C.; Abbasi, N.; Abbastabar, H.; Abd-Allah, F.; Abdela, J.; Abdelalim, A.; et al. Global, regional, and national age-sex-specific mortality for 282 causes of death in 195 countries and territories, 1980–2017: A systematic analysis for the Global Burden of Disease Study 2017. *Lancet* **2018**, *392*, 1736–1788. [CrossRef]
2. World Health Organization. Global Status Report on Noncommunicable Diseases 2014. Available online: http://apps.who.int/iris/bitstream/handle/10665/148114/9789241564854_eng.pdf;jsessionid=0712F3C8D5628766B719D32D4DB31713?sequence=1 (accessed on 19 September 2021).
3. Benjamin, E.J.; Virani, S.S.; Callaway, C.W.; Chamberlain, A.M.; Chang, A.R.; Cheng, S.; Chiuve, S.E.; Cushman, M.; Delling, F.N.; Deo, R.; et al. Heart disease and stroke statistics-2018 update: A report from the American Heart Association. *Circulation* **2018**, *137*, 67–492. [CrossRef] [PubMed]
4. Busch, M.; Kuhnert, R. 12-Month prevalence of coronary heart disease in Germany. *J. Health Monit.* **2017**, *2*, 58–63. [CrossRef]
5. Timmis, A.; Townsend, N.; Gale, C.; Grobbee, R.; Maniadakis, N.; Flather, M.; Wilkins, E.; Vos, R.; Bax, J.; Blum, M.; et al. European society of cardiology: Cardiovascular disease statistics 2017. *Eur. Hear. J.* **2018**, *39*, 508–579. [CrossRef]
6. Robert Koch-Institut Prävalenz der Koronaren Herzerkrankung. Available online: https://www.rki.de/DE/Content/Gesundheitsmonitoring/Gesundheitsberichterstattung/GBEDownloadsB/Geda09/koronare_herzerkrankung.pdf?__blob=publicationFile (accessed on 19 September 2021).
7. Khot, U.N.; Khot, M.B.; Bajzer, C.T.; Sapp, S.K.; Ohman, E.M.; Brener, S.J.; Ellis, S.G.; Lincoff, A.M.; Topol, E.J. Prevalence of conventional risk factors in patients with coronary heart disease. *J. Am. Med. Assoc.* **2003**, *290*, 898–904. [CrossRef] [PubMed]
8. MacMahon, S.; Peto, R.; Collins, R.; Godwin, J.; MacMahon, S.; Cutler, J.; Sorlie, P.; Abbott, R.; Collins, R.; Neaton, J.; et al. Blood pressure, stroke, and coronary heart disease. Part 1, prolonged differences in blood pressure: Prospective observational studies corrected for the regression dilution bias. *Lancet* **1990**, *335*, 765–774. [CrossRef]
9. Chauvet-Gelinier, J.C.; Bonin, B. Stress, anxiety and depression in heart disease patients: A major challenge for cardiac rehabilitation. *Ann. Phys. Rehabil. Med.* **2017**, *60*, 6–12. [CrossRef]
10. Gan, Y.; Gong, Y.; Tong, X.; Sun, H.; Cong, Y.; Dong, X.; Wang, Y.; Xu, X.; Yin, X.; Deng, J.; et al. Depression and the risk of coronary heart disease: A meta-analysis of prospective cohort studies. *BMC Psychiatry* **2014**, *14*, 1–11. [CrossRef]
11. Hale, K.; Capra, S.; Bauer, J. A framework to assist health professionals in recommending high-quality apps for supporting chronic disease self-management: Illustrative assessment of type 2 diabetes apps. *JMIR mHealth uHealth.* **2015**, *3*, 1–12. [CrossRef]
12. Piette, J.D.; List, J.; Rana, G.K.; Townsend, W.; Striplin, D.; Heisler, M. Mobile health devices as tools for worldwide cardiovascular risk reduction and disease management. *Circulation* **2015**, *132*, 2012–2027. [CrossRef]
13. Park, L.G.; Beatty, A.; Stafford, Z.; Whooley, M.A. Mobile phone interventions for the secondary prevention of cardiovascular disease. *Prog. Cardiovasc. Dis.* **2016**, *58*, 639–650. [CrossRef]
14. Piepoli, M.F.; Corrà, U.; Dendale, P.; Frederix, I.; Prescott, E.; Schmid, J.P.; Cupples, M.; Deaton, C.; Doherty, P.; Giannuzzi, P.; et al. Challenges in secondary prevention after acute myocardial infarction: A call for action. *Eur. J. Prev. Cardiol.* **2016**, *23*, 1994–2006. [CrossRef]
15. Frederix, I.; Vanhees, L.; Dendale, P.; Goetschalckx, K. A review of telerehabilitation for cardiac patients. *J. Telemed. Telecare* **2015**, *21*, 45–53. [CrossRef]
16. Frederix, I.; Caiani, E.G.; Dendale, P.; Anker, S.; Bax, J.; Böhm, A.; Cowie, M.; Crawford, J.; de Groot, N.; Dilaveris, P.; et al. ESC e-Cardiology Working Group Position Paper: Overcoming challenges in digital health implementation in cardiovascular medicine. *Eur. J. Prev. Cardiol.* **2019**, *26*, 1166–1177. [CrossRef]
17. Harrison, V.; Proudfoot, J.; Wee, P.P.; Parker, G.; Pavlovic, D.H.; Manicavasagar, V. Mobile mental health: Review of the emerging field and proof of concept study. *J. Ment. Health* **2011**, *20*, 509–524. [CrossRef]
18. Pejovic, V.; Mehrotra, A.; Musolesi, M. Anticipatory mobile digital health: Towards personalised proactive therapies and prevention strategies. *Anticip. Med.* **2015**, 253–267.
19. Burke, L.E.; Ma, J.; Azar, K.M.J.; Bennett, G.G.; Peterson, E.D.; Zheng, Y.; Riley, W.; Stephens, J.; Shah, S.H.; Suffoletto, B.; et al. Current science on consumer use of mobile health for cardiovascular disease prevention: A scientific statement from the American heart association. *Circulation* **2015**, *132*, 1157–1213. [CrossRef]
20. Sheridan, S.L.; Viera, A.J.; Krantz, M.J.; Ice, C.L.; Steinman, L.E.; Peters, K.E.; Kopin, L.A.; Lungelow, D. The effect of giving global coronary risk information to adults: A systematic review. *Arch. Intern. Med.* **2010**, *170*, 230–239. [CrossRef] [PubMed]
21. Honeyman, E.; Ding, H.; Varnfield, M.; Karunanithi, M. Mobile health applications in cardiac care. *Interv. Cardiol.* **2014**, *6*, 227–240. [CrossRef]
22. Cowie, M.R.; Bax, J.; Bruining, N.; Cleland, J.G.F.; Koehler, F.; Malik, M.; Pinto, F.; Van Der Velde, E.; Vardas, P. E-Health: A position statement of the European Society of Cardiology. *Eur. Heart J.* **2016**, *37*, 63–66. [CrossRef] [PubMed]
23. Athilingam, P.; Jenkins, B. Mobile phone apps to support heart failure self-care management: Integrative review. *JMIR Cardio* **2018**, *2*. [CrossRef]
24. Creber, R.M.M.; Maurer, M.S.; Reading, M.; Hiraldo, G.; Hickey, K.T.; Iribarren, S. Review and analysis of existing mobile phone apps to support heart failure symptom monitoring and self-care management using the mobile application rating scale (MARS). *JMIR mHealth uHealth.* **2016**, *4*. [CrossRef]
25. Ayyaswami, V.; Padmanabhan, D.L.; Crihalmeanu, T.; Thelmo, F.; Prabhu, A.V.; Magnani, J.W. Mobile health applications for atrial fibrillation: A readability and quality assessment. *Int. J. Cardiol.* **2019**, *293*, 288–293. [CrossRef] [PubMed]

26. Jamaladin, H.; van de Belt, T.H.; Luijpers, L.C.H.; de Graaff, F.R.; Bredie, S.J.H.; Roeleveld, N.; van Gelder, M.M.H.J. Mobile apps for blood pressure monitoring: Systematic search in app stores and content analysis. *JMIR mHealth uHealth.* **2018**, *6*. [CrossRef] [PubMed]
27. Zang, J.; Dummit, K.; Graves, J.; Lisker, P.; Sweeney, L. Who Knows What About Me? A Survey of Behind the Scenes Personal Data Sharing to Third Parties by Mobile Apps. Available online: https://techscience.org/a/2015103001/ (accessed on 19 September 2021).
28. Nguyen, H.H.; Silva, J.N.A. Use of smartphone technology in cardiology. *Trends Cardiovasc. Med.* **2016**, *26*, 376–386. [CrossRef]
29. Martínez-Pérez, B.; de la Torre-Díez, I.; López-Coronado, M.; Herreros-González, J. Mobile apps in cardiology: Review. *JMIR mHealth uHealth* **2013**, *1*, e15. [CrossRef] [PubMed]
30. Treskes, R.W.; Wildbergh, T.X.; Schalij, M.J.; Scherptong, R.W.C. Expectations and perceived barriers to widespread implementation of e-Health in cardiology practice: Results from a national survey in the Netherlands. *Netherlands Hear. J.* **2019**, *27*, 18–23. [CrossRef]
31. Stach, M.; Kraft, R.; Probst, T.; Messner, E.-M.; Terhorst, Y.; Baumeister, H.; Schickler, M.; Reichert, M.; Sander, L.B.; Pryss, R. Mobile health app database-a repository for quality ratings of mHealth Apps. *Int. Symp. Comput. Med. Syst.* **2020**, 427–432.
32. Messner, E.M.; Terhorst, Y.; Barke, A.; Baumeister, H.; Stoyanov, S.; Hides, L.; Kavanagh, D.; Pryss, R.; Sander, L.; Probst, T. The german version of the mobile app rating scale (MARS-G): Development and validation study. *JMIR mHealth uHealth.* **2020**, *8*, e14479. [CrossRef]
33. Stoyanov, S.R.; Hides, L.; Kavanagh, D.J.; Zelenko, O.; Tjondronegoro, D.; Mani, M. Mobile app rating scale: A new tool for assessing the quality of health mobile apps. *JMIR mHealth uHealth.* **2015**, *3*, 1–9. [CrossRef]
34. Terhorst, Y.; Philippi, P.; Sander, L.B.; Schultchen, D.; Paganini, S.; Bardus, M.; Santo, K.; Knitza, J.; Machado, G.C.; Schoeppe, S.; et al. Validation of the mobile application rating scale (MARS). *PLoS ONE.* **2020**, *15*, e0241480. [CrossRef]
35. Fleiss, J.L. *The Design and Analysis of Clinical Experiments*; John Wiley & Sons: Hoboken, NJ, USA, 1999.
36. Albrecht, U.-V.; Höhn, M.; von Jan, U. *Chancen und Risiken von Gesundheits-Apps (CHARISMHA)*; Albrecht, U.-V., Ed.; Medizinische Hochschule Hannover: Hannover, Germany, 2016.
37. Kumar Rathi, R.; Kalantri, A.; Kalantri, S.; Rathi, V.; Gandhi, M. Symptom-Based smartphone app for detecting acute coronary syndrome: A diagnostic accuracy study. *JACC* **2016**, *67*, 632. [CrossRef]
38. Sander, L.B.; Schorndanner, J.; Terhorst, Y.; Spanhel, K.; Pryss, R.; Baumeister, H.; Messner, E.M. 'Help for trauma from the app stores?' A systematic review and standardised rating of apps for Post-Traumatic Stress Disorder (PTSD). *Eur. J. Psychotraumatol.* **2020**, *11*, 1. [CrossRef]
39. Terhorst, Y.; Rathner, E.M.; Baumeister, H.; Sander, L. "Help from the App Store?": A systematic review of depression apps in german app stores. *Verhaltenstherapie* **2018**, *28*, 101–112. [CrossRef]
40. Madalina, S.; Pim, C.; Frederick, M.; Roxana, C.; Radu, S.; Anca, D.; Patriciu, A.-C.; Daniel, D. Anxiety: There is an app for that. A systematic review of anxiety apps. *Depress. Anxiety* **2017**, *34*, 518–525. [CrossRef]
41. Schultchen, D.; Terhorst, Y.; Holderied, T.; Stach, M.; Messner, E.-M.; Baumeister, H.; Sander, L.B. Stay present with your phone: A systematic review and standardized rating of mindfulness apps in European App Stores. *Int. J. Behav. Med.* **2021**, *28*, 1–9. [CrossRef] [PubMed]
42. Schoeppe, S.; Alley, S.; Rebar, A.L.; Hayman, M.; Bray, N.A.; Van Lippevelde, W.; Gnam, J.P.; Bachert, P.; Direito, A.; Vandelanotte, C. Apps to improve diet, physical activity and sedentary behaviour in children and adolescents: A review of quality, features and behaviour change techniques. *Int. J. Behav. Nutr. Phys. Act.* **2017**, *14*, 1–10. [CrossRef]
43. Carlo, A.D.; Hosseini Ghomi, R.; Renn, B.N.; Areán, P.A. By the numbers: Ratings and utilization of behavioral health mobile applications. *NPJ Digit. Med.* **2019**, *2*, 54. [CrossRef] [PubMed]
44. Larsen, M.E.; Nicholas, J.; Christensen, H. Quantifying app store dynamics: Longitudinal tracking of mental health apps. *JMIR mHealth uHealth* **2016**, *4*, e6020. [CrossRef]
45. Baumel, A.; Faber, K.; Mathur, N.; Kane, J.M.; Muench, F. Enlight: A comprehensive quality and therapeutic potential evaluation tool for mobile and web-based eHealth interventions. *J. Med. Internet Res.* **2017**, *19*, e82. [CrossRef]

International Journal of
Environmental Research and Public Health

Review

An Updated Meta-Analysis of Remote Blood Pressure Monitoring in Urban-Dwelling Patients with Hypertension

Sang-Hyun Park [1,†], Jong-Ho Shin [1,†], Joowoong Park [2] and Woo-Seok Choi [3,4,*]

[1] Department of Internal Medicine, Daejeon Eulji Medical Center, Eulji University School of Medicine, Daejeon 35233, Korea; psh@eulji.ac.kr (S.-H.P.); redsea98@eulji.ac.kr (J.-H.S.)
[2] Research Strategy Division, Korea Aerospace Research Institute (KARI), Daejeon 34133, Korea; park@kari.re.kr
[3] Moon Soul Graduate School of Future Strategy, Korea Advanced Institute of Science and Technology (KAIST), Daejeon 34141, Korea
[4] Keyu Internal Medicine Clinic, Daejeon 35250, Korea
* Correspondence: keyuim@kaist.ac.kr; Tel.: +82-42-483-7554; Fax: +82-42-485-7554
† These authors contributed equally to this work.

Abstract: Following the coronavirus disease-2019 pandemic, this study aimed to evaluate the overall effects of remote blood pressure monitoring (RBPM) for urban-dwelling patients with hypertension and high accessibility to healthcare and provide updated quantitative summary data. Of 2721 database-searched articles from RBPM's inception to November 2020, 32 high-quality studies (48 comparisons) were selected as primary data for synthesis. A meta-analysis was undertaken using a random effects model. Primary outcomes were changes in office systolic blood pressure (SBP) and diastolic blood pressure (DBP) following RBPM. The secondary outcome was the BP control rate. Compared with a usual care group, there was a decrease in SBP and DBP in the RBPM group (standardized mean difference 0.507 (95% confidence interval [CI] 0.339–0.675, $p < 0.001$; weighted mean difference [WMD] 4.464 mmHg, $p < 0.001$) and 0.315 (CI 0.209–0.422, $p < 0.001$; WMD 2.075 mmHg, $p < 0.001$), respectively). The RBPM group had a higher BP control rate based on a relative ratio (RR) of 1.226 (1.107–1.358, $p < 0.001$). RBPM effects increased with increases in city size and frequent monitoring, with decreases in intervention duration, and in cities without medically underserved areas. RBPM is effective in reducing BP and in achieving target BP levels for urban-dwelling patients with hypertension.

Keywords: blood pressure; remote monitoring; hypertension; telemedicine; urban

Citation: Park, S.-H.; Shin, J.-H.; Park, J.; Choi, W.-S. An Updated Meta-Analysis of Remote Blood Pressure Monitoring in Urban-Dwelling Patients with Hypertension. *Int. J. Environ. Res. Public Health* **2021**, *18*, 10583. https://doi.org/10.3390/ijerph182010583

Academic Editors: Irene Torres-Sanchez and Marie Carmen Valenza

Received: 23 September 2021
Accepted: 5 October 2021
Published: 9 October 2021

Publisher's Note: MDPI stays neutral with regard to jurisdictional claims in published maps and institutional affiliations.

1. Introduction

Hypertension is widely recognized as the most important risk factor for cardiovascular disease (CVD), which is a major cause of total mortality [1]. A 2 mmHg fall in systolic blood pressure (SBP) has been reported to reduce the incidence of ischemic CVD and stroke by 7% [2]. However, even in advanced countries, target blood pressure (BP) is achieved in <50% of patients with hypertension [3,4]. The 2017 American College of Cardiology/American Heart Association (ACC/AHA) and 2018 European Society of Cardiology/European Society of Hypertension (ESC/ESH) treatment recommendations state that BP must be controlled to stricter levels [5,6].

Remote BP monitoring (RBPM) has been recommended for hypertension diagnosis and treatment [5,6], as it has been reported to predict CVD morbidity and mortality with higher accuracy than office BP monitoring [7]. As a method of telemedicine, RBPM is known to be an effective tool to enhance drug adherence and BP control in patients with hypertension [8–12]. RBPM has been suggested as a potential solution to overcome the geographical limitations of healthcare services [13], with significant effects shown in randomized controlled trials (RCTs) and meta-analysis studies [10,14–16]. The 2017

Int. J. Environ. Res. Public Health **2021**, *18*, 10583. https://doi.org/10.3390/ijerph182010583   https://www.mdpi.com/journal/ijerph

ACC/AHA guidelines also recommended RBPM for hypertension diagnosis and control, and for enhancing patients' drug adherence [6].

According to the United Nations, approximately 68% of the human population is predicted to dwell in urban settings by 2050 [17]. Urbanization is a rapidly growing 21st century trend, with significant effects on human health. However, despite increased interest in new health technologies, several studies have reported that remote monitoring has limited application in urban settings where high-quality face-to-face care is possible and healthcare accessibility is high [18,19]. Moreover, there is no comprehensive evidence concerning the effect of RBPM in improving clinical outcomes of urban-dwelling patients with hypertension or whether RBPM can become a standard treatment for hypertension management.

In a previous meta-analysis of RCTs using the Jovell/Navarro-Rubio classification system to determine the strength of evidence, RBPM showed statistically significant reductions in SBP (3.48 mmHg) and diastolic BP (DBP, 1.64 mmHg) compared with usual care (UC) after an average of 7.6 months for patients dwelling in an urban setting. In terms of CVD prevention, however, RBPM induced <0.5% of CVD prevention in low-risk patients with hypertension. Therefore, some studies have concluded that RBPM is of little practical significance to policy-makers [20,21]. The coronavirus disease-2019 (COVID-19) pandemic resulted in a steeply increased demand for telemedicine, even in urban settings, for those otherwise having adequate availability and accessibility to healthcare services. More generally, characteristically dense populations in cities have resulted in the rapid spread of infectious diseases, leading to the expansion of infrastructure for non-face-to-face care in line with a rapid increase in the use of the internet and mobile devices.

Considering the global rate of BP control, according to 2017 ACC/AHA guidelines for hypertension diagnosis and control, which is the latest strict guideline for hypertension diagnosis and control, the proportion of patients achieving the target BP is predicted to decrease further. The use of remote medical care services suddenly increased during the COVID-19 pandemic [14,22,23], and its use needs to be verified based on the integration of previous findings, given that hypertension is a chronic disease requiring long-term management for CVD prevention and for efficient healthcare policies to be implemented in urban settings. Therefore, relevant studies need to be extended through an updated compilation of BP data. The objective of our study is to evaluate whether RBPM could be utilized as an alternative to standard treatment for urban-dwelling patients with hypertension during the COVID-19 pandemic. Thus, this study aimed to determine the relative effects of RBPM compared with UC based on outcomes including SBP, DBP, and BP control rates. Intervention duration, city size, setting, frequency of remote transmission of BP data, and the presence of medically underserved areas (MUAs) in the city were analyzed as secondary factors to evaluate the effects of RBPM. We hypothesized that the effects of RBPM were equivalent to those of UC. To test this hypothesis, relevant, up-to-date RCTs were systematically reviewed and transparent and reliable quantitative data synthesis was performed.

2. Materials and Methods

2.1. Searching for Eligible Studies

This study followed the Preferred Reporting Items for Systematic Reviews and Meta-Analyses (PRISMA) guidelines of the Cochrane Collaboration and a checklist was provided [Supplementary Materials] [24]. To identify eligible studies, two investigators (SHP and JHS) independently searched the following electronic databases: PubMed, EBSCOhost, Embase, and the Cochrane Library, from RBPM's inception to November 30, 2020. Free terms were used, including and related to *urban*, *hypertension*, and *remote monitoring*, along with medical subject heading (MeSH) terms. Truncation and phrasing methods were applied to derive a structured search formula [20] (Appendix A). The formula was first applied to the Cochrane Library and then converted to suit each database for the subsequent search. Articles written in English were retrieved. To include as many relevant articles as possible, all systematic reviews and meta-analyses related to the search themes were collected from each database and Google Scholar, and their reference lists were reviewed. To identify gray literature, relevant websites were used, and all studies including those in which the city area was not clearly defined were identified through a manual search.

2.2. Inclusion and Exclusion Criteria

All included studies were blinded RCT studies with random and uniform allocation of patients with hypertension into an RBPM group and a traditional face-to-face UC group. Articles reporting pre- and post-intervention data were targeted, with participants satisfying the following criteria: (1) patients with hypertension under management through regular visits to an urban medical institution; (2) patients able to measure their own BP at home; (3) patients able to transmit their BP data to the physician via post, phone, Bluetooth device, mobile phone, web, or computer (wired or wireless); (4) adults aged $\geq$18 years; (5) BP measurement through ambulatory monitoring; and (6) various transmission methods from real-time or a stored and forward method to an automatic or manual method. Exclusion criteria comprised the following: (1) sudden BP changes due to an acute CVD or cerebrovascular accident (CVA); (2) patients undergoing hemodialysis due to acute or chronic renal disease; (3) female patients before and after pregnancy; (4) cases not reported for urban areas or cases for urban and rural areas reported together; (5) cases from unclear target areas; (6) cases where monitoring was aided by medical staff at a nursing management unit or care facility; and (7) cluster trials or cross-over studies.

2.3. Study Selection

The citations retrieved from each database were exported to EndNote X8.2, and two investigators (SHP and JHS) independently eliminated those not satisfying the criteria to confirm the reliability of identification. First, the title and abstract were screened, and for studies satisfying the criteria, full texts were obtained and scrutinized. Primary studies were selected independently, and their reference lists were reviewed. Final articles for data synthesis were determined after discussion with the senior author (WSC).

2.4. Data Extraction and Coding

For the selected studies, data extraction was performed independently by two investigators (JHS and WSC), and relevant values were coded in an electronic sheet. The extracted data included demographic and pre- and post-intervention SBP and DBP data. BP data were mostly obtained using an automated device and, in the case of ambulatory BP monitoring (ABPM), the mean of each group was calculated and coded. If an article did not report BP values or standard deviations (SDs), preventing calculations with a 95% confidence interval, the values were first checked on the trial registries website and, in cases where the required information could not be obtained, an attempt was made to contact the author of the article [25,26]. Articles that satisfied the inclusion criteria but did not report the main BP data were excluded from the final data synthesis. For some studies with missing SDs, data imputation was performed using a simple method [27,28]. The mean of all other studies, excluding those with missing data, was obtained. Regarding the rate of BP control, the number of patients satisfying the level of normal BP, determined during the final follow-up period of comparison in each study, was calculated and compared between the two groups. If a single primary study included several different follow-up periods for comparison [26–35]; applied a different, additional intervention [25,34]; or had multiple varying sample sizes and thus reported varying results, each result was included in the analysis as an independent study. Disagreements between investigators were resolved through consultation with the senior author (WSC).

2.5. Quality Assessment and Publication Bias

The quality assessment of the primary studies included evaluating the risk of bias (RoB) and was performed independently by two investigators (SHP and JP). Using the

Review Manager program (RevMan, version 5.3.5, Copenhagen, Denmark) software from the Cochrane Collaboration, the evaluation was performed according to the Cochrane Handbook for Systematic Reviews of Interventions guidelines [24,36]. Disagreements were resolved through discussion among investigators. To identify publication bias, Egger's regression, classic fail-safe N, Duval and Tweedie's trim-and-fill method, and funnel plots were used.

2.6. Statistical Analysis

To ensure the reliability of the analysis, coded data were analyzed by two investigators (SHP & JHS) using Comprehensive Meta-Analysis version 2 (CMA, Biostat, Englewood, NJ, USA) software. For primary outcomes, continuous variables comprised the weighted mean difference (WMD) and the standardized mean difference (SMD) obtained from the mean SBP and DBP values measured at baseline and during follow-up in the office. Despite divided opinions regarding the use of continuous variables, SMD has shown a trend of higher statistically significant generalizability and percentage agreement than the WMD in a random effects model (REM) and a fixed effects model (FEM) [37,38]. Therefore, SMD was used in this study to report the results of the data synthesis for continuous variables. Considering the generalizability of each result, the WMD was additionally estimated for comparing the subgroup results [38]. Based on Cohen's general rule of thumb, the effect size was set as follows: SMD 0.2 (small effect); SMD 0.5 (medium effect), and SMD 0.8 (large effect) [39]. Accordingly, when the SMD was $\geq$0.5, we considered the effect size to be significant in this study. The rate of BP control was a dichotomous variable, for which BP normalization data were extracted from each study, and effect size based on relative risk (RR) was used. A 95% confidence interval (CI) was used for all data. To analyze the inter-rater difference, a χ^2 test was used and the level of significance was set to $p < 0.10$. The model of analysis was applied after assessing the enrolled population of each study and the heterogeneity among research centers. Between-study heterogeneity was presented using Tau-squared (τ^2) and I-squared (I^2) indices, and the adequacy of results was determined based on Cohen's general rule of thumb [40]. Therefore, in this study, $30 \leq I^2 \leq 60$ indicated moderate heterogeneity and $50 \leq I^2 \leq 90$ indicated substantial heterogeneity [39]. To assess the quality of each trial and the consequent impact on the overall effect size, sensitivity was tested using the "one study removed" method (Appendix B, Figure A1). A cumulative analysis was run for a total of 48 comparisons, and the range of summary effect sizes at each step according to temporal progression was determined. *p*-values and the presence of outliers affecting the overall effect size were also determined (Appendix C, Figure A2). An additional sensitivity test was performed to determine differences between the data before and after imputing the missing values.

3. Results

3.1. Study Characteristics

Through an initial search of available databases, reference to trial registries, and a manual search of reference lists, a total of 2721 citations were retrieved (Figure 1). Of these, 992 duplicates were removed, leaving 1729 citations to be identified. Next, titles and abstracts for each identified citation were screened, and 1217 irrelevant citations were excluded. For the remaining 512 articles, the full text was obtained and scrutinized, and studies without available data (n = 206), studies not performed in an urban area, studies either reporting combined results of urban and rural areas or not reporting the area (n = 192), studies conducted on patients with CVD or CVA that may induce a sudden change in BP, studies conducted on patients undergoing hemodialysis or including patients with chronic renal disease, and studies involving female patients before or after pregnancy (n = 46) or patients aged <18 years (n = 21) were excluded. In total, 32 independent studies (48 comparisons) satisfying the inclusion criteria were used in the final data synthesis (Table 1).

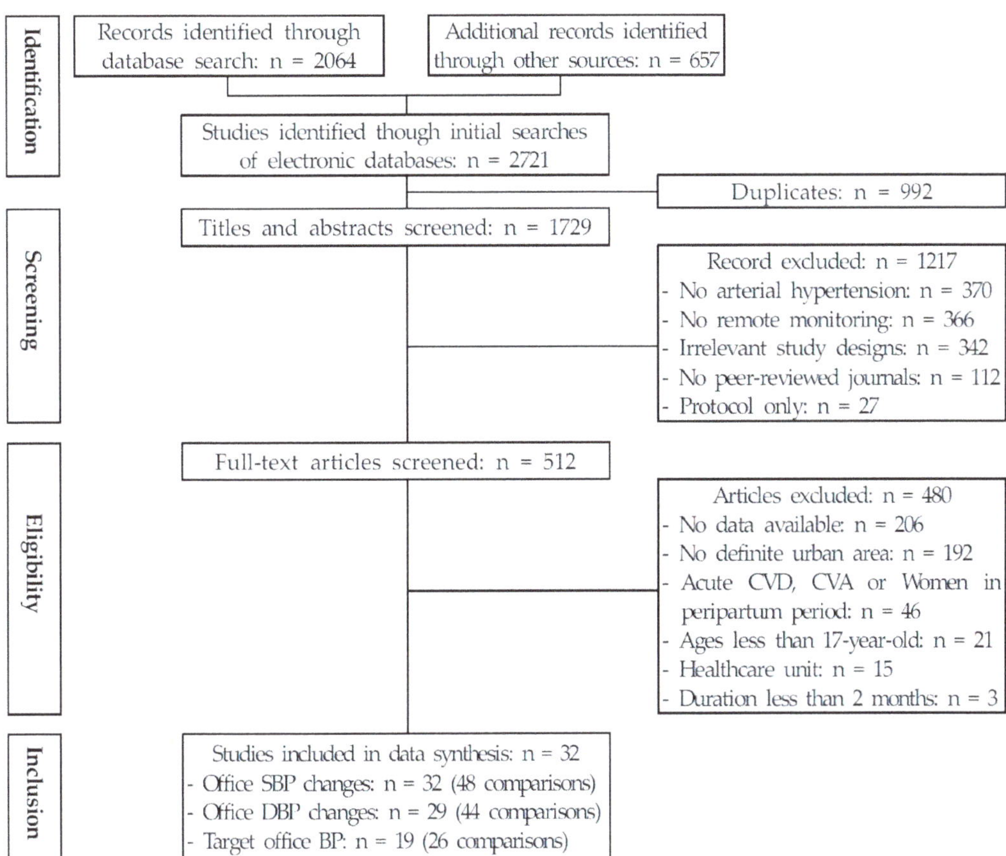

Figure 1. PRISMA flow of study. Abbreviations: BP, blood pressure; CVD, cardiovascular disease; CVA, cerebro-vascular accident; DBP, diastolic blood pressure; SBP, systolic blood pressure.

For the primary studies included in the meta-analysis in this study, the duration of RBPM was 2–18 months (mean, 7.37 months), and the number of participants in the UC and RBPM groups was 5666 and 5729, respectively. The mean age of participants in the UC and RBPM groups was 52.63 and 52.17 years, respectively. No significant intergroup differences were found in terms of sex and baseline BP. No differences in ethnicity were observed. Fourteen studies were conducted in primary medical institutions, 12 in community healthcare centers, and 22 in hospitals or higher-level institutions. The completion dates were in or prior to the year 2000 for two studies [41,42], between 2001 and 2010 for 14 studies [25,29,43–50], and between 2011 and 2020 for 32 studies. Seven studies had used mean values for ABPM [47,48,50–54].

Table 1. Characteristics of individual primary studies.

Study	Included Participants	Participants Number		Participants' Age Interval (Years)			Duration (Months)	City Name (Country)	Population of City	Setting	Description of Intervention	Intervention Frequency	Outcomes
		UC	RBPM	Age Interval	UC	RBPM							
Bosworth (2007) [25]	Treated hypertensive patients	150	150	Child, Adult, Older Adult	Not re-ported	Not re-ported	18	Durham (USA)	232,299 in 2005	Durham VA general internal medicine clinics (Not underserved)	Nurse-administered tailored behavioral intervention with telemedicine device connected to telephone	Once a day	1. Primary: BP control. 2. Secondary: knowledge and perceived risks related with hyper-tension
Bosworth (2007) [25]	Treated hypertensive patients	150	150	Child, Adult, Older Adult	Not re-ported	Not re-ported	18	Durham (USA)	232,299 in 2005	Durham VA general internal medicine clinics (Not underserved)	Nurse-administered medication management	Once a day	1. Primary: BP control 2. Secondary: knowledge and perceived risks related with hyper-tension
Bosworth (2007) [25]	Treated hypertensive patients	150	150	Child, Adult, Older Adult	Not re-ported	Not re-ported	18	Durham (USA)	232,299 in 2005	Durham VA general internal medicine clinics (Not underserved)	Nurse-administered tailored behavioral intervention and medication management	Once a day	1. Primary: BP control 2. Secondary: knowledge and perceived risks related with hyper-tension

Table 1. *Cont.*

Study	Included Participants	Participants Number		Participants' Age Interval (Years)			Duration (Months)	City Name (Country)	Population of City	Setting	Description of Intervention	Intervention Frequency	Outcomes
		UC	RBPM	Age Interval	UC	RBPM							
Kerry (2013) [26]	Hypertensive patients with history of stroke or transient ischemic attack	169	168	16 or older (Child, Adult, Older Adult) Average: 71.9	72.6 ± 11.4	71.1 ± 12.6	6	London (UK)	6,984,772 in 2007	Community healthcare center (Not underserved)	Home BP monitoring with nurse-led support through telephone	Twice a week	Reduction of SBP
Kerry (2013) [26]	Hypertensive patients with history of stroke or transient ischemic attack	169	168	16 or older (Child, Adult, Older Adult) Average: 71.9	72.6 ± 11.4	71.1 ± 12.6	12	London (UK)	6,984,772 in 2007	Community healthcare center (Not underserved)	Home BP monitoring with nurse-led support through telephone	Twice a week	Reduction of SBP
Pan (2018) [27]	Patients diagnosed hypertension	55	52	Between 35 and 75. Average: 57.2	56.55 ± 9.80	57.8 ± 10.87	3	Beijing (China)	11,895,973 in 2016	Fangzhuang Community Health Center (Not underserved)	Mobile phone-linked computer system	Once a day	BP control
Pan (2018) [27]	Patients diagnosed hypertension	55	52	Between 35 and 75. Average: 57.2	56.55 ± 9.80	57.8 ± 10.87	6	Beijing (China)	11,895,973 in 2016	Fangzhuang Community Health Center (Not underserved)	Mobile phone-linked computer system	Once a day	BP control

Table 1. *Cont.*

Study	Included Participants	Participants Number		Participants' Age Interval (Years)			Duration (Months)	City Name (Country)	Population of City	Setting	Description of Intervention	Intervention Frequency	Outcomes
		UC	RBPM	Age Interval	UC	RBPM							
Zha (2020) [28]	Uncontrolled hypertensive patients	13	12	Between 18 and 64. Average: 52.3	55.5 ± 5.2	48.9 ± 8.0	3	Newark (USA)	278,366 in 2016	Jordan and Harris Community Health Center (Local community health center) (Underserved)	Smartphone-linked system by nurse	Visit office once a week. Instant feedback after all measurements.	BP control (Changes in SBP and DBP), perceived self-efficacy, HRQOL
Zha (2020) [28]	Uncontrolled hypertensive patients	13	12	Between 18 and 64. Average: 52.3	55.5 ± 5.2	48.9 ± 8.0	6	Newark (USA)	278,366 in 2016	Jordan and Harris Community Health Center (Local community health center) (Underserved)	Smartphone-linked system by nurse	Visit office once a week. Instant feedback after all measurements.	BP control (Changes in SBP and DBP), perceived self-efficacy, HRQOL
Artinian (2007) [29]	African American hypertensive patients	157	164	18 or more	60.2 ± 12.3	59.1 ± 13.0	3	Detroit (USA)	594,562 in 2002	Family community center (Underserved)	Telephonic transmission with BP monitoring device linked to telephone	Once a week	Office BP changes (SBP, DBP)
Artinian (2007) [29]	African American hypertensive patients	163	168	18 or more	60.2 ± 12.3	59.1 ± 13.0	6	Detroit (USA)	594,562 in 2002	Family community center (Undeserved)	Telephonic transmission with BP monitoring device linked to telephone	Once a month	Office BP changes (SBP, DBP)

Table 1. *Cont.*

Study	Included Participants	Participants Number		Participants' Age Interval (Years)			Duration (Months)	City Name (Country)	Population of City	Setting	Description of Intervention	Intervention Frequency	Outcomes
		UC	RBPM	Age Interval	UC	RBPM							
Artinian (2007) [29]	African American hypertensive patients	169	167	18 or more	60.2 ± 12.3	59.1 ± 13.0	12	Detroit (USA)	594,562 in 2002	Family community center (Undeserved)	Telephonic transmission with BP monitoring device linked to telephone	Once a month	Office BP changes (SBP, DBP)
Cicolini (2013) [30]	Treated or untreated hypertensive patients	98	100	Between 18 and 80. (Adult, Older Adult) Average: 59.1	58.3 ± 13.9	59.8 ± 15.0	3	Chieti (Italy)	43,824 in 2011	Italian Hypertension Primary Care Center (Not underserved)	Nurse-led reminder through e-mail	Once a week	1. BP changes 2. BMI, alcohol consumption, cigarette smoking, adherence to therapy
Cicolini (2013) [30]	Treated or untreated hypertensive patients	98	100	Between 18 and 80. (Adult, Older Adult) Average: 59.1	58.3 ± 13.9	59.8 ± 15.0	6	Chieti (Italy)	43,824 in 2011	Italian Hypertension Primary Care Center (Not underserved)	Nurse-led reminder through e-mail	Once a week	1. BP changes 2. BMI, alcohol consumption, cigarette smoking, adherence to therapy

Table 1. *Cont.*

Study	Included Participants	Participants Number		Participants' Age Interval (Years)			Duration (Months)	City Name (Country)	Population of City	Setting	Description of Intervention	Intervention Frequency	Outcomes
		UC	RBPM	Age Interval	UC	RBPM							
Hebert (2012) [31]	Uncontrolled hypertensive patients	83	85	18 or more. Average: 60.8	(61.3 ± 11.7)	61.3 ± 11.7	9	New York (USA)	8,174,959 in 2010	One academic medical center, two medium-sized hospitals, one community hospital (Underserved)	Telephone	Once a week (Meetings: once in two weeks)	Blood pressure reduction
Hebert (2012) [31]	Uncontrolled hypertensive patients	78	79	18 or more. Average: 60.8 Average: 60.8	(61.3 ± 11.7)	61.3 ± 11.7	18	New York (USA)	7,721,457 in 2010	One academic medical center, two medium-sized hospitals, one community hospital (Underserved)	Telephone	Once a week (Meetings: once in two weeks)	Blood pressure reduction
Kim (2014) [32]	Uncontrolled Korean–American hypertensive seniors	192	191	60 or older adult. Average: 70.9	71.2 ± 5.6	70.6 ± 5.0	6	Ellicott City (USA)	60,489 in 2007	Korean Resource Center (Hospital) (Not Undeserved)	Telephone-monitoring system and telephone counseling	At least once a week (Measure-ment: at least twice a day, Monthly telephone counseling)	Changes in SBP and DBP

Table 1. *Cont.*

Study	Included Participants	Participants Number		Participants' Age Interval (Years)			Duration (Months)	City Name (Country)	Population of City	Setting	Description of Intervention	Intervention Frequency	Outcomes
		UC	RBPM	Age Interval	UC	RBPM							
Kim (2014) [32]	Uncontrolled Korean–American hypertensive seniors	185	187	60 or older adult. Average: 70.9	71.2 ± 5.6	70.6 ± 5.0	12	Ellicott City (USA)	60,489 in 2007	Korean Resource Center (Hospital) (Not Undeserved)	Telephone-monitoring system and telephone counseling	At least once a week (Measurement: at least twice a day, Monthly telephone counseling,	Changes in SBP and DBP
Kim (2014) [32]	Uncontrolled Korean–American hypertensive seniors	185	184	60 or older adult. Average: 70.9	71.2 ± 5.6	70.6 ± 5.0	18	Ellicott City (USA)	60,489 in 2007	Korean Resource Center (Hospital) (Not Undeserved)	Telephone-monitoring system and telephone counseling	At least once a week (Measurement: at least twice a day, Monthly telephone counseling,	Changes in SBP and DBP
Mohsen (2020) [33]	Treated hypertensive patients with antihypertensive medication	50	50	Between 35 and 65. Average: 56.41	55.01 ± 7.50	57.81 ± 9.52	3	Shibin El Kom (Egypt)	190.064 in 2019	Medical outpatient clinic of Menoufia University Hospital (Not Undeserved)	Tele-nursing intervention with telephone support	Twice a week (Measurement: every day)	1. Reduction of SBP and DBP 2. BMI difference

Table 1. *Cont.*

Study	Included Participants	Participants Number		Participants' Age Interval (Years)			Duration (Months)	City Name (Country)	Population of City	Setting	Description of Intervention	Intervention Frequency	Outcomes
		UC	RBPM	Age Interval	UC	RBPM							
Mohsen (2020) [33]	Treated hypertensive patients with medication	50	50	Between 35 and 65. Average: 56.41	55.01 ± 7.50	57.81 ± 9.52	6	Shibin El Kom (Egypt)	190.064 in 2019	Medical outpatient clinic of Menoufia University Hospital (Not Undeserved)	Tele-nursing intervention with telephone support	Twice a week. (Measurement: every day)	1. Reduction of SBP and DBP 2. BMI difference
Pour (2020) [34]	Treated hypertensive patients with medication	21	21	Between 35 and 64. Average: 55.7	56.71 ± 5.73	54.71 ± 6.11	3	Tehran (Iran)	7,250,693 in 2019	Military hospital (Not underserved)	Interactive SMS	Once a week	BP control (Changes in SBP and DBP),
Pour (2020) [34]	Treated hypertensive patients with medication	21	21	Between 35 and 64. Average: 55.7	56.71 ± 5.73	54.71 ± 6.11	4	Tehran (Iran)	7,250,693 in 2019	Military hospital (Not underserved)	Interactive SMS	Once a week	BP control (Changes in SBP and DBP),
Pour (2020) [34]	Treated hypertensive patients with medication	21	21	Between 35 and 64. Average: 55.7	56.71 ± 5.73	54.71 ± 6.11	3	Tehran (Iran)	7,250,693 in 2019	Military hospital (Not underserved)	Non-Interactive SMS	Once a week	BP control (Changes in SBP and DBP),
Pour (2020) [34]	Treated hypertensive patients with medication	21	21	Between 35 and 64. Average: 55.7	56.71 ± 5.73	54.71 ± 6.11	4	Tehran (Iran)	7,250,693 in 2019	Military hospital (Not underserved)	Non-Interactive SMS	Once a week	BP control (Changes in SBP and DBP),

Table 1. *Cont.*

Study	Included Participants	Participants Number		Participants' Age Interval (Years)			Duration (Months)	City Name (Country)	Population of City	Setting	Description of Intervention	Intervention Frequency	Outcomes
		UC	RBPM	Age Interval	UC	RBPM							
Rubinstein (2016) [35]	Untreated pre-hypertensive patients	276	270	Between 30 and 60. Average: 43.4	43.2 ± 8.4	43.6 ± 8.4	6	Buenos Aires (Argentina) and Guatemala City (Guatemala) and Lima (Peru)	12,271,254 (Buenos Aires) and 880,893 (Guatemala City) and 7,136,586 (Lima) in 2012	Institute for Clinical Effectiveness and Health Policy (Buenos Aires, Argentina), Institute of Nutrition of Central America and Panama (Guatemala City, Guatemala), Universidad Peruana Cayetano Heredia (Lima, Peru) (Underserved)	Mobile phone transmission	Once a month	Mean changes in SBP and DBP

Table 1. *Cont.*

Study	Included Participants	Participants Number		Participants' Age Interval (Years)			Duration (Months)	City Name (Country)	Population of City	Setting	Description of Intervention	Intervention Frequency	Outcomes
		UC	RBPM	Age Interval	UC	RBPM							
Rubinstein (2016) [35]	Untreated pre-hypertensive patients	287	266	Between 30 and 60. Average: 43.4	43.2 ± 8.4	43.6 ± 8.4	12	Buenos Aires (Ar-gentina) and Guatemala City (Guatemala) and Lima (Peru)	12,271,254 (Buenos Aires) and 880,893 (Guatemala city) and 7,136,586 (Lima) in 2012	Institute for Clinical Effectiveness and Health Policy (Buenos Aires, Argentina), Insitute of Nutrition of Central America and Panama (Guatemala City, Guatemala), Universidad Peruana Cayetano Heredia (Lima, Peru) (Underserved)	Mobile phone transmission	Once a month	Mean changes in SBP and DBP
Hill (1999) [41]	Black or African American hypertensive young male residents within hospital catchment area	77	78	Between 22 and 49 Average: 39.0			12	Baltimore (USA)	503,998 in 1995	Johns Hopkins Hospital Outpatient General Clinical Research Center (Underserved)	Telephone	Once a month	Office BP changes

Table 1. *Cont.*

Study	Included Participants	Participants Number		Participants' Age Interval (Years)			Duration (Months)	City Name (Country)	Population of City	Setting	Description of Intervention	Intervention Frequency	Outcomes
		UC	RBPM	Age Interval	UC	RBPM							
Friedman (1996) [42]	Treated hypertensive patients	134	133	Over 60 Average: 76.5	77	76	6	Boston (USA)	534,743 in 1994	Senior centers in 29 different communities(Not underserved)	Telephone-linked computer system	Once a week	Office BP changes
McMahon (2005) [43]	Poorly controlled diabetics and hypertensive patients	35	37	Older than 18. Average: 63.5	63 $\pm$ 7	64 $\pm$ 7	12	Boston (USA)	580,352 in 2001	Hospital (Not underserved)	Web-base	At least three times a week	Changes in A$_{1c}$, BP, lipid profiles
Shea (2006) [44]	Diabetic hypertensive patients	347	333	55 or older (Adult, Older Adult) Average: 70.8 $\pm$ 6.7	70.9 $\pm$ 6.8	70.8 $\pm$ 6.5	12	Syracuse (USA)	129,966 in 2005	SUNY Upstate Medical University hospital, (Underserved)	Telephone-linked web system	Regularly	Changes in hemoglobin A$_{1c}$, BP, cholesterol level
Carrasco (2008) [45]	Treated or untreated hypertensive patients	142	131	Average age: 62.5	62.8 $\pm$ 12.5	62.1 $\pm$ 11.9	3	Madrid (Spain)	3,116,909 in 2006	21 regional public health centers (the corporative network of the "Servicio Madrileno de Salud") (Not underserved)	Mobile phone transmission	During the six-month follow-up, four times a week (Monday and Thursday, morning and night)	1. BP control 2. the impact on patient QoL and anxiety, and economic aspects concerning the viability of the telemedicine system

Table 1. *Cont.*

Study	Included Participants	Participants Number		Participants' Age Interval (Years)			Duration (Months)	City Name (Country)	Population of City	Setting	Description of Intervention	Intervention Frequency	Outcomes
		UC	RBPM	Age Interval	UC	RBPM							
Green (2008) [46]	Treated hypertensive patients	247	246	Between 25 and 75. (Adult, Older Adult) Average: 59.1	58.6 ± 8.5	59.5 ± 8.3	12	Seattle, USA	622,927 in 2006	10 medical centers within Group Health Research Institute (Not underserved)	Home BP monitors, instruction on their use, and proficiency training on web-based communication	Report once every two weeks (measurement at least twice a week)	Office SBP and DBP changes and control of BP
Madsen (2008) [47]	Treated or untreated hypertensive patients	123	113	Between 20 and 80. Average Age: 55.9	56.7 ± 11.6	55.0 ± 11.7	6	Holstebro (Denmark)	29,888 in 2004	Holstebro Hospital (Not underserved)	PDA-embedded mobile-web phone (mobile)	Three times a week during the first 3 months and once a week during the last 3 months	Difference in systolic daytime ABPM change
Parati (2009) [48]	Uncontrolled hypertensive patients	111	187	Between 17 and 75. Average age: 57.5	58.1 ± 10.8	57.2 ± 10.7	6	Milan (Italy)	1,198,182 in 2006	Primary care units in Milan (Not underserved)	Telephone-linked computer system	Regularly	Percentage of patients who reached normalization of BP
Park (2009) [49]	Obese hypertensive patients	21	28	Average age: 53.8	54.6 ± 11.0	53.2 ± 6.9	2	Seoul (S. Korea)	9,828,102 in 2007	University-affiliated tertiary care hospital (Not underserved)	Telephone and internet transmission	Once a week	Change in blood pressure, body weight, waist circumference, and serum lipid profile

Table 1. *Cont.*

Study	Included Participants	Participants Number		Participants' Age Interval (Years)			Duration (Months)	City Name (Country)	Population of City	Setting	Description of Intervention	Intervention Frequency	Outcomes
		UC	RBPM	Age Interval	UC	RBPM							
Varis (2010) [50]	Untreated hypertensive patients	68	89	Between 40 and 80	Not reported	Not reported	13	Helsinki and Tampere and Turku (Finland)	536,160 and 194,594 and 168,920 In 2007	Not underserved	Letter to physician	Every five weeks (measurement every day)	Changes in BP and target BP
Hoffmann-Petersen (2017) [51]	Treated un-complicated hypertensive patients	181	175	Between 55 and 64 Average: 60.4	60.4 ± 2.9	60.5 ± 2.6	3	Holstebro (Denmark)	30,885 in 2011	Holstebro Regional Hospital (Not underserved)	Telephone and e-mail communication (Telephone-linked computer system)	Once every two weeks	Daytime ABPM reduction and percentage of target BP
Ionov (2020) [52]	Uncontrolled hypertension patients	80	160	Between 18 and 78	49 (20 to 77)	47 (18 to 78)	3	Saint-Petersburg, (Russia)	5,076,520 in 2019	Federal Medical Research Center Hospital (Not underserved)	Mobile phone communication	Once a week (Measurement: twice a day)	Change of SBP and rate of BP control.
Logan (2012) [53]	Uncontrolled hypertensive and diabetic patients	51	54	30 or more Average: 62.9	62.7 ± 7.8	63.1 ± 9.0	12	Toronto (Canada)	2,423,221 in 2011	Mount Sinai Hospital (Not underserved)	Bluetooth-enabled BP device paired with smartphone (mobile-web)	Twice a day	Changes in ambulatory BP
Neumann (2011) [54]	Inadequately treated hypertensive patients	29	28	Between 18 and 80. Average age: 55.5	56.2 ± 17.4	54.7 ± 17.9	3	Göttingen (Germany)	119,161 in 2009	Not underserved	Mobile phone-linked computer system	Once a Day	BP Control

Table 1. *Cont.*

Study	Included Participants	Participants Number		Participants' Age Interval (Years)			Duration (Months)	City Name (Country)	Population of City	Setting	Description of Intervention	Intervention Frequency	Outcomes
		UC	RBPM	Age Interval	UC	RBPM							
Wakefield (2011) [55]	Type 2 diabetics and hypertensive patients	97	83	Between 40 and 89. Average: 48.1	67.9 ± 9.9	68.4 ± 9.5	6	Iowa City (USA)	67,548 in 2006	Iowa City VA Health Care System (Not underserved)	Telephonic transmission	Every day	Changes in hemoglobin A_{1c} and SBP
Bosworth (2011) [56]	Treated hypertensive patients	137	127	Child, Adult, Older Adult Average Age: 63.5	64 ± 10	63 ± 11	12	Durham (USA)	234,477 in 2006	Durham VA Medical Center (Not underserved)	Telephonic transmission	Once a day	1. BP control 2. SBP and DBP change
Migneault (2012) [57]	African American hypertensive patients	140	125	35 or more. Average age: 56.5	56.8 ± 11.4	56.3 ± 10.6	8	Boston (USA)	590,971, in 2003	Boston Medical Center primary care practices of a large, safety-net hospital and four affiliated community health centers. (Underserved)	Automated, computer-based, interactive telephone counseling system	Once a week	Change in diet quality, leisure time physical activity of moderate-or-greater intensity, and adherence to the antihypertensive medication regimen and change in BP.

Table 1. *Cont.*

Study	Included Participants	Participants Number		Participants' Age Interval (Years)			Duration (Months)	City Name (Country)	Population of City	Setting	Description of Intervention	Intervention Frequency	Outcomes
		UC	RBPM	Age Interval	UC	RBPM							
Park (2012) [58]	Post-menopausal obese hypertensive patients	33	34	Average age: 56.7	57.6 ± 5.5	55.8 ± 5.7	3	Seoul (S. Korea)	9,828,102 in 2007	University medical center (Not underserved)	Reporting on website. Mobile and internet transmission	Once a week.	Change in waist circumference, body weight, and blood pressure, fasting plasma glucose, and serum lipid levels
Bove (2013) [59]	Systolic hypertensive patients	107	99	Between 18 and 85 (Adult, Older Adult) Average: 59.6	58.2 ± 13.5	61.0 ± 13.6	6	Philadelphia/ Wilmington, USA	1,480,457/ 109,499 in 2010	University hospital (Underserved)	Telephone and internet-based System	Once a day	BP control at 6 months
Wakefield (2014) [60]	Type 2 diabetics and uncontrolled hypertensive patients	43	40	18 or more. Average: 60.0	62.5 ± 10.9	57.7 ± 10.8	3	Columbia (USA)	112,498 in 2010	University hospital (Not underserved)	Web System through mobile phone or personal computer	Twice a week (Measurement: every day)	Changes in hemoglobin A_{1c} and SBP

Table 1. *Cont.*

Study	Included Participants	Participants Number		Participants' Age Interval (Years)			Duration (Months)	City Name (Country)	Population of City	Setting	Description of Intervention	Intervention Frequency	Outcomes
		UC	RBPM	Age Interval	UC	RBPM							
Yi (2015) [61]	Uncontrolled hypertensive patients	332	329	18 or more. Average: 61.3	61.3 ± 12.2	61.3 ± 11.9	9	Bronx and Brooklyn and New York (USA)	1,308,242 and 2,172,989 and 7,721,458 in 2010	Riverdale Family Practice (Bronx), Lutheran Family Health Centers (Brooklyn), New York City Department of Health and Mental Hygiene (New York City), Heritage Health Care (New York City) (Underserved)	Telephone-linked computer system	Once a month (Measurement: every day)	Change in SBP and DBP and achievement of BP control

Abbreviations: ABPM, ambulatory blood pressure monitoring; A_{1c}, glycated hemoglobin; BMI, body mass index; BP, blood pressure; DBP, diastolic blood pressure; HRQOL, health-related quality of life; QoL, quality of life; RBPM, remote blood pressure monitoring; SBP, systolic blood pressure; SMS, short message service; UC, usual care.

3.2. Risk Assessment

To check for bias in RCT studies, the Cochrane Group's RoB tool of the Cochrane group was used for domain analysis based on a checklist. Across seven domains, a low risk of selection bias related to sequence generation or allocation concealment was shown. Similarly, the risk of detection bias related to blinding of personnel and patients was appropriately reported. Concerning attrition bias (incomplete outcome data), an unclear or sufficiently high risk was shown that raised concern in a number of studies; however, as most studies showed a low risk (≥ 4) across the seven domains, the overall RoB was deemed to be low [62].

Egger's regression intercept was 4.516 (1.363–7.669; $p = 0.005$) in two-tailed 95% CIs [37]. The number of studies needed to attain $p > 0.05$ for a classic fail-safe N was 5085. The point estimate of SBP in Duval and Tweedie's trim-and-fill analysis (SMD, 0.507 mmHg (0.339–0.645, $p < 0.001$); WMD, 4.464 mmHg ($p < 0.001$)) coincided with the summary effect size, while no imputed study was found in the funnel plot (Figure 2) [63]. The SMD of DBP was 0.253 (0.215–0.292), and no study was trimmed (Figure 3). In the analysis of the rate of target BP achievement, RR was 1.237 (1.107–1.381), three studies were imputed, and the adjusted value was 1.161 (1.032–1.306, Figure 4). Although RoB assessment detected a certain level of publication bias, the overall data were statistically significant and the analysis results were not rejected.

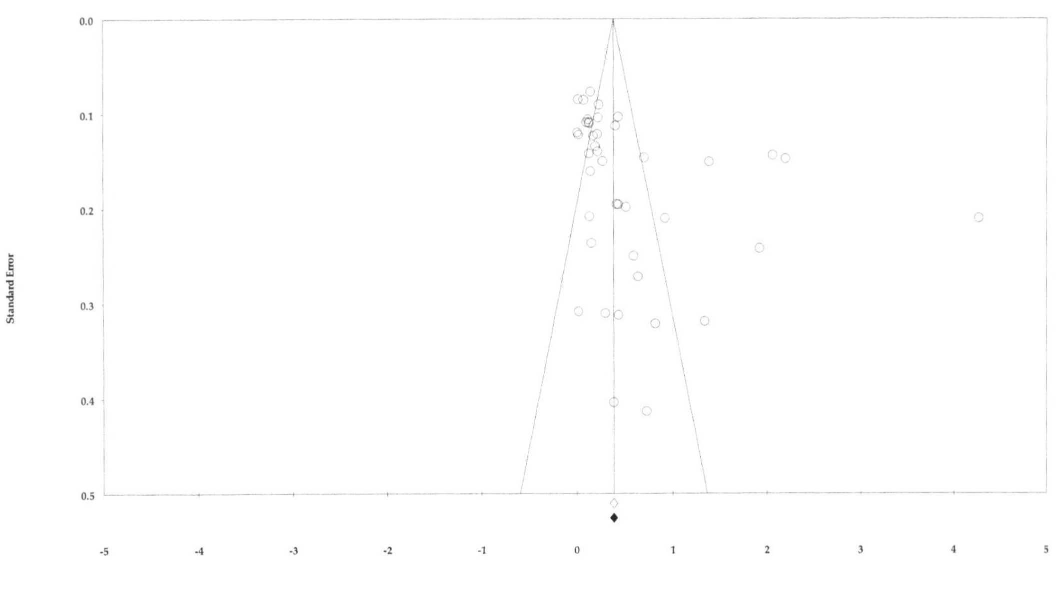

Figure 2. A funnel plot of the standardized mean difference in systolic blood pressure. Note: summary effect size (◇), summary effect size of imputed studies (◆), individual study (○).

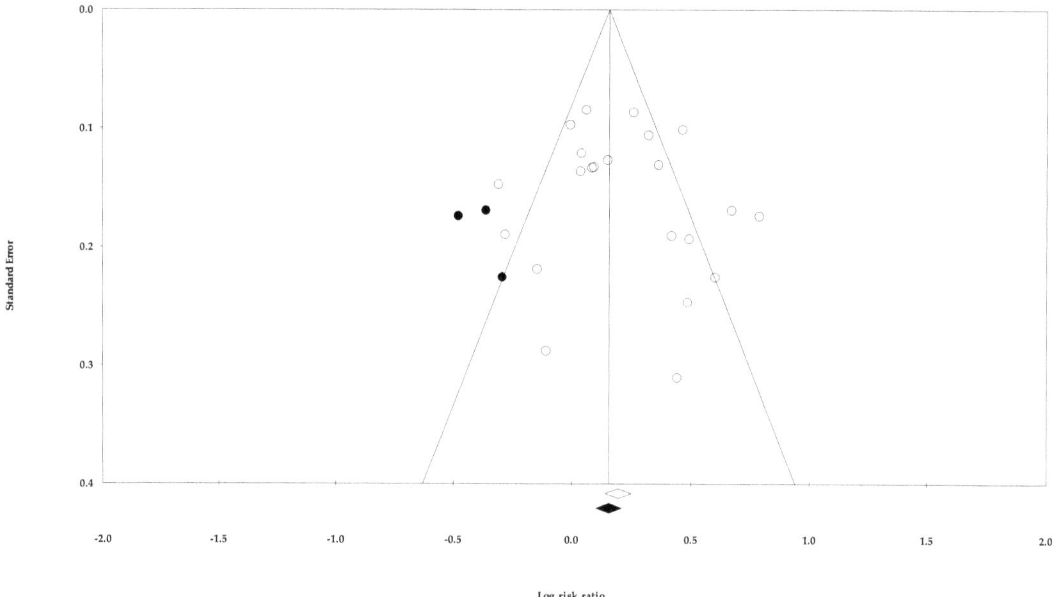

Figure 3. A funnel plot of standardized mean difference in diastolic blood pressure. Note: summary effect size (◇), summary effect size of imputed studies (◆), individual study (○).

Figure 4. A funnel plot of relative risk for the target blood pressure rate. Note: summary effect size (◇), imputed study (●), summary effect size of imputed studies (◆), individual study (○).

A sensitivity test was performed for studies that had been included to prevent small-study effects, excluding those with a sample size of ≤50 for the RBPM group [64]. The test results showed an SMD of 0.501 mmHg (0.313–0.689, $p < 0.001$) and a WMD of 4.238 mmHg ($p < 0.001$), indicating that the difference from the overall summary effect size was not clinically significant and that the potential small-study effect was not significant in this study.

3.3. Primary Outcomes

3.3.1. Systolic Blood Pressure

Across 32 independent studies (48 comparisons), 11,395 patients (UC group, $n = 5666$; RBPM group, $n = 5,729$) were analyzed for SBP [25–35,41–61]. The summary SMD was 0.507 (0.339–0.675, $p < 0.001$), showing an above moderate effect size, and the WMD after conversion was 4.464 mmHg (3.371–5.556, $p < 0.001$; Figure 5). The between-group heterogeneity was significant ($I^2 = 70.908\%$, $p < 0.001$). To determine the effect of individual studies on the total summary effect size, a sensitivity test was performed using the "one study removed" method, whereby each study was sequentially omitted (Appendix B). Here, the point estimate of the summary effect size showed no significant difference and no outliers were detected.

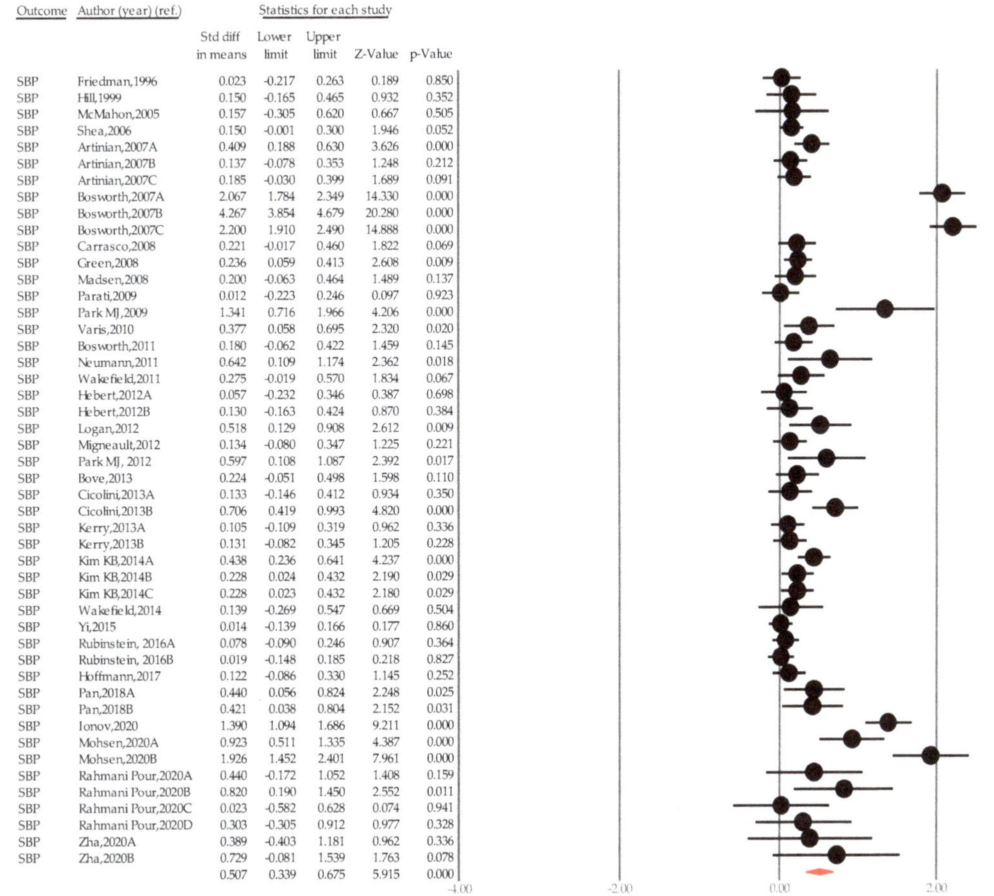

Outcome	Author (year) (ref.)	Std diff in means	Lower limit	Upper limit	Z-Value	p-Value
SBP	Friedman,1996	0.023	-0.217	0.263	0.189	0.850
SBP	Hill,1999	0.150	-0.165	0.465	0.932	0.352
SBP	McMahon,2005	0.157	-0.305	0.620	0.667	0.505
SBP	Shea,2006	0.150	-0.001	0.300	1.946	0.052
SBP	Artinian,2007A	0.409	0.188	0.630	3.626	0.000
SBP	Artinian,2007B	0.137	-0.078	0.353	1.248	0.212
SBP	Artinian,2007C	0.185	-0.030	0.399	1.689	0.091
SBP	Bosworth,2007A	2.067	1.784	2.349	14.330	0.000
SBP	Bosworth,2007B	4.267	3.854	4.679	20.280	0.000
SBP	Bosworth,2007C	2.200	1.910	2.490	14.888	0.000
SBP	Carrasco,2008	0.221	-0.017	0.460	1.822	0.069
SBP	Green,2008	0.236	0.059	0.413	2.608	0.009
SBP	Madsen,2008	0.200	-0.063	0.464	1.489	0.137
SBP	Parati,2009	0.012	-0.223	0.246	0.097	0.923
SBP	Park MJ,2009	1.341	0.716	1.966	4.206	0.000
SBP	Varis,2010	0.377	0.058	0.695	2.320	0.020
SBP	Bosworth,2011	0.180	-0.062	0.422	1.459	0.145
SBP	Neumann,2011	0.642	0.109	1.174	2.362	0.018
SBP	Wakefield,2011	0.275	-0.019	0.570	1.834	0.067
SBP	Hebert,2012A	0.057	-0.232	0.346	0.387	0.698
SBP	Hebert,2012B	0.130	-0.163	0.424	0.870	0.384
SBP	Logan,2012	0.518	0.129	0.908	2.612	0.009
SBP	Migneault,2012	0.134	-0.080	0.347	1.225	0.221
SBP	Park MJ, 2012	0.597	0.108	1.087	2.392	0.017
SBP	Bove,2013	0.224	-0.051	0.498	1.598	0.110
SBP	Cicolini,2013A	0.133	-0.146	0.412	0.934	0.350
SBP	Cicolini,2013B	0.706	0.419	0.993	4.820	0.000
SBP	Kerry,2013A	0.105	-0.109	0.319	0.962	0.336
SBP	Kerry,2013B	0.131	-0.082	0.345	1.205	0.228
SBP	Kim KB,2014A	0.438	0.236	0.641	4.237	0.000
SBP	Kim KB,2014B	0.228	0.024	0.432	2.190	0.029
SBP	Kim KB,2014C	0.228	0.023	0.432	2.180	0.029
SBP	Wakefield,2014	0.139	-0.269	0.547	0.669	0.504
SBP	Yi,2015	0.014	-0.139	0.166	0.177	0.860
SBP	Rubinstein, 2016A	0.078	-0.090	0.246	0.907	0.364
SBP	Rubinstein, 2016B	0.019	-0.148	0.185	0.218	0.827
SBP	Hoffmann,2017	0.122	-0.086	0.330	1.145	0.252
SBP	Pan,2018A	0.440	0.056	0.824	2.248	0.025
SBP	Pan,2018B	0.421	0.038	0.804	2.152	0.031
SBP	Ionov,2020	1.390	1.094	1.686	9.211	0.000
SBP	Mohsen,2020A	0.923	0.511	1.335	4.387	0.000
SBP	Mohsen,2020B	1.926	1.452	2.401	7.961	0.000
SBP	Rahmani Pour,2020A	0.440	-0.172	1.052	1.408	0.159
SBP	Rahmani Pour,2020B	0.820	0.190	1.450	2.552	0.011
SBP	Rahmani Pour,2020C	0.023	-0.582	0.628	0.074	0.941
SBP	Rahmani Pour,2020D	0.303	-0.305	0.912	0.977	0.328
SBP	Zha,2020A	0.389	-0.403	1.181	0.962	0.336
SBP	Zha,2020B	0.729	-0.081	1.539	1.763	0.078
		0.507	0.339	0.675	5.915	0.000

Figure 5. A forest plot of standardized mean difference in systolic blood pressure. Note: point estimate of individual study (●), summary effect size (◆); SBP, systolic blood pressure; UC, usual care; RBPM, remote blood pressure monitoring.

When the average effect of RBPM was chronologically divided into three time-frames and compared with the UC group (Phase I, inception of RBPM to 2000; phase II, 2001–2010; phase III, 2011–2020), the WMD was 1.515 mmHg ($n = 2$, −4.031–7.061, $p = 0.592$; $I^2 = 0.000\%$, $p = 0.478$) in phase I [41,42], 4.333 mmHg ($n = 14$, 2.338–6.328, $p < 0.001$; $I^2 = 38.554$, $p < 0.001$) in phase II [25,29,43–50], and 4.719 mmHg ($n = 32$, 3.343–6.094, $p < 0.001$; $I^2 = 77.361\%$, $p < 0.001$) in phase III [26–28,30–35,51–61].

3.3.2. Diastolic Blood Pressure

To determine the effect of RBPM on DBP, data concerning 10,482 patients (UC group, $n = 5192$; RBPM group, $n = 5290$) were analyzed across 29 studies (44 comparisons) [25,27–35,41–54,56–59,61]. Compared with the UC group, the RBPM group showed greater BP reduction (SMD, 0.315 mmHg (0.209–0.402), $p < 0.001$; WMD, 2.075 mmHg (1.399–2.750) $p < 0.001$) after conversion (Figure 6). The between-study heterogeneity was substantial (I^2, 68.021%; $p < 0.001$). No outliers were detected in the sensitivity test performed through sequentially omitting each study.

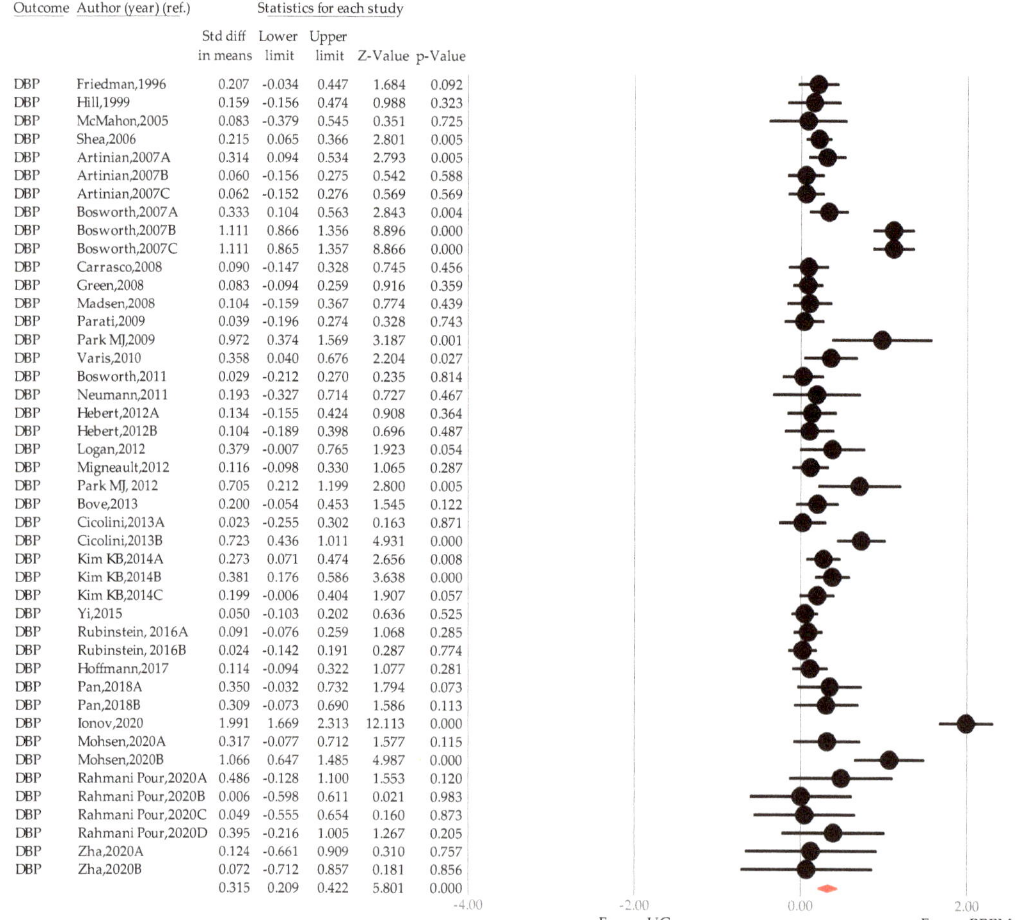

Outcome	Author (year) (ref.)	Std diff in means	Lower limit	Upper limit	Z-Value	p-Value
DBP	Friedman,1996	0.207	-0.034	0.447	1.684	0.092
DBP	Hill,1999	0.159	-0.156	0.474	0.988	0.323
DBP	McMahon,2005	0.083	-0.379	0.545	0.351	0.725
DBP	Shea,2006	0.215	0.065	0.366	2.801	0.005
DBP	Artinian,2007A	0.314	0.094	0.534	2.793	0.005
DBP	Artinian,2007B	0.060	-0.156	0.275	0.542	0.588
DBP	Artinian,2007C	0.062	-0.152	0.276	0.569	0.569
DBP	Bosworth,2007A	0.333	0.104	0.563	2.843	0.004
DBP	Bosworth,2007B	1.111	0.866	1.356	8.896	0.000
DBP	Bosworth,2007C	1.111	0.865	1.357	8.866	0.000
DBP	Carrasco,2008	0.090	-0.147	0.328	0.745	0.456
DBP	Green,2008	0.083	-0.094	0.259	0.916	0.359
DBP	Madsen,2008	0.104	-0.159	0.367	0.774	0.439
DBP	Parati,2009	0.039	-0.196	0.274	0.328	0.743
DBP	Park MJ,2009	0.972	0.374	1.569	3.187	0.001
DBP	Varis,2010	0.358	0.040	0.676	2.204	0.027
DBP	Bosworth,2011	0.029	-0.212	0.270	0.235	0.814
DBP	Neumann,2011	0.193	-0.327	0.714	0.727	0.467
DBP	Hebert,2012A	0.134	-0.155	0.424	0.908	0.364
DBP	Hebert,2012B	0.104	-0.189	0.398	0.696	0.487
DBP	Logan,2012	0.379	-0.007	0.765	1.923	0.054
DBP	Migneault,2012	0.116	-0.098	0.330	1.065	0.287
DBP	Park MJ, 2012	0.705	0.212	1.199	2.800	0.005
DBP	Bove,2013	0.200	-0.054	0.453	1.545	0.122
DBP	Cicolini,2013A	0.023	-0.255	0.302	0.163	0.871
DBP	Cicolini,2013B	0.723	0.436	1.011	4.931	0.000
DBP	Kim KB,2014A	0.273	0.071	0.474	2.656	0.008
DBP	Kim KB,2014B	0.381	0.176	0.586	3.638	0.000
DBP	Kim KB,2014C	0.199	-0.006	0.404	1.907	0.057
DBP	Yi,2015	0.050	-0.103	0.202	0.636	0.525
DBP	Rubinstein, 2016A	0.091	-0.076	0.259	1.068	0.285
DBP	Rubinstein, 2016B	0.024	-0.142	0.191	0.287	0.774
DBP	Hoffmann,2017	0.114	-0.094	0.322	1.077	0.281
DBP	Pan,2018A	0.350	-0.032	0.732	1.794	0.073
DBP	Pan,2018B	0.309	-0.073	0.690	1.586	0.113
DBP	Ionov,2020	1.991	1.669	2.313	12.113	0.000
DBP	Mohsen,2020A	0.317	-0.077	0.712	1.577	0.115
DBP	Mohsen,2020B	1.066	0.647	1.485	4.987	0.000
DBP	Rahmani Pour,2020A	0.486	-0.128	1.100	1.553	0.120
DBP	Rahmani Pour,2020B	0.006	-0.598	0.611	0.021	0.983
DBP	Rahmani Pour,2020C	0.049	-0.555	0.654	0.160	0.873
DBP	Rahmani Pour,2020D	0.395	-0.216	1.005	1.267	0.205
DBP	Zha,2020A	0.124	-0.661	0.909	0.310	0.757
DBP	Zha,2020B	0.072	-0.712	0.857	0.181	0.856
		0.315	0.209	0.422	5.801	0.000

Figure 6. A forest plot of standardized mean difference in diastolic blood pressure. Note: point estimate of individual study (●), summary effect size (◆); DBP, diastolic blood pressure; UC, usual care; RBPM, remote blood pressure monitoring.

The WMD according to time interval was 2.059 mmHg in phase I ($n = 2$, −1.143–5.262, $p = 0.208$; $I^2 = 0.000\%$, $p = 0.45$)[41,42], 1.587 mmHg in phase II ($n = 14$, 0.421–2.753, $p < 0.001$; $I^2 = 17.407\%$, $p < 0.001$) [25,29,43–50], and 2.348 mmHg in phase III ($n = 28$; 1.480–3.216, $p < 0.001$; $I^2 = 76.230\%$, $p < 0.001$) [26–28,30–35,51–61].

3.3.3. Target Blood Pressure Rate

To determine the effect of RBPM, the rate of BP control was estimated based on BP normalization criteria defined in each primary study. Across 16 studies (25 comparisons), the data of 2655 patients in the UC group and 2816 patients in the RBPM group were comprehensively analyzed [13,25,30–33,45–47,50–53,59,61]. Compared with the UC group, the RBPM group showed a significant effect, with an approximately 23.7% higher improvement in BP control based on RR (RR= 1.226 (1.107–1.358), $p < 0.001$; Figure 7). The between-study heterogeneity was substantial ($I^2 = 70.656\%$; $p < 0.001$). No significant difference in summary effect size was found in the sensitivity test.

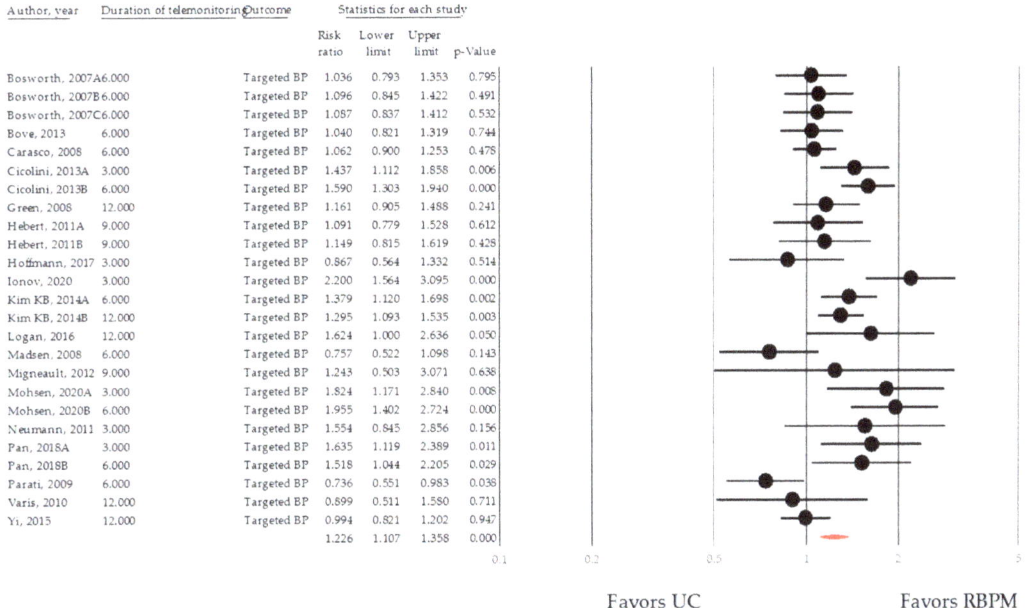

Figure 7. Risk ratio of target blood pressure using remote blood pressure monitoring. Note: point estimate of individual study (•), summary effect size (♦); BP, blood pressure; UC, usual care; RBPM, remote blood pressure monitoring.

3.4. Subgroup Analysis

3.4.1. City Size

Generally accepted international criteria define city size according to population size in a given area. In this study, a metropolitan city was defined as a city with a population of at least one million. Thus, the RCT studies included in this study were categorized based on city size as either a small-to-medium-sized city study or a large city study, and the two categories were analyzed separately. Population size was estimated from the data of the latest international population survey performed in the nearest period of time to this study. Of the 48 studies, 22 were conducted in small-to-medium cities [25,28,29,32,41–43,46,47,50, 51,54–57] and 26 were conducted in large cities [26,27,30,31,33–35,44,45,48,49,52,53,58–61]. For the former, the SBP showed a WMD of 3.860 mmHg (2.271–5.450, $p < 0.001$) without between-study heterogeneity ($I^2 = 0.000$, $p = 0.478$; Tau^2 = 0.000). For the latter, the SBP showed a WMD of 5.056 mmHg (3.503–6.609, $p < 0.001$) with a significant level of between-study heterogeneity ($I^2 = 82.177\%$, $p < 0.001$, $Tau^2 = 17.368$); the magnitude of the effect size was above moderate.

3.4.2. Medically Underserved Areas

The presence of MUAs for each group was reflected in the analysis only if the study clearly indicated the respective area. As a result, 17 studies were categorized as MUAs [28,29,31,32,35,41,44,57,59,61] and 31 as non-MUAs [25–27,30,33,34,42,43,45–56,58,60]. In terms of MUAs, the effect of RBPM on SBP showed a WMD of 3.213 mmHg (1.521–4.905, $p < 0.001$), with $I^2 = 48.904\%$ ($p = 0.012$, $Tau^2 = 2.793$), indicating a moderate degree of between-study heterogeneity based on Cohen's rule of thumb. In contrast, in non-MUAs, the effect of RBPM on SBP showed a WMD of 5.224 mmHg (3.878–6.569; $p < 0.001$), with $I^2 = 73.152\%$ ($p < 0.001$, $Tau^2 = 12.943$).

3.4.3. Duration of Intervention

The effect of reduced SBP based on the WMD varied according to the duration of the intervention. For an intervention duration ≤ 3 months [27–30,33,34,49,51,52,54,58,60], the effect was a WMD of 6.219 mmHg ($n = 15$, 3.970–8.468, $p < 0.001$; $I^2 = 70.060$, $p < 0.001$). For 6 months [26–30,32,33,35,42,45,47,48,55,59], the effect was a WMD of 4.491 mmHg ($n = 14$, 2.461–6.521, $p < 0.001$; $I^2 = 84.562$, $p < 0.001$). For 12 months, the effect was a WMD of 3.446 mmHg ($n = 12$, 1.209–5.683, $p = 0.003$; $I^2 = 34.656$, $p = 0.113$). The rate of BP control had an RR of 1.540 ($n = 6$, 1.223–1.939, $p < 0.001$) after 3 months [27,30,33,51,52,54], an RR of 1.159 ($n = 11$, 1.002–1.341, $p = 0.047$) after 6 months [25,27,30,32,33,45,47,48,59], and an RR of 1.167 ($n = 5$, 0.930–1.464, $p = 0.183$) after 12 months [32,46,50,53,61] (Appendix D, Figure A3).

3.4.4. Setting

The BP reducing effect was analyzed according to the size of the medical institution where RBPM was mainly performed. In primary care clinics, the WMD was 2.981 mmHg ($n = 14$, 1.323–4.639, $p < 0.001$; $I^2 = 45.343$, $p = 0.034$) [25,35,44,45,47,48,50,51,53,54,56]. In community health centers, the WMD was 3.512 mmHg ($n = 12$, 1.651–5.373, $p < 0.001$; $I^2 = 31.670$, $p < 0.001$) [27–30,42,57,61], and the WMD at hospital level was 6.333 mmHg ($n = 22$, 4.750–7.917, $p < 0.001$; $I^2 = 73.401$, $p < 0.001$) [26,31–34,41,43,46,49,52,55,58–60].

3.4.5. Frequency of Remote Transmission of Blood Pressure Data

In the primary studies in which the frequency of remote BP transmission was reported, when BP information was transmitted daily, the WMD was 5.881 mmHg ($n = 13$, 3.898–7.864, $p < 0.001$; $I^2 = 14.635$, $p < 0.001$) [27,34,49,53–55,59–61]. For weekly BP transmission, the WMD was 4.024 ($n = 15$, 2.641–5.406, $p < 0.001$; $I^2 = 54.610$, $p < 0.001$) [28,30,32,42,43,45,47,52,56–58]. For biweekly BP transmission, the WMD was 3.941 mmHg ($n = 4$, 1.428–6.454, $p < 0.001$; $I^2 = 0.000$). For monthly BP transmission, the WMD was 1.803 mmHg ($n = 6$, -0.234–3.841, $p = 0.083$; $I^2 = 21.639$, $p = 0.056$) [26,35,41,50].

4. Discussion

The development of healthcare infrastructure and physicians' preference for practice in an urban setting implies higher accessibility to healthcare and higher patient satisfaction regarding healthcare [65]. However, the COVID-19 pandemic has raised concerns regarding face-to-face care in cities being a potential infection route between healthcare professionals and patients. In this study, data published since September 2018 were included and integrated with data from previous studies to undertake an updated analysis.

Compared with UC, RBPM for urban-dwelling patients with hypertension was found to significantly reduce SBP and DBP in both statistical and clinical terms, while improving the rate of BP control. Following RBPM, SBP and DBP WMDs decreased by 4.464 mmHg and 2.075 mmHg, respectively, compared with UC. This change, observed through quantitative data, showed a greater margin of decrease than reported in a previous meta-analysis

(SBP, 3.482 mmHg; DBP, 1.638 mmHg) [20]. Moreover, according to the temporal interval, the decrease in SBP (1.515 vs. 4.719 mmHg) and DBP (2.059 vs. 2.438 mmHg) in phase III was significantly greater than that in phase I. Therefore, we consider that the demand for RBPM has increased in line with technological advancements, the increased use of mobile devices, and the acceptance of new technologies [66].

RBPM is frequently used in pilot projects preceding the full launch of telemedicine, as it is relatively simple and cost-effective compared with other types of telemedicine. However, reports on the effect of RBPM on the rate of BP control have been inconsistent across numerous previous studies [14]. In this study, where additional data were comprehensively analyzed to extend the meta-analysis, RBPM led to an approximately 20% higher rate of BP control than UC. This is a greater magnitude of improvement than the 13% figure reported in a previous analysis [20]. Considering that the rate of BP control is <50% in traditional face-to-face care, even in countries with advanced healthcare systems, an improvement of 20% is indicative of a highly significant contribution to the prevention of CVD [67].

The ultimate objective behind attempts to lower and control BP in patients with hypertension and to bring it closer to a target BP is to reduce the incidence of CVD. However, in the meta-analysis in this study, data were not analyzed in relation to cardiovascular (CV) events because the included RCTs primarily showed outcomes that targeted changes in BP or the rate of BP control, not CV events. Nevertheless, the effect of RBPM on CV events in urban-dwelling patients with hypertension can be conjectured based on the results of previous studies. In a previous large-scale meta-analysis on prospective monitoring, including randomized, controlled, placebo trials or anti-hypertensive studies, a decrease of 2–3 mmHg in SBP in patients with a moderate risk of CVD was shown to cause a 10% reduction in CV mortality and a 20–30% reduction in major adverse CV events [2,68–70]. Thus, the observed decrease in SBP of 4.464 mmHg in this study, when the WMD was compared between UC and RBPM, is clinically significant and potentially contributes to reducing CV events.

The effect size of the primary outcomes was set as the SMD and, as it showed moderate-to-high heterogeneity (I^2 = 70.908%; $p < 0.001$), a subgroup analysis was performed (Appendix E, Table A1). First, the analysis according to city size (based on population size) showed that the effect of RBPM was greater in cities with a population of ≥ 1 million (SBP, 3.860 mmHg, $p < 0.001$; I^2 = 0.000, $p = 0.478$) than in small-to-medium cities with a smaller population, although within-study heterogeneity was high (I^2 = 82.177, $p < 0.001$). The effect of RBPM in reducing SBP was statistically significant compared with UC, irrespective of city size. The rate of BP control also showed greater effects in large cities (RR, 1.268; $p < 0.001$) than in small-to-medium-sized cities (RR, 1.157; $p = 0.094$). In a previous literature review, the intervention effect was found to be smaller in larger cities (large city, 3.229 mmHg vs. small-to-medium city, 3.765 mmHg), where the difference was considered to be associated with the difference in technological utility based on acceptance [66]. In particular, there was a sudden rise in demand for telemedicine to avoid the transmission of infectious diseases in response to the COVID-19 pandemic in 2020 [71,72].

Second, subgroup analysis was also performed according to urban MUAs in terms of healthcare accessibility. The decrease in BP in relation to RBPM in non-MUAs was 5.224 mmHg (I^2 = 73.152%, $p < 0.001$), indicating a greater effect of RBPM in reducing SBP compared with MUAs (3.213 mmHg, $p < 0.001$; I^2 = 48.904%, $p = 0.012$). The extent to which the level of within-study heterogeneity affects the summary effect size remains unclear, but the results of the analysis provided supporting evidence for determining the overall effect. Although a precise reason for this result could not be identified in this study, the following factors may be considered: changes in attitudes towards the use of mobile devices and chronic disease management and changes in economic lifestyle related to reduced opportunities for healthcare. These results may be used as evidence by healthcare policy-makers to support the need for differentiated policies for the supply of telemedicine in urban settings.

Third, a subgroup analysis was also performed concerning the duration of intervention. No optimal schedule has been established for the period of management of hypertension based on RBPM and the frequency of remote transmission of data [47,73]. Despite slight differences in the magnitude of reduction in SBP, RBPM in this study showed a consistent effect of reducing SBP, regardless of duration. Nonetheless, as the intervention duration increased, the level of BP reduction decreased. The reason for such a decrease could not be clearly identified, but possible causes may be fatigue, indifference, and inadequate level of perceived utility due to the prolonged performance of the intervention [73,74]. However, considering that it is essential to achieve a target BP as early as possible in patients with hypertension to prevent CVD, the effect of RBPM on the early outcome of BP reduction may be emphasized for its use in practice. The optimal duration of RBPM should be limited to a short period of time due to hypertension being a chronic disease requiring long-term management.

Fourth, in this updated study, subgroup analysis was undertaken according to the setting where RBPM was mainly implemented. Accordingly, when the intervention was performed at a tertiary hospital, RBPM had a significant reduction in BP (6.33 mmHg, $p < 0.001$; $I^2 = 73.401\%$, $p < 0.001$). The same numerical comparison was not compared in each group and, in the case of hospitals, its size was not analyzed separately; however, the results were statistically significant and included a sufficient number of studies to support the results; therefore, the significance of the results should not be ignored. The reason that RBPM had a higher BP lowering effect in tertiary medical institutions than in primary medical institutions may be due to the greater financial and human resource capacity in tertiary medical institutions [75].

Finally, this study observed the effect of RBPM with respect to the frequency of transmission of BP data. In the case of daily transmission, the WMD decreased by 5.881 mmHg. In contrast, in the case of monthly transmission, a decrease of 1.803 mmHg was observed. Some conflicting studies show that the higher the frequency of remote transmission, the lower the BP reduction effect [60,61]. However, in our study on cities, the longer the transmission interval, the lower the effect.

In previous meta-analyses, the number of studies conducted in urban settings was insufficient, and no study showed a change according to temporal progression. In this updated research, we included a comparison of the average effect over time, which was not covered in previous studies, and the effect according to the frequency of setting and data teletransmission. In particular, in our previous meta-analysis, it was reported that the effect of RBPM on patients with hypertension in metropolitan cities was not as large as that in small and medium cities. However, in this updated study, we found that the decrease in SBP and DBP was large in cities with a population of ≥ 1 million. Therefore, this study addressed the limitations of previous studies. Advancements in telecommunication technology have led to increased use of remote monitoring in healthcare [76]. In situations where physical distancing is emphasized, such as in the case of COVID-19, it is essential to assess the effects of RBPM in an urban setting [77]. To our knowledge, this study is the first meta-analysis to assess the effects of RBPM in urban-dwelling patients with hypertension from RBPM inception to the end of November 2020, including during the COVID-19 pandemic period, and these comprehensive results may provide a clinical basis for developing future healthcare policies.

In this study, a structured formula was applied, and a transparent process was followed to analyze RCTs with a high level of evidence. However, this study had some limitations. First, although the final studies were selected through a structured search using reliable databases, there may have been a language barrier. No outlier was found to have an influence on the summary effect size through the "one-study removed" sensitivity test method and a cumulative meta-analysis; however, selecting articles in different languages may have prevented adequate accounting for errors. Although most abstracts included in the search were written in English, the collected data may not have been sufficient. To overcome this limitation, multiple languages need to be set in the search with a wider scope to include gray literature. Second, the number of small-sized articles was insufficient to test for publication errors. Egger's test for the results in this study was used to determine combined two-tailed p-value significance, and the number of articles with a nil result in terms of a 95% CI was as high as 2898, which increased reliability. Nevertheless, there remained the possibility of publication errors. This limitation could be addressed through including a larger number of small-sized articles. Third, as the studies included in this meta-analysis varied in terms of the period when they were conducted, the criteria for target BP reflected in the rate of BP control may also have varied. Thus, further studies should set a clear BP target for collecting and synthesizing the data to produce more accurate results. Fourth, the authors categorized time intervals to compare the average SBP according to time interval and to quantify the results, which involved dividing the studies according to time based on the year 2000, when internet use expanded globally, and making simple comparisons at 10-year intervals thereafter. However, distinctions between time intervals may have been unclear. Although it is not possible to clearly divide the development time of telemedicine technology, we consider that the timeframe could be set more precisely based on historical developments in mobile communication technology and telemedicine. Finally, we examined trends in the effect of RBPM over time through categorizing studies based on their publication dates to indicate the temporal association with COVID-19. However, since differences between the actual dates of research and publication dates are possible, a future study should clarify the dates during which studies were conducted or include more studies published after the onset of the COVID-19 pandemic to address this limitation.

5. Conclusions

Our study findings indicated that RBPM for urban-dwelling patients with hypertension was a practical and clinically effective means of reducing office BP. As the cumulative analysis shows, a consistent and clear effect was found in terms of reduction in office SBP following RBPM according to the temporal progress of the primary studies included in this study; an identical trend was found for 2020.

Based on the primary findings, the effects were classified according to intervention duration, city size, setting, frequency of remote monitoring of BP data, and urban MUAs, and it is anticipated that the implementation of specific policies in relation to these factors would more effectively guide the application of efficient and successful urban remote monitoring. Future studies should analyze more specific variables and include a greater number of studies to obtain more reliable results.

Supplementary Materials: The following are available online at https://www.mdpi.com/article/10.3390/ijerph182010583/s1, Table S1: PRISMA 2020 Checklist.

Author Contributions: Conceptualization, W.-S.C. and J.-H.S.; methodology, W.-S.C.; software, W.-S.C.; validation, W.-S.C., S.-H.P., J.P. and J.-H.S.; formal analysis, W.-S.C.; investigation, S.-H.P. and J.-H.S.; resources, W.-S.C.; data curation, J.P.; writing—original draft preparation, S.-H.P. and J.-H.S.; writing—review and editing, W.-S.C. and S.-H.P.; visualization, J.P. and W.-S.C.; supervision, W.-S.C. All authors have read and agreed to the published version of the manuscript.

Funding: This research received no external funding.

Institutional Review Board Statement: Not applicable.

Informed Consent Statement: Not applicable.

Data Availability Statement: Not applicable.

Conflicts of Interest: The authors declare no conflict of interest.

Appendix A. Searching Strategy via Cochrane Library

1. MeSH descriptor: [Hypertension] explode all trees
2. hypertensi* OR high blood pressure
3. OR/1,2
4. MeSH descriptor: [Urban population] explode all trees
5. MeSH descriptor: [Urban health] explode all trees urban health [Mesh]
6. MeSH descriptor: [Urban health services] explode all trees
7. MeSH descriptor: [Cities] explode all trees
8. urban* OR city OR cities OR central cit*
9. OR/4–8
10. AND/3,10
11. MeSH descriptor: [Telemedicine] explode all trees
12. MeSH descriptor: [Telemetry] explode all trees
13. MeSH descriptor: [Blood pressure monitoring, ambulatory] explode all trees
14. telemedicine OR telemetry OR telenurs* OR telemonitor* OR eHealth OR telehealth OR remote monitor* OR technolog* OR telephone OR smartphone OR internet
15. OR/12–15
16. AND/11,16
17. randomised controlled trial OR randomized controlled
18. controlled clinical trial
19. randomised [tiab] OR randomized [tiab]
20. 2placebo [tiab]
21. drug therapy [sh]
22. groups [tiab]
23. clinical trials as topic [tiab]
24. randomly [tiab]
25. trial [tiab]
26. OR/18–26
27. 27 NOT cluster randomized controlled trials
28. 28 NOT cross over study
29. AND/17,29

Appendix B. Sensitivity Test Based on a "One-Study Removed" Approach

Author (year) (ref.)	Outcome	Statistics with study removed					
		Lower limit	Upper limit	Z-Value	Point	p-Value	
Artinian,2007A	SBP	0.367	0.779	5.456	0.573	0.000	
Artinian,2007B	SBP	0.375	0.786	5.536	0.580	0.000	
Bosworth,2007A	SBP	0.340	0.715	5.519	0.528	0.000	
Bosworth,2007B	SBP	0.314	0.630	5.864	0.472	0.000	
Bosworth,2007C	SBP	0.339	0.709	5.541	0.524	0.000	
Bosworth,2011	SBP	0.374	0.784	5.537	0.579	0.000	
Bove,2013	SBP	0.373	0.782	5.539	0.578	0.000	
Carrasco,2008	SBP	0.373	0.783	5.522	0.578	0.000	
Cicolini,2013A	SBP	0.376	0.784	5.570	0.580	0.000	
Cicolini,2013B	SBP	0.361	0.769	5.436	0.565	0.000	
Friedman,1996	SBP	0.379	0.787	5.595	0.583	0.000	
Green,2008	SBP	0.370	0.786	5.457	0.578	0.000	
Hill,1999	SBP	0.376	0.783	5.575	0.579	0.000	
Hoffmann,2017	SBP	0.375	0.786	5.536	0.581	0.000	
Ionov,2020	SBP	0.348	0.746	5.391	0.547	0.000	
Kerry,2013A	SBP	0.376	0.786	5.547	0.581	0.000	
Kerry,2013B	SBP	0.375	0.786	5.536	0.580	0.000	
Kim KB,2014A	SBP	0.366	0.779	5.432	0.572	0.000	
Kim KB,2014B	SBP	0.372	0.784	5.492	0.578	0.000	
Logan,2012	SBP	0.367	0.773	5.497	0.570	0.000	
Madsen,2008	SBP	0.374	0.783	5.542	0.578	0.000	
McMahon,2005	SBP	0.376	0.781	5.593	0.578	0.000	
Mohsen,2020A	SBP	0.358	0.762	5.424	0.560	0.000	
Mohsen,2020B	SBP	0.338	0.734	5.308	0.536	0.000	
Neumann,2011	SBP	0.365	0.769	5.490	0.567	0.000	
Pan,2018A	SBP	0.369	0.775	5.514	0.572	0.000	
Pan,2018B	SBP	0.369	0.776	5.518	0.572	0.000	
Parati,2009	SBP	0.379	0.788	5.597	0.583	0.000	
Park MJ, 2012	SBP	0.365	0.771	5.495	0.568	0.000	
Park MJ,2009	SBP	0.351	0.753	5.378	0.552	0.000	
Rahmani Pour,2020A	SBP	0.369	0.774	5.538	0.572	0.000	
Rahmani Pour,2020B	SBP	0.361	0.765	5.466	0.563	0.000	
Rahmani Pour,2020C	SBP	0.378	0.783	5.630	0.581	0.000	
Rahmani Pour,2020D	SBP	0.372	0.777	5.567	0.575	0.000	
Rubinstein, 2016A	SBP	0.375	0.789	5.514	0.582	0.000	
Rubinstein, 2016B	SBP	0.377	0.790	5.546	0.584	0.000	
Shea,2006	SBP	0.372	0.789	5.448	0.580	0.000	
Wakefield,2011	SBP	0.372	0.781	5.532	0.576	0.000	
Wakefield,2014	SBP	0.376	0.782	5.593	0.579	0.000	
Zha,2020A	SBP	0.370	0.774	5.557	0.572	0.000	
Zha,2020B	SBP	0.364	0.768	5.498	0.566	0.000	
		0.369	0.768	5.585	0.568	0.000	

Favors UC — Favors RBPM (scale: −5.00, −2.50, 0.00, 2.50)

Figure A1. Note: point estimate of individual study (•), summary effect size (◆); SBP, systolic blood pressure; UC, usual care; RBPM, remote blood pressure monitoring.

Appendix C. Cumulative Meta-Analysis of RBPM According to the SMD of SBP

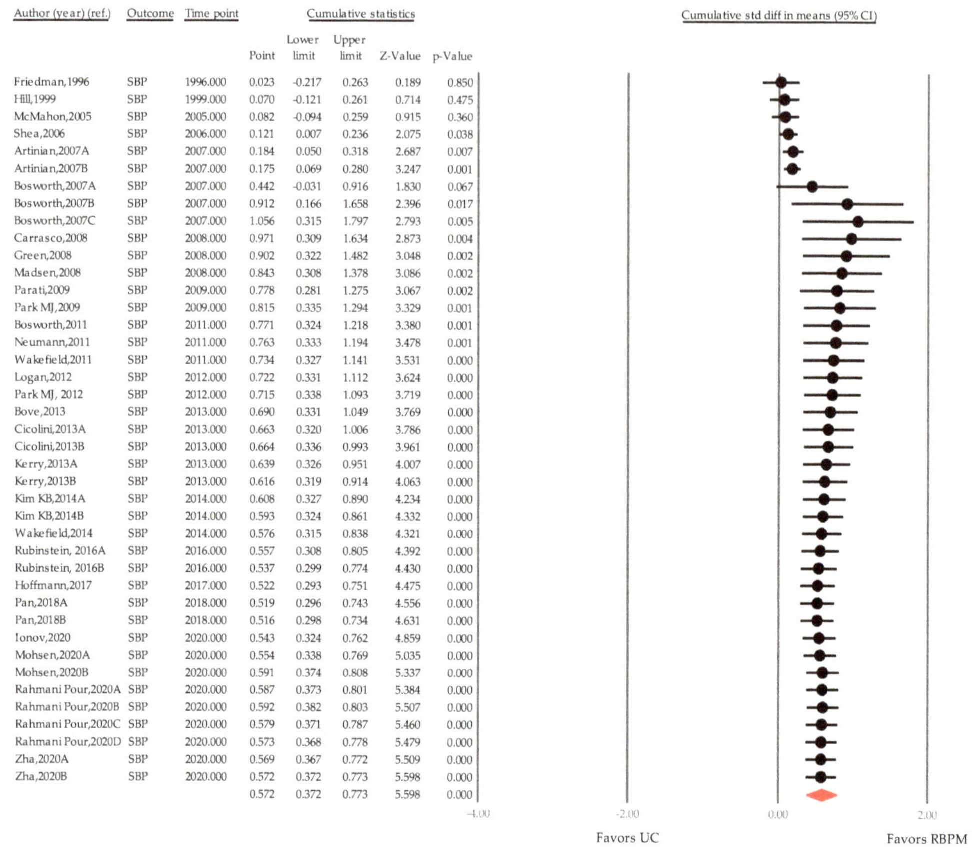

Author (year) (ref.)	Outcome	Time point	Point	Lower limit	Upper limit	Z-Value	p-Value
Friedman,1996	SBP	1996.000	0.023	-0.217	0.263	0.189	0.850
Hill,1999	SBP	1999.000	0.070	-0.121	0.261	0.714	0.475
McMahon,2005	SBP	2005.000	0.082	-0.094	0.259	0.915	0.360
Shea,2006	SBP	2006.000	0.121	0.007	0.236	2.075	0.038
Artinian,2007A	SBP	2007.000	0.184	0.050	0.318	2.687	0.007
Artinian,2007B	SBP	2007.000	0.175	0.069	0.280	3.247	0.001
Bosworth,2007A	SBP	2007.000	0.442	-0.031	0.916	1.830	0.067
Bosworth,2007B	SBP	2007.000	0.912	0.166	1.658	2.396	0.017
Bosworth,2007C	SBP	2007.000	1.056	0.315	1.797	2.793	0.005
Carrasco,2008	SBP	2008.000	0.971	0.309	1.634	2.873	0.004
Green,2008	SBP	2008.000	0.902	0.322	1.482	3.048	0.002
Madsen,2008	SBP	2008.000	0.843	0.308	1.378	3.086	0.002
Parati,2009	SBP	2009.000	0.778	0.281	1.275	3.067	0.002
Park MJ,2009	SBP	2009.000	0.815	0.335	1.294	3.329	0.001
Bosworth,2011	SBP	2011.000	0.771	0.324	1.218	3.380	0.001
Neumann,2011	SBP	2011.000	0.763	0.333	1.194	3.478	0.001
Wakefield,2011	SBP	2011.000	0.734	0.327	1.141	3.531	0.000
Logan,2012	SBP	2012.000	0.722	0.331	1.112	3.624	0.000
Park MJ, 2012	SBP	2012.000	0.715	0.338	1.093	3.719	0.000
Bove,2013	SBP	2013.000	0.690	0.331	1.049	3.769	0.000
Cicolini,2013A	SBP	2013.000	0.663	0.320	1.006	3.786	0.000
Cicolini,2013B	SBP	2013.000	0.664	0.336	0.993	3.961	0.000
Kerry,2013A	SBP	2013.000	0.639	0.326	0.951	4.007	0.000
Kerry,2013B	SBP	2013.000	0.616	0.319	0.914	4.063	0.000
Kim KB,2014A	SBP	2014.000	0.608	0.327	0.890	4.234	0.000
Kim KB,2014B	SBP	2014.000	0.593	0.324	0.861	4.332	0.000
Wakefield,2014	SBP	2014.000	0.576	0.315	0.838	4.321	0.000
Rubinstein, 2016A	SBP	2016.000	0.557	0.308	0.805	4.392	0.000
Rubinstein, 2016B	SBP	2016.000	0.537	0.299	0.774	4.430	0.000
Hoffmann,2017	SBP	2017.000	0.522	0.293	0.751	4.475	0.000
Pan,2018A	SBP	2018.000	0.519	0.296	0.743	4.556	0.000
Pan,2018B	SBP	2018.000	0.516	0.298	0.734	4.631	0.000
Ionov,2020	SBP	2020.000	0.543	0.324	0.762	4.859	0.000
Mohsen,2020A	SBP	2020.000	0.554	0.338	0.769	5.035	0.000
Mohsen,2020B	SBP	2020.000	0.591	0.374	0.808	5.337	0.000
Rahmani Pour,2020A	SBP	2020.000	0.587	0.373	0.801	5.384	0.000
Rahmani Pour,2020B	SBP	2020.000	0.592	0.382	0.803	5.507	0.000
Rahmani Pour,2020C	SBP	2020.000	0.579	0.371	0.787	5.460	0.000
Rahmani Pour,2020D	SBP	2020.000	0.573	0.368	0.778	5.479	0.000
Zha,2020A	SBP	2020.000	0.569	0.367	0.772	5.509	0.000
Zha,2020B	SBP	2020.000	0.572	0.372	0.773	5.598	0.000
			0.572	0.372	0.773	5.598	0.000

Figure A2. Note: point estimate of individual study excluding each individual study (●), summary effect size (◆); SBP, systolic blood pressure; SMD, standardized mean difference; UC, usual care; RBPM, remote blood pressure monitoring.

Appendix D. Meta-Regression of Risk Ratio According to RBPM Duration

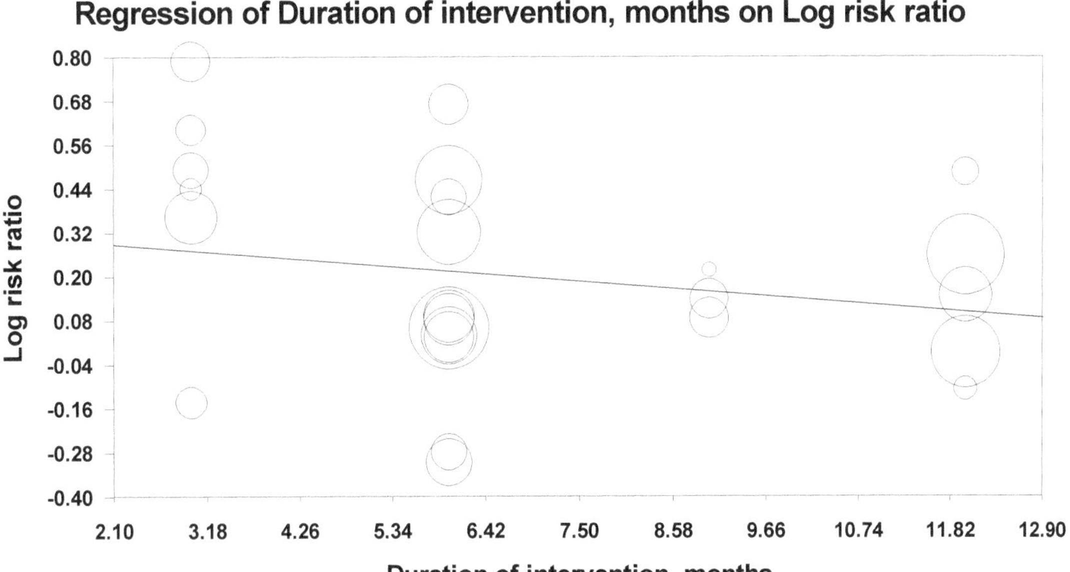

Figure A3. Note: point estimate of individual study (○); RBPM, remote blood pressure monitoring.

Appendix E. Subgroup Analysis

Table A1. CI, confidence interval; FEM, fixed effects model; SBP, systolic blood pressure; WMD, weighted mean difference.

Category	Number of Studies	Summary WMD of SBP, mmHg (95% CI)	Heterogeneity, I^2 (%) Using an FEM (p-Value)	Heterogeneity, Tau-Squared (τ^2) Using an FEM
Overall	48	4.464 (3.371–5.556)	70.908 ($p < 0.001$)	9.200
City size (population)				
<1 million	22	3.860 (2.271–5.450)	0.000 ($p = 0.478$)	0.000
>1 million	26	5.056 (3.503–6.609)	82.177 ($p < 0.001$)	17.368
Medically underserved areas				
Underserved	17	3.213 (1.521–4.905)	48.904 ($p = 0.012$)	2.793
Not underserved	31	5.224 (3.878–6.569)	73.152 ($p < 0.001$)	12.943
Duration (month)				
≤3	15	6.198 (4.019–8.377)	70.060 ($p < 0.001$)	14.069
6	14	4.479 (2.524–6.433)	84.562 ($p < 0.001$)	17.240
9	4	2.116 (-1.816–6.048)	0.000 ($p = 0.752$)	0.000
12	12	3.436 (1.281–5.591)	34.656 ($p = 0.113$)	1.646
Setting				
Primary care clinic	14	2.981 (1.323–4.639)	45.243 ($p = 0.034$)	1.989
Community health center	12	3.512 (1.651–5.373)	31.670 ($p = 0.138$)	1.883

Table A1. *Cont.*

Category	Number of Studies	Summary WMD of SBP, mmHg (95% CI)	Heterogeneity, I^2 (%) Using an FEM (p-Value)	Heterogeneity, Tau-Squared (τ^2) Using an FEM
Hospital	22	6.333 (4.750–7.917)	73.401 ($p < 0.001$)	17.133
Frequency of data transmission				
Daily	13	5.881 (3.898–7.864)	14.635 ($p = 0.297$)	1.637
Weekly	15	4.024 (2.641–5.406)	53.610 ($p = 0.007$)	4.505
Bi-weekly	4	3.941 (1.428–6.454)	0.000 ($p = 0.622$)	0.000
Monthly	6	1.803 (-0.234–3.841)	21.639 ($p = 0.271$)	0.552

References

1. Kishore, S.P.; Heller, D.J.; Vasan, A. Beyond hypertension: Integrated cardiovascular care as a path to comprehensive primary care. *B. World Health Organ.* **2018**, *96*, 219–221. [CrossRef] [PubMed]
2. Lewington, S.; Clarke, R.; Qizilbash, N.; Peto, R.; Collins, R. Age-specific relevance of usual blood pressure to vascular mortality: A meta-analysis of individual data for one million adults in 61 prospective studies. *Lancet* **2002**, *360*, 1903–1913. [PubMed]
3. Chow, C.K.; Teo, K.K.; Rangarajan, S.; Islam, S.; Gupta, R.; Avezum, A.; Bahonar, A.; Chifamba, J.; Dagenais, G.; Diaz, R.; et al. Prevalence, awareness, treatment, and control of hypertension in rural and urban communities in high-, middle-, and low-income countries. *JAMA* **2013**, *310*, 959–968. [CrossRef]
4. Egan, B.M.; Li, J.; Hutchison, F.N.; Ferdinand, K.C. Hypertension in the United States, 1999 to 2012: Progress toward Healthy People 2020 goals. *Circulation* **2014**, *130*, 1692–1699. [CrossRef]
5. Williams, B.; Mancia, G.; Spiering, W.; Rosei, E.A.; Azizi, M.; Burnier, M.; Clement, D.L.; Coca, A.; de Simone, G.; Dominiczak, A.; et al. 2018 ESC/ESH Guidelines for the management of arterial hypertension. *Eur. Heart J.* **2018**, *39*, 3021–3104. [CrossRef]
6. Whelton, P.K.; Carey, R.M.; Aronow, W.S.; Casey, D.E., Jr.; Collins, K.J.; Himmelfarb, C.D.; DePalma, S.M.; Gidding, S.; Jamerson, K.A.; Jones, D.W.; et al. 2017 ACC/AHA/AAPA/ABC/ACPM/AGS/APhA/ASH/ASPC/NMA/PCNA Guideline for the Prevention, Detection, Evaluation, and Management of High Blood Pressure in Adults: A Report of the American College of Cardiology/American Heart Association Task Force on Clinical Practice Guidelines. *Circulation* **2018**, *138*, e484–e594. [PubMed]
7. Ward, A.M.; Takahashi, O.; Stevens, R.; Heneghan, C. Home measurement of blood pressure and cardiovascular disease: Systematic review and meta-analysis of prospective studies. *J. Hypertens.* **2012**, *30*, 449–456. [CrossRef]
8. McManus, R.J.; Mant, J.; Bray, E.P.; Holder, R.; Jones, M.I.; Greenfield, S.; Kaambwa, B.; Banting, M.; Bryan, S.; Little, P.; et al. Telemonitoring and self-management in the control of hypertension (TASMINH2): A randomised controlled trial. *Lancet* **2010**, *376*, 163–172. [CrossRef]
9. McManus, R.; Mant, J.; Haque, M.; Bray, E.P.; Bryan, S.; Greenfield, S.; Jones, M.I.; Jowett, S.; Little, P.; Peñaloza, C.; et al. Effect of self-monitoring and medication self-titration on systolic blood pressure in hypertensive patients at high risk of cardiovascular disease: The TASMIN-SR randomized clinical trial. *JAMA* **2014**, *312*, 799–808. [CrossRef]
10. Tucker, K.L.; Sheppard, J.P.; Stevens, R.; Bosworth, H.B.; Bove, A.; Bray, E.P.; Earle, K.; George, J.; Godwin, M.; Green, B.B.; et al. Self-monitoring of blood pressure in hypertension: A systematic review and individual patient data meta-analysis. *PLoS Med.* **2017**, *14*, e1002389. [CrossRef]
11. Kim, Y.N.; Shin, D.G.; Park, S.; Lee, C.H. Randomized clinical trial to assess the effectiveness of remote patient monitoring and physician care in reducing office blood pressure. *Hypertens. Res.* **2015**, *38*, 491–497. [CrossRef]
12. Green, B.B.; Anderson, M.L.; Cook, A.J.; Catz, S.; Fishman, P.A.; McClure, J.B.; Reid, R.J. e-Care for heart wellness: A feasibility trial to decrease blood pressure and cardiovascular risk. *Am. J. Prev. Med.* **2014**, *46*, 368–377. [CrossRef] [PubMed]
13. Parati., G.; Ochoa, J.E.; Postel-Vinay, N.; Pellegrini, D.; Torlasco, C.; Omboni, S.; McManus, R. Home Blood Pressure Telemonitoring: Conventional Approach and Perspectives from Mobile Health Technology. In *Home Blood Pressure Monitoring*; Springer: Cham, Switzerland, 2019; pp. 103–119. [CrossRef]
14. Omboni, S.; McManus, R.J.; Bosworth, H.B.; Chappell, L.C.; Green, B.B.; Kario, K.; Logan, A.G.; Magid, D.J.; McKinstry, B.; Margolis, K.L.; et al. Evidence and Recommendations on the Use of Telemedicine for the Management of Arterial Hypertension: An International Expert Position Paper. *Hypertension* **2020**, *76*, 1368–1383. [CrossRef] [PubMed]
15. Lu, X.; Yang, H.; Xia, X.; Lu, X.; Lin, J.; Liu, F.; Gu, D. Interactive Mobile Health Intervention and Blood Pressure Management in Adults. *Hypertension* **2019**, *74*, 697–704. [CrossRef] [PubMed]
16. Duan, Y.; Xie, Z.; Dong, F.; Wu, Z.; Lin, Z.; Sun, N.; Xu, J. Effectiveness of home blood pressure telemonitoring: A systematic review and meta-analysis of randomised controlled studies. *J. Hum. Hypertens.* **2017**, *31*, 427–437. [CrossRef]
17. United Nations Department of Economic and Social Affairs. *World Urbanization Prospects: The 2018 Revision*; online edition; United Nations Department of Economic and Social Affairs: New York, NY, USA, 2018.

18. Omboni, S.; Gazzola, T.; Carabelli, G.; Parati, G. Clinical usefulness and cost effectiveness of home blood pressure telemonitoring: Meta-analysis of randomized controlled studies. *J. Hypertens.* **2013**, *31*, 455–467, discussion 67–68. [CrossRef]

19. Choi, W.S.; Park, J.; Choi, J.Y.B.; Yang, J.S. Stakeholders' resistance to telemedicine with focus on physicians: Utilizing the Delphi technique. *J. Telemed. Telecare* **2019**, *25*, 378–385. [CrossRef]

20. Choi, W.S.; Choi, J.H.; Oh, J.; Shin, I.S.; Yang, J.S. Effects of Remote Monitoring of Blood Pressure in Management of Urban Hypertensive Patients: A Systematic Review and Meta-Analysis. *Telemed. J. e-Health* **2020**, *26*, 744–759. [CrossRef]

21. Jovell, A.J.; Navarro-Rubio, M.D. [Evaluation of scientific evidence]. *Med. Clin.* **1995**, *105*, 740–743.

22. Wosik, J.; Fudim, M.; Cameron, B.; Gellad, Z.F.; Cho, A.; Phinney, D.; Curtis, S.; Roman, M.; Poon, E.G.; Ferranti, J.; et al. Telehealth transformation: COVID-19 and the rise of virtual care. *J. Am. Med. Inform. Assoc.* **2020**, *27*, 957–962. [CrossRef]

23. Cowie, M.R.; Lam, C.S.P. Remote monitoring and digital health tools in CVD management. *Nat. Rev. Cardiol.* **2021**, *18*, 457–458. [CrossRef]

24. Higgins, J.P.T.; Green, S. *Cochrane Handbook for Systematic Reviews of Interventions*; Version 502 [updated September 2009]; Wiley: New York, NY, USA, 2009.

25. Bosworth, H.B.; Olsen, M.K.; McCant, F.; Harrelson, M.; Gentry, P.; Rose, C.; Goldstein, M.K.; Hoffman, B.B.; Powers, B.; Oddone, E.Z. Hypertension Intervention Nurse Telemedicine Study (HINTS): Testing a multifactorial tailored behavioral/educational and a medication management intervention for blood pressure control. *J. Am. Heart Assoc.* **2007**, *153*, 918–924. [CrossRef]

26. Kerry, S.; Markus, H.; Khong, T.; Cloud, G.; Tulloch, J.; Coster, D.; Ibison, J.; Oakeshott, P. Home blood pressure monitoring with nurse-led telephone support among patients with hypertension and a history of stroke: A community-based randomized controlled trial. *CMAJ* **2012**, *185*, 23–31. [CrossRef]

27. Pan, F.; Wu, H.; Liu, C.; Zhang, X.; Peng, W.; Wei, X.; Gao, W. Effects of home telemonitoring on the control of high blood pressure: A randomised control trial in the Fangzhuang Community Health Center, Beijing. *Aust. J. Prim. Health* **2018**, *24*, 398–403. [CrossRef]

28. Zha, P.; Qureshi, R.; Porter, S.; Chao, Y.Y.; Pacquiao, D.; Chase, S.; O'Brien-Richardson, P. Utilizing a Mobile Health Intervention to Manage Hypertension in an Underserved Community. *West. J. Nurs. Res.* **2020**, *42*, 201–209. [CrossRef]

29. Artinian, N.T.; Flack, J.M.; Nordstrom, C.K.; Hockman, E.M.; Washington, O.G.; Jen, K.L.; Fathy, M. Effects of nurse-managed telemonitoring on blood pressure at 12-month follow-up among urban African Americans. *Nurs. Res.* **2007**, *56*, 312–322. [CrossRef] [PubMed]

30. Cicolini, G.; Simonetti, V.; Comparcini, D.; Celiberti, I.; Di Nicola, M.; Capasso, L.M.; Flacco, M.E.; Bucci, M.; Mezzetti, A.; Manzoli, L. Efficacy of a nurse-led email reminder program for cardiovascular prevention risk reduction in hypertensive patients: A randomized controlled trial. *Int. J. Nurs. Stud.* **2014**, *51*, 833–843. [CrossRef] [PubMed]

31. Hebert, P.; Sisk, J.; Tuzzio, L.; Casabianca, J.; Pogue, V.; Wang, J.; Chen, Y.; Cowles, C.; McLaughlin, M.A. Nurse-led disease management for hypertension control in a diverse urban community: A randomized trial. *J. Gen. Intern. Med.* **2012**, *27*, 630–639. [CrossRef] [PubMed]

32. Kim, K.B.; Han, H.R.; Huh, B.; Nguyen, T.; Lee, H.; Kim, M.T. The effect of a community-based self-help multimodal behavioral intervention in Korean American seniors with high blood pressure. *Am. J. Hypertens.* **2014**, *27*, 1199–1208. [CrossRef] [PubMed]

33. Magda, M.M.; Riad, N.A.; Badawy, A.E.; El Gafar, S.E.A.; El-Hammed, B.M.A.; Eltomy, E.M. Tele-nursing versus Routine Outpatient Teaching for Improving Arterial Blood Pressure and Body Mass Index for Hypertensive Patients. *Am. J. Nurs. Res.* **2020**, *8*, 18–26.

34. Pour, E.R.; Aliyari, S.; Farsi, Z.; Ghelich, Y. Comparing the effects of interactive and noninteractive education using short message service on treatment adherence and blood pressure among patients with hypertension. *Nurs. Midwifery Stud.* **2020**, *9*, 68–76.

35. Rubinstein, A.; Miranda, J.J.; Beratarrechea, A.; Diez-Canseco, F.; Kanter, R.; Gutierrez, L.; Bernabé-Ortiz, A.; Irazola, V.; Fernandez, A.; Letona, P.; et al. Effectiveness of an mHealth intervention to improve the cardiometabolic profile of people with prehypertension in low-resource urban settings in Latin America: A randomised controlled trial. *Lancet Diabetes Endocrinol.* **2016**, *2016 4*, 52–63. [CrossRef]

36. The Cochrane Collaboration. *Review Manager (RevMan)[Computer Program]*, Version 5.4 for Windows; The Cochrane Collaboration: Oxford, UK, 2020.

37. Egger, M.; Davey-Smith, G.; Altman, D. (Eds.) *Systematic Reviews in Healthcare: Meta-Analysis in Context*; John Wiley & Sons: New York, NY, USA, 2008.

38. Takeshima, N.; Sozu, T.; Tajika, A.; Ogawa, Y.; Hayasaka, Y.; Furukawa, T.A. Which is more generalizable, powerful and interpretable in meta-analyses, mean difference or standardized mean difference? *BMC Med. Res. Methodol.* **2014**, *14*, 30. [CrossRef]

39. Cohen, J. *Statistical Power Analysis for the Behavioral Sciences*; Lawrence Erlbaum Associates: New York, NY, USA, 1988.

40. Higgins, J.P.; Thompson, S.G.; Deeks, J.J.; Altman, D.G. Measuring inconsistency in meta-analyses. *BMJ* **2003**, *327*, 557–560. [CrossRef]

41. Hill, M.; Bone, L.; Hilton, S.; Roary, M.; Kelen, G.; Levine, D. A clinical trial to improve high blood pressure care in young urban black men: Recruitment, follow-up, and outcomes. *Am. J. Hypertens.* **1999**, *12*, 548–554. [CrossRef]

42. Friedman, R.H.; Kazis, L.E.; Jette, A.; Smith, M.B.; Stollerman, J.; Torgerson, J.; Carey, K. A telecommunications system for monitoring and counseling patients with hypertension: Impact on medication adherence and blood pressure control. *Am. J. Hypertens.* **1996**, *9*, 285–292. [CrossRef]

43. McMahon, G.; Gomes, H.; Hickson, H.S.; Hu, T.; Levine, B.; Conlin, P. Web-based care management in patients with poorly controlled diabetes. *Diabetes Care* **2005**, *28*, 1624–1629. [CrossRef]

44. Shea, S.; Weinstock, R.; Starren, J.; Teresi, J.; Palmas, W.; Field, L.; Morin, P.; Goland, R.; Izquierdo, R.E.; Wolff, L.T.; et al. A randomized trial comparing telemedicine case management with usual care in older, ethnically diverse, medically underserved patients with diabetes mellitus. *J. Am. Med. Inform. Assoc.* **2006**, *13*, 40–51. [CrossRef] [PubMed]

45. Carrasco, M.P.; Salvador, C.H.; Sagredo, P.G.; Marquez-Montes, J.; de Mingo, M.A.G.; Fragua, J.A.; Rodríguez, M.C.; García-Olmos, L.M.; García-López, F.; Muñoz Carrero, A.; et al. Impact of patient-general practitioner short-messages-based interaction on the control of hypertension in a follow-up service for low-to-medium risk hypertensive patients: A randomized controlled trial. *IEEE Trans. Inf. Technol. Biomed.* **2008**, *12*, 780–791. [CrossRef]

46. Green, B.B.; Cook, A.J.; Ralston, J.D.; Fishman, P.A.; Catz, S.L.; Carlson, J. Effectiveness of home blood pressure monitoring, Web communication, and pharmacist care on hypertension control: A randomized controlled trial. *JAMA* **2008**, *299*, 2857–2867. [CrossRef]

47. Madsen, L.B.; Kirkegaard, P.; Pedersen, E.B. Blood pressure control during telemonitoring of home blood pressure. A randomized controlled trial during 6 months. *Blood Press.* **2008**, *17*, 78–86. [CrossRef] [PubMed]

48. Parati, G.; Omboni, S.; Albini, F.; Piantoni, L.; Giuliano, A.; Revera, M.; Illyes, M.; Mancia, G.; Tele, B.S.G. Home blood pressure telemonitoring improves hypertension control in general practice. *The TeleBPCare study. J. Hypertens.* **2009**, *27*, 198–203. [CrossRef] [PubMed]

49. Park, M.; Kim, H.; Kim, K. Cellular phone and Internet-based individual intervention on blood pressure and obesity in obese patients with hypertension. *Int. J. Med. Inform.* **2009**, *78*, 704–710. [CrossRef] [PubMed]

50. Varis, J.; Kantola, I. The choice of home blood pressure result reporting method is essential: Results mailed to physicians did not improve hypertension control compared with ordinary office-based blood pressure treatment. *Blood Press.* **2010**, *19*, 319–324. [CrossRef]

51. Hoffmann-Petersen, N.; Lauritzen, T.; Bech, J.N.; Pedersen, E.B. Short-term telemedical home blood pressure monitoring does not improve blood pressure in uncomplicated hypertensive patients. *J. Hum. Hypertens.* **2017**, *31*, 93–98. [CrossRef] [PubMed]

52. Ionov, M.V.; Zhukova, O.V.; Yudina, Y.S.; Avdonina, N.G.; Emelyanov, I.V.; Kurapeev, D.I.; Zvartau, N.E.; Konradi, A.O. Value-based approach to blood pressure telemonitoring and remote counseling in hypertensive patients. *Blood Press.* **2020**, *30*, 1–11. [CrossRef]

53. Logan, A.G.; Irvine, M.J.; McIsaac, W.J.; Tisler, A.; Rossos, P.G.; Easty, A.; Feig, D.S.; Cafazzo, J.A. Effect of Home Blood Pressure Telemonitoring With Self-Care Support on Uncontrolled Systolic Hypertension in Diabetics. *Hypertension* **2012**, *60*, 51–57. [CrossRef]

54. Neumann, C.L.; Menne, J.; Rieken, E.M.; Fischer, N.; Weber, M.H.; Haller, H.; Schulz, E.G. Blood pressure telemonitoring is useful to achieve blood pressure control in inadequately treated patients with arterial hypertension. *J. Hum. Hypertens.* **2011**, *25*, 732–738. [CrossRef] [PubMed]

55. Wakefield, B.J.; Holman, J.E.; Ray, A.; Scherubel, M.; Adams, M.R.; Hillis, S.L.; Rosenthal, G.E. Effectiveness of home telehealth in comorbid diabetes and hypertension: A randomized, controlled trial. *Telemed. J. e-Health* **2011**, *17*, 254–261. [CrossRef]

56. Bosworth, H.B.; Powersm, B.J.; Olsen, M.K.; McCant, F.; Grubber, J.; Smith, V.; Gentry, P.W.; Rose, C.; Van Houtven, C.; Wang, V.; et al. Home blood pressure management and improved blood pressure control: Results from a randomized controlled trial. *Arch. Intern. Med.* **2011**, *171*, 1173–1180. [CrossRef]

57. Migneault, J.; Dedier, J.; Wright, J.; Heeren, T.; Campbell, M.; Morisky, D.; Rudd, P.; Friedman, R.H. A culturally adapted telecommunication system to improve physical activity, diet quality, and medication adherence among hypertensive African-Americans: A randomized controlled trial. *Ann. Behav. Med.* **2012**, *43*, 62–73. [CrossRef] [PubMed]

58. Park, M.; Kim, H. Evaluation of mobile phone and Internet intervention on waist circumference and blood pressure in post-menopausal women with abdominal obesity. *Int. J. Med. Eng. Inform.* **2012**, *81*, 388–394. [CrossRef] [PubMed]

59. Bove, A.; Homko, C.; Santamore, W.; Kashem, M.; Kerper, M.; Elliott, D. Managing hypertension in urban underserved subjects using telemedicine–a clinical trial. *Am. Heart J.* **2013**, *165*, 615–621. [CrossRef]

60. Wakefield, B.J.; Koopman, R.J.; Keplinger, L.E.; Bomar, M.; Bernt, B.; Johanning, J.L.; Kruse, R.L.; Davis, J.W.; Wakefield, D.S.; Mehr, D.R. Effect of home telemonitoring on glycemic and blood pressure control in primary care clinic patients with diabetes. *Telemed. J. e-Health* **2014**, *20*, 199–205. [CrossRef]

61. Yi, S.; Tabaei, B.; Angell, S.; Rapin, A.; Buck, M.; Pagano, W.; Maselli, F.J.; Simmons, A. Self-blood pressure monitoring in an urban, ethnically diverse population: A randomized clinical trial utilizing the electronic health record. *Circ. Cardiovasc. Qual. Outcomes* **2015**, *8*, 138–145. [CrossRef]

62. Higgins, J.P.T.; Savović, J.; Page, M.J.; Elbers, R.G.; Sterne, J.A.C. Chapter 8: Assessing risk of bias in a randomized trial. In *Cochrane Handbook for Systematic Reviews of Interventions*; Higgins, J., Thomas, J., Eds.; Cochrane: London, UK, 2021; Available online: www.training.cochrane.org/handbook (accessed on 5 August 2021).

63. Duval, S.; Tweedie, R. Trim and fill: A simple funnel-plot–based method of testing and adjusting for publication bias in meta-analysis. *Biometrics* **2000**, *56*(2), 455–463. [CrossRef]

64. Sterne, J.A.; Egger, M. Funnel plots for detecting bias in meta-analysis: Guidelines on choice of axis. *J. Clin. Epidemiol.* **2001**, *54*, 1046–1055. [CrossRef]

65. OECD Indicators. *Health at a Glance 2019: OECD Indicators*; OECD Publishing: Paris, French, 2019.

66. Hwang, J.Y.; Kim, K.Y.; Lee, K.H. Factors that influence the acceptance of telemetry by emergency medical technicians in ambulances: An application of the extended technology acceptance model. *Telemed. J. e-Health* **2014**, *20*, 1127–1134. [CrossRef]
67. National Center for Health Statistics (US). Health, United States, 2013: With Special Feature on Prescription Drugs. National Center for Health Statistics (US): Hyattsville, MD, USA, 2014.
68. Neal, B.; MacMahon, S.; Chapman, N. Effects of ACE inhibitors, calcium antagonists, and other blood-pressure-lowering drugs: Results of prospectively designed overviews of randomised trials. Blood Pressure Lowering Treatment Trialists' Collaboration. *Lancet* **2000**, *356*, 1955–1964. [PubMed]
69. Cook, N.R.; Cohen, J.; Hebert, P.R.; Taylor, J.O. Hennekens, C.H. Implications of small reductions in diastolic blood pressure for primary prevention. *Arch Intern. Med.* **1995**, *155*, 701–709. [CrossRef]
70. Law, M.R.; Morris, J.K.; Wald, N.J. Use of blood pressure lowering drugs in the prevention of cardiovascular disease: Meta-analysis of 147 randomised trials in the context of expectations from prospective epidemiological studies. *BMJ* **2009**, *338*, b1665. [CrossRef]
71. Hammersley, V.; Parker, R.; Paterson, M.; Hanley, J.; Pinnock, H.; Padfield, P.; Stoddart, A.; Park, H.G.; Sheikh, A.; McKinstry, B. Telemonitoring at scale for hypertension in primary care: An implementation study. *PLoS Med.* **2020**, *17*, e1003124. [CrossRef]
72. Omboni, S. Telemedicine During the COVID-19 in Italy: A Missed Opportunity? *Telemed. J. e-Health* **2020**, *26*, 973–975. [CrossRef]
73. Margolis, K.L.; Asche, S.E.; Dehmer, S.P.; Bergdall, A.R.; Green., B.B.; Sperl-Hillen, J.M.; Nyboer, R.A.; Pawloski, P.A.; Maciosek, M.V.; Trower, N.K.; et al. Long-term Outcomes of the Effects of Home Blood Pressure Telemonitoring and Pharmacist Management on Blood Pressure Among Adults With Uncontrolled Hypertension: Follow-up of a Cluster Randomized Clinical Trial. *JAMA Netw. Open* **2018**, *1*, e181617. [CrossRef] [PubMed]
74. Williams, B.; Mancia, G.; Spiering, W.; Rosei, E.A.; Azizi, M.; Burnier, M.; Clement, D.; Coca, A.; De Simone, G.; Dominiczak, A.; et al. 2018 Practice Guidelines for the management of arterial hypertension of the European Society of Hypertension and the European Society of Cardiology: ESH/ESC Task Force for the Management of Arterial Hypertension. *J. Hypertens.* **2018**, *36*, 2284–2309. [CrossRef] [PubMed]
75. Hjelm, N.M. Benefits and drawbacks of telemedicine. *J. Telemed. Telecare* **2005**, *11*, 60–70. [CrossRef] [PubMed]
76. Filchev, R.; Pavlova, D.; Dimova, R.; Dovramadjiev, T. Healthcare System Sustainability by Application of Advanced Technologies in Telemedicine and eHealth. In *International Conference on Human Interaction and Emerging Technologies*; Springer: Cham, Switzerland, 2021; pp. 1011–1017.
77. Chu, D.K.; Akl, E.A.; Duda, S.; Solo, K.; Yaacoub, S.; Schünemann, H.J. Physical distancing, face masks, and eye protection to prevent person-to-person transmission of SARS-CoV-2 and COVID-19: A systematic review and meta-analysis. *Lancet* **2020**, *395*, 1973–1987. [CrossRef]

International Journal of
Environmental Research and Public Health

MDPI

Systematic Review

Virtual Reality in the Treatment of Adults with Chronic Low Back Pain: A Systematic Review and Meta-Analysis of Randomized Clinical Trials

Beatriz Brea-Gómez, Irene Torres-Sánchez *, Araceli Ortiz-Rubio, Andrés Calvache-Mateo, Irene Cabrera-Martos, Laura López-López and Marie Carmen Valenza

Physical Therapy Department, Faculty of Health Sciences, University of Granada, 18016 Granada, Spain;
beatrizbreagomez@gmail.com (B.B.-G.); aortiz@ugr.es (A.O.-R.); andrescalvache@ugr.es (A.C.-M.);
irenecm@ugr.es (I.C.-M.); lauralopez@ugr.es (L.L.-L.); cvalenza@ugr.es (M.C.V.)
* Correspondence: irenetorres@ugr.es

Citation: Brea-Gómez, B.;
Torres-Sánchez, I.; Ortiz-Rubio, A.;
Calvache-Mateo, A.;
Cabrera-Martos, I.; López-López, L.;
Valenza, M.C. Virtual Reality in the
Treatment of Adults with Chronic
Low Back Pain: A Systematic Review
and Meta-Analysis of Randomized
Clinical Trials. *Int. J. Environ. Res.
Public Health* **2021**, *18*, 11806. https://
doi.org/10.3390/ijerph182211806

Academic Editor: Paul B. Tchounwou

Received: 21 October 2021
Accepted: 7 November 2021
Published: 11 November 2021

Abstract: Virtual reality (VR) can present advantages in the treatment of chronic low back pain. The objective of this systematic review and meta-analysis was to analyze the effectiveness of VR in chronic low back pain. This review was designed according to PRISMA and registered in PROSPERO (CRD42020222129). Four databases (PubMed, Cinahl, Scopus, Web of Science) were searched up to August 2021. Inclusion criteria were defined following PICOS recommendations. Methodological quality was assessed with the Downs and Black scale and the risk of bias with the Cochrane Risk of Bias Assessment Tool. Fourteen studies were included in the systematic review and eleven in the meta-analysis. Significant differences were found in favor of VR compared to no VR in pain intensity postintervention (11 trials; n = 569; SMD = -1.92; 95% CI = -2.73, -1.11; $p < 0.00001$) and followup (4 trials; n = 240; SDM = -6.34; 95% CI = -9.12, -3.56; $p < 0.00001$); and kinesiophobia postintervention (3 trials; n = 192; MD = -8.96; 95% CI = -17.52, -0.40; $p = 0.04$) and followup (2 trials; n = 149; MD = -12.04; 95% CI = -20.58, -3.49; $p = 0.006$). No significant differences were found in disability. In conclusion, VR can significantly reduce pain intensity and kinesiophobia in patients with chronic low back pain after the intervention and at followup. However, high heterogeneity exists and can influence the consistency of the results.

Keywords: chronic low back pain; virtual reality; videogames; horse simulator riding; rehabilitation; physical therapy

1. Introduction

Chronic low back pain (CLBP) is one of the main causes of pain, dysfunction, and disability [1,2]. It is one of the most common reasons for which patients require medical attention [3]. Furthermore, it is the world's leading cause of years of life lived with disability [4]. In most cases, it is not possible to identify the specific nociceptive cause of CLBP and therefore, it is classified as nonspecific (pain not caused by a specific pathology such as infection, tumor, fracture, or inflammation) [2]. CLBP affects the physical, psychological, and social areas and carries a great socioeconomic burden, as it is the main cause of work absenteeism and the excessive use of therapeutic services [5]. For all these reasons, it is essential to establish an effective treatment.

There are many ways to treat CLBP in the clinical environment, such as surgery, medication, or physical therapy. In addition to analgesic treatment with drugs, manual therapy, pain management, and early physical exercise (coordination, strengthening, and resistance exercises) have been recommended with a strong level of evidence, as they can be beneficial in reducing pain and achieve a functional improvement [5,6]. However, in many cases the main limitation of physical exercise is lack of motivation and adherence [7]. Virtual reality (VR) can present some advantages in the face of these problems, since it contributes the motivational component and interactivity to the treatment [8]. The patient is involved in their recovery in a fun and attractive way and the interactive elements and feedback offered by the virtual environment can increase adherence to the exercises [9,10]. Negative thoughts and beliefs about pain experienced by some patients can lead to pain avoidance behaviors, causing inactivity, and preventing recovery and pain reduction [11]. VR treatment is a powerful pain distraction mechanism by focusing on an external stimulus and not on body movement, reducing attention to pain by dividing attention tasks [9,12]. Furthermore, compared to traditional methods, VR is considered a cost-effective and efficient tool [13].

In the current scientific literature, we found different reviews about VR in the treatment of pain in various areas. Gumaa et al. [14] explored VR effectiveness in orthopedic rehabilitation, showing inconclusive results in low back pain. In addition, they referred to the need for higher quality studies to establish more solid conclusions. In another review, VR in spinal pain was investigated [15]. Due to the low quality of the included studies, Ahern et al. [15] concluded that higher quality studies were necessary. A recent review published by Bordeleau et al. [16] concluded that while the specific set of studies showed high heterogeneity across several methodological factors, a tentative conclusion could be drawn that VR is effective improving back pain intensity and tends to have a positive effect on improving other pain outcomes and motion function. Bordeleau et al. [16] highlights that methodology framework for the development of VR treatments should be considered.

Since the completion of the search of the review of Bordeleau et al. [16], several new randomized clinical trials have been published on this topic, so there is new evidence to contribute to this issue. Additionally, a subgroup analysis of the different interventions is needed. Whether VR is applied alone or added to a physical therapy intervention could produce different results; furthermore, the comparison should also be taken into account.

Additionally, an analysis comparing the effects of the different VR interventions, the different durations of the interventions and the effects of VR at followup should be useful. It would also be of interest to explore other variables related to pain, in addition to pain intensity.

Therefore, the objective of this systematic review and meta-analysis of randomized clinical trials was to analyze the effectiveness of VR interventions in the treatment of CLBP. Implications and considerations may arise regarding the characteristics of the intervention programs.

2. Materials and Methods

2.1. Design

A systematic review was performed to identify randomized clinical trials exploring the effects of VR on the treatment of CLBP. The guidelines of the Preferred Reporting Items for Systematic Reviews and Meta-Analyses (PRISMA) [17] was used to carry out this systematic review. This systematic review was previously registered at the International prospective register of systematic reviews (PROSPERO) with number CRD42020222129. Available from: https://www.crd.york.ac.uk/prospero/display_record.php?RecordID=222129.

2.2. Search Strategy

Four databases were searched from their inception up to August 2021 without language restrictions. We used PubMed, Cinahl, Scopus, and Web of Science. The full search strategy is described in Appendix A. In order to find other relevant articles to the study, the reference list of other reviews and related articles were reviewed.

Additionally, a search was conducted for ongoing randomized clinical trials, which have not yet been published, to find out if they could be included in our review. The clinical trials registries ClinicalTrials.gov, the International Standard Randomized Controlled Trial Number (ISRCTN) Registry, and the International Clinical Trials Registry Platform (ICTRP) were used. Appendix B describes the search strategy used in each database.

2.3. Study Selection

The selection of studies was conducted systematically based on the prespecified PICOS (participants, interventions, comparisons, outcome, and study design) eligibility criteria: (1) Participants: adults ($\geq$18 years) with CLBP (12 weeks or more) [18]; (2) Interventions: interventions based on VR; length of intervention of at least four weeks; (3) Comparisons: no intervention, interventions without VR, standard treatment, usual care, placebo or control; (4) Outcomes: pain intensity and other outcomes related to pain; (5) Study design: randomized clinical trials.

For the first screening title and abstract of each article was evaluated. We excluded those that did not meet the inclusion criteria defined with the PICOS strategy. After, the full text of relevant studies was assessed to check if they met the inclusion criteria. The list of excluded studies in the last screening and reason for exclusion is described in Appendix C.

When full text was not available, we contacted the corresponding author of the study via email. Two reviewers (BBG and ITS) independently carried out the search and selection of studies. If needed, disagreements were resolved with a third reviewer.

2.4. Data Extraction

The following data were recorded from the included articles: author, year of publication, country, disease, sample size, age (years), gender (percentage of males), outcome measures, main results (outcomes that showed significant differences ($p \leq 0.05$)), measuring instrument, and time point assessment. This information is summarized in Table 1. In addition, the score obtained on the Downs and Black methodological quality scale [19] was added. Table 2 shows the characteristics of interventions: experimental group intervention, control group interventions, session duration, frequency, program duration, supervision, and adverse events.

When the information was insufficient or unclear, we contacted the corresponding author of the study via email. If the data were still unclear after contacting the corresponding author or if contact was not possible, it was analyzed using the available data. Two reviewers (BBG and ITS) independently carried out the data extraction. If needed, disagreements were resolved with a third reviewer.

2.5. Methodological Quality of Included Studies

Downs and Black quality assessment method [19] was used to assess the methodological quality of included studies in this review. This scale is one of the six best quality assessment scales [20–22]. This method contains 27 items divided into 5 sections: study quality (10 items), external validity (3 items), study bias (7 items), confounding and selection bias (6 items), and study power (1 item). In this review, we used the modified Downs and Black scale. The scoring for item 27 was simplified to a choice of 0 ("no"/"unable to determine") or 1 point ("yes"). These scores will be the same for the rest of the items except item 5 which is valued as 0 ("no"/"unable to determine"), 1 ("partially"), or 2 ("yes"). Therefore, the scores range from 0 to 28 and the higher ones indicate a better methodological quality of the study [22,23]. According to their quality, studies can be categorized as excellent (26–28), good (20–25), fair (15–19), and poor ($\leq$14) [22–24].

2.6. Risk of Bias of Included Studies

The Cochrane Risk of Bias Assessment Tool [25] was used to assess the risk of bias of included studies. This tool assesses seven domains: random sequence generation, allocation concealment, blinding of participants and personnel, blinding of outcome assessment, incomplete outcome data, selective reporting, and other bias. For each study, the different domains were scored as "high risk of bias", "low risk of bias", or "unclear".

Two reviewers (BBG and ITS) carried out the assessment of risk of bias, as well as the assessment of methodological quality independently, and in case of doubt or disagreement a third reviewer was consulted.

2.7. Statistical Analysis

We used the Review Manager (RevMan) software version 5.4 to perform statistical analysis and used forest plots to display the results. Analysis was performed for those outcomes repeated at least in three comparisons or studies. Regarding the period of time, the analysis was carried out after the intervention and at 6 months followup. Mean, standard deviation (SD) and sample size were extracted from included studies to estimate the overall effect. For continuous outcomes, the mean difference (MD) and the 95% confidence intervals (CI) were used when the outcomes were evaluated with the same scale and the standardized mean difference (SMD) when the scales were different. The method utilized was inverse variance. The fixed effects model was used and the random effects model was applied when heterogeneity was greater than 75%. A value of $p \leq 0.05$ was considered statistically significant. Heterogeneity between studies was assessed using the I^2 test. The degree of heterogeneity was categorized as low ($I^2 < 25\%$), moderate ($I^2 = 25$–75%), and high ($I^2 > 75\%$). In order to explore possible causes of heterogeneity among study results we conducted a subgroup analysis. Subgroups were performed according to the comparisons (VR vs. no intervention, VR vs. placebo, VR vs. oral treatment, VR vs. physiotherapy, VR + physiotherapy vs. physiotherapy, and VR + physiotherapy vs. no VR exercise + physiotherapy); the type of intervention with VR (Nintendo consoles, Horse Simulator Riding, and Prokin System), and the duration of the intervention (4, 8, or 12 weeks).

3. Results

3.1. Search Selection

After the initial search in the databases and reference lists, we found 1363 manuscripts. After removing duplicates, we obtained 838 potentially eligible records. After screening by title and abstract, 58 articles remained, of which the full text was assessed. Of those 58 studies, 14 randomized clinical trials met the inclusion criteria, and finally 11 were included in quantitative synthesis.

In addition, we searched for ongoing randomized clinical trials. Of the 63 studies found in the three clinical trial registries consulted, 17 finally met the inclusion criteria. Figure 1 shows the flow diagram of the articles during the study selection process in the databases and clinical trial registries. The list of ongoing randomized clinical trials that could be included in the review is shown in Appendix D. None of the ongoing randomized clinical trials were included in this review.

3.2. Characteristics of Included Studies

Table 1 shows the characteristics of included studies in this review. All studies were randomized clinical trials and are arranged chronologically from oldest to newest. The included studies were published between 2013 and 2021.

Six studies were conducted in South Korea [26–31], one in Brazil [32], one in the USA [33], one in Australia [7], three in Saudi Arabia [34–36], one in Turkey [37], and one in Japan [38]. The total number of participants was 765. The mean age of the participants was 40.04 with 62.08% men. All studies measured pain intensity. It was measured using VAS in nine studies [26–29,31,34–36,38], 11-point Numeric Pain Rating Scale (11-NPRS) in four studies [7,30,32,37], and The Defense and Veterans Pain Rating Scale (DVPRS) in one study [33]. Four studies measured disability associated with low back pain using the Oswestry Disability Index (ODI) [29–31,37], four studies measured kinesiophobia using the 17-item Tampa Scale of Kinesiophobia (17-TSK) [7,35,36,38], and four studies measured body composition using bioelectrical impedance analysis method [27,28,31,38]. Other variables assessed more frequently were severity of disability with Roland Morris Disability Questionnaire (RMDQ) in two studies [7,30], isokinetic trunk flexion/extension with a dynamometer in three studies [27,28,31], pain self-efficacy with the 10-item Pain Self-Efficacy (10-PSEQ) in two studies [7,38], pain catastrophizing with Pain Catastrophizing Scale (PCS) in two studies [33,38], and blood serum levels of stress hormones in two studies [35,36]. Variables were assessed before and after the intervention in all articles. Five studies included followup, one at 8 weeks and 6 months [34], other at 3 and 6 months [7], and three only at 6 months [30,35,36]. In addition, one of these studies included a midterm assessment after 4 weeks [30], and one study assessed outcomes during intervention [33].

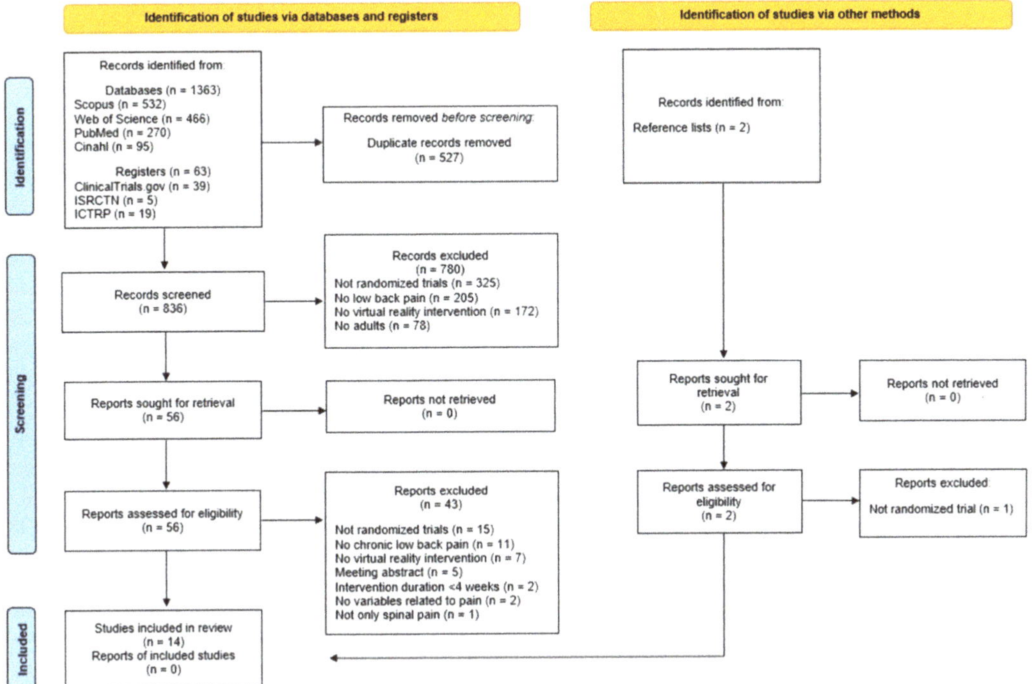

Figure 1. PRISMA flow diagram: database and clinical trials register search and other sources.

3.3. Characteristics of Interventions

Characteristics of the interventions of the included studies are described in Table 2.

Regarding the interventions, three studies compared VR with no intervention [7,27,28], two studies with a placebo [31,33], and other study with oral treatment (Nonsteroidal Anti-Inflammatory Drugs (NSAIDs), tramadol, and duloxetine) [38]. In two studies, comparisons consisted of VR versus physiotherapy [30,37]. In addition, three studies combined VR + physiotherapy versus physiotherapy alone [26,29,32], and four studies combined those interventions and compared them with no VR exercises and physiotherapy [26,34–36].

Four of the fourteen studies used Nintendo programs and consoles [7,26,32,38], whereas three studies used other types of video games with sensors and a monitor (Prokin System) [34–36]. Five studies used a horse simulator riding [27–31]. One study used a system similar to VR, but without video games, based on biofeedback [37], and in one study, the intervention was a behavioral skills-based VR program with VR glasses [33].

Table 1. Characteristics of included studies.

Author (Year) [Ref.]	Country	Sample Size	Age (Years) Mean ± SD	Gender (% Males)	Outcome Measures	Measuring Instrument	Time Points Assessment	Quality
Park et al. (2013) [26]	South Korea	n = 24 EG1:8 EG2: 8 CG: 8	EG1: 44.12 ± 5.48 EG2: 43.37 ± 5.42 CG: 45.50 ± 5.34	100	Pain intensity Back strength Functional balance Health status	VAS Isometric lifting strength One-legged stand test RAND-36	Pre-intervention Post-intervention	16
Oh et al. (2014) [27]	South Korea	n = 37 EG1: 10 EG2: 9 EG3: 9 CG: 9	EG1: 20.56 ± 0.69 EG2: 20.33 ± 0.52 EG3: 20.44 ± 0.27 CG: 20.70 ± 0.37	100	**Pain intensity** Body composition **Isokinetic trunk flexion/extension Isokinetic hip flexion/extension and ABD/ADD**	VAS Bioelectrical impedance analysis method and anthropometer Isokinetic dynamometer Isokinetic dynamometer	Pre-intervention Post-intervention	15
Yoo et al. (2014) [28]	South Korea	n = 47 EG: 24 CG: 23	EG: 20.44 ± 1.33 CG: 20.70 ± 1.45	100	**Pain intensity Body composition** **Isokinetic trunk flexion/extension**	VAS Bioelectrical impedance analysis method Isokinetic dynamometer	Pre-intervention Post-intervention	17
Monteiro-Junior et al. (2015) [32]	Brazil	n = 30 EG: 16 CG: 14	68 ± 4	0	Pain intensity Static balance Functional capacity Mood	11-NPRS Wii balance board Sit-to-stand test POMS	Pre-intervention Post-intervention	21
Chen et al. (2016) [29]	South Korea	n = 19 EG: 10 CG: 9	19-30	-	Pain intensity Disability associated with low back pain Dynamic balance	VAS KODI LoS test with Biorescue	Pre-intervention Post-intervention	13
Zadro et al. (2019) [7]	Australia	n = 60 EG: 30 CG: 30	EG: 68.8 ± 5.5 CG: 67.8 ± 6	EG: 20 CG: 28.3	**Pain self-efficacy** Care-seeking **Physical activity** Pain intensity **Function changes** Severity of disability Kinesiophobia Falls efficacy	10-PSEQ 3-items questionnaire The Rapid Assessment of Physical Activity Questionnaire 11-NPRS PSFS RMDQ 17-TSK FEQ-I	Pre-intervention Post-intervention 3 months follow-up 6 months follow-up	26
Kim et al. (2020) [30]	South Korea	n = 48 EG: 24 CG: 24	EG: 26 ± 3.82 CG: 28.79 ± 9.05	EG: 68.2 CG: 42.3	Pain intensity Disability associated with low back pain Severity of disability **Fear and avoidance beliefs**	11-NPRS ODI RMDQ FABQ	Pre-intervention 4 weeks Post-intervention 6 months follow-up	22
Nambi et al. (2020) A [34]	Saudi Arabia	n = 45 EG1: 15 EG2: 15 CG: 15	EG1: 21.25 ± 1.2 EG2: 20.23 ± 1.6 CG: 20.78 ± 1.6	100	**Pain intensity Player wellness Sprint performance:** - 40 m sprint and 4 × 5 m sprint - **Submaximal shuttle running** **Jump performance: CJ and SJ**	VAS Player wellness questionnaire Photocell timer MEMS device Optical timing system	Pre-intervention Post-intervention 8 weeks follow-up 6 months follow-up	25

Table 1. *Cont.*

Author (Year) [Ref.]	Country	Sample Size	Age (Years) mean ± SD	Gender (% Males)	Outcome Measures	Measuring Instrument	Time Points Assessment	Quality
Nambi et al. (2020) B [35]	Saudi Arabia	n = 60 EG1: 20 EG2:20 CG:20	EG1: 23.2 ± 1.5 EG2: 22.8 ± 1.6 CG: 23.3 ± 1.5	100	**Pain intensity** **Kinesiophobia** **Blood serum levels of stress hormones**	VAS 17-TSK 20 ml venous blood sample	Pre-intervention Post-intervention 6 months follow-up	24
Park et al. (2020) [31]	South Korea	n = 80 EG: 40 CG:40	EG: 71.50 ± 6.34 CG: 72.05 ± 6.82	0	**Pain intensity** **Disability associated with low back pain** **Isokinetic trunk flexion/extension** **Body composition** **Spinal alignment**	VAS ODI Isokinetic dynamometer Bioelectrical impedance analysis method Raster stereography	Pre-intervention Post-intervention	21
Tomruk et al. (2020) [37]	Turkey	n = 42 EG: 21 CG: 21	Median (IQR) EG: 46 (40.05-50.50) CG: 45 (44-48)	-	**Pain intensity** **Disability associated with low back pain** **Postural control** Physical activity	11-NPRS ODI LoS PS tests by Biodex Balance System SenseWear Armband	Pre-intervention Post-intervention	18
García et al. 2021) [33]	USA	n = 179 EG: 89 CG: 90	EG: 51.5 ± 13.5 CG: 51.4 ± 12.9	EG: 25 CG: 21	**Pain intensity** **Pain interference** Patient's global impression of change **Physical function and sleep disturbances** Pain catastrophizing Pain self-efficacy Chronic pain acceptance Satisfaction with treatment VR device use System usability Over-the-counter analgesic medication use Opioid use data	DVPRS DVPRS-II Question and 7-point scale ranging PROMIS PCS PSEQ-2 CPAG-8 6-point scale Device SUS Question yes/no Self-reported	Pre-intervention During intervention Post-intervention	27
Nambi et al. (2021) [36]	Saudi Arabia	n = 54 EG1: 18 EG2:18 CG:18	EG1: 22.3 ± 1.6 EG2: 21.4 ± 1.8 CG: 21.9 ± 1.8	100	**Pain intensity** **Kinesiophobia** **Blood serum levels of hormones**	VAS 17-TSK 20 ml venous blood sample	Pre-intervention Post-intervention 6 months follow-up	24
Sato et al. (2021) [38]	Japan	n = 40 EG: 20 CG: 20	EG: 49.31 ± 12.59 CG: 55.61 ± 10.96	EG: 45 CG: 40	**Pain intensity** **Buttock pain** Leg numbness Body composition Left and right GS **Pain self-efficacy** Pain catastrophizing Kinesiophobia	VAS VAS VAS Bioelectrical impedance analysis method Dynamometer 10-PSEQ PCS 17-TSK	Pre-intervention Post-intervention	22

SD: Standard Deviation; EG: experimental group; CG: control group; LBP: Low Back Pain; VAS: Visual Analogue Scale; RAND-36: RAND-36 Health Status Inventory; CLBP: Chronic Low Back Pain; ABD: Abduction; ADD: adduction; 11-NPRS: Numeric Pain Rating Scale; POMS: Profile of Mood States; KODI: the Korean Oswestry Disability Index; LoS: Limits of Stability; 10-PSEQ: 10 items Pain-Self Efficacy; Questionnaire; PSFS: Patients Specific Functional Scale; RMSQ: Roland Morris Disability Questionnaire;17-TSK: 17 item Tampa Scale of Kinesiophobia; FEQ-I: Falls Efficacy Scale Questionnaire International, ODI: Oswestry Disability Index; FABQ: Fear-avoidance Beliefs Questionnaire; CJ: Countermovement jump, SJ: Squat jump; PS: Postural Stability; DVPRS: The Defense and Veterans Pain Rating Scale; DVPRS-II: The Defense and Veterans Pain Rating Scale interference; PROMIS: The NIH Physical Function and Sleep Disturbance; PCS: Pain Catastrophizing Scale; PSEQ-2: 2 items Pain-Self Efficacy Questionnaire; CPAG8: 8 item Chronic Pain Acceptance Questionnaire; SUS: System Usability Scale; GS: Grip Strength. Outcomes with significant differences ($p < 0.05$) between groups are presented in bold.

Table 2. Characteristics of interventions.

Author (Year) [Ref.]	Interventions	Session Duration	Frequency	Program Duration	Supervision	Adverse Events
Park et al. (2013) [26]	**EG1: Physical therapy (50 min) + Nintendo Wii exercises (30 min)** *Game*: Nintendo Wii sports. *VR program*: wakeboard, frisbee dog, jet ski, and canoe games. Participants controlled a virtual character on the screen by swinging, rowing, and tilting remote controllers with motion sensors. Participants chose which program they performed. *Time using videogame*: 30 min (2 min break every 10 min).	80 min	3 sessions per week	8 weeks	-	-
	EG2: Physical therapy (50 min) + Lumbar stabilization exercises (30 min) 7 positions based on the back bridge, hands and knees, and side bridge. Maintain each position for 15 s for 3 sets.					
	CG: Physical therapy (50 min) Hot pack (30 min), interferential current therapy (15 min), and deep heat with ultrasound (5 min).	50 min				
Oh et al. (2014) [27]	**EG1: Horse simulator riding (10 min)** *VR system*: Horse simulator machine (HJL Co. Ltd., Korea). *VR program*: warmup (stretching 5 min + ordinary walking on the horse simulator 5 min) + work-out (sitting trotting and rising trotting 10 min) + cool-down (supine stretching 10 min). *Time using videogame*: 15 min.	30 min	5 sessions per week	8 weeks	Supervised	-
	EG2: Horse simulator riding (20 min) *VR system*: Horse simulator machine (HJL Co. Ltd., Korea). *VR program*: warmup (stretching 5 min + ordinary walking on the horse simulator 5 min) + work-out (sitting trotting and rising trotting 20 min) + cool-down (supine stretching 10 min). *Time using videogame*: 25 min.	40 min				
	EG3: Horse simulator riding (30 min) *VR system*: Horse simulator machine (HJL Co. Ltd., Korea). *VR program*: warmup (stretching 5 min + ordinary walking on the horse simulator 5 min) + work-out (sitting trotting and rising trotting 30 min) + cool-down (supine stretching 10 min). *Time using videogame*: 35 min.	50 min				
	CG: No intervention.	-	-		-	
Yoo et al. (2014) [28]	**EG: Horse simulator riding** *VR system*: Horse simulator machine (HJL Co. Ltd., Korea). *VR program*: warmup (stretching 10 min) + work-out + cool-down (stretching 10 min). *Workout*: ordinary walking and sitting trotting (week 1), increase riding time and riding trotting (weeks 2–3), increase riding time and intensity (weeks 4–5), and increase riding time and intensity (weeks 6–8). *Time using videogame*: 10 min (week 1), 20 min (weeks 2–3), 30 min (weeks 4–5) and 40 min (weeks 6–8).	Week 1: 30 min Week 2–3: 40 min Week 4–5: 50 min Week 6–8: 60 min	3 sessions per week	8 weeks	Supervised	-
	CG: No intervention.	-	-		-	
Monteiro-Junior et al. (2015) [32]	**EG: Core exercises and strength training + 8 Wii Fit Plus workout** *VR system*: Nintendo Wii Motion and Wii Balance Board Games. *Games*: chair, tightrope walk, ski slalom, balance bubble, table tilt, sideways, rowing squat, lunge. *VR program*: familiarization, play games 2 times (3 initial sessions). Only one attempt for each game (from 4 session). Verbal stimulus and rest periods (1–2 min). *Time using videogame*: 30 min.	90 min	3 sessions per week	8 weeks	Supervised by a physiotherapist	- Vertigo
	CG: Core exercises and strength training Core exercises: postures (15–30 s, 3 series, rest 10–15 s) + Strength training: exercises (10 reps, 3 series).					

Table 2. *Cont.*

Author (Year) [Ref.]	Interventions	Session Duration	Frequency	Program Duration	Supervision	Adverse Events
Chen et al. (2016) [29]	**EG: Lumbar strengthening exercise (15 min) + Horse simulator riding (15 min)** *VR system:* indoor riding machine (Hongjin Leports, South Korea). *VR program:* the horse simulator riding used in this study simulated riding a real horse through the visual information that appeared on the front screen by the virtual environment. *Time using videogame:* 15 min. **CG: Lumbar strengthening exercise** The exercise program consisted of 6 movements for 1 set. Each movement was held for 5s with 8 reps. All the programs were repeated for 5 times for 1 set.	30 min	3 sessions per week	4 weeks	-	-
Zadro et al. (2019) [7]	**EG: Wii Fit U exercises at home** *VR system:* Nintendo Wii U console and Wii Fit U software. *VR program:* booklet with exercises preselected by the research team (flexibility, strengthening, and aerobic exercises). *Categories:* yoga, muscle, aerobic, and balance. *Intensity:* 13 on the Borg rating scale + symptoms 24 h after exercise. 1 day of rest between exercise sessions. *Learning session:* 1 or 2 h. *Time using videogame:* 60 min. **CG: No intervention.**	60 min -	3 sessions per week -	8 weeks	Unsupervised: EG received fortnightly calls from a physiotherapist -	No adverse events were reported -
Kim et al. (2020) [30]	**EG: Simulated horseback riding** *VR system:* Horse simulator (FORTIS-102; Daewon FORTIS, Republic of Korea). *VR program:* stretching and cooldown (10 min) + workout (30 min) + rest time (6 min). *Workout:* consisted of walking, slow trotting, and fast trotting of a real horse gait. *Time using videogame:* 30 min. **CG: Stabilization exercises** Stretching and cooldown (10 min) + workout (30 min) + rest time (6 min). The stabilization exercises with suspension (Redcord AS, Norway) consisted of a supine pelvic lift, supine and prone bridging exercise, and side-lying hip abduction. Each movement was performed forabout 10 s.	46 min	2 sessions per week	8 weeks	Supervised by practitioner	No adverse events were reported
Nambi et al. (2020) A [34]	**EG1: VR training + Physiotherapy** *VR system:* ProKin system PK252N (Pelvic Module balance trunk MF; TechnoBody, Lanusei, Italy). *Game:* VR balance training with shooting game. The game is controlled by moving the trunk back and forth and left and right according to the signs. The activities were made gradually more difficult with more participant muscle activity and movement. Level of difficulty was defined by the number of enemies, angle of throw, frequency of shoot, frequency of flashing of enemies, and number of balls appearing around the participant. *Time using video game:* 30 min. **EG2: Isokinetic training + Physiotherapy** Warmup: stretching (5 min) + Isokinetic dynamometer: exercise at an angular speed of 60°, 90°, and 120° (15 reps of 3 sets and rest between sets 30 s and between pace 60 s). **CG: Conventional balance training + Physiotherapy** Active isotonic exercise and isometric exercise (10–15 reps/day) + stretching (3 reps for 10 s). **Physiotherapy:** Hot pack therapy (20 min) and ultrasound (5 min) + home-based exercise protocol.	30 min + 25 min — + 25 min — + 25 min 25 min	5 sessions per week	4 weeks	Supervisor	-

Table 2. *Cont.*

Author (Year) [Ref.]	Interventions	Session Duration	Frequency	Program Duration	Supervision	Adverse Events
Nambi et al. (2020) B [35]	**EG1: VR training + Physiotherapy** *VR system*: ProKin system PK252N (Pelvic Module balance trunk MF; TechnoBody, Lanusei, Italy). *Game*: VR balance training with shooting game. The game is controlled by moving the trunk back and forth and left and right according to the signs. The activities were made gradually more difficult with more participant muscle activity and movement. Level of difficulty was defined by the number of enemies, angle of throw, frequency of shoot, frequency of flashing of enemies, and number of balls appearing around the participant. *Time using video game*: 30 min.	30 min + 25 min	5 sessions per week	4 weeks	Supervisor	-
	EG2: Isokinetic training + Physiotherapy Warmup: stretching (5 min) + Isokinetic dynamometer: exercise at an angular speed of 60°, 90°, and 120° (15 reps of 3 sets and rest between sets 30 s and between pace 60 s).	— + 25 min				
	CG: Conventional balance training + Physiotherapy Active isotonic exercise and isometric exercise (10–15 reps/day) + stretching (3 reps for 10 s).	— + 25 min				
	Physiotherapy: Hot pack therapy (20 min) and ultrasound (5 min) + home-based exercise protocol.	25 min				
Park et al. (2020) [31]	**EG: Equine riding simulator** *VR system*: Horse simulator (FORTIS 101, Daewon, Corp., Seoul, Korea). *VR program*: warmup (8 min) + workout (15 min) + cooldown (7 min). *Work-out*: walking (weeks 1–4), walking 10 min + trotting 5 min (weeks 5–8) and trotting 10 min + cantering 5 min (weeks 9–12). *Time using videogame*: 15 min.	30 min	3 sessions per week	12 weeks	Supervised by researcher	-
	CG: Watching video Participants sat on the horse and watched the video from the monitor					
Tomruk et al. (2020) [37]	**EG: Computer-based stability training** *VR system*: Biodex Balance System. *VR program*: postural stability training, limits of stability training, weight shift training, and maze control training. 12 stability levels, 3 trials with 10 s rest in each condition. *Time using videogame*: 30 min.	30 min	2 sessions per week	12 weeks	Supervised by physiotherapist	-
	CG: Traditional training Traditional postural control exercises by giving them visual, vestibular, or proprioceptive stimulus under the cues of a physiotherapist.					
Garcia et al. (2021) [33]	**EG: EaseVRx at home** *VR system*: Pico G2 4K all-in-one head-mounted VR device. *VR program*: the program delivers a multifaceted combination of pain relief skills training through a prescribed sequence of daily immersive experiences (3D images). Each VR experiences lasts between 2–16 min (average 6 min). *Categories*: pain education, relaxation/interoception, mindful escapes, pain distraction games and dynamic breathing. *Time using video game*: 2–16 min depending on the experience.	2–16 min	7 sessions per week	8 weeks	Unsupervised	- Nausea - Motion sickness
	CG: Sham VR at home *VR system*: Pico G2 4K all-in-one head-mounted VR device. *VR program*: sham VR headset displayed 2D nature footage with neutral music that was selected to be neither overly relaxing, aversive, nor distracting. The experience of Sham VR is similar to viewing nature scenes on a large-screen television and is not interactive. Twenty videos were rotated over the 56 sessions, with average duration of sessions closely matching those of EaseVRx.					

Table 2. *Cont.*

Author (Year) [Ref.]	Interventions	Session Duration	Frequency	Program Duration	Supervision	Adverse Events
Nambi et al. (2021) [36]	**EG1: VR training + Physiotherapy** *VR system*: Pro-Kin system PK 252N (Pelvic Module balance trunk MF; TechnoBody, Lanusei, Italy). *Game*: shooting game. The subjects are sitting on the virtual platform and visualizing the game on the computer display screen. The game was executed and controlled by moving the trunk back and forth and left and right according to the signs. The activities were made gradually more difficult with more participant muscle activity and movement. The level of difficulty was defined by the number of enemies, angle of throw, frequency of shoot, frequency of flashing of enemies, and number of balls appearing around the participant. *Time using videogame*: 30 min.	30 min + 25 min	5 sessions per week	4 weeks	Supervised by physiotherapist	-
	EG2: Combined physical rehabilitation + Physiotherapy The participant received balance training through a Swiss ball for core muscles. *Exercises*: wall squat, Russian twist, leg lift, plank saw, cobra and hip raise on the Swiss ball, 15 times per set for 3 sets. Maintain each position 10 s with a 3 s break between repetitions.	— + 25 min				
	CG: Conventional balance training + Physiotherapy Active isotonic and isometric exercises for abdominal muscles, deep abdominal muscles, and back muscles. 10 to 15 reps. Stretching was focused on each muscle group for 3 reps for 10 s per muscle group.	— + 25 min				
	Physiotherapy: Hot pack therapy (20 min) + ultrasound (5 min) + home-based exercise protocol.	25 min				
Sato et al. (2021) [38]	**EG: Nintendo Ring Fit Adventure Exergame** *VR system*: Ring Fit Adventure of Nintendo Switch. *VR program*: Adventure mode (30 min) + low back pain improvement program (10 min). Ring Fit Adventure is a fitness role-playing game that uses a ring-shaped controller as a device for resistance training. The player advances the story while exercising as the movement of the player is linked to the main character on the screen. *Time using videogame*: 40 min.	40 min	1 session per week	8 weeks	-	-
	CG: Oral treatment *Drugs*: Nonsteroidal Anti-Inflammatory Drugs, Tramadol, and Duloxetine. Each drug was started at the standard dose, with patients coming in every 2 weeks to be interviewed for pain, and if pain relief was not adequate, then the dose was gradually increased to its highest recommended level. If pain relief was still inadequate, the next drug was added.	-	-			

EG: Experimental Group; min: minutes; s: seconds; CG: Control Group; VR: Virtual Reality; reps: repetitions.

The mean time using VR was 28.29 min and the mean session duration was 46.21 min. Regarding the frequency of the sessions, it varied from one weekly session [38] to seven sessions per week [33]. The duration of the program in the different studies ranged from 4 [29,34–36] to 12 weeks [31,37]. In nine studies, the interventions were supervised [27,28,30–32,34–37]. In one article, participants were contacted by phone calls [7], and one did not include any type of supervision [33]. Three studies did not report on supervision of the intervention [26,29,38]. Of all the articles, only two reported adverse events derived from the intervention with VR (e.g., nausea, motion sickness, vertigo, etc.) [32,33]; in two articles no adverse events were reported [7,30], and in the rest no information was provided.

3.4. Methodological Quality of Included Studies

Downs and Black quality assessment method [19] was used to assess the methodological quality of included studies in this review. The total score for each study is shown in Table 1, and the score for each item is summarized in Appendix E. According to their score, of the 14 articles evaluated, two were classified as excellent (26–28), seven as good (20–25), four as fair (19–15), and one as poor (≤14). The mean score of the included studies was 20.79 (range: 13–27).

3.5. Risk of Bias of Included Studies

The Cochrane Risk of Bias Assessment Tool [25] was used to assess the risk of bias of the articles included in this review. Figures 2 and 3 show the summary and the graph of the risk of bias assessment, respectively. Random sequence generation, allocation concealment, incomplete outcome data, and selective reporting did not obtain a high risk of bias in any study. In addition, other bias obtained unclear risk of bias in all of the included studies. Blinding of participants and personnel and blinding of outcome assessment was evaluated as a high risk of bias in four [7,30,32,38] and two [33,38] studies, respectively. Two studies obtained unclear risk of bias in all items [29,37] and other two studies obtained unclear risk of bias in all items, except in incomplete outcome data [26,28].

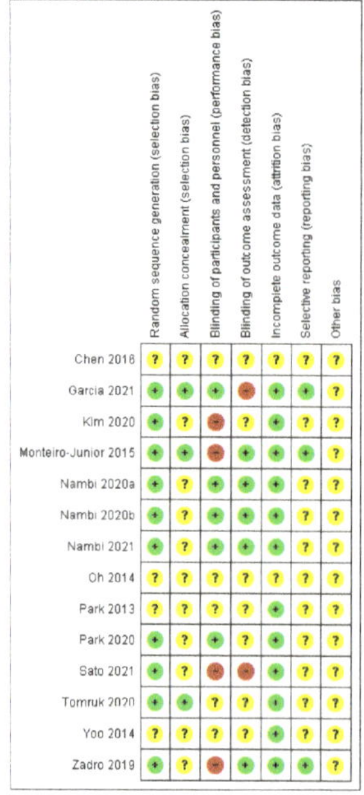

Figure 2. Risk of bias summary.

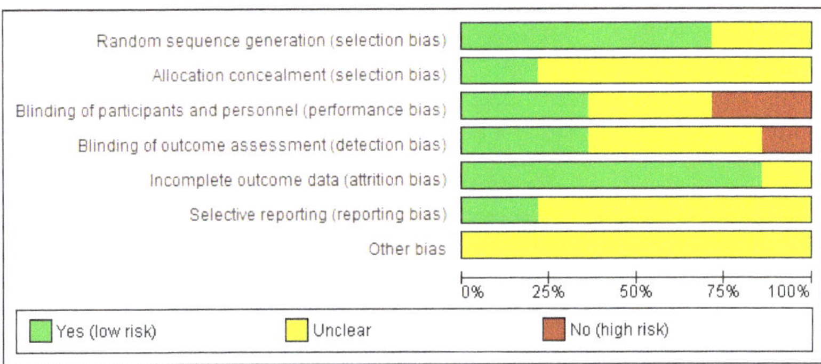

Figure 3. Risk of bias graph.

3.6. Effects of Virtual Reality vs. No Virtual Reality in Chronic Low Back Pain

For meta-analysis, we only considered the outcome pain intensity and outcomes related to pain.

Eleven studies were included in the meta-analysis. All of them were included for pain intensity postintervention; four for pain intensity at the 6 month followup; three for disability postintervention; three for kinesiophobia postintervention, and two (four comparisons) for kinesiophobia at the six months followup. Two articles were excluded from the meta-analysis because they did not express data in mean $\pm$ SD [33,37]. In addition, Yoo et al. [28] was excluded because the SD was 0, and it was not estimable by RevMan.

3.6.1. Subgroup Based on Intervention Comparisons: Virtual Reality Alone or Combined with Physiotherapy vs. Control Group Interventions

Firstly, a subgroup analysis of the different interventions was performed to know if VR applied alone or added to a physical therapy intervention could produce different results, and if it differed depending on the type of intervention of the control group. We analyzed pain intensity, disability, and kinesiophobia postintervention; and pain intensity and kinesiophobia at the 6 months followup. The Visual Analog Scale (VAS) to evaluate pain intensity was adjusted to a scale of 0–10 cm when it was expressed in millimeters.

In Figure 4a, the results show significant differences (SMD = -1.92; 95% CI = -2.73, -1.11; $p < 0.00001$) in favor of VR compared to no VR in pain intensity postintervention. When VR was compared with no intervention (SMD = -1.84; 95% CI = -3.48, -0.21; $p = 0.03$), placebo (SMD = -2.71; 95% CI = -3.33, -2.10; $p < 0.00001$), or oral treatment (SMD = -0.78; 95% CI = -1.42, -0.13; $p = 0.02$), the subgroup analysis showed significant differences in favor of VR. In addition, when VR + physiotherapy were compared with no VR exercise + physiotherapy, the subgroup analysis showed significant differences (SMD = -3.26; 95% CI = -5.08. -1.44; $p = 0.0004$) in favor of VR too. However, no significant differences were observed between VR and physiotherapy (SMD = -0.28; 95% CI = -0.85, 0.28; $p = 0.33$) or VR + physiotherapy and physiotherapy (SMD = 0.08; 95% CI = -0.42, 0.59; $p = 0.75$). Heterogeneity was high in overall effect (I^2 = 93%; $p < 0.00001$) and in two subgroups, VR versus no intervention (I^2 = 90%; $p < 0.00001$) and VR + physiotherapy versus no VR exercise + physiotherapy (I^2 = 95%; $p < 0.00001$). According to the I^2 statistic, 0% of variation across studies was due to heterogeneity ($p = 0.98$) in VR + physiotherapy versus the physiotherapy subgroup.

In Figure 4b, the results show significant differences (SDM = -6.34; 95% CI = -9.12, -3.56; $p < 0.00001$) in pain intensity at the six month followup in favor of VR compared to no VR. When VR was compared with physiotherapy, the subgroup analysis showed no significant differences (SDM = 0.17; 95% CI = -0.54, 0.87; $p = 0.64$). However, when VR + physiotherapy were compared with no VR exercise + physiotherapy, the subgroup analysis showed significant differences in favor of VR (SDM = -7.56; 95% CI = -10.79, -4.32; $p < 0.00001$). Heterogeneity was high in overall effect (I^2 = 97%; $p < 0.00001$)

and in VR + physiotherapy versus no VR exercise + physiotherapy subgroup (I^2 = 96%; p < 0.00001).

As shown in Figure 5, no significant differences were found between VR interventions and other interventions without VR (MD = 10.46; 95% CI = −30.02, 9.09; p = 0.29) in disability postintervention. Subgroup analysis did not show significant differences between VR and physiotherapy (MD = −3.26; 95% CI = −8.44, 1.92; p = 0.22) or between VR + physiotherapy and physiotherapy (MD = −0.10; 95% CI = −3.47, 3.27; p = 0.95). However, when VR was compared with the placebo, the subgroup analysis showed significant differences in favor of VR (MD = −27.89; 95% CI = −30.77, −25.01; p < 0.00001). Heterogeneity between studies was high (I^2 = 99%; p < 0.00001).

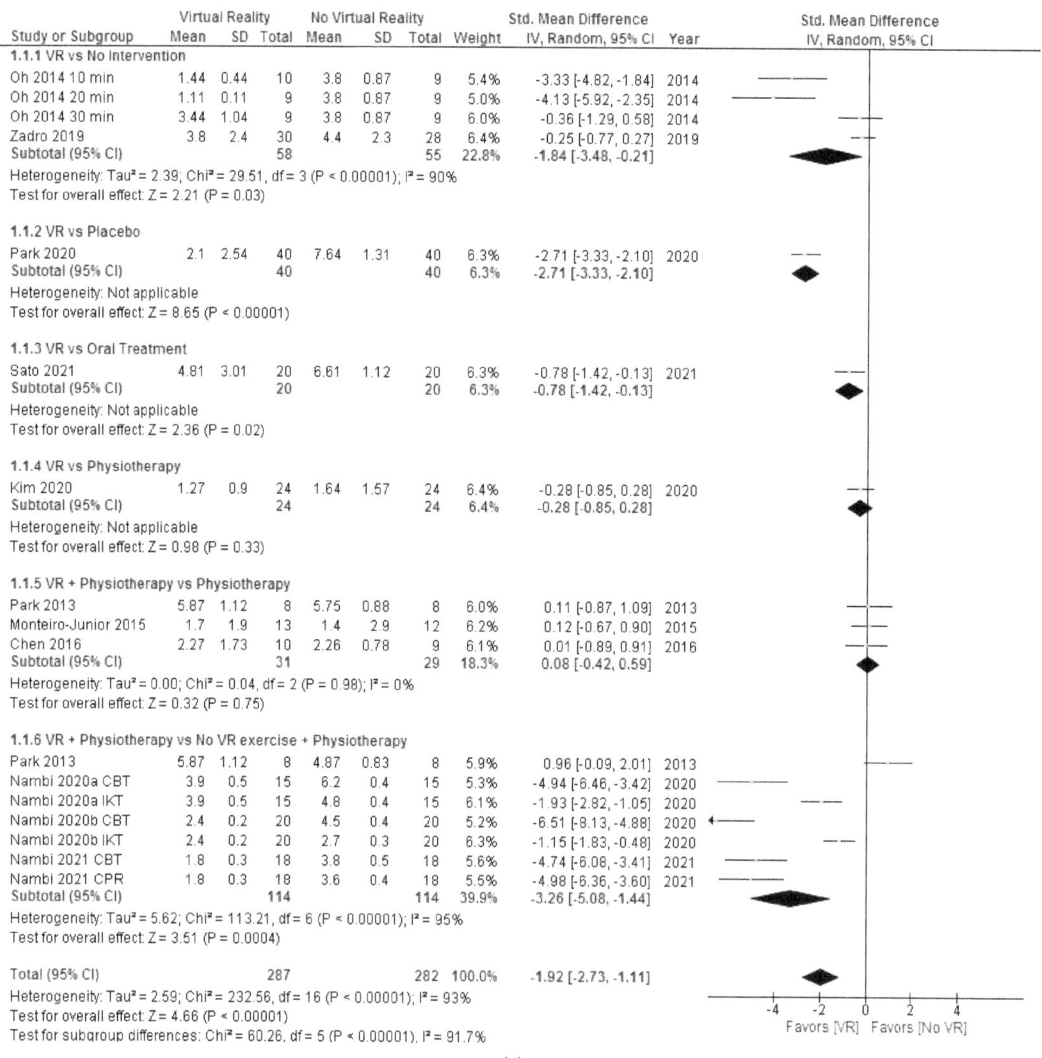

Figure 4. *Cont.*

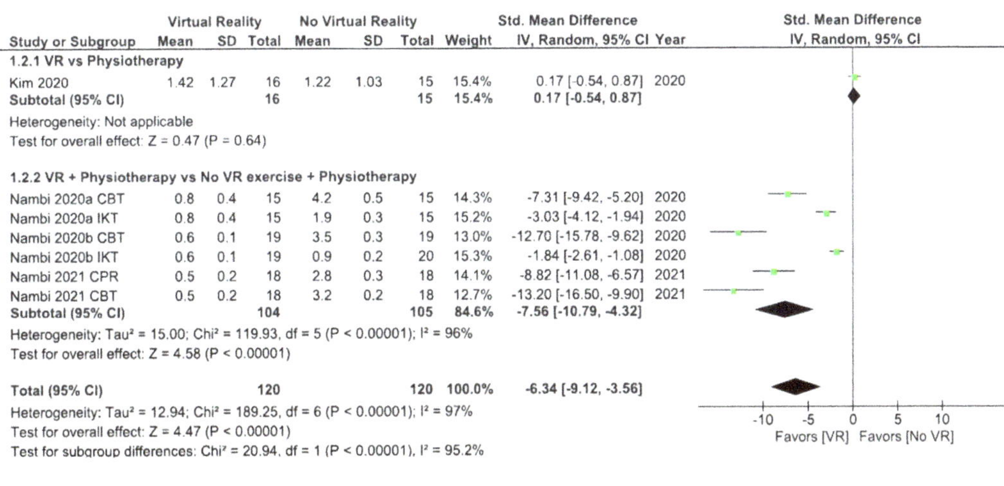

(**b**)

Figure 4. Effect of virtual reality versus no virtual reality in chronic low back pain for pain intensity postintervention (**a**) and at the six month followup (**b**) based on the type of intervention. CBT: conventional balance training; IKT: isokinetic training; CPR: combined physical rehabilitation; VR: virtual reality.

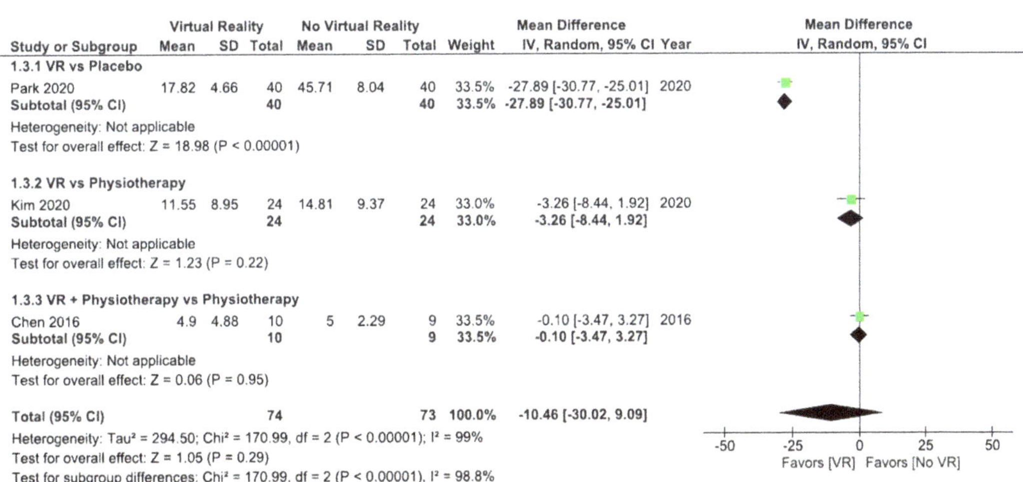

Figure 5. Effect of virtual reality versus no virtual reality in chronic low back pain for disability postintervention based on the type of intervention. VR: virtual reality.

As shown in Figure 6a, the results showed significant differences (MD = −8.96; 95% CI = −17.52, −0.40; p = 0.04) in favor of VR in total comparison in kinesiophobia postintervention. When VR was compared with oral treatment, the subgroup analysis showed significant differences in favor of oral treatment (MD = 3.47; 95% CI = 1.00, 5.94; p = 0.006). However, when VR + physiotherapy were compared with no VR exercises + physiotherapy, the subgroup analysis showed significant differences in favor of VR (MD = −12.05; 95% CI = −20.13, −3.98; p = 0.003). Heterogeneity was high in overall effect (I^2 = 99%; p < 0.00001) and in VR + physiotherapy versus no VR exercise + physiotherapy subgroup (I^2 = 98%; p < 0.00001).

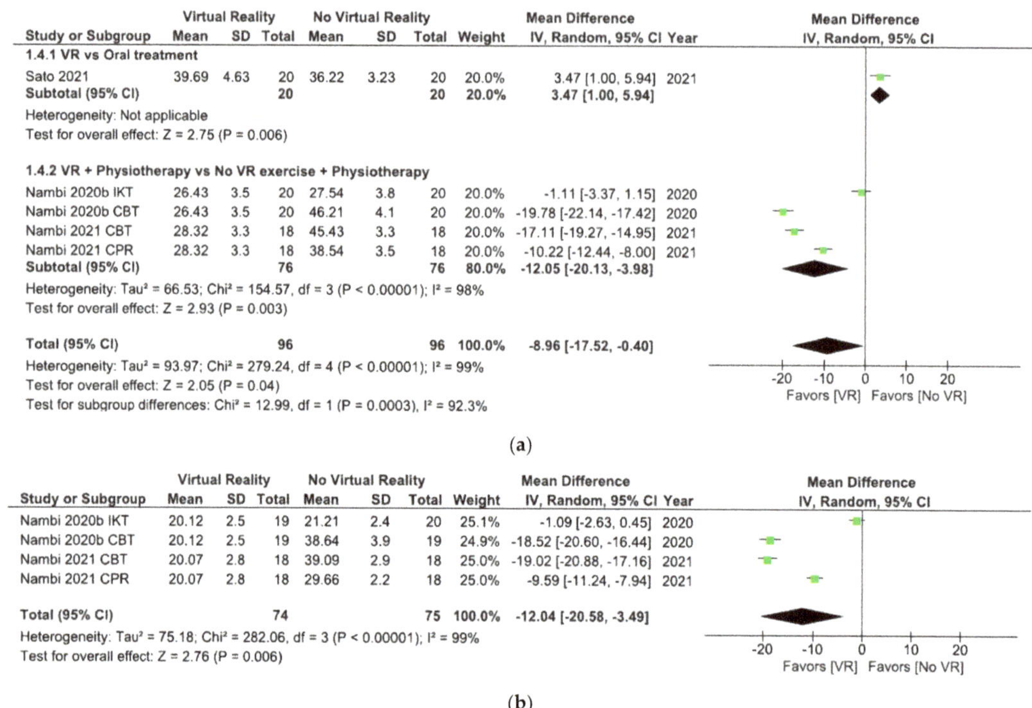

Figure 6. Effect of virtual reality versus no virtual reality in chronic low back pain for kinesiophobia postintervention (**a**) and at the six month followup (**b**) based on the type of intervention. IKT: isokinetic training; CBT: conventional balance training; CPR: combined physical rehabilitation; VR: virtual reality.

All studies in this meta-analysis (Figure 6b) compared VR + physiotherapy versus no VR exercise + physiotherapy. The results showed significant differences (MD = −12.04; 95% CI = −20.58, –3.49; p = 0.006) in favor of VR in kinesiophobia at the 6 month followup. Heterogeneity between studies was high (I^2 = 99% p < 0.00001).

3.6.2. Subgroups Based on Virtual Reality Interventions

Other subgroup analysis was based on the type of VR intervention. The studies were divided into three subgroups: Nintendo consoles, Horse Simulator Riding, or Prokin System. We analyzed pain intensity, disability, and kinesiophobia postintervention and pain intensity and kinesiophobia at the 6 months followup.

As shown in Figure 7a, the results showed significant differences (SMD = −1.92; 95% CI = −2.73, −1.11; p < 0.00001) in favor of VR versus no VR in pain intensity postintervention. When Nintendo consoles were compared with interventions without VR, the subgroup analysis showed no significant differences (SMD = −0.07; 95% CI = −0.57, 0.43; p = 0.78). However, when horse simulator riding (SMD = −1.68; 95% CI = −2.95, –0.41; p = 0.009) or Prokin System (SMD = −3.96; 95% CI = −5.71, −2.21; p < 0.00001) were compared with interventions without VR, the subgroup analysis showed significant differences in favor of VR. Heterogeneity was high in overall effect (I^2 = 93%; p < 0.00001) and in two subgroups, horse simulator riding (I^2 = 92%; p < 0.00001) and Prokin System (I^2 = 93%; p < 0.00001). According to the I^2 statistic, 54% of variation across studies was due to heterogeneity (p = 0.07) in the Nintendo consoles subgroup.

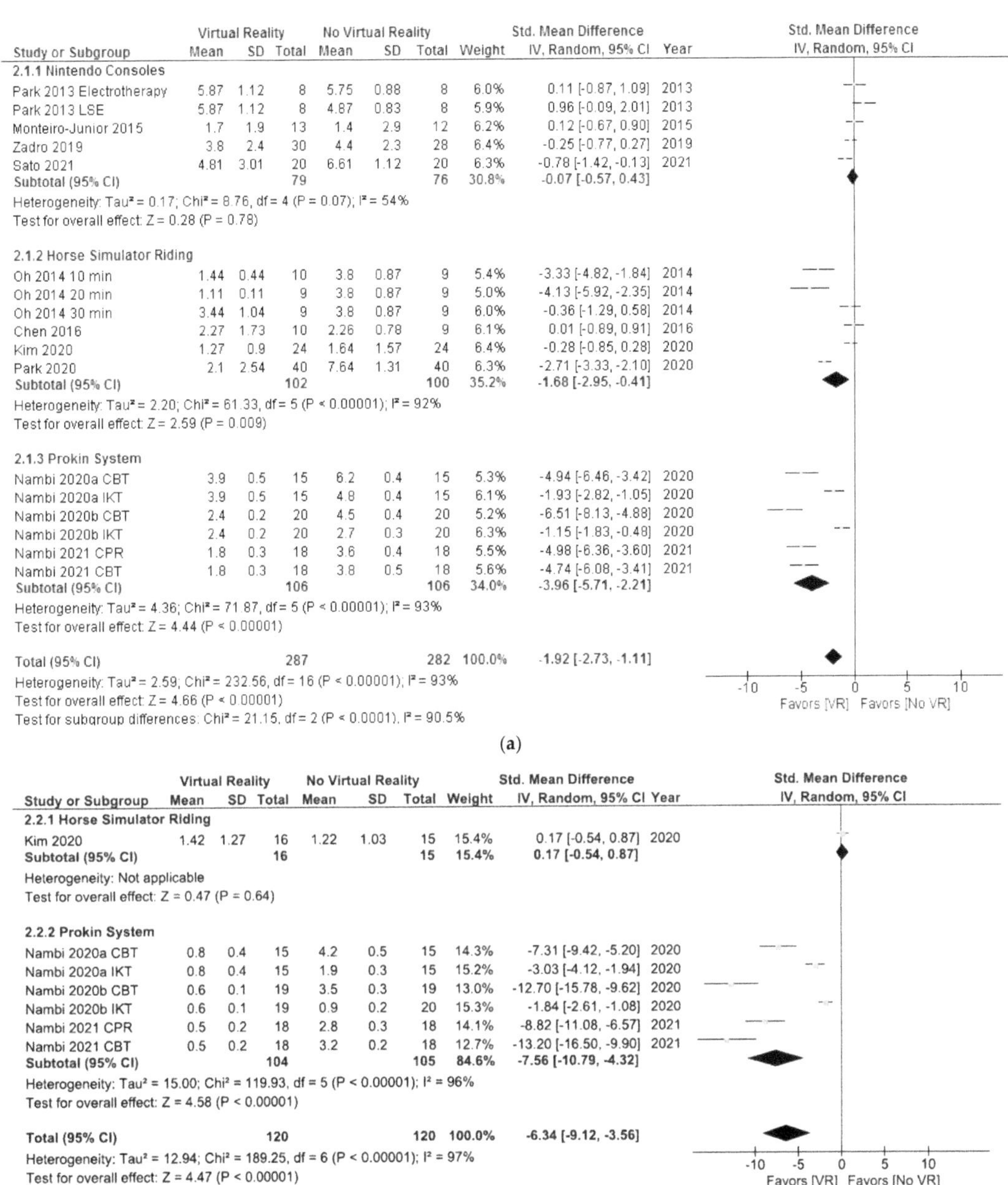

Figure 7. Effect of virtual reality versus no virtual reality in chronic low back pain for pain intensity postintervention (**a**) and at the six months followup (**b**) based on the type of virtual reality intervention. CBT: conventional balance training; IKT: isokinetic training; CPR: combined physical rehabilitation; VR: virtual reality.

As shown in Figure 7b the results showed significant differences (SDM = -6.34; 95% CI = $-9.12, -3.56$; $p < 0.00001$) in favor of VR in total comparison in pain intensity at the 6 month followup. Regarding subgroup analysis, no significant differences were found between horse simulator riding and no VR interventions (SDM = 0.17; 95% CI = $-0.54, 0.87$; $p = 0.64$). However, significant differences in favor of VR were found in the Prokin System subgroup (SDM = -7.56; 95% CI = $-10.79, -4.32$; $p < 0.00001$). Heterogeneity was high in overall effect ($I^2 = 97\%$; $p < 0.00001$) and in Prokin System versus no VR interventions subgroup ($I^2 = 96\%$; $p < 0.00001$).

All studies in Figure 8 compared horse simulator riding interventions versus other interventions without VR. No significant differences were found between VR and no VR (MD = -10.46; 95% CI = $-30.02, 9.09$; $p = 0.29$) in disability post-intervention. Heterogeneity between studies was high ($I^2 = 99\%$; $p < 0.00001$).

Study or Subgroup	Virtual Reality Mean	SD	Total	No Virtual Reality Mean	SD	Total	Weight	Mean Difference IV, Random, 95% CI	Year
Chen 2016	4.9	4.88	10	5	2.29	9	33.5%	-0.10 [-3.47, 3.27]	2016
Park 2020	17.82	4.66	40	45.71	8.04	40	33.5%	-27.89 [-30.77, -25.01]	2020
Kim 2020	11.55	8.95	24	14.81	9.37	24	33.0%	-3.26 [-8.44, 1.92]	2020
Total (95% CI)			**74**			**73**	**100.0%**	**-10.46 [-30.02, 9.09]**	

Heterogeneity: Tau² = 294.50; Chi² = 170.99, df = 2 (P < 0.00001); I² = 99%
Test for overall effect: Z = 1.05 (P = 0.29)

Figure 8. Effect of virtual reality versus no virtual reality in chronic low back pain for disability postintervention based on the type of virtual reality intervention. VR: virtual reality.

As shown in Figure 9a, the results showed significant differences (MD = -8.96; 95% CI = $-17.52, -0.40$; $p = 0.04$) in favor of VR in total comparison in kinesiophobia postintervention. The results showed significant differences in favor of interventions without VR versus interventions with Nintendo consoles (MD = 3.47; 95% CI = 1.00, 5.94; $p = 0.006$). However, when the Prokin System was compared with interventions without VR significant differences were found in favor of the Prokin System subgroup (MD = -12.05; 95% CI = $-20.13, -3.98$; $p = 0.003$). Heterogeneity was high in overall effect ($I^2 = 99\%$; $p < 0.00001$) and in Prokin System versus interventions without VR ($I^2 = 98\%$; $p < 0.00001$).

All studies in Figure 9b compared Prokin System versus interventions without VR. The results showed significant differences (MD = -12.04; 95% CI = $-20.58, -3.49$; $p = 0.006$) in favor of VR in kinesiophobia at the 6 month followup. Heterogeneity between studies was high ($I^2 = 99\%$; $p < 0.00001$).

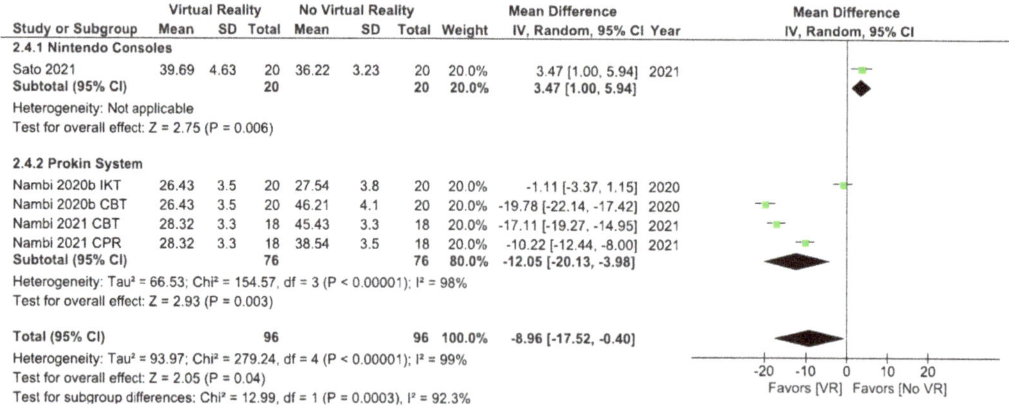

Study or Subgroup	Virtual Reality Mean	SD	Total	No Virtual Reality Mean	SD	Total	Weight	Mean Difference IV, Random, 95% CI	Year
2.4.1 Nintendo Consoles									
Sato 2021	39.69	4.63	20	36.22	3.23	20	20.0%	3.47 [1.00, 5.94]	2021
Subtotal (95% CI)			20			20	20.0%	3.47 [1.00, 5.94]	
Heterogeneity: Not applicable									
Test for overall effect: Z = 2.75 (P = 0.006)									
2.4.2 Prokin System									
Nambi 2020b IKT	26.43	3.5	20	27.54	3.8	20	20.0%	-1.11 [-3.37, 1.15]	2020
Nambi 2020b CBT	26.43	3.5	20	46.21	4.1	20	20.0%	-19.78 [-22.14, -17.42]	2020
Nambi 2021 CBT	28.32	3.3	18	45.43	3.3	18	20.0%	-17.11 [-19.27, -14.95]	2021
Nambi 2021 CPR	28.32	3.3	18	38.54	3.5	18	20.0%	-10.22 [-12.44, -8.00]	2021
Subtotal (95% CI)			76			76	80.0%	-12.05 [-20.13, -3.98]	
Heterogeneity: Tau² = 66.53; Chi² = 154.57, df = 3 (P < 0.00001); I² = 98%									
Test for overall effect: Z = 2.93 (P = 0.003)									
Total (95% CI)			96			96	100.0%	-8.96 [-17.52, -0.40]	

Heterogeneity: Tau² = 93.97; Chi² = 279.24, df = 4 (P < 0.00001); I² = 99%
Test for overall effect: Z = 2.05 (P = 0.04)
Test for subgroup differences: Chi² = 12.99, df = 1 (P = 0.0003), I² = 92.3%

(**a**)

Figure 9. *Cont.*

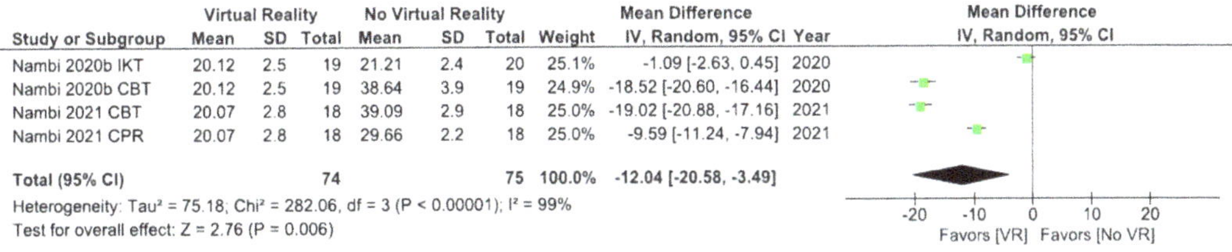

(**b**)

Figure 9. Effect of virtual reality versus no virtual reality in chronic low back pain for kinesiophobia postintervention (**a**) and at the six month followup (**b**) based on the type of virtual reality intervention. IKT: isokinetic training; CBT: conventional balance training; CPR: combined physical rehabilitation; VR: virtual reality.

3.6.3. Subgroups Based on the Duration of the Intervention

The last subgroup analysis was based on the duration of the intervention. The studies were divided into three subgroups: four weeks, eight weeks, or twelve weeks of intervention. We analyzed pain intensity, disability, and kinesiophobia postintervention and pain intensity and kinesiophobia at the 6 month followup.

As shown in Figure 10a, the results showed significant differences (SMD = -1.92; 95% CI = -2.73, -1.11; $p < 0.00001$) in favor of VR versus no VR in pain intensity postintervention. Subgroup analysis showed significant differences in favor of VR after 4 weeks of intervention (SMD = -3.38; 95% CI = -5.06, -1.70; $p < 0.0001$), 8 weeks of intervention (SMD = -0.65; 95% CI = -1.29, -0.00; $p = 0.05$), and 12 weeks of intervention (SMD = -2.71; 95% CI = -3.33, -2.10; $p < 0.00001$). Heterogeneity was high in overall effect ($I^2 = 93\%$; $p < 0.00001$) and in all subgroups ($I^2 = 94\%$; $p < 0.00001$) ($I^2 = 81\%$; $p < 0.00001$).

As shown in Figure 10b, the results showed significant differences (SDM = -6.34; 95% CI = -9.12, -3.56; $p < 0.00001$) in favor of VR in total comparison in pain intensity at the 6 month followup. Regarding subgroup analysis, no significant differences were found between VR versus no VR after 8 weeks of intervention (SDM = 0.17; 95% CI = -0.54, 0.87; $p = 0.64$). However, significant differences in favor of VR were found after 4 weeks of intervention (SDM = -7.56; 95% CI = -10.79, -4.32; $p < 0.00001$). Heterogeneity was high in overall effect ($I^2 = 97\%$; $p < 0.00001$) and in the 4 weeks of intervention subgroup ($I^2 = 96\%$; $p < 0.00001$).

No significant differences were found between VR interventions and other interventions without VR (MD = -10.46; 95% CI = -30.02, 9.09; $p = 0.29$) in disability postintervention. Subgroup analysis did not show significant differences between VR and no VR after 4 weeks (MD = -0.10; 95% CI = -3.47, 3.27; $p = 0.95$) or 8 weeks of intervention (MD = -3.26; 95% CI = -8.44, 1.92; $p = 0.22$). However, significant differences were found in favor of VR after 12 weeks of intervention (MD = -27.89; 95% CI = -30.77, -25.01; $p < 0.00001$). Heterogeneity between studies was high ($I^2 = 99\%$; $p < 0.00001$). Figure 11 shows these results.

As shown in Figure 12a, the results showed significant differences (MD = -8.96; 95% CI = -17.52, -0.40; $p = 0.04$) in favor of VR in total comparison in kinesiophobia postintervention. After 8 weeks of intervention, the results showed significant differences in favor of no VR intervention (MD = 3.47; 95% CI = 1.00, 5.94; $p = 0.006$). However, significant differences in favor of VR were observed after 4 weeks of intervention (MD = -12.05; 95% CI = -20.13, -3.98; $p = 0.003$). Heterogeneity was high in overall effect ($I^2 = 99\%$; $p < 0.00001$) and in the 4 weeks of intervention subgroup ($I^2 = 98\%$; $p < 0.00001$).

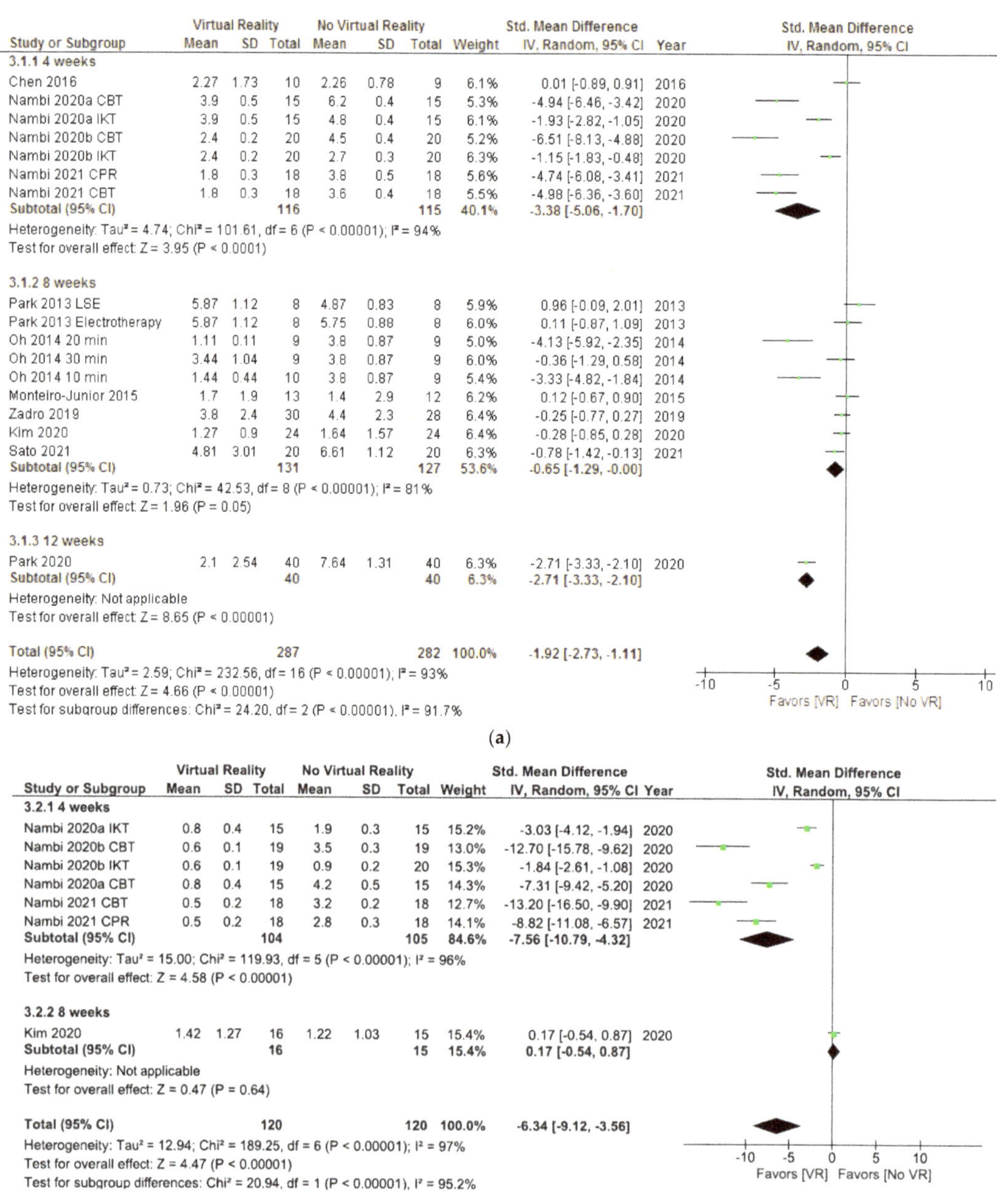

(a)

(b)

Figure 10. Effect of virtual reality versus no virtual reality in chronic low back pain for pain intensity postintervention (**a**) and at the six month followup (**b**) based on the duration of the intervention. CBT: conventional balance training; IKT: isokinetic training; CPR: combined physical rehabilitation; VR: virtual reality.

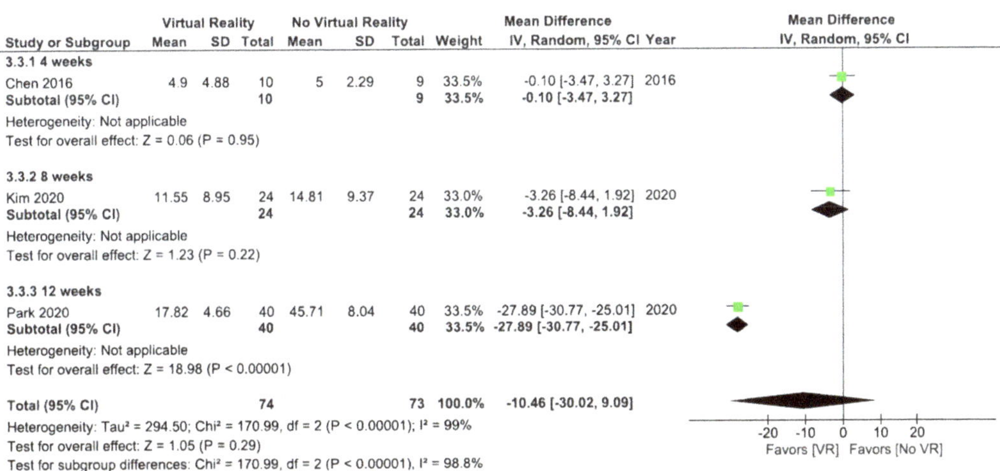

Figure 11. Effect of virtual reality versus no virtual reality in chronic low back pain for disability postintervention based on the duration of the intervention. VR: virtual reality.

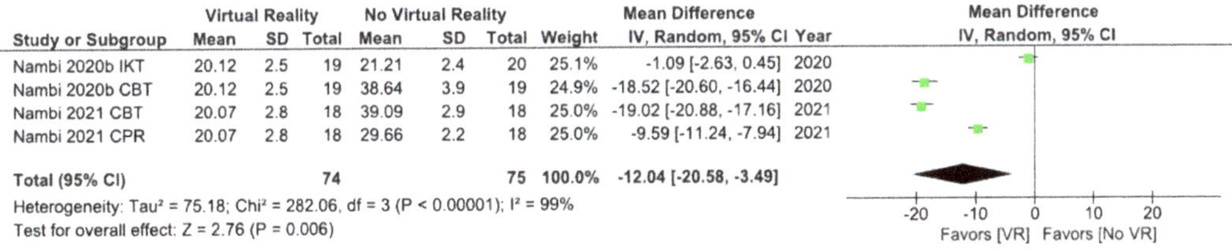

(a)

(b)

Figure 12. Effect of virtual reality versus no virtual reality in chronic low back pain for kinesiophobia postintervention (**a**) and at the six month followup (**b**) based on the duration of the intervention. IKT: isokinetic training; CBT: conventional balance training; CPR: combined physical rehabilitation; VR: virtual reality.

All studies shown in Figure 12b conducted a 4-week intervention. The results showed significant differences (MD = −12.04; 95% CI = −20.58, −3.49; *p* = 0.006) in favor of VR in kinesiophobia at the 6 month followup. Heterogeneity between studies was high (I^2 = 99%; *p* < 0.00001).

4. Discussion

The objective of this systematic review and meta-analysis was to analyze the effectiveness of VR interventions in the treatment of CLBP. Fourteen studies were included in this review and eleven of them in the meta-analysis. The results showed significant differences in favor of VR interventions in pain intensity and kinesiophobia postintervention and at the six month followup. However, no significant differences were found in disability postintervention.

4.1. Pain Intensity

Pain intensity was assessed in all of the studies included in meta-analysis. The meta-analysis showed significant differences in favor of interventions with VR versus interventions without VR in pain intensity postintervention and at the six month followup. On the one hand, the effect of VR was superior to no intervention [7,27], placebo [31], and oral treatment [38] in pain intensity postintervention, but it should be noted that there was only a study in two of these subgroups. Significant differences in favor of VR + physiotherapy were observed when we compared with no VR exercise + physiotherapy [26,34–36] in pain intensity postintervention and at the six month follow-up. Most of the studies included in this subgroup had good methodological quality and obtained significant differences in favor of VR in the rest of the variables not included in the meta-analysis. However, it must be taken into account that these results have been obtained from studies that only included young athletic men and cannot be generalized [34–36]. On the other hand, VR was not superior to physiotherapy in pain intensity postintervention or at the six month followup. Nevertheless, there was only one article (with young adults and a high dropout rate) in this subgroup [30]. Neither were significant differences found between VR + physiotherapy versus physiotherapy in pain intensity postintervention. It should be noted that these studies had a small sample size and some of them had low methodological quality [26,29,32].

Regarding the type of VR, horse simulator riding and Prokin System were superior to interventions without VR in pain intensity postintervention. However, in the horse simulator riding subgroup, most of the studies compared this type of VR with no intervention [27] or placebo [31], which can explain the good results in the analysis. Nintendo consoles did not show significant differences. This may be because the Prokin System and horse simulator riding are specialized VR devices compared to Nintendo consoles. At the six month followup, the results showed significant differences in favor of the Prokin System but not in favor of horse simulator riding. It must be taken into account that there was only one study (with young adults and a high dropout rate) in this subgroup [30]. The results showed significant differences between VR and no VR in pain intensity postintervention after 4 weeks, 8 weeks, or 12 weeks of intervention. At the six month followup, significant differences in favor of VR were found after 4 weeks of intervention but not after 8 weeks of intervention. It should be noted that there was only one study in this subgroup [30].

4.2. Disability

No significant differences were found between VR interventions (horse simulator riding) and no VR interventions in disability postintervention. However, when VR was compared with placebo and when the intervention lasted 12 weeks, the different subgroups analysis showed significant differences in favor of VR. This can be explained because the same article was included in the subgroups [31]. No significant differences were observed between VR and physiotherapy [30] or VR + physiotherapy and physiotherapy [29], or after four [29] or eight weeks of intervention [30]. It should be noted that there was only one study in each subgroup. In addition, these studies had a small sample size, poor methodological quality [29], and some limitations, such as a high dropout rate [30].

4.3. Kinesiophobia

The results showed significant differences in favor of VR when compared with no VR in kinesiophobia postintervention and at the six month followup. When VR + physiotherapy were compared with no VR exercises + physiotherapy, the subgroup analysis showed significant differences in favor of VR in kinesiophobia postintervention and at the six month followup. These studies used Prokin System as the VR intervention, so significant differences in favor of Prokin System also were observed in this subgroup. The intervention lasted 4 weeks in all these articles, thus, the same results were found in subgroups based on duration of intervention. Although these articles had a good methodological quality, they only included young athletic men and their results cannot be generalized [35,36]. The other study that assessed kinesiophobia postintervention, Sato et al. [38] compared Nintendo Switch with oral treatment and the duration of the intervention was 8 weeks. In all of the different subgroups realized significant differences were found in favor of oral treatment. It must be taken into account that there was only this study in each subgroup (Nintendo and 8 weeks).

4.4. Virtual Reality in Other Populations

Other studies explored the effects of VR in different populations (such as, patients with chronic neck pain, fibromyalgia, acute pain, Parkinson's disease, stroke, etc.) and the results differ in part from ours.

In similar chronic pathologies, for example chronic neck pain, VR showed significant differences compared with no VR in pain intensity postintervention, which is in line with our results [39,40]. In this case, interventions consisted of VR compared with physiotherapy and in our review this type of comparison did not obtain significant results. No significant differences were found between VR + physiotherapy versus physiotherapy in pain intensity postintervention [41], which coincides with our review. However, in patients with fibromyalgia, VR combined with exercises showed significant improvement compared to exercises [42,43], although, results in pain intensity are not always conclusive [42]. In addition, in chronic neck pain, significant differences were observed in favor of VR in disability postintervention [39,40] which differs with the results found in our studies. These differences can be explained; the VR interventions in chronic neck pain articles were immersive, and the articles had better methodological quality.

In acute pain pathologies, VR has proven to be an adjuvant tool that can reduce procedural pain [44], burn pain, and anxiety [45]. In addition, it can reduce the use of medication [45]. As we have mentioned previously, in our review the studies that made a comparison between VR + another treatment versus same treatment did not obtain significant differences. In this case, this may be due to differences in the duration of pain and its origin.

Regarding neurological pathologies, such as Parkinson's disease, VR rehabilitation showed better results in overall improvement than conventional rehabilitation [46]. However, in another study, VR combined with exercises was statistically as effective as each intervention alone [47]. In any case, these results do not agree with ours, since no significant differences were found for these comparisons in CLBP. In stroke patients, VR combined with conventional physical therapy obtained significantly higher improvements than conventional physical therapy [48–50]. These results are also not in line with the current review.

4.5. Discussion with Other Reviews

The results obtained in our meta-analysis differ partially from those found in other reviews. Bordeleau et al. [16] found significant differences in favor of VR versus no VR interventions for pain intensity postintervention, which is in line with our results. Nevertheless, there are differences in subgroup analysis. When we compared VR with no intervention, the subgroup analysis showed significant differences in favor of VR but in Bordeleau et al. [16] significant differences were not found. The differences found between the meta-analysis may be due to the different articles included in each one and how they was carried out. In addition, they included studies with back pain, whereas we only included studies limited to CLBP patients. In Gumaa et al. [14] the results of the meta-analysis did not show significant differences between VR interventions compared to no intervention in pain intensity postintervention. It should be noted that in one of the studies there was an intervention, since there was electrotherapy [26], and another had a short intervention [51] compared to the others, so we did not consider it comparable. Our results showed significant differences in favor of VR versus no intervention. This can be explained by the greater number of articles included in our meta-analysis and by the different comparisons realized. However, most of the studies included in this meta-analysis had a small sample size, fair methodological quality, and unclear risk of bias.

Bordeleau et al. [16] did not observe significant differences between horse simulator riding and interventions without VR, whereas in our meta-analysis significant differences in favor of horse simulator riding were obtained in pain intensity postintervention. In addition, this is consistent with the results found in two reviews. Collado-Mateo et al. [52] concluded that horse-riding simulators are a promising tool to reduce pain intensity in low back patients, but the interpretation of the results must be performed with extreme caution due to the large heterogeneity, the low number of studies, and the potential risk of bias. Ren et al. [53] also found significant differences in favor of horse simulator riding compared with control in pain intensity postintervention and severity of disability in people with CLBP. However, Ren el al. included another type of VR in addition to horse simulator riding and patients with subacute low back pain.

In Bordeleau et al. [16] the results showed that the potential beneficial effect of VR was more important when more than 12 sessions were performed. In our review, the interventions of included articles lasted 4, 8, or 12 weeks. In all of these cases significant differences in favor of VR were found in pain intensity postintervention, but it should be noted that the best results were obtained in the 12 weeks of intervention subgroup. However, only one study was included [31].

4.6. Strengths and Limitations

This review represents an update in the knowledge about the effects of VR treatment in CLBP, incorporating a meta-analysis of outcomes that could not be performed before.

The strengths of the current systematic review included following the PRISMA guidelines [17] for implementation and the use of the PICOS strategy to define the inclusion criteria. Another strength was the performance of meta-analysis. The assessment of methodological quality was carried out with the Downs and Black scale [19], one of the six best scales of methodological quality [21]. Additionally, the risk of bias was assessed with the Cochrane Risk of Bias Assessment Tool [25]. Furthermore, the review was previously registered in PROSPERO with registration number CRD42020222129.

However, although PRISMA guidelines were adhered to and the methodology was strictly followed, completely accounting for the limitations of the included studies was impossible. One of the main limitations was the high heterogeneity between included studies and the difficulties found in making comparable subgroups in order to draw solid conclusions. There were also differences in the age ranges and in the clinical profile of the participants. Regarding the characteristics of the patients, in four studies pain was defined as nonspecific [7,29,30,32], in another study pain was related to work [26]. Four studies [27,35,36,38] made reference to nonspecific pain, however, pain was not defined as nonspecific in the inclusion criteria of these studies. Finally, in three studies pain may have been related to sports practice [34–36], as the participants were football players. These differences in the origin of pain must be taken into account, because they could influence the results. Furthermore, the sample size of the included studies was relatively small in some of the studies (19 to 179) and there are no data on long-term outcomes. Finally, three studies compared VR with no intervention and it was expected that results in favor of VR would be observed.

4.7. Clinical Implications for Practice

VR interventions could be integrated into clinical practice to reduce pain intensity and kinesiophobia in patients with CLBP, with good results in the short and midterm followup. However, its effects on midterm followup have only been analyzed in a specific population of young sportsmen and cannot be generalized to the general population. Evidence for the efficacy of VR in disability associated with low back pain remains limited. Of the different types of VR, the Prokin System and horse simulator riding have obtained the best short-term results. However, only the studies using the Prokin System showed significant differences at midterm followup (6 months). In addition, this type of VR was combined with physiotherapy. Regarding the duration of the program, an intervention of 12 weeks showed the best results. However, interventions of 4 or 8 weeks also obtained significant results in favor of VR.

4.8. Future Research

None of the included studies assessed the variables at long-term followup so future research needs to focus on long-term effects. It may be interesting to conduct more studies comparing VR and physiotherapy versus physiotherapy due to the results obtained in other populations and the low quality of the studies included in this review. Prokin System and horse simulator riding showed good results in the treatment of CLBP. However, these devices are sophisticated and specialized and can be difficult to obtain for a clinic. Therefore, more studies would be necessary to explore the effects of Nintendo consoles in the treatment of CLBP. Although its results are inconclusive, it is commercially available and easier to implement in clinical practice. In addition, there is the possibility of it being used at home. Most of the studies included in this review have been conducted in adult patients under 30 years of age, and our best results were obtained in studies that only included young sportsmen. So, studies of similar quality in other types of populations are needed.

5. Conclusions

The results suggest that VR interventions can significantly reduce pain intensity and kinesiophobia in patients with CLBP after the intervention and at the 6 month followup. However, these studies showed high heterogeneity among them, influencing the consistency of the results. VR treatment showed the best results when it was compared with no intervention, placebo, or oral treatment in pain intensity postintervention. VR combined with physiotherapy versus no VR exercise and physiotherapy obtained significant differences in pain intensity and kinesiophobia postintervention and at the six month followup. Regarding VR systems, the Prokin System and horse simulator riding were the most effective short-term. Evidence of Nintendo consoles is still inconclusive, but they present some advantages, so more research is necessary. In terms of the duration of the program, 4, 8, or 12 week interventions showed good results. Studies are needed to evaluate the long-term effects of these interventions.

Author Contributions: Conceptualization, idea, and research design: B.B.-G., I.T.-S., A.O.-R., A.C.-M., L.L.-L., I.C.-M., M.C.V.; data collection: B.B.-G., I.T.-S.; data analysis: B.B.-G., I.T.-S.; writing—original draft preparation: B.B.-G., I.T.-S., A.O.-R., A.C.-M., L.L.-L., I.C.-M., M.C.V.; writing—review and editing: B.B.-G., I.T.-S., A.O.-R., A.C.-M., L.L.-L., I.C.-M., M.C.V. All authors have read and agreed to the published version of the manuscript.

Funding: This work was supported by the Spanish Ministry of Education by a FPU ("Formación Profesorado Universitario") grant for the authors Andrés Calvache-Mateo and Laura López-López (grant number, FPU:19/02609, FPU:17/00408).

Institutional Review Board Statement: Not applicable.

Informed Consent Statement: Not applicable.

Data Availability Statement: The data presented in this study are available in selected articles in the reference list.

Conflicts of Interest: The authors declare no conflict of interest.

Appendix A

Table A1. Search Strategy Studies.

Database	PubMed
Date	05/08/2021
Strategy	#1 AND #2
#1	("back pain"[Mesh] OR "back pain" OR "low back pain"[Mesh] OR "backache" OR "spine pain" OR "spinal pain" OR "lumbago" OR "sciatica")
#2	("Video Games"[Mesh] OR "video game*" OR "videogame*" OR "Gaming" OR "Game" OR "games" OR "Wii" OR "Nintendo" OR "Kinect" OR "Xbox" OR "PlayStation" OR "Virtual Reality"[Mesh] OR "virtual reality" OR "Virtual Reality Exposure Therapy"[Mesh] OR "exergame*" OR "gamification" OR "virtual" OR "computer-based" OR "augmented reality" OR "horse riding" OR "horseback" OR "hippotherapy simulator" OR "equine simulator")
Database	**Web of Science**
Date	05/08/2021
Strategy	#1 AND #2
#1	TS = ("back pain"[Mesh] OR "back pain" OR "low back pain"[Mesh] OR "backache" OR "spine pain" OR "spinal pain" OR "lumbago" OR "sciatica")
#2	TS = ("Video Games"[Mesh] OR "video game*" OR "videogame*" OR "Gaming" OR "Game" OR "games" OR "Wii" OR "Nintendo" OR "Kinect" OR "Xbox" OR "PlayStation" OR "Virtual Reality"[Mesh] OR "virtual reality" OR "Virtual Reality Exposure Therapy"[Mesh] OR "exergame*" OR "gamification" OR "virtual" OR "computer-based" OR "augmented reality" OR "horse riding" OR "horseback" OR "hippotherapy simulator" OR "equine simulator")
Database	**Scopus**
Date	07/08/2021
Strategy	#1 AND #2
#1	TITLE-ABS-KEY ("back pain" OR "low back pain" OR "backache" OR "spine pain" OR "spinal pain" OR "lumbago" OR "sciatica")
#2	TITLE-ABS-KEY ("video game*" OR "videogame*" OR "Gaming" OR "Game" OR "games" OR "Wii" OR "Nintendo" OR "Kinect" OR "Xbox" OR "PlayStation" OR "virtual reality" OR "Virtual Reality Exposure Therapy" OR "exergame*" OR "gamification" OR "virtual" OR "computer-based" OR "augmented reality" OR "horse riding" OR "horseback" OR "hippotherapy simulator" OR "equine simulator")
Database	**Cinahl**
Data	06/08/2021
Strategy	#1 AND #2
#1	AB ("back pain"[Mesh] OR "back pain" OR "low back pain"[Mesh] OR "backache" OR "spine pain" OR "spinal pain" OR "lumbago" OR "sciatica")
#2	AB ("Video Games"[Mesh] OR "video game*" OR "videogame*" OR "Gaming" OR "Game" OR "games" OR "Wii" OR "Nintendo" OR "Kinect" OR "Xbox" OR "PlayStation" OR "Virtual Reality"[Mesh] OR "virtual reality" OR "Virtual Reality Exposure Therapy"[Mesh] OR "exergame*" OR "gamification" OR "virtual" OR "computer-based" OR "augmented reality" OR "horse riding" OR "horseback" OR "hippotherapy simulator" OR "equine simulator")

Appendix B

Table A2. Search Strategy Ongoing Trials.

Database	ClinicalTrials.gov
Date	17/08/2021
Strategy	("back pain" OR "low back pain") AND ("video games" OR "virtual reality" OR "virtual reality exposure therapy") Filter: study type → interventional (clinical trial)
Database	ISRCTN registry
Date	17/08/2021
Strategy	"back pain" AND "virtual reality" "back pain" AND "virtual reality exposure therapy" "back pain" AND "video games" "low back pain" AND "virtual reality" "low pain" AND "virtual reality exposure therapy" "low back pain" AND "video games"
Database	ICTRP
Date	28/08/2021
Strategy	"back pain" AND "virtual reality" "back pain" AND "virtual reality exposure therapy" "back pain" AND "video games" "low back pain" AND "virtual reality" "low back pain" AND "virtual reality exposure therapy" "low back pain" AND "video games"

Appendix C

Table A3. Excluded Studies in the Last Screening with Reasons for Exclusion ($n = 44$).

Article	Reason for Exclusion
Virtual Environment Rehabilitation for Patients with Motor Neglect Trial (VERMONT): A Single-Center Randomized Controlled Feasibility Trial	No chronic low back pain
Home-Based Balance Training Using the Wii Balance Board	No chronic low back pain
Interactive Sections of an Internet-Based Intervention Increase Empowerment of Chronic Back Pain Patients: Randomized Controlled Trial	No chronic low back pain
Response latencies to postural disturbances when using a virtual reality balance trainer or wobble board in persons with low back pain	No chronic low back pain treatment
Feasibility, Acceptability and Effects of a Home-Based Exercise Program Using a Gerontechnology on Physical Capacities After a Minor Injury in Community-Living Older Adults: A Pilot Study	No chronic low back pain
Effectiveness of Trunk Balance Exercises and Wii Fit TM Balance Exercises in Managing Disability and Pain in Patients with Chronic Low Back Pain	Not randomized trial
Serious Gaming During Multidisciplinary Rehabilitation for Patients With Chronic Pain or Fatigue Symptoms: Mixed Methods Design of a Realist Process Evaluation	Not randomized trial
Examining virtual reality gaming for pain-related fear and disability in chronic low back pain	Meeting abstract
Using Virtual Reality to Treat Chronic Pain: Virtual Graded Exposure for Chronic Low Back Pain and Virtual Walking for Persistent Neuropathic Pain in Spinal Cord Injury	Meeting abstract
Cost effectiveness of virtual reality game versus clinic based mckenzie extension therapy for chronic non specific low back pain	Meeting abstract
Modulating body-image in people with chronic back pain using virtual reality	Meeting abstract
Preliminary Feasibility of a Graded, Locomotor-Enabled, Whole-Body Virtual Reality Intervention for Individuals with Chronic Low Back Pain	Not randomized trial
RabbitRun: An Immersive Virtual Reality Game for Promoting Physical Activities Among People with Low Back Pain dagger	Not randomized trial
Virtual Reality Serious Game for Musculoskeletal Disorder Prevention	Not randomized trial
Exploring the role of pain-related fear and catastrophizing in response to a virtual reality gaming intervention for chronic low back pain	Meeting abstract

Table A3. *Cont.*

Article	Reason for Exclusion
Effects of a Nintendo Wii exercise program versus Tai Chi Chuan on standing balance in older adults: a preliminary study	Not randomized trial
A novel, web-enabled multimedia approach, with 3D virtual reality internal and external human body tours, to support low back pain diagnosis	Not randomized trial
Low Back Pain Attenuation Employing Virtual Reality Physiotherapy	Not randomized trial
Mindfulness-based cognitive-behavior therapy (MCBT) versus virtual reality (VR) enhanced CBT, versus treatment as usual for chronic back pain. A clinical trial	Not randomized trial
A Portable Wireless Solution for Back Pain Telemonitoring: A 3D-Based, Virtual Reality Approach	Not randomized trial
Assessing the Perception of Trunk Movements in Military Personnel with Chronic Non-Specific Low Back Pain Using a Virtual Mirror	Not randomized trial
ALFRED Back Trainer: Conceptualization of a Serious Game-Based Training System for Low Back Pain Rehabilitation Exercises	Not randomized trial
A Virtual Reality Lower-Back Pain Rehabilitation Approach: System Design and User Acceptance Analysis	Not randomized trial
Efficacy of virtual reality to reduce chronic low back pain: Proof-of-concept of a non-pharmacological approach on pain, quality of life, neuropsychological and functional outcome	Not randomized trial
Proposed Game for Promoting Physical Activities among People with Low Back Pain using Virtual Reality	Not randomized trial
The influence of a biopsychosocial educational internet-based intervention on pain, dysfunction, quality of life, and pain cognition in chronic low back pain patients in primary care: a mixed methods approach	Not randomized trial
Tailored, multimedia versus traditional educational interventions for patients with low back pain: a randomized clinical trial.	No virtual reality intervention
Seeing It Helps: Movement-related Back Pain Is Reduced by Visualization of the Back During Movement	No virtual reality intervention
New exercise system for waist and back and its effect detection	No virtual reality intervention
Tele-rehabilitation for back pain in Korean farmers	No virtual reality intervention
Body schema acuity training and Feldenkrais RTM movements compared to core stabilization biofeedback and motor control exercises: Comparative effects on chronic non-specific low back pain in an outpatient clinical setting: A randomized controlled comparative study	No virtual reality intervention
Effect of Motor Control Training on Muscle Size and Football Games Missed from Injury	No virtual reality intervention
Randomized trial comparing interferential therapy with motorized lumbar traction and massage in the management of low back pain in a primary care setting	No virtual reality intervention
Self-Administered Skills-Based Virtual Reality Intervention for Chronic Pain: Randomized Controlled Pilot Study	Not only spinal pain
Effects of physiotherapy associated to virtual games in pain perception and heart rate variability in cases of low back pain	No chronic low back pain
Adherence to home exercises in non-specific low back pain. A randomised controlled pilot trial.	No chronic low back pain
Radiological (Magnetic Resonance Image and Ultrasound) and biochemical effects of virtual reality training on balance training in football players with chronic low back pain: A randomized controlled study.	No variables related to pain
The Effects of VR-based Wii Fit Yoga on PhysicalFunction in Middle-aged Female LBP Patients	No chronic low back pain
Evaluation of biofeedback based bridging exercises on older adults with low back pain: A randomized controlled trial	No chronic low back pain
Is physiotherapy integrated virtual walking effective on pain, function, and kinesiophobia in patients with non-specific low-back pain? Randomied controlled trial	No chronic low back pain
Effect of hippotherapy simulator on pain, disability and range of motion of the spinal column in subjects with mechanical low back pain: A randomized single-blind clinical trial	No chronic low back pain
A change in the size of the abdominal muscles and balance ability after virtual reality exercise in the elderly with chronic low back pain	No variables related to pain
Feasibility and Safety of a Virtual Reality Dodgeball Intervention for Chronic Low Back Pain: A Randomized Clinical Trial	Intervention duration < 4 weeks
Virtual reality distraction induces hypoalgesia in patients with chronic low back pain: a randomized controlled trial	Intervention duration < 4 weeks

Appendix D

Table A4. Characteristics of Included Registry Entries or Ongoing Trials (*n* = 17).

Number	Article	Recruitment Status
NCT02125968	Therapeutic Effects of Video Game Play Therapy on Patients With Chronic Low Back Pain	Unknown status
NCT03819907	The Use of Virtual Reality for Lumbar Pain Management in an Outpatient Spine Clinic	Completed
NCT04468074	Virtual Reality Treatment for Adults With Chronic Back Pain	Completed
NCT04042090	The Efficacy, Acceptability, Tolerability and Feasibility of a Therapeutic Virtual Reality Application	Active, not recruiting
NCT04273919	Virtual Reality for the Treatment of Chronic Low Back Pain	Recruiting
NCT04236804	Implementing TMC-CP01 Treatment Based on the Virtual Autonomic Neuromodulation Induced Systemic Healing System in Reducing Pain and Opioid Requirement in Subjects Suffering From Chronic Low Back Pain	Recruiting
NCT04307446	Immersive Virtual Reality and Chronic Back Pain	Recruiting
NCT04139564	EaseVRx for the Reduction of Chronic Pain and Opioid Use	Recruiting
NCT04225884	Digital Therapeutics (DTx) for Pain: Pilot Study of a Virtual Reality Software for Chronic Pain	Recruiting
NCT04609787	Immersive Virtual Reality and Central Sensitization in People With Chronic Pain	Recruiting
NCT03909048	Chronic Low Back Pain Graded - Exposure Psychoeducation Intervention	Completed
PACTR202010569932287	Comparative effects of augmented, virtual and mixed reality on pain characteristics and health-related quality of life of patients with chronic non-specific low-back pain.	Not recruiting
PACTR202007533977502	Comparative effects of clinic and virtual reality-based McKenzie extension therapy in chronic non-specific low-back pain.	Not recruiting
IRCT20200330046895N1	Effects of virtual reality exercises on clinical outcomes in patients with chronic low back pain: a randomized controlled trial.	Recruiting
ACTRN12619001776190	Altering body image in chronic low back pain using virtual reality: A proof of concept randomised clinical trial.	Not recruiting
PACTR201907749053096	Effects of Core Stability Exercise Combined with Virtual Reality in Collegiate Athletes with Nonspecific Low Back Pain: A Randomized Clinical Trial.	Not recruiting
JPRN-UMIN000035505	Effect of exercise by Virtual Reality in patients with chronic low back pain.	Not recruiting

Appendix E

Table A5. Methodological Quality of Included Studies.

Author (Year) [Ref.]	Study Quality										External Validity				Study Bias							Confounding and Selection Bias			Study Power			Total	Quality
	1	2	3	4	5	6	7	8	9	10	11	12	13	14	15	16	17	18	19	20	21	22	23	24	25	26	27		
Park et al. (2013) [26]	1	1	1	1	2	1	1	0	1	0	0	0	0	0	0	1	1	1	1	1	1	0	1	0	0	0	0	16	FAIR
Oh et al. (2014) [27]	1	1	1	1	2	1	0	0	0	1	1	0	0	0	0	1	1	1	0	1	1	0	1	0	0	0	0	15	FAIR
Yoo et al. (2014) [28]	1	1	1	1	2	1	0	0	1	1	1	0	0	0	0	1	1	0	1	1	1	0	1	0	0	1	0	17	FAIR
Monteiro-Junior et al. (2015) [32]	1	1	1	1	1	1	1	1	1	0	0	0	1	0	1	1	1	1	1	1	1	0	1	1	0	1	1	21	GOOD
Chen et al. (2016) [29]	1	1	1	1	0	0	1	0	0	1	0	0	0	0	0	1	1	1	0	1	1	0	1	0	0	0	1	13	POOR
Zadro et al. (2019) [7]	1	1	1	1	1	1	1	1	1	1	1	1	1	0	1	1	1	1	1	1	1	1	1	1	1	1	1	26	EXCELLENT
Kim et al. (2020) [30]	1	1	1	1	0	1	1	1	1	1	1	1	1	0	0	1	1	1	1	1	1	1	1	0	1	0	1	22	GOOD
Nambi et al. A (2020) [34]	1	1	1	1	2	1	1	0	1	1	1	1	0	1	1	1	1	1	1	1	1	0	1	1	1	1	1	25	GOOD
Nambi et al. B (2020) [35]	1	1	1	1	2	1	1	0	1	1	1	1	0	1	1	1	1	1	0	1	1	0	1	1	1	1	1	24	GOOD
Park et al. (2020) [31]	1	1	1	1	2	1	1	0	1	1	0	1	0	1	0	1	1	1	1	1	0	0	1	0	1	1	1	21	GOOD
Tomruk et al. (2020) [37]	1	1	1	1	2	1	1	0	1	1	0	0	0	0	0	1	1	1	1	1	1	0	1	0	0	0	1	18	FAIR
Garcia et al. (2021) [33]	1	1	1	1	2	1	1	1	1	1	1	1	1	1	0	1	1	1	1	1	1	1	1	1	1	1	1	27	EXCELLENT
Nambi et al. (2021) [36]	1	1	1	1	2	1	1	0	1	1	0	1	0	1	1	1	1	1	1	1	1	0	1	1	1	1	1	24	GOOD
Sato et al. (2021) [38]	1	1	1	1	2	1	1	0	1	1	1	0	1	0	0	1	1	1	1	1	1	1	0	0	1	1	1	22	GOOD

References

1. Hoy, D.; Bain, C.; Williams, G.; March, L.; Brooks, P.; Blyth, F.; Woolf, A.; Vos, T.; Buchbinder, R. A systematic review of the global prevalence of low back pain. *Arthritis Rheum.* **2012**, *64*, 2028–2037. [CrossRef] [PubMed]
2. Hartvigsen, J.; Hancock, M.; Kongsted, A.; Louw, Q.; Ferreira, M.L.; Genevay, S.; Hoy, D.; Karppinen, J.; Pransky, G.; Sieper, J.; et al. What low back pain is and why we need to pay attention. *Lancet* **2018**, *391*, 2356–2367. [CrossRef]
3. Hoy, D.; Brooks, P.; Blyth, F.; Buchbinder, R. The Epidemiology of low back pain. *Best Pr. Res. Clin. Rheumatol.* **2010**, *24*, 769–781. [CrossRef]
4. GBD 2015 Disease and Injury Incidence and Prevalence Collaborators. Global, regional, and national incidence, prevalence, and years lived with disability for 310 diseases and injuries, 1990–2015: A systematic analysis for the Global Burden of Disease Study 2015. *Lancet* **2016**, *388*, 1545–1602. [CrossRef]
5. Delitto, A.; George, S.Z.; Van Dillen, L.R.; Whitman, J.M.; Sowa, G.; Shekelle, P.; Denninger, T.R.; Godges, J.J. Low back pain. *J. Orthop. Sports Phys. Ther.* **2012**, *42*, 1–57. [CrossRef] [PubMed]
6. Chou, R.; Qaseem, A.; Snow, V.; Casey, D.; Cross, J.T.; Shekelle, P.; Owens, D.K. American Pain Society Low Back Pain Guidelines for the Clinical Efficacy Assessment Subcommittee of the American College of Physicians and the American College of Physicians/American Pain Society Low Back Pain Guidelines Panel. Diagnosis and Treatment of Low Back Pain: A Joint Clinical Practice Guideline from the American College of Physicians and the American Pain Society. *Ann. Intern. Med.* **2007**, *147*, 478–491. [CrossRef]
7. Zadro, J.; Shirley, D.; Simic, M.; Mousavi, S.J.; Ceprnja, D.; Maka, K.; Sung, J.; Ferreira, P. Video-Game–Based Exercises for Older People with Chronic Low Back Pain: A Randomized Controlledtable Trial (GAMEBACK). *Phys. Ther.* **2018**, *99*, 14–27. [CrossRef]
8. Kato, P.M. Video Games in Health Care: Closing the Gap. *Rev. Gen. Psychol.* **2010**, *14*, 113–121. [CrossRef]
9. Pereira, M.F.; Prahm, C.; Kolbenschlag, J.; Oliveira, E.; Rodrigues, N.F. Application of AR and VR in hand rehabilitation: A systematic review. *J. Biomed. Inform.* **2020**, *111*, 103584. [CrossRef]
10. De Miguel-Rubio, A.; Rubio, M.D.; Alba-Rueda, A.; Salazar, A.; Moral-Munoz, J.A.; Lucena-Anton, D. Virtual Reality Systems for Upper Limb Motor Function Recovery in Patients with Spinal Cord Injury: Systematic Review and Meta-Analysis. *JMIR mHealth uHealth* **2020**, *8*, e22537. [CrossRef]
11. Vlaeyen, J.W.; Linton, S.J. Fear-avoidance model of chronic musculoskeletal pain: 12 years on. *Pain* **2012**, *153*, 1144–1147. [CrossRef]
12. Matheve, T.; Bogaerts, K.; Timmermans, A. Virtual reality distraction induces hypoalgesia in patients with chronic low back pain: A randomized controlled trial. *J. Neuroeng. Rehabil.* **2020**, *17*, 1–12. [CrossRef]
13. Li, L.; Yu, F.; Shi, D.; Shi, J.; Tian, Z.; Yang, J.; Wang, X.; Jiang, Q. Application of virtual reality technology in clinical medicine. *Am. J. Transl. Res.* **2017**, *9*, 3867–3880.
14. Gumaa, M.; Youssef, A.R. Is Virtual Reality Effective in Orthopedic Rehabilitation? A Systematic Review and Meta-Analysis. *Phys. Ther.* **2019**, *99*, 1304–1325. [CrossRef] [PubMed]

15. Ahern, M.M.; Dean, L.V.; Stoddard, C.C.; Agrawal, A.; Kim, K.; Cook, C.E.; Garcia, A.N. The Effectiveness of Virtual Reality in Patients with Spinal Pain: A Systematic Review and Meta-Analysis. *Pain Pr.* **2020**, *20*, 656–675. [CrossRef] [PubMed]
16. Bordeleau, M.; Stamenkovic, A.; Tardif, P.-A.; Thomas, J. The Use of Virtual Reality in Back Pain Rehabilitation: A Systematic Review and Meta-Analysis. *J. Pain* **2021**, in press. [CrossRef] [PubMed]
17. Page, M.J.; McKenzie, J.E.; Bossuyt, P.M.; Boutron, I.; Hoffmann, T.C.; Mulrow, C.D.; Shamseer, L.; Tetzlaff, J.M.; Akl, E.A.; Brennan, S.E. The PRISMA 2020 statement: An updated guideline for reporting systematic reviews. *BMJ* **2021**, *372*, n71. [CrossRef]
18. Furlan, A.; Malmivaara, A.; Chou, R.; Maher, C.; A Deyo, R.; Schoene, M.L.; Bronfort, G.; van Tulder, M. 2015 Updated Method Guideline for Systematic Reviews in the Cochrane Back and Neck Group. *Spine* **2015**, *40*, 1660–1673. [CrossRef]
19. Downs, S.H.; Black, N. The feasibility of creating a checklist for the assessment of the methodological quality both of randomised and non-randomised studies of health care interventions. *J. Epidemiol. Community Health* **1998**, *52*, 377–384. [CrossRef]
20. Deeks, J.J.; Dinnes, J.; D'Amico, R.; Sowden, A.J.; Sakarovitch, C.; Song, F.; Petticrew, M.; Altman, D.G. Evaluating non-randomized intervention studies. *Health Technol. Assess.* **2003**, *7*, iii–173. [CrossRef]
21. Saunders, L.D.; Soomro, G.M.; Buckingham, J.; Jamtvedt, G.; Raina, P. Assessing the Methodological Quality of Nonrandomized Intervention Studies. *West. J. Nurs. Res.* **2003**, *25*, 223–237. [CrossRef] [PubMed]
22. Sánchez, I.T.; Salmerón, Y.M.; López, L.L.; Ortiz-Rubio, A.; Torres, J.R.; Valenza, M.C. Videogames in the Treatment of Obstructive Respiratory Diseases: A Systematic Review. *Games Health J.* **2019**, *8*, 237–249. [CrossRef]
23. Silverman, S.R.; Schertz, L.A.; Yuen, H.K.; Lowman, J.D.; Bickel, C.S. Systematic review of the methodological quality and outcome measures utilized in exercise interventions for adults with spinal cord injury. *Spinal Cord* **2012**, *50*, 718–727. [CrossRef] [PubMed]
24. Hooper, P.; Jutai, J.; Strong, G.; Russell-Minda, E. Age-related macular degeneration and low-vision rehabilitation: A systematic review. *Can. J. Ophthalmol.* **2008**, *43*, 180–187. [CrossRef]
25. Higgins, J.P.T.; Altman, D.G.; Gøtzsche, P.C.; Jüni, P.; Moher, D.; Oxman, A.D.; Savović, J.; Schulz, K.F.; Weeks, L.; Sterne, J.A.C.; et al. The Cochrane Collaboration's tool for assessing risk of bias in randomised trials. *BMJ* **2011**, *343*, d5928. [CrossRef]
26. Park, J.-H.; Lee, S.-H.; Ko, D.-S. The Effects of the Nintendo Wii Exercise Program on Chronic Work-related Low Back Pain in Industrial Workers. *J. Phys. Ther. Sci.* **2013**, *25*, 985–988. [CrossRef]
27. Oh, H.-W.; Lee, M.-G.; Jang, J.-Y.; Jin, J.-J.; Cha, J.-Y.; Jin, Y.-Y.; Jee, Y.-S. Time-effects of horse simulator exercise on psychophysiological responses in men with chronic low back pain. *Isokinet. Exerc. Sci.* **2014**, *22*, 153–163. [CrossRef]
28. Yoo, J.-H.; Kim, S.-E.; Lee, M.-G.; Jin, J.-J.; Hong, J.; Choi, Y.-T.; Kim, M.-H.; Jee, Y.-S. The effect of horse simulator riding on visual analogue scale, body composition and trunk strength in the patients with chronic low back pain. *Int. J. Clin. Pr.* **2014**, *68*, 941–949. [CrossRef] [PubMed]
29. Chen, S.-Y.; Kim, S.-K.; Kim, K.-H.; Lee, I.-S.; Hwangbo, G. Effects of Horse Riding Simulator on Pain, Oswestry Disability Index and Balance in Adults with Nonspecific Chronic Low Back Pain. *J. Korean Soc. Phys. Med.* **2016**, *11*, 79–84. [CrossRef]
30. Kim, T.; Lee, J.; Oh, S.; Kim, S.; Yoon, B. Effectiveness of Simulated Horseback Riding for Patients with Chronic Low Back Pain: A Randomized Controlled Trial. *J. Sport Rehabil.* **2020**, *29*, 179–185. [CrossRef]
31. Park, S.; Park, S.; Min, S.; Kim, C.-J.; Jee, Y.-S. A Randomized Controlled Trial Investigating the Effects of Equine Simulator Riding on Low Back Pain, Morphological Changes, and Trunk Musculature in Elderly Women. *Medicina* **2020**, *56*, 610. [CrossRef]
32. Monteiro-Junior, R.S.; De Souza, C.P.; Lattari, E.; Rocha, N.; Mura, G.; Machado, S.; Da Silva, E.B. Wii-Workouts on Chronic Pain, Physical Capabilities and Mood of Older Women: A Randomized Controlled Double Blind Trial. *CNS Neurol. Disord. Drug Targets* **2015**, *14*, 1157–1164. [CrossRef]
33. Garcia, L.M.; Birckhead, B.J.; Krishnamurthy, P.; Sackman, J.; Mackey, I.G.; Louis, R.G.; Salmasi, V.; Maddox, T.; Darnall, B.D. An 8-Week Self-Administered At-Home Behavioral Skills-Based Virtual Reality Program for Chronic Low Back Pain: Double-Blind, Randomized, Placebo-Controlled Trial Conducted During COVID-19. *J. Med Internet Res.* **2021**, *23*, e26292. [CrossRef] [PubMed]
34. Nambi, G.; AbdelBasset, W.K.; Elsayed, S.H.; Alrawaili, S.M.; Abodonya, A.M.; Saleh, A.K.; Elnegamy, T.E. Comparative Effects of Isokinetic Training and Virtual Reality Training on Sports Performances in University Football Players with Chronic Low Back Pain-Randomized Controlled Study. *Evid. Based Complement. Altern. Med.* **2020**, *2020*, 1–10. [CrossRef]
35. Nambi, G.; Abdelbasset, W.K.; Alrawaili, S.M.; Alsubaie, S.F.; Abodonya, A.M.; Saleh, A.K. Virtual reality or isokinetic training; its effect on pain, kinesiophobia and serum stress hormones in chronic low back pain: A randomized controlled trial. *Technol. Health Care* **2021**, *29*, 155–166. [CrossRef] [PubMed]
36. Nambi, G.; Abdelbasset, W.K.; Alsubaie, S.F.; Saleh, A.K.; Verma, A.; Abdelaziz, M.A.; Alkathiry, A.A. Short-Term Psychological and Hormonal Effects of Virtual Reality Training on Chronic Low Back Pain in Soccer Players. *J. Sport Rehabil.* **2021**, *30*, 884–893. [CrossRef] [PubMed]
37. Tomruk, M.S.; Kara, B.; Erbayraktar, R.S. The Effect of Computer-Based Training on Postural Control in Patients with Chronic Low Back Pain: A Randomized Controlled Trial. *J. Basic Clin. Health Sci.* **2020**, *4*, 329–334. [CrossRef]
38. Sato, T.; Shimizu, K.; Shiko, Y.; Kawasaki, Y.; Orita, S.; Inage, K.; Shiga, Y.; Suzuki, M.; Sato, M.; Enomoto, K.; et al. Effects of Nintendo Ring Fit Adventure Exergame on Pain and Psychological Factors in Patients with Chronic Low Back Pain. *Games Health J.* **2021**, *10*, 158–164. [CrossRef]
39. Bahat, H.S.; Croft, K.; Carter, C.; Hoddinott, A.; Sprecher, E.; Treleaven, J. Remote kinematic training for patients with chronic neck pain: A randomized controlled trial. *Eur. Spine J.* **2018**, *27*, 1309–1323. [CrossRef]
40. Rezaei, I.; Razeghi, M.; Ebrahimi, S.; Kayedi, S.; Zadeh, A.R. A Novel Virtual Reality Technique (Cervigame®) Compared to Conventional Proprioceptive Training to Treat Neck Pain: A Randomized Controlled Trial. *J. Biomed. Phys. Eng.* **2019**, *9*, 355–366. [CrossRef]

41. Bahat, H.S.; Takasaki, H.; Chen, X.; Bet-Or, Y.; Treleaven, J. Cervical kinematic training with and without interactive VR training for chronic neck pain—A randomized clinical trial. *Man. Ther.* **2015**, *20*, 68–78. [CrossRef] [PubMed]
42. Polat, M.; Kahveci, A.; Muci, B.; Günendi, Z.; Karataş, G.K. The Effect of Virtual Reality Exercises on Pain, Functionality, Cardiopulmonary Capacity, and Quality of Life in Fibromyalgia Syndrome: A Randomized Controlled Study. *Games Health J.* **2021**, *10*, 165–173. [CrossRef] [PubMed]
43. Gulsen, C.; Soke, F.; Eldemir, K.; Apaydin, Y.; Ozkul, C.; Guclu-Gunduz, A.; Akcali, D.T. Effect of fully immersive virtual reality treatment combined with exercise in fibromyalgia patients: A randomized controlled trial. *Assist. Technol.* **2020**, 1–8. [CrossRef]
44. Joo, Y.; Kim, E.-K.; Song, H.-G.; Jung, H.; Park, H.; Moon, A.J.Y. Effectiveness of virtual reality immersion on procedure-related pain and anxiety in outpatient pain clinic: An exploratory randomized controlled trial. *Korean J. Pain* **2021**, *34*, 304–314. [CrossRef]
45. McSherry, T.; Atterbury, M.; Gartner, S.; Helmold, E.; Searles, D.M.; Schulman, C. Randomized, Crossover Study of Immersive Virtual Reality to Decrease Opioid Use during Painful Wound Care Procedures in Adults. *J. Burn. Care Res.* **2017**, *39*, 278–285. [CrossRef]
46. Pazzaglia, C.; Imbimbo, I.; Tranchita, E.; Minganti, C.; Ricciardi, D.; Monaco, R.L.; Parisi, A.; Padua, L. Comparison of virtual reality rehabilitation and conventional rehabilitation in Parkinson's disease: A randomised controlled trial. *Physiotherapy* **2020**, *106*, 36–42. [CrossRef] [PubMed]
47. Santos, P.; Machado, T.; Santos, L.; Ribeiro, N.; Melo, A. Efficacy of the Nintendo Wii combination with Conventional Exercises in the rehabilitation of individuals with Parkinson's disease: A randomized clinical trial. *NeuroRehabilitation* **2019**, *45*, 255–263. [CrossRef]
48. Marques-Sule, E.; Arnal-Gómez, A.; Buitrago-Jiménez, G.; Suso-Martí, L.; Cuenca-Martínez, F.; Espí-López, G.V. Effectiveness of Nintendo Wii and Physical Therapy in Functionality, Balance, and Daily Activities in Chronic Stroke Patients. *J. Am. Med. Dir. Assoc.* **2021**, *22*, 1073–1080. [CrossRef]
49. Park, Y.-S.; An, C.-S.; Lim, C.-G. Effects of a Rehabilitation Program Using a Wearable Device on the Upper Limb Function, Performance of Activities of Daily Living, and Rehabilitation Participation in Patients with Acute Stroke. *Int. J. Environ. Res. Public Health* **2021**, *18*, 5524. [CrossRef]
50. Saeedi, S.; Ghazisaeedi, M.; Rezayi, S. Applying Game-Based Approaches for Physical Rehabilitation of Poststroke Patients: A Systematic Review. *J. Health Eng.* **2021**, *2021*, 1–27. [CrossRef] [PubMed]
51. Thomas, J.S.; France, C.R.; Applegate, M.E.; Leitkam, S.T.; Walkowski, S. Feasibility and Safety of a Virtual Reality Dodgeball Intervention for Chronic Low Back Pain: A Randomized Clinical Trial. *J. Pain* **2016**, *17*, 1302–1317. [CrossRef] [PubMed]
52. Collado-Mateo, D.; Lavín-Pérez, A.; García, J.F.; García-Gordillo, M.; Villafaina, S. Effects of Equine-Assisted Therapies or Horse-Riding Simulators on Chronic Pain: A Systematic Review and Meta-Analysis. *Medicina* **2020**, *56*, 444. [CrossRef] [PubMed]
53. Ren, C.; Liu, T.; Zhang, J. Horse-riding simulators in treatment of chronic low back pain: A meta-analysis. *Int. J. Clin. Pr.* **2021**, *75*, e14198. [CrossRef]

International Journal of
Environmental Research and Public Health

Review

Methods Used to Evaluate mHealth Applications for Cardiovascular Disease: A Quasi-Systematic Scoping Review

Felix Holl [1,2,*], Jennifer Kircher [1], Walter J. Swoboda [1] and Johannes Schobel [1]

[1] DigiHealth Institute, Neu-Ulm University of Applied Sciences, 89231 Neu-Ulm, Germany;
Jennifer.Kircher@student.hnu.de (J.K.); walter.swoboda@hnu.de (W.J.S.); johannes.schobel@hnu.de (J.S.)

[2] Institute for Medical Information Processing, Biometry, and Epidemiology, Ludwig Maximilian University of Munich, 81377 Munich, Germany

* Correspondence: felix.holl@hnu.de; Tel.: +49-731-9762-1613

Abstract: In the face of demographic change and constantly increasing health care costs, health care system decision-makers face ever greater challenges. Mobile health applications (mHealth apps) have the potential to combat this trend. However, in order to integrate mHealth apps into care structures, an evaluation of such apps is needed. In this paper, we focus on the criteria and methods of evaluating mHealth apps for cardiovascular disease and the implications for developing a widely applicable evaluation framework for mHealth interventions. Our aim is to derive substantiated patterns and starting points for future research by conducting a quasi-systematic scoping review of relevant peer-reviewed literature published in English or German between 2000 and 2021. We screened 4066 articles and identified $n = 38$ studies that met our inclusion criteria. The results of the data derived from these studies show that usability, motivation, and user experience were evaluated primarily using standardized questionnaires. Usage protocols and clinical outcomes were assessed primarily via laboratory diagnostics and quality-of-life questionnaires, and cost effectiveness was tested primarily based on economic measures. Based on these findings, we propose important considerations and elements for the development of a common evaluation framework for professional mHealth apps, including study designs, data collection tools, and perspectives.

Keywords: mobile health; cardiovascular diseases; evaluation methods

Citation: Holl, F.; Kircher, J.; Swoboda, W.J.; Schobel, J. Methods Used to Evaluate mHealth Applications for Cardiovascular Disease: A Quasi-Systematic Scoping Review. *Int. J. Environ. Res. Public Health* **2021**, *18*, 12315. https://doi.org/10.3390/ijerph182312315

Academic Editors: Irene Torres-Sanchez and Marie Carmen Valenza

Received: 20 October 2021
Accepted: 20 November 2021
Published: 23 November 2021

Publisher's Note: MDPI stays neutral with regard to jurisdictional claims in published maps and institutional affiliations.

1. Introduction

In 2019, over 331,000 deaths in Germany were attributed to cardiovascular disease (CVD) [1], the treatment of which generates higher medical costs to the German healthcare system than any other single illness, estimated at € 46.4 billion in 2015 [2]. Similarly, in the US, CVD is among the most expensive and most frequent causes of death among the population [3]. Kvedar et al. [4] pointed out the urgent need to develop, optimize, and evaluate programs and technologies that ensure more effective care for patients, where mobile health (mHealth) concepts are likely to play a significant role [5]. The World Health Organization defines mHealth as "Medical and public health practice supported by mobile devices, such as mobile phones, patient monitoring devices, personal digital assistants (PDAs), and other wireless devices" [6].

The 2019 German Digital Healthcare Act (DVG) permitted mobile health applications (mHealth apps)that meet specific requirements to be included the list of reimbursable digital health applications (DiGA list) [7]. Germany is one of the first countries to introduce a standardized mechanism for reimbursing digital health services and its healthcare and medical insurance policy-makers are still working through several challenges. For example, the DiGA list only includes mHealth apps classified as medical devices as defined in the Medical Devices Act administered by the German Federal Institute for Drugs and Medical Devices (BfArM) [8]. While other professional mHealth apps, such as medication reminders or prevention apps, demonstrate both medical benefit and positive care effects, they remain ineligible for reimbursement.

Beyond narrowly defined medical devices, the data and treatment results provided by other professional mHealth apps require equally stringent assessment to ensure reliably high-quality care. Notably, there is currently no established and broadly applicable framework for evaluating mHealth interventions [9].

As a step toward filling this gap, this study examines the criteria and methods for evaluating mHealth interventions for cardiovascular disease discussed in the published literature as a basis for developing a more broadly applicable framework.

2. Materials and Methods

In this study, we conducted a quasi-systematic scoping review of methods and criteria used to evaluate cardiovascular disease mHealth apps in the published literature. In a preliminary scoping review, we identified gaps in the literature and synthesized key concepts in a narrative review [10]. Then, in an iterative process, we scoped the literature with refined search terms, performing a final quasi-systematic search with fixed search terms [11].

2.1. Preliminary Scoping Review

We conducted a preliminary scoping review of articles of mHealth apps for CVD through an unstructured and open search to generate an overview of existing methods of evaluating mHealth apps for CVD [12] and to confirm the validity of our research objective. The results of this review informed the development of our final search strategy and analysis.

2.2. Inclusion and Exclusion Criteria

Our preliminary scoping review revealed various apps designed to reduce the users' risk of developing cardiovascular disease. These apps focus mainly on reduction and control of risk factors for CVD, such as diabetes, hypertension, chronic obstructive pulmonary disease, nutrition, and physical activity. Based on these results, we derived inclusion and exclusion criteria for the subsequent quasi-systematic scoping review of publications in German and English evaluating mHealth apps designed for adult patients diagnosed with acquired cardiovascular disease. Table A1 in the Appendix A provides a complete overview of our inclusion and exclusion criteria.

2.3. Search Strategy

Our final search followed a quasi-systematic approach. We searched the "PubMed", "Livivo", and "ProQuest" databases to identify relevant literature published between 2000 and the beginning of April 2021. The last search took place on 6 April 2021. Using keywords and index terms relevant to cardiovascular disease, mHealth, and evaluation, we developed search strings, which we adjusted for each database. Table A2 in the Appendix A provides a list of our search terms.

2.4. Literature Selection

In selecting suitable literature, we applied the Preferred Reporting Items for Systematic Reviews and Meta-Analyses (PRISMA) scheme [13]. The process steps and the results of the study selection are illustrated in Figure 1 below.

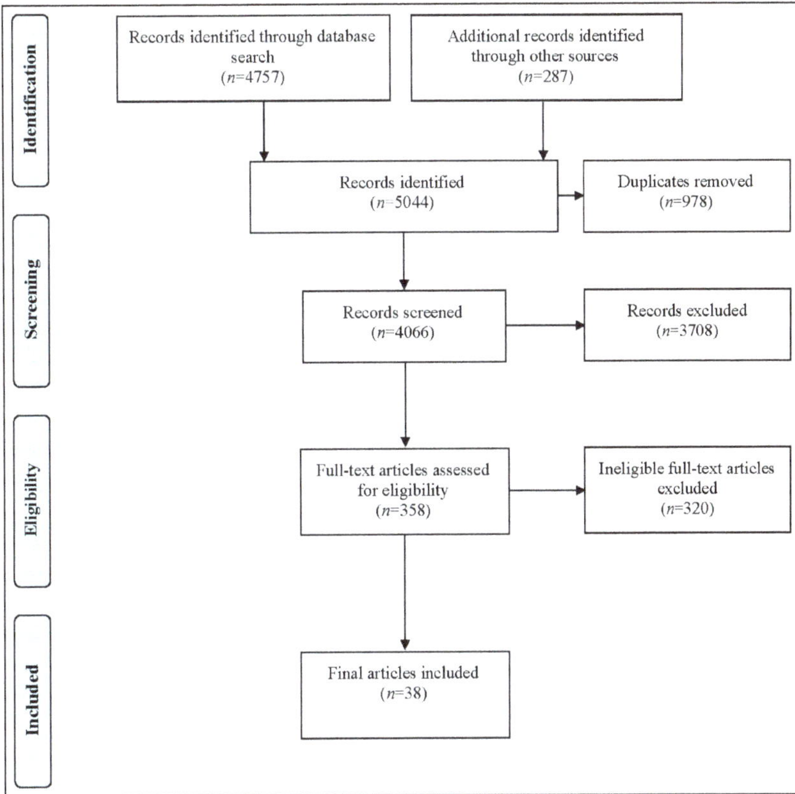

Figure 1. PRISMA flow diagram of the study.

After importing our 5044 records into *Covidence*, we excluded 978 duplicates. Then, two scholars independently screened the titles and abstracts of the remaining 4066 entries to identify adherence to previously defined inclusion and exclusion criteria. After resolving inconsistencies by consensus, 3708 studies were excluded. We then undertook a full-text review of the remaining 358 articles, excluding an additional 320 studies because they failed to meet our inclusion criteria. Many of the articles we excluded were study protocols, focused on apps designed only to prevent risk factors, such as high blood pressure or diabetes apps, or assessed apps that rely on implanted sensor technology. Our final sample of $n = 38$ articles was included in the scoping review and approved for data extraction.

2.5. Data Extraction and Analysis

In a next step, we extracted data from the studies according to variables, in order to sort and map the literature to reveal patterns, key information, and research gaps in a data chart for subsequent evaluation. The data extraction sheet was developed by two authors based on the findings of the preliminary scoping review and adapted as part of the iterative process to ensure all relevant information from the studies were captured and included in the analysis. To identify evaluation approaches and criteria, we classified the studies into three categories. Interventions carried out using only an app are classified as "mHealth app"; interventions using an app plus at least one additional device, such as an electrocardiogram or smartwatch, are classified as "mHealth system"; and interventions using only text messages are classified as "mHealth text messaging". Table A3 in the Appendix A summarizes the extracted information as a data chart.

3. Results

3.1. Characteristics of the Identified Studies

All articles included in our study were published between 2012 and 2020, even though our search spanned 2000 to April 2021. One-third of the articles were published by scholars in the US ($n = 13$), 13% by scholars in Australia, and 10% by scholars in China. Studies with quantitative and qualitative research designs were included in our review. The largest proportion ($n = 18$) consists of randomized controlled trials (RCTs), followed by single-arm prospective studies and mixed-methods studies (each $n = 7$). Figure 2 illustrates the frequency of study designs.

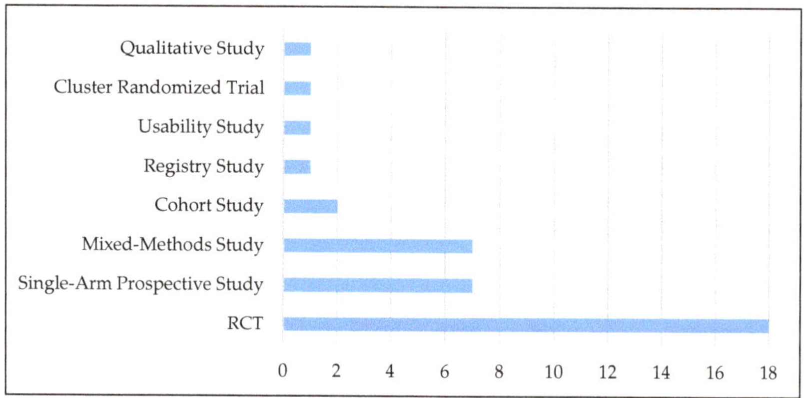

Figure 2. Study designs of the studies identified.

Four of the studies [14–17] lasted over 12 months, while the shortest study lasted 2 weeks [18]. The largest study had 767 participants [18], while the smallest study had 8 participants [19]. Just over half (57.9%) of the studies reported a retention rate (RR) (the percentage of study participants who remained in the study until the defined end of the study process) of between 90% and 100%, while only four studies [20–23] reported an RR of below 50%. For analysis purposes, we also tracked the corresponding loss to follow-up (LTFU) (the percentage of study participants who drop out of a study before the defined end of the study process) figure for each study.

Just over half (52.6%) of the studies focused on mHealth systems (app plus device). The context includes applications for telemonitoring ($n = 12$) as well as for cardiac rehabilitation (CR) ($n = 8$). Seven studies in the mHealth apps (app only) category focused on self-management applications and five focused on CR. In contrast, the smallest share (15.7%) of studies focused on text messaging for self-management purposes (mHealth text messaging category).

3.2. Methods and Measurements for Evaluating mHealth Technologies

The studies followed qualitative, quantitative, and mixed-methods designs and the great majority ($n = 31$) analyzed data collected through standardized questionnaires. In most cases ($n = 33$), the overall aim of the research was to assess participants' perceptions of treatment and subjective health. In addition to general questionnaires on quality of life (e.g., "EQ-5D" [15], "health-related quality of life" [15], illness (e.g., "Self-Care of Heart Failure Index" [24]) or the psychological well-being of the patients (e.g., "8-item Morisky Medication Adherence Scale" [25,26], "Hospital Anxiety and Depression Scale" [20]), specific question sets for digital applications were also used. The Mobile Application Rating Scale (MARS) was frequently applied in assessing mHealth apps [27]. The "Perceived Health Web Site Usability Questionnaire" (PHWSUQ) [28] specifically addresses assessing the usability of websites among elderly participants [29]. Each questionnaire appeared once in the analysis [18,28]. In addition to standardized question sets, self-defined questionnaires (*n* = 3), interviews (*n* = 5), and open-feedback rounds (*n* = 7) were conducted to determine perceptions.

A large proportion of the publications (63%) evaluated mHealth interventions using medical measurements (e.g., blood pressure, pulse, weight), comparing health parameters before and after the intervention. The results were often compared directly between the standard of care and the mHealth intervention (*n* = 15). The medical outcomes were used to assess, among others, the feasibility of the intervention (*n* = 16) and physical activity (*n* = 21). The measurements were either documented by the participants using the mHealth device or determined by healthcare providers using monitoring data or laboratory diagnostics.

Interactions with the mHealth app on the part of patients (*n* = 19) and health care providers (*n* = 2) were often recorded in usage protocols (*n* = 19) used to draw conclusions about participants' motivation (*n* = 17), adherence (*n* = 18), and self-efficacy (*n* = 14). In mHealth apps for CR, usage data and logging activities related to login-ins, training, or learning modules were analyzed [30,31]. In one study of an mHealth system for medication adherence [32], the number of times two electronic pill bottles were opened was documented using timestamps.

The usability of mHealth interventions (*n* = 14) was evaluated using several measurement methods and instruments, such as the PHWSUQ and the "System Usability Scale" [33]. A theoretical basis was used in two studies [34,35] to develop the intervention and measure usability. One study adapted the Unified Theory of Acceptance and Use of Technology 2 (UTAUT2) to measure various factors influencing mHealth intervention technology use behavior [36]. In another study [34], the practice of mHealth was prompted by the responsible intervention team as part of a usability test.

Over one-third of the studies (*n* = 14) investigated the effectiveness and efficiency of mHealth for new clinical treatments. Several studies relied on various key performance indicators (KPIs) in assessing mHealth effectiveness (*n* = 11), including, most frequently, hospital readmission, length of hospital stay, number of doctor visits, and hospital admittance due to heart defects. Less attention was paid to mortality and personnel resources required for monitoring. Two studies [37,38] undertook cost-effectiveness analyses. A small number of studies used application-specific indicators, such as data management [38,39], communication between users [38,40], app features [18,41], design characteristics [42], or technology and algorithm analyses [43].

4. Discussion

The integration of mHealth apps into healthcare structures is a relatively young field of investigation: the analysis shows that the oldest two studies [14,24] date back less than 10 years, probably due to relatively recent and rapid developments in mobile technologies. The relevance of the research topic of mHealth systems and their evaluation is supported by the large number of publications that we found, and a large body of research exists for health applications for certain manageable illnesses and conditions, such as diabetes, high blood pressure, and obesity-related health problems. Most of the studies included in the analysis were randomized controlled trials, thus providing high-quality evidence-based results and high proof of efficacy [44].

4.1. Patient Empowerment in mHealth Interventions for CR

Overall, our results show that mHealth interventions for cardiac rehabilitation (CR) can be used to reduce or manage coronary heart disease (CHD) and potentially contribute to secondary prevention by empowering heart attack survivors to monitor their risk factors themselves and act accordingly. We find that by using self-management functions, patients can participate actively in their care process and take more responsibility for their health [45]. We thus identify self-efficacy and motivation as key indicators for evaluating mHealth interventions and in an evaluation framework. This recommendation underscores Schwab et al.'s discussion of approaches to developing mHealth applications and the importance they attribute to increasing awareness and empowerment among patients and healthcare professionals [46].

4.2. Usage Behavior and Motivation

Our results show that the retention rate and LTFU are suitable measures of motivation and commitment among mHealth intervention users. The fact that more than half of the studies identified had a very high retention rate indicates an overall positive perception of mHealth interventions among users. Our results indicate that usage protocols provide reliable insights into usability, acceptance, and user motivation levels. We also identify the benefits of adapting the Unified Theory of Acceptance and Use of Technology 2" (UTAUT2) to fit the mHealth application use context: the modified construct includes seven factors influencing intention to use a telemonitoring system, together with the independent variables age, gender, and experience influencing the factors.

4.3. Quantitative and Qualitative Research Methods

While both quantitative and qualitative research methods can be used to collect data, almost all included studies use standardized validated questionnaires and scales, enabling the analysis and comparison of large samples and yielding comparable quantifiable results. Using validated tools is cost and time efficient [47]. Since quantitative research methods often allow little room to interpret the questions, the research framework should include open questions, such as semi-structured interviews or focus groups [48]. Our results illustrate the benefits of combing quantitative and qualitative research methods, particularly in assessing patient satisfaction with the intervention.

4.4. Quality Assessment

The Mobile Application Rating Scale (MARS) [18] has been used as an instrument to assess the quality of mHealth apps according to the following quality indicators: engagement, functionality, aesthetics, information quality, and subjective app quality [27]. Terhorst et al. [49] demonstrated the suitability and validity of these indicators and recommended using the instrument to increase transparency for stakeholders and patients. While an mHealth intervention evaluation framework should include app quality criteria, the quality assessment should not be limited to subjective user feedback but rather should include data quality and interoperability with other devices and interfaces.

4.5. Privacy and Data Security

Data security and privacy are important to patients and legally protected. Schnall et al. [50] found a decrease in trust in mHealth solutions and data transfer over time and Zhou et al. [51] showed that some patients refuse to use mHealth applications because of security concerns, loss of interest, or hidden costs. Despite these concerns, our results show that little attention has been paid to data management, such as data transfer between health care providers and participants, data privacy, and data security. An mHealth app evaluation framework should assess the app's data protection systems carefully and communicate the results transparently.

4.6. Economic Evaluation

Performance measures, such as hospital readmissions, are an important indicator of the effectiveness and efficiency of mHealth systems and should be included in an evaluation framework as well. In the CR mHealth intervention context, our results show that mHealth apps can reduce heart failure-related hospital days and studies conducting cost-effectiveness analysis underscore that shortening out- and inpatient stays also cuts healthcare costs [52]. Similarly, Maddison et al.'s [37] post-hoc economic evaluation assessed the costs of implementing and delivering the intervention to calculate the incremental cost-effectiveness ratio (ICER) between costs and quality-adjusted life years (QALYs) gained and to compare the health benefit gains of switching from standard in- and outpatient care to mHealth-supported care. The authors found that mHealth interventions are more cost-effective compared to the standard care and can improve health-related quality of life in an ongoing program. Martín et al. applied a "Hidden Markov Model" to measure cost-effectiveness. Long-term costs and outcomes associated with an illness and a particular health intervention can be estimated over multiple cycles, based on resource use and health outcomes [53]. Martín et al.'s [38] study modeled the different disease states of patients during the mHealth intervention, using economic parameters for the outcome analysis and aligning participants' health-specific and follow-up data with healthcare costs published by the health care system. Their cost-effectiveness analysis model showed that introducing an mHealth app lowered the overall cost of disease management by 33% of the total cost of disease management [38]. Pavlović et al.'s [54] results are equally striking: introducing mHealth apps can reduce the total expenses related to data collection in medical scenarios by 50%.

5. Conclusions

Our scoping review of scholarly articles including criteria and methods of evaluating mHealth apps for cardiovascular disease makes recommendations for developing an evaluation framework for mHealth interventions. In keeping with recent research on the health benefits of active patient involvement in their treatment process, we recommend adopting a user perspective. While various methods and criteria have been used, we recommend quantitative methods using validated standardized questionnaires to generate comparable quantifiable results with a reasonable effort in terms of time commitment and cost. In addition to considering the overall effects of mHealth apps on mental and physical health, we recommend that mHealth intervention evaluations apply usage protocols to understand the patients' interaction with the application and assess their motivation, engagement, and acceptance of integrating the interventions into healthcare processes sustainably. We also recommend including the retention rate and LTFUs, and adapting use and acceptance constructs, such as UTAUT2, into the mHealth technology use setting by modifying its assessment dimensions accordingly.

In terms of the overall scope and considerations for the development of an mHealth app evaluation framework, we recommend focusing on the added value of an mHealth intervention. Specifically, we recommend laboratory diagnostics and physical tests to assess objective physical health, standardized surveys and semi-structured interviews to assess subjective quality of life, and economic performance and efficiency KPIs, such as hospital readmission data and incremental cost-effectiveness ratios between costs and quality-adjusted life years. Heterogeneity of results by using different standardized surveys and questionnaires could be a major challenge for the analysis and comparisons of the results from such a framework. Therefore, the selection of data collection tools needs to be made carefully.

mHealth app providers, patients, healthcare providers, healthcare systems, and society at large will benefit by applying these recommendations when developing a holistic framework to evaluate mHealth apps and interventions to ensure that they are effective, efficient, empowering, accurate, sustainable, and safe. Such a framework will enable an informed decision when choosing an mHealth app.

Author Contributions: Conceptualization, F.H., methodology, J.K. and F.H.; software, J.K. and F.H.; validation, F.H.; formal analysis, J.K.; investigation, F.H.; resources, J.K.; data curation, J.K.; writing—original draft preparation, J.K.; writing—review and editing, F.H., W.J.S. and J.S.; visualization, J.K.; supervision, J.S. and W.J.S. All authors have read and agreed to the published version of the manuscript.

Funding: This research received no external funding.

Institutional Review Board Statement: Not applicable.

Informed Consent Statement: Not applicable.

Data Availability Statement: All relevant data can be found in the Appendix A.

Acknowledgments: The authors would like to thank Timo Guter for his guidance and support during the literature search.

Conflicts of Interest: The authors declare no conflict of interest.

Appendix A

Table A1. Inclusion and exclusion criteria based on the PCC elements.

PCC Elements	Inclusion Criteria	Exclusion Criteria
Population	Patients (>18 years) with a diagnosed CHD No limitation of the number of participants, origin, gender of the study participants	Patients who are at risk of coronary heart disease Relatives of cardiovascular patients, e.g., children Comorbid heart disease (e.g., congenital heart defect, heart transplant) Healthy, voluntary study participants
Concept	*mHealth Application* Wearable mHealth applications for patients with CHD Studies using qualitative or quantitative methods to evaluate mHealth applications (e.g., standardized questionnaires, quality guidelines, device data sets, usage logs) No limitation of the evaluation parameters Fully developed mHealth applications *Study Design* Single study designs for evaluating a mHealth intervention for patients with CHD Written in English or German	mHealth applications for the use of exclusively: Risk factors (e.g., high blood pressure) Diabetes Chronic Obstructive Pulmonary Disease Pregnancy Nutrition assessment (e.g., food tracking) Sport and Wellness Sensor technology (e.g., implanted sensor) Applications that are only designed for health care providers, e.g., Clinical Assessment Tool A risk screening tool of CHD for the population Pure descriptions of the apps (e.g., system, technical, program, algorithm description) Study protocols Preliminary studies (e.g., for the development of the app) Reviews (e.g., systematic reviews, scoping reviews) Case studies
Context	No limitation of cultural parameters (e.g., geographical location, social origin, gender-specific interests) No restriction of the setting, e.g., acute care, primary care, rehabilitation facilities Full texts	Unpublished literature

Table A2. Search strings and number of results.

Database	Search String	Search Date	Results (*n*)
PubMed	Heart Disease* OR Cardiovascular Disease* AND "Mobile Health" OR "mHealth" OR Smartphone App* AND Evaluation	5 January 2021	2916
Livivo	Cardiovascular disease AND mHealth OR mobile health app AND evaluation	13 January 2021	485
Proquest	(mHealth OR "mobile health" app) AND Evaluation AND cardiovascular disease	13 January 2021	1356
Total records			**4757**
Pubmed	+ Additional studies from reference lists of 37 systematic reviews	6 April 2021	287
Total records generated by search			**5044**

Table A3. Extracted data of the 38 studies included in the analysis.

Country [Ref]	Setting	Type of Intervention	Study Design	Type(s) of Evaluation	Evaluation Indicators	Evaluation Methods
Canada [24]	Home-based and hospital	mHealth system devices: mobile phone, weight scale, blood pressure monitor, ECG recordings	RCT Sample size $n = 100$ Duration: 6 months Retention rate: 94% Loss to follow-up: 6	Feasibility Medical Outcomes Comparison with standard of care Utilization Clinical Management Quality of Life Effectiveness/ Efficiency	Clinical endpoints Physical well-being Health parameters (BP, weight, ECG) Hospital KPIs application: Patient perception /feedback Clinicians' interaction	Medical measurements Standardized questionnaires Collection of hospital KPI data
USA [34]	Home-based	mHealth app	Usability study Sample: $n = 15$ Duration: - Retention rate: 87% Loss to follow-up: 2	Acceptability Usability Medical outcome Self-efficacy	Clinical endpoints: Physical activity Application: Task completion success Mobile technology use Patients' interaction	Interviews Standardized questionnaires Open feedback Usability testing Guidance by UTAUT2 construct
USA [30]	Home-based and cardiac rehabilita-tion	mHealth system devices: app, monitoring dashboard	Single-arm prospective study Sample: $n = 18$ Duration: 3 months Retention rate: 72% Loss to follow-up: 5	Feasibility Engagement Acceptability Medical outcome	Clinical endpoints: Health parameters (BP, functional capacity, safety) Application: Patients' interaction with app Patient percep-tion/feedback	Open feedback Usage logs
Belgium [31]	Home-based and cardiac rehabilita-tion	mHealth app	Mixed-methods study Sample: $n = 32$ Duration: 4 months Retention rate: 88% Loss to follow-up: 4	Comparison of usual care Engagement Effectiveness Usefulness Medical outcome Quality of life	Clinical endpoints: Physical activity Health parameters Application: Patients' percep-tion/feedback Patients' interaction	Interviews Standardized questionnaires Medical measurements Usage logs
China [39]	Home-based	mHealth app	Cluster randomized trial Sample: $n = 209$ Duration: 3 months Retention rate: 80% Loss to follow-up: 42	Usability Feasibility Acceptability Medical outcome Safety accuracy/consistency Quality of life Self-efficacy	Clinical endpoints: Health parameters Psychological well-being Application: Patients' percep-tion/feedback Knowledge Data management	Open Feedback Medical measurements Standardized questionnaires Questionnaires (self-defined) Collection of cointervention data (medical outcome data)
USA [55]	Home-based and hospital	mHealth system devices: wireless ECG, app	Cohort study Sample: $n = 46$ Duration: 6 months Retention rate: 76% Loss to follow-up: 11	Comparison of usual care Feasibility Quality of life Medical outcome Self-efficacy	Clinical endpoints: Physical and psychological well-being Health parameters (ECG) Application: Patient percep-tion/feedback	Standardized questionnaires Medical measurements Usability testing
USA [40]	Home-based and hospital	mHealth system devices: tablet, Bluetooth-weight scale, pulse wave blood pressure wrist monitor	Mixed-methods study Sample: $n = 28$ Duration: 3 months Retention rate: 89% Loss to follow-up: 3	Feasibility Comparison of usual care Usability Acceptability Medical outcome Clinical management Self-efficacy	Clinical endpoints: Health parameters Physical well-being Physical activity Application: Adherence Patients' percep-tion/feedback Clinicians' interaction	Standardized questionnaires Medical measurements Interviews

Table A3. *Cont.*

Country [Ref]	Setting	Type of Intervention	Study Design	Type(s) of Evaluation	Evaluation Indicators	Evaluation Methods
USA [41]	Home-based	mHealth app	RCT Sample: $n = 60$ Duration: one month Retention rate: 92% Loss to follow-up: 5	Comparison of telehealth Medication adherence Feasibility Quality of life Acceptability Self-efficacy	Clinical endpoints: psychological and physical well-being Application: App features Patients' interaction	Questionnaires (self-defined) Usage logs
New Zealand [56]	Home-based	mHealth system devices: mobile phone, device for internet support	RCT Sample: $n = 171$ Duration: 6 months Retention rate: 92% Loss to follow-up: 14	Medical outcome Self-efficacy	Clinical endpoints: Physical well-being Physical activity (leisure-time and walking) Health parameters	Standardized questionnaires Medical measurements
USA [42]	Home-based	mHealth app	Mixed-methods study Sample: $n = 12$ Duration: one month Retention rate: 92% Loss to follow-up: 1	Feasibility Usability Quality of life Self-efficacy Acceptability Effectiveness/efficacy Medical outcome	Clinical endpoints: Health parameters Hospital KPIs Application: Patient perception/feedback Message characteristics	Open feedback Medical measurements Standardized questionnaires Collection of hospital KPI data
Australia [35]	Home-based	mHealth system devices: app, tracking tools (accelerometer, wrist-worn Fitbit Flex), web-based program	Cohort Study Sample: $n = 21$ Duration: 4 months Retention rate: 62% Loss to follow-up: 8	Feasibility Usability Medical outcome Self-efficacy Quality of life Medical outcome	Clinical endpoints: Health parameters Physical activity Psychological well-being Application: Mobile Technology Use Patient perception/Feedback Resource Requirements Patients' interaction	Medical measurements Standardized questionnaires Usage logs
USA [16]	Home-based	mHealth—Text messaging	RCT Sample: $n = 84$ Duration: 12 months Retention rate: 99% Loss to follow-up: 1	Comparison of usual care Medication adherence	Clinical endpoints: Physical well- Physical activity Application: Patients' interaction	Usage logs Medical measurements Questionnaire
USA [57]	Home-based and hospital	mHealth system devices: apps, bp cuff, scale, dashboard, medicine software platform	Registry study Sample: $n = 60$ Duration: one month Retention rate: 97% Loss to follow-up: 2	Feasibility Acceptability Effectiveness/efficacy Medical outcome	Clinical endpoints: Health parameters Hospital KPIs Application: Patients' interaction	Collection of hospital KPI data Usage logs
Australia [23]	Home-based	mHealth app	RCT Sample: $n = 166$ Duration: 3 months Retention rate: 92% Loss to follow-up: 14	Medication adherence Feasibility Comparison of usual care Adherence Acceptability Medical outcome	Clinical endpoints: Health parameters Application: Patient perception/feedback	Standardized questionnaires Open feedback Medical measurements
Malaysia [25]	Home-based	mHealth -text messaging	RCT Sample: $n = 62$ Duration: 2 months Retention rate: 97% Loss to follow-up: 2	Medication adherence Medical outcome Effectiveness/efficacy	Clinical endpoints: Health parameters Hospital KPIs Application: Patient perception/feedback	Medical measurements Standardized Questionnaires Collection of Hospital KPIs data

Table A3. *Cont.*

Country [Ref]	Setting	Type of Intervention	Study Design	Type(s) of Evaluation	Evaluation Indicators	Evaluation Methods
USA [32]	Home-based	mHealth system devices: mobile phone, electronic pillbox, web-based platform	RCT Sample: n = 90 Duration: one month Retention rate: 93% Loss to follow-up: 6	Medication adherence Feasibility Acceptability Comparison of usual care Usability	Application: Patient perception/feedback Patients' interaction	Standardized Questionnaires Usage logs
USA [58]	Home-based and hospital	mHealth system devices: tablet, blood pressure cuff, weight scale, web-based platform	Single-arm prospective study Sample: n = 21 Duration: 3.2 months Retention rate: 95% Loss to follow-up: 1	Engagement Effectiveness/efficacy Acceptability Feasibility Usability (incl. ease of use) Quality of life Medical outcome	Clinical endpoints: Health parameters Hospital KPIs Application: Patient perception/feedback Patients' interaction	Questionnaires (self-defined) Medical measurements Usage logs Collection of hospital KPIs data Standardized questionnaires
Norway [33]	Home-based and cardiac rehabilitation	mHealth app	Single-arm prospective study Sample: n = 14 Duration: 3 months Retention rate: 100% Loss to follow-up: 0	Feasibility Quality of life Usability Effectiveness/efficacy	Clinical endpoints: Physical well-being Hospital KPIs Application: Patient perception/feedback Patient satisfaction Adherence Patients' interaction	Standardized questionnaires Open feedback Usage logs Collection of hospital KPIs data
New Zealand [37]	Home-based	mHealth System Devices: mobile phone, web-based platform	RCT Sample: n = 171 Duration: 6 months Retention rate: 89% Loss to follow-up: 18	Comparison of usual care Effectiveness Self-efficacy Engagement Medical outcome Quality of life Economic outcome	Clinical endpoints: Physical activity Health parameters Application: Patient perception/feedback Cost and Cost-effectiveness	Medical measurements Standardized questionnaires Economic measurements
Norway [15]	Home-based and cardiac rehabilitation	mHealth app	RCT Sample: n = 113 Duration: 12 months Retention rate: 98% Loss to follow-up: 2	Comparison of usual care Medical outcome Quality of life	Clinical endpoints: Health parameters Application: Patient perception/feedback Patient satisfaction	Medical measurements Standardized questionnaires
France [59]	Home-based	mHealth—text messaging	RCT Sample: n = 521 Duration: one month Retention rate: 96% Loss to follow-up: 22	Medication adherence Comparison of usual care	Clinical endpoints: Health parameters Application: Patient perception/feedback	Open feedback Medical measurements
China [28]	Home-based and hospital	mHealth system devices: apps, smart tracking devices (bp cuff, weight scale, wearable ECG), remote monitoring service platform	Single-arm prospective study Sample: n = 70 Duration: 4 months Retention rate: 94% Loss to follow-up: 4	Usability Medical outcome Satisfaction Engagement Feasibility	Clinical endpoints: Physical activity Health parameters Application: Mobile Technology Use Patient perception/feedback Health care provider experience Relatives' experience Patients' interaction	Interviews Standardized questionnaires Usage logs Medical record entries Medical measurements

Table A3. *Cont.*

Country [Ref]	Setting	Type of Intervention	Study Design	Type(s) of Evaluation	Evaluation Indicators	Evaluation Methods
Netherlands [60]	Home-based and hospital	mHealth system devices: app, weight scale, blood pressure monitor, rhythm monitor, step counter	RCT Sample: $n = 200$ Duration: - Retention rate: 90% Loss to follow-up: 20	Medical outcome Feasibility Satisfaction Effectiveness/efficacy Comparison of usual care	Clinical endpoints: Health parameters Hospital KPIs Application: Patients' interaction Patient perception/feedback	Medical measurements Standardized questionnaires Collection of hospital KPIs data Medical record entries Usage logs
Canada [61]	Home-based and hospital	mHealth system devices: app, weight scales, blood pressure monitors	Single-arm prospective study Sample: $n = 315$ Duration: 6 months Retention rate: 90% Loss to follow-up: 30	Quality of life Effectiveness/efficacy Medical outcome Self-care	Clinical endpoints: Hospital KPIs Health parameters Application: Patient perception/feedback	Collection of hospital KPIs data Standardized questionnaires Medical measurements
USA [21]	Home-based and cardiac rehabilitation	mHealth app	Qualitative Study Sample: $n = 16$ Duration: 2.2 months Retention rate: 25% Loss to follow-up: 12	Feasibility Acceptability Medical outcome Medication adherence Engagement Effectiveness/efficacy	Clinical endpoints: Health parameters Physical activity Hospital KPIs Application: Patients' interaction Patient perception/feedback	Medical measurement Usage logs Collection of hospital KPIs data
China [19]	Home-based	mHealth—text messaging	RCT Sample: $n = 767$ Duration: 6.4 months Retention rate: 95% Loss to follow-up: 37	Effectiveness/Efficacy Quality of life Self-efficacy Medication adherence	Clinical endpoints: Hospital KPIs Health parameters Application: Patient perception/feedback	Collection of hospital KPIs data Standardized questionnaires
USA [62]	Home-based	mHealth system	RCT Sample: $n = 90$ Duration: one month Retention rate: 93% Loss to follow-up: 6	Medication adherence Self-efficacy	Clinical endpoints: Psychological well-being Application: Patients' interaction Patient perception/feedback	Standardized questionnaires Usage logs
Spain [38]	Home-based	mHealth app	RCT Sample: $n = 630$ Duration: - Retention rate: 86% Loss to follow-up: 86	Economic outcome Engagement Quality of life Efficacy	Application: Cost-effectiveness Patient satisfaction Data management Communication	Economic measurements
Australia [18]	Home-based	mHealth app	Mixed-methods study Sample: $n = 8$ Duration: between 2 and 4 weeks Retention rate: 75% Loss to follow-up: 2	Usability	Clinical endpoints: Physical activity Application: Patient perception/feedback App features Mobile technology use	Standardized questionnaires Interviews
Canada [17]	Home-based and hospital	mHealth system devices: app, weight scales, blood pressure monitors	Mixed-methods study Sample: $n = 231$ Duration: 12 months Retention rate: 87% Loss to follow-up: 30	Usability Adherence Engagement Medical outcome	Clinical endpoints: Health parameters Application: Mobile technology use Adherence Patients' interaction Patient perception/Feedback	Guidance by UTAUT2 construct interviews Usage logs Standardized questionnaire Medical measurements

Table A3. *Cont.*

Country [Ref]	Setting	Type of Intervention	Study Design	Type(s) of Evaluation	Evaluation Indicators	Evaluation Methods
China [63]	Home-based and hospital	mHealth—text messaging	Mixed-methods study Sample: $n = 190$ Duration: 3 months Retention rate: 93% Loss to follow-up: 13	Feasibility Usability Acceptability Medication adherence Economic outcome	Clinical endpoints: Physical activity Application: Patient satisfaction Patient perception/feedback costs	Standardized questionnaires Open feedback Economic measurements
Israel [64]	Home-based and cardiac rehabilitation	mHealth system devices: mobile phone, smartwatch, monitoring system	Single-arm prospective study Sample: $n = 22$ Duration: 6 months Retention rate: 100% Loss to follow-up: 0	Feasibility Safety Adherence Effectiveness/efficacy Medical outcome Usability	Clinical endpoints: Physical activity Hospital KPIs Health parameters Application: Patient satisfaction Patients' interaction Patient perception/Feedback	Collection of hospital KPIs data Medical measurements Usage logs Standardized questionnaires
Norway [20]	Home-based	mHealth system devices: mobile phone, web-based platform	RCT Sample: $n = 69$ Duration: 3 months Retention rate: 28% Loss to follow-up: 50	Comparison of usual care Usability Self-efficacy Adherence	Clinical endpoints: Physical activity Psychological well-being Application: Patients' interaction Patient perception/Feedback	Standardized questionnaires Usage logs
Australia [43]	Home-based and cardiac rehabilitation	mHealth system devices: app, blood pressure monitor, weight scale, web-based platform	RCT Sample: $n = 66$ Duration: 6 months Retention rate: 77% Loss to follow-up: 15	Medical outcome Feasibility Security	Clinical endpoints: Physical activity Health parameters Psychological well-being Application: Technology and algorithm	Medical measurement Standardized questionnaires
New Zealand [65]	Home-based and cardiac rehabilitation	mHealth system devices: mobile phone, web-based platform, pedometer	RCT Sample: $n = 123$ Duration: 6 months Retention rate: 94% Loss to follow-up: 7	Comparison of usual care Medical outcome Medication adherence Self-efficacy Acceptancy	Clinical endpoints: Physical activity Psychological well-being Health parameters Application: Patient perception/feedback	Standardized questionnaire Open feedback Guidance following on the mHealth development and evaluation framework
Australia [23]	Home-based	mHealth App	Mixed-methods study Sample: $n = 58$ Duration: 3 months Retention rate: 26% Loss to follow-up: 43	Comparison of usual care Medication adherence Acceptability Utilization Engagement	Clinical endpoints: Health parameters Application: Patient perception/feedback Patients' interaction	Standardized questionnaire Usage logs Open feedback
Spain [14]	Home-based and cardiac rehabilitation	mHealth system devices: mobile phone, web-based platform, sphygmomanometer, glucose, and lipid meter	RCT Sample: $n = 203$ Duration: 12 months Retention rate: 90% Loss to follow-up: 21	Usefulness Medical outcome Quality of life	Clinical endpoints: Health parameters Psychological well-being Application: Patient perception/feedback	Medical measurement Standardized questionnaires
USA [22]	Home-based	mHealth—text messaging	Single-arm prospective study Sample: $n = 15$ Duration: one month Retention rate: 40% Loss to follow-up: 9	Feasibility Acceptability Medication adherence Adherence Engagement	Application: Patient perception/feedback Patient satisfaction Patients' interaction	Usage logs Standardized questionnaires

References

1. Statistisches Bundesamt (Destatis). Gestorbene: Deutschland, Jahre, Todesursachen, Geschlecht. Available online: https://www-genesis.destatis.de/genesis/online?sequenz=tabelleErgebnis&selectionname=23211-0002#abreadcrumb (accessed on 12 July 2021).
2. Nier, H. Herz-Kreislauf-Erkrankungen Verursachen Höchste Kosten. Available online: https://de.statista.com/infografik/11301/herz-kreislauf-erkrankungen-verursachen-hoechste-kosten/ (accessed on 12 July 2021).
3. Virani, S.S.; Alonso, A.; Aparicio, H.J.; Benjamin, E.J.; Bittencourt, M.S.; Callaway, C.W.; Carson, A.P.; Chamberlain, A.M.; Cheng, S.; Delling, F.N.; et al. Heart Disease and Stroke Statistics-2021 Update: A Report from the American Heart Association. *Circulation* **2021**, *143*, e254–e743. [CrossRef] [PubMed]
4. Kvedar, J.; Coye, M.J.; Everett, W. Connected health: A review of technologies and strategies to improve patient care with telemedicine and telehealth. *Health Aff.* **2014**, *33*, 194–199. [CrossRef] [PubMed]
5. Tomlinson, M.; Rotheram-Borus, M.J.; Swartz, L.; Tsai, A.C. Scaling up mHealth: Where is the evidence? *PLoS Med.* **2013**, *10*, e1001382. [CrossRef] [PubMed]
6. World Health Organization (WHO). mHealth: New Horizons for Health through Mobile Technologies: Second Global Survey on eHealth, Global Observatory for eHealth. World Health Organization. Available online: https://apps.who.int/iris/handle/10665/44607 (accessed on 12 July 2021).
7. Das Fast-Track-Verfahren Für Digitale Gesundheitsanwendungen (Diga) Nach § 139e SGB V Ein Leitfaden Für Hersteller, Leistungserbringer Und Anwender. Available online: https://www.bfarm.de/SharedDocs/Downloads/DE/Medizinprodukte/diga_leitfaden.pdf?__blob=publicationFile (accessed on 12 July 2021).
8. Verordnung (EU) 2017/745 Des Europäischen Parlaments Und Des Rates. Available online: https://lexparency.de/eu/MDR/ (accessed on 12 July 2021).
9. Holl, F.; Swoboda, W. Methods to Measure the Impact of mHealth Applications: Preliminary Results of a Scoping Review. *Stud. Health Technol. Inform.* **2018**, *251*, 285–288.
10. Anderson, S.; Allen, P.; Peckham, S.; Goodwin, N. Asking the right questions: Scoping studies in the commissioning of research on the organisation and delivery of health services. *Health Res. Policy Syst.* **2008**, *6*, 7. [CrossRef]
11. Arksey, H.; O'Malley, L. Scoping studies: Towards a methodological framework. *Int. J. Soc. Res. Methodol.* **2005**, *8*, 19–32. [CrossRef]
12. Holl, F.; Kircher, J.; Swoboda, W. Evaluation Methods Used to Assess mHealth Applications for Cardiovascular Disease: First Results of a Scoping Review. *Stud. Health Technol. Inform.* **2021**, *281*, 1083–1084. [CrossRef]
13. Moher, D.; Liberati, A.; Tetzlaff, J.; Altman, D.G. Preferred reporting items for systematic reviews and meta-analyses: The PRISMA statement. *PLoS Med.* **2009**, *6*, e1000097. [CrossRef]
14. Blasco, A.; Carmona, M.; Fernández-Lozano, I.; Salvador, C.H.; Pascual, M.; Sagredo, P.G.; Somolinos, R.; Muñoz, A.; García-López, F.; Escudier, J.M.; et al. Evaluation of a telemedicine service for the secondary prevention of coronary artery disease. *J. Cardiopulm. Rehabil. Prev.* **2012**, *32*, 25–31. [CrossRef]
15. Lunde, P.; Bye, A.; Bergland, A.; Grimsmo, J.; Jarstad, E.; Nilsson, B.B. Long-term follow-up with a smartphone application improves exercise capacity post cardiac rehabilitation: A randomized controlled trial. *Eur. J. Prev. Cardiol.* **2020**, *27*, 1782–1792. [CrossRef]
16. Pandey, A.; Krumme, A.; Patel, T.; Choudhry, N. The impact of text messaging on medication adherence and exercise among postmyocardial infarction patients: Randomized controlled pilot trial. *J. Med. Internet Res. mHealth uHealth* **2017**, *5*, e7144. [CrossRef]
17. Ware, P.; Dorai, M.; Ross, H.J.; Cafazzo, J.A.; Laporte, A.; Boodoo, C.; Seto, E. Patient adherence to a Mobile phone–based heart failure Telemonitoring program: A longitudinal mixed-methods study. *J. Med. Internet Res. mHealth uHealth* **2019**, *7*, e13259. [CrossRef]
18. Woods, L.S.; Duff, J.; Roehrer, E.; Walker, K.; Cummings, E. Patients' experiences of using a consumer mhealth app for self-management of heart failure: Mixed-methods study. *J. Med. Internet Res. Hum. Factors* **2019**, *6*, e13009. [CrossRef]
19. Chen, C.; Li, X.; Sun, L.; Cao, S.; Kang, Y.; Hong, L.; Liang, Y.; You, G.; Zhang, Q. Post-discharge short message service improves short-term clinical outcome and self-care behaviour in chronic heart failure. *Eur. Soc. Cardiol. Heart Fail.* **2019**, *6*, 164–173. [CrossRef]
20. Antypas, K.; Wangberg, S.C. An Internet-and mobile-based tailored intervention to enhance maintenance of physical activity after cardiac rehabilitation: Short-term results of a randomized controlled trial. *J. Med. Internet Res.* **2014**, *16*, e77. [CrossRef]
21. Layton, A.M.; Whitworth, J.; Peacock, J.; Bartels, M.N.; Jellen, P.A.; Thomashow, B.M. Feasibility and acceptability of utilizing a smartphone based application to monitor outpatient discharge instruction compliance in cardiac disease patients around discharge from hospitalization. *Int. J. Telemed. Appl.* **2014**, *2014*, 17. [CrossRef]
22. Nundy, S.; Razi, R.R.; Dick, J.J.; Smith, B.; Mayo, A.; O'Connor, A.; Meltzer, D.O. A text messaging intervention to improve heart failure self-management after hospital discharge in a largely African-American population: Before-after study. *J. Med. Internet Res.* **2013**, *15*, e53. [CrossRef]
23. Santo, K.; Singleton, A.; Chow, C.K.; Redfern, J. Evaluating reach, acceptability, utility, and engagement with an app-based intervention to improve medication adherence in patients with coronary heart disease in the MedApp-CHD study: A mixed-methods evaluation. *Med. Sci.* **2019**, *7*, 68. [CrossRef]

24. Seto, E.; Leonard, K.J.; Cafazzo, J.A.; Barnsley, J.; Masino, C.; Ross, H.J. Mobile phone-based telemonitoring for heart failure management: A randomized controlled trial. *J. Med. Internet Res.* **2012**, *14*, e31. [CrossRef]
25. Khonsari, S.; Subramanian, P.; Chinna, K.; Latif, L.A.; Ling, L.W.; Gholami, O. Effect of a reminder system using an automated short message service on medication adherence following acute coronary syndrome. *Eur. J. Cardiovasc. Nurs.* **2015**, *14*, 170–179. [CrossRef]
26. Santo, K.; Singleton, A.; Rogers, K.; Thiagalingam, A.; Chalmers, J.; Chow, C.K.; Redfern, J. Medication reminder applications to improve adherence in coronary heart disease: A randomised clinical trial. *Heart* **2019**, *105*, 323–329. [CrossRef] [PubMed]
27. Stoyanov, S.R.; Hides, L.; Kavanagh, D.J.; Zelenko, O.; Tjondronegoro, D.; Mani, M. Mobile App Rating Scale: A New Tool for Assessing the Quality of Health Mobile Apps. *JMIR mHealth uHealth* **2015**, *3*, e27. [CrossRef] [PubMed]
28. Guo, X.; Gu, X.; Jiang, J.; Li, H.; Duan, R.; Zhang, Y.; Sun, L.; Bao, Z.; Shen, J.; Chen, F. A Hospital-Community-Family–Based Telehealth Program for Patients With Chronic Heart Failure: Single-Arm, Prospective Feasibility Study. *J. Med. Internet Res. mHealth uHealth* **2019**, *7*, e13229. [CrossRef] [PubMed]
29. Nahm, E.S.; Resnick, B.; Mills, M.E. Development and pilot-testing of the perceived health Web Site usability questionnaire (PHWSUQ) for older adults. *Stud. Health Technol. Inform.* **2006**, *122*, 38–43.
30. Harzand, A.; Witbrodt, B.; Davis-Watts, M.L.; Alrohaibani, A.; Goese, D.; Wenger, N.K.; Shah, A.J.; Zafari, A.M. Feasibility of a smartphone-enabled cardiac rehabilitation program in male veterans with previous clinical evidence of coronary heart disease. *Am. J. Cardiol.* **2018**, *122*, 1471–1476. [CrossRef]
31. Sankaran, S.; Dendale, P.; Coninx, K. Evaluating the impact of the HeartHab app on motivation, physical activity, quality of life, and risk factors of coronary artery disease patients: Multidisciplinary crossover study. *J. Med. Internet Res. mHealth uHealth* **2019**, *7*, e10874. [CrossRef]
32. Park, L.G.; Howie-Esquivel, J.; Chung, M.L.; Dracup, K. A text messaging intervention to promote medication adherence for patients with coronary heart disease: A randomized controlled trial. *Patient Educ. Couns.* **2014**, *94*, 261–268. [CrossRef]
33. Lunde, P.; Nilsson, B.B.; Bergland, A.; Bye, A. Feasibility of a mobile phone app to promote adherence to a heart-healthy lifestyle: Single-arm study. *J. Med. Internet Res. Form. Res.* **2019**, *3*, e12679. [CrossRef]
34. Beatty, A.L.; Magnusson, S.L.; Fortney, J.C.; Sayre, G.G.; Whooley, M.A. VA FitHeart, a mobile app for cardiac rehabilitation: Usability study. *J. Med. Internet Res. Hum. Factors* **2018**, *5*, e3. [CrossRef]
35. Freene, N.; van Berlo, S.; McManus, M.; Mair, T.; Davey, R. A behavioral change smartphone APP and program (ToDo-CR) to decrease sedentary behavior in cardiac rehabilitation participants: Prospective feasibility cohort study. *J. Med. Internet Res. Form. Res.* **2020**, *4*, e17359. [CrossRef]
36. Venkatesh, V.; Thong, J.Y.L.; Xu, X. Consumer Acceptance and Use of Information Technology: Extending the Unified Theory of Acceptance and Use of Technology. *MIS Q.* **2012**, *36*, 157–178. [CrossRef]
37. Maddison, R.; Pfaeffli, L.; Whittaker, R.; Stewart, R.; Kerr, A.; Jiang, Y.; Kira, G.; Leung, W.; Dalleck, L.; Carter, K.; et al. A mobile phone intervention increases physical activity in people with cardiovascular disease: Results from the HEART randomized controlled trial. *Eur. J. Prev. Cardiol.* **2015**, *22*, 701–709. [CrossRef]
38. Martín, J.A.C.; Martínez-Pérez, B.; de la Torre-Díez, I.; López-Coronado, M. Economic impact assessment from the use of a mobile app for the self-management of heart diseases by patients with heart failure in a Spanish region. *J. Med. Syst.* **2014**, *38*, 1–7. [CrossRef]
39. Guo, Y.; Chen, Y.; Lane, D.A.; Liu, L.; Wang, Y.; Lip, G.Y.H. Mobile health technology for atrial fibrillation management integrating decision support, education, and patient involvement: mAF app trial. *Am. J. Med.* **2017**, *130*, 1388–1396.e6. [CrossRef]
40. Lefler, L.L.; Rhoads, S.J.; Harris, M.; Funderburg, A.E.; Lubin, S.A.; Martel, I.D.; Faulkner, J.L.; Rooker, J.L.; Bell, D.K.; Marshall, H.; et al. Evaluating the use of mobile health technology in older adults with heart failure: Mixed-methods study. *J. Med. Internet Res. Aging* **2018**, *1*, e12178. [CrossRef]
41. Goldstein, C.M.; Gathright, E.C.; Dolansky, M.A.; Gunstad, J.; Sterns, A.; Redle, J.D.; Josephson, R.; Hughes, J.W. Randomized controlled feasibility trial of two telemedicine medication reminder systems for older adults with heart failure. *J. Telemed. Telecare* **2014**, *20*, 293–299. [CrossRef]
42. Heiney, S.P.; Donevant, S.B.; Adams, S.A.; Parker, P.D.; Chen, H.; Levkoff, S. A smartphone app for self-management of heart failure in older African Americans: Feasibility and usability study. *J. Med. Internet Res. Aging* **2020**, *3*, e17142. [CrossRef]
43. Del Rosario, M.B.; Lovell, N.H.; Fildes, J.; Holgate, K.; Yu, J.; Ferry, C.; Schreier, G.; Ooi, S.Y.; Redmond, S.J. Evaluation of an mHealth-based adjunct to outpatient cardiac rehabilitation. *Inst. Electr. Electron. Eng. J. Biomed. Health Inform.* **2017**, *22*, 1938–1948. [CrossRef]
44. Evidence-Based Medicine Working Group. Evidence-based medicine. A new approach to teaching the practice of medicine. *JAMA* **1992**, *268*, 2420–2425. [CrossRef]
45. Albrecht, U.-V. Chancen und Risiken von Gesundheits-Apps. In *Recht & Netz*, 1st ed.; Albers, M., Katsivelas, I., Eds.; Nomos Verlagsgesellschaft mbH & Co. KG: Baden-Baden, Germany, 2018; pp. 417–430.
46. Schwab, J.D.; Schobel, J.; Werle, S.D.; Fürstberger, A.; Ikonomi, N.; Szekely, R.; Thiam, P.; Hühne, R.; Jahn, N.; Schuler, R.; et al. Perspective on mHealth Concepts to Ensure Users' Empowerment–From Adverse Event Tracking for COVID-19 Vaccinations to Oncological Treatment. *IEEE Access* **2021**, *9*, 83863–83875. [CrossRef]
47. SurveyMonkey. Der standardisierte Fragebogen SurveyMonkey. Available online: https://www.surveymonkey.de/mp/der-standardisierte-fragebogen/ (accessed on 13 July 2021).

48. SurveyMonkey. Der Unterschied zwischen quantitativer und qualitativer Forschung. SurveyMonkey. Available online: https://www.surveymonkey.de/mp/quantitative-vs-qualitative-research/ (accessed on 13 July 2021).
49. Terhorst, Y.; Philippi, P.; Sander, L.B.; Schultchen, D.; Paganini, S.; Bardus, M.; Santo, K.; Knitza, J.; Machado, G.C.; Schoeppe, S.; et al. Validation of the Mobile Application Rating Scale (MARS). *PLoS ONE* **2020**, *15*, e0241480. [CrossRef] [PubMed]
50. Schnall, R.; Higgins, T.; Brown, W.; Carballo-Dieguez, A.; Bakken, S. Trust, Perceived Risk, Perceived Ease of Use and Perceived Usefulness as Factors Related to mHealth Technology Use. *Stud. Health Technol. Inform.* **2015**, *216*, 467–471. [PubMed]
51. Zhou, L.; Bao, J.; Watzlaf, V.; Parmanto, B. Barriers to and Facilitators of the Use of Mobile Health Apps From a Security Perspective: Mixed-Methods Study. *JMIR Mhealth Uhealth* **2019**, *7*, e11223. [CrossRef] [PubMed]
52. Carbo, A.; Gupta, M.; Tamariz, L.; Palacio, A.; Levis, S.; Nemeth, Z.; Dang, S. Mobile Technologies for Managing Heart Failure: A Systematic Review and Meta-analysis. *Telemed. J. E Health* **2018**, *24*, 958–968. [CrossRef]
53. Briggs, A.; Sculpher, M. An introduction to Markov modelling for economic evaluation. *Pharmacoeconomics* **1998**, *13*, 397–409. [CrossRef]
54. Pavlović, I.; Kern, T.; Miklavcic, D. Comparison of paper-based and electronic data collection process in clinical trials: Costs simulation study. *Contemp. Clin. Trials* **2009**, *30*, 300–316. [CrossRef]
55. Hickey, K.T.; Biviano, A.B.; Garan, H.; Sciacca, R.R.; Riga, T.; Warren, K.; Frulla, A.P.; Hauser, N.R.; Wang, D.Y.; Whang, W. Evaluating the utility of mHealth ECG heart monitoring for the detection and management of atrial fibrillation in clinical practice. *J. Atr. Fibrillation* **2017**, *9*, 1546. [CrossRef]
56. Maddison, R.; Pfaeffli, L.; Stewart, R.; Kerr, A.; Jiang, Y.; Rawstorn, J.; Carter, K.; Whittaker, R. The HEART mobile phone trial: The partial mediating effects of self-efficacy on physical activity among cardiac patients. *Front. Public Health* **2014**, *2*, 56. [CrossRef]
57. Park, C.; Otobo, E.; Ullman, J.; Rogers, J.; Fasihuddin, F.; Garg, S.; Kakkar, S.; Goldstein, M.; Chandrasekhar, S.V.; Pinney, S.; et al. Impact on readmission reduction among heart failure patients using digital health monitoring: Feasibility and adoptability study. *J. Med. Internet Res. Med. Inform.* **2019**, *7*, e13353. [CrossRef]
58. Zan, S.; Agboola, S.; Moore, S.A.; Parks, K.A.; Kvedar, J.C.; Jethwani, K. Patient engagement with a mobile web-based telemonitoring system for heart failure self-management: A pilot study. *J. Med. Internet Res. mHealth uHealth* **2015**, *3*, e3789. [CrossRef]
59. Quilici, J.; Fugon, L.; Beguin, S.; Morange, P.E.; Bonnet, J.L.; Alessi, M.C.; Carrieri, P.; Cuisset, T. Effect of motivational mobile phone short message service on aspirin adherence after coronary stenting for acute coronary syndrome. *Int. J. Cardiol.* **2013**, *168*, 568–569. [CrossRef]
60. Treskes, R.W.; van Winden, L.A.; van Keulen, N.; van der Velde, E.T.; Beeres, S.L.; Atsma, D.E.; Schalij, M.J. Effect of smartphone-enabled health monitoring devices vs regular follow-up on blood pressure control among patients after myocardial infarction: A randomized clinical trial. *J. Am. Med. Assoc. Netw. Open* **2020**, *3*, e202165. [CrossRef]
61. Ware, P.; Ross, H.J.; Cafazzo, J.A.; Boodoo, C.; Munnery, M.; Seto, E. Outcomes of a heart failure telemonitoring program implemented as the standard of care in an outpatient heart function clinic: Pretest-posttest pragmatic study. *J. Med. Internet Res.* **2020**, *22*, e16538. [CrossRef]
62. Park, L.G.; Howie-Esquivel, J.; Whooley, M.A.; Dracup, K. Psychosocial factors and medication adherence among patients with coronary heart disease: A text messaging intervention. *Eur. J. Cardiovasc. Nurs.* **2015**, *14*, 264–273. [CrossRef]
63. Chen, S.; Gong, E.; Kazi, D.S.; Gates, A.B.; Bai, R.; Fu, H.; Peng, W.; De La Cruz, G.; Chen, L.; Liu, X.; et al. Using mobile health intervention to improve secondary prevention of coronary heart diseases in China: Mixed-methods feasibility study. *J. Med. Internet Res. mHealth uHealth* **2018**, *6*, e7849. [CrossRef]
64. Nabutovsky, I.; Ashr, S.; Nachshon, A.; Tesler, R.; Shapiro, Y.; Wright, E.; Vadasz, B.; Offer, A.; Grosman-Rimon, L.; Klempfner, R. Feasibility, Safety, and Effectiveness of a Mobile Application in Cardiac Rehabilitation. *Isr. Med. Assoc. J. IMAJ* **2020**, *22*, 357–363.
65. Dale, L.P.; Whittaker, R.; Jiang, Y.; Stewart, R.; Rolleston, A.; Maddison, R. Text message and internet support for coronary heart disease self-management: Results from the Text4Heart randomized controlled trial. *J. Med. Internet Res.* **2015**, *17*, e4944.

International Journal of
*Environmental Research
and Public Health*

MDPI

Review

Open Issues and Practical Suggestions for Telemedicine in Chronic Pain

Marco Cascella [1], Franco Marinangeli [2], Alessandro Vittori [3,*], Cristina Scala [4], Massimo Piccinini [5], Alessandro Braga [6], Luca Miceli [7] and Renato Vellucci [8]

1. Division of Anesthesia and Pain Medicine, Istituto Nazionale Tumori—IRCCS—Fondazione Pascale, 80131 Napoli, Italy; m.cascella@istitutotumori.na.it
2. Department of Life, Health and Environmental Sciences, University of L'Aquila, 67100 L'Aquila, Italy; francomarinangeli@gmail.com
3. Department of Anesthesia and Critical Care, ARCO, Ospedale Pediatrico Bambino Gesù IRCCS, 00165 Rome, Italy
4. UOC Anesthesia, Intensive Care and Pain Therapy, Senigallia Hospital, 60123 Ancona, Italy; scalacri@virgilio.it
5. Anesthesia, Critical Care, Palliative Medicine and Pain Therapy Service, L'Aquila ASL1 Abruzzo, 67100 L'Aquila, Italy; massimo.piccinini@yahoo.it
6. Grunenthal, Italia srl, 20019 Milan, Italy; alessandro.braga@grunenthal.com
7. Department of Clinical and Experimental Pain Medicine, IRCCS CRO of Aviano, 33081 Aviano, Italy; luca.miceli@cro.it
8. Pain and Palliative Care Clinic, University Hospital of Careggi, 50121 Florence, Italy; renato.vellucci@gmail.com
* Correspondence: alexvittori@libero.it or alexvittori82@gmail.com; Tel.: +39-06-68592397

Citation: Cascella, M.; Marinangeli, F.; Vittori, A.; Scala, C.; Piccinini, M.; Braga, A.; Miceli, L.; Vellucci, R. Open Issues and Practical Suggestions for Telemedicine in Chronic Pain. *Int. J. Environ. Res. Public Health* **2021**, *18*, 12416. https://doi.org/10.3390/ijerph182312416

Academic Editors: Marie Carmen Valenza and Irene Torres-Sanchez

Received: 27 October 2021
Accepted: 24 November 2021
Published: 25 November 2021

Publisher's Note: MDPI stays neutral with regard to jurisdictional claims in published maps and institutional affiliations.

Abstract: Telemedicine represents a major opportunity to facilitate continued assistance for patients with chronic pain and improve their access to care. Preliminary data show that an improvement can be expected of the monitoring, treatment adherence, assessment of treatment effect including the emotional distress associated with pain. Moreover, this approach seems to be convenient and cost-effective, and particularly suitable for personalized treatment. Nevertheless, several open issues must be highlighted such as identification of assessment tools, implementation of monitoring instruments, and ability to evaluate personal needs and expectations. Open questions exist, such as how to evaluate the need for medical intervention and interventional procedures, and how to define when a clinical examination is required for certain conditions. In this context, it is necessary to establish dynamic protocols that provide the right balance between face-to-face visits and telemedicine. Useful tips are provided to start an efficient experience. More data are needed to develop precise operating procedures. In the meantime, the first experiences from such settings can pave the way to initiate effective care pathways in chronic pain.

Keywords: chronic pain; telemedicine; healthcare delivery; health-related quality of life; functionality

1. Introduction

Telemedicine has been defined as the use of electronic technologies for communication and information of patients, to provide the public with remote healthcare services [1]. Although it has existed for more than two decades, its implementation has been limited for many years, until the emergency during the COVID-19 pandemic. This catastrophic event has promoted attempts to provide chronic patients with adequate care despite restrictions to in-presence activities [2]. The definition of adequate care pathways for chronic pain will need long clinical trials as the condition is complex and many different cases must be faced. As an example, patients with chronic post-surgical pain who have a certain diagnosis, a complex condition, and need careful and prolonged follow-up, and can benefit from the assistance through remote services. Recent experiences have suggested that telemedicine

341

can improve access to care, facilitate continuity of care, allow better resource efficiency, and lower costs, compared with traditional in-person hospital or ambulatory visits [1,3,4]. Improved access to care is expected to enhance timely adjustment of therapy, and improved adherence, which could reduce the progression toward reduced functionality of patients with chronic pain [5]. Appropriate uses of telemedicine for patients with chronic pain have recently been described. It has opened a promising field of activity, although a business case analysis would be needed in each prospective application [2,6].

On these premises, best practice approaches for telemedicine programs in chronic pain need to be suggested, to enable clinicians to provide and patients to benefit from remote assistance.

This article is based on the direct experience of a group of clinicians and attempts to provide a framework to prepare physicians, patients with chronic pain, and caregivers to use telemedicine with satisfactory results.

2. The Complexity of the Patient with Chronic Pain

According to the International Association for the Study of Pain (IASP), chronic pain is commonly defined as persistent or recurrent pain lasting more than 3 months or beyond normal tissue healing [7]. It has been recognized as a real disease associated with multiple adaptations in the nervous, endocrine, and immune systems [8]. Consequently, chronic pain is a complex multidimensional experience severely compromising the patient's health-related quality of life (HRQoL), often limiting the ability to work, sleep, and affecting social interactions with friends and family [2,9]. Reduced functionality, emotional imbalances, and social isolation are frequently associated complaints and may exacerbate each other in a vicious circle that compromises the HRQoL and induces a progression toward disability [10,11]. Furthermore, chronic pain is often associated with multimorbidity. In particular, many patients with chronic pain have other comorbidities, such as depression, cardiovascular and pulmonary diseases, diabetes mellitus, and cancer [12,13]. Notably, proper pain control may be extremely challenging in multi-morbid patients as comorbidities and their treatments can increase the risk of side effects of analgesics and thus limit the applicability of disease-specific clinical guidelines [14]. On the other hand, chronic pain is an independent risk factor for mortality in people with other co-morbidities [15]. Moreover, patients with chronic pain require multidisciplinary, continuous, and skilled management, which may challenge healthcare system organizations. Difficulties with the traditional models of care, with in-person patient visits to their physicians inherently leading to delayed care, have further cumulated during the COVID-19 pandemic and the need to structure new care pathways to ensure appropriate treatment of chronic pain patients, became even more prominent.

The first and necessary steps for the management of chronic pain are diagnosis and assessment of all pain dimensions. Given the 11th International Classification of Diseases (ICD-11) introduction, chronic pain will be classified into primary and secondary pain. Chronic primary pain can be conceived as a disease. This is a definition, which applies to chronic pain syndromes that are best conceived as health conditions in their own right [7]. Chronic secondary pain syndromes are linked to other diseases as the underlying cause, for which pain may initially be regarded as a symptom. Secondary pain includes cancer-related, post-traumatic and postsurgical, neuropathic, visceral, musculoskeletal, and headache/orofacial pain [7]. In many cases, the secondary chronic pain may continue beyond the successful treatment of the initial cause; in such cases, the pain diagnosis will remain, even after the diagnosis of the underlying disease is no longer relevant. This distinction is particularly important because it conditions the clinical, diagnostic, and therapeutic approaches. For example, in the cancer patient, close monitoring with close controls is mandatory when opioids are used. Moreover, in primary chronic pain (e.g., fibromyalgia) a combined approach with the collaboration of different professional figures may be necessary.

3. Approaches of Telemedicine for Patients with Chronic Pain

Telemedicine has been defined as the use of electronic technologies for communication and information to provide the public with healthcare services at a distance. More recently, the American Medical Association defined "telehealth" as a general group of modalities allowing: (1) real-time audio and visual connections between patients and physicians in different locations; (2) image and data collection storage and sharing for later interpretation; (3) remote patient monitoring tools, including mobile health (mHealth) tools, wearables, and devices; and (4) virtual check-ins through voice-only patient portals, messaging technologies [16]. The Italian Ministry of Health produced guidelines for telemedicine and defined the televisit as "a health act in which the doctor interacts remotely with the patient". The definition specifies that "the health act of diagnosis that arises from the visit may give rise to the prescription of drugs or treatments" [17].

Thus, the terms "telemedicine" and "telehealth" are commonly used interchangeably and encompassed in the set of tools called "telecare" [18]. Telecare includes several possible modalities and activities, such as archiving and sharing medical images or biosignals (e.g., in the fields of radiology, or dermatology), telemonitoring, and real-time interactive services. The latter modalities include a variety of services, such as telenursing, telepharmacy, telerehabilitation, emergency counseling, and mostly online consultation via remote visits and/or multi-professional teleconsulting.

4. Online Consultation Pathway

A remote system is a great opportunity to improve access to care and continuing assistance, which may help to personalize treatments and to increase adherence. To fulfil these objectives, it is necessary to structure a defined pathway. It is divided into a series of technical processes (information technology infrastructure) and operational phases (preparation, execution, scheduling of controls) (Figure 1).

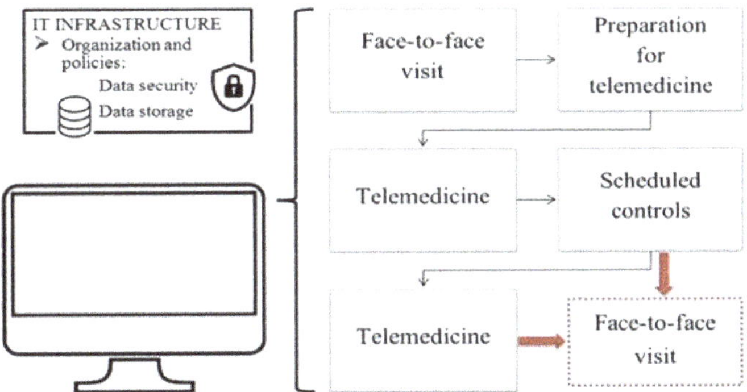

Figure 1. Telemedicine pathway for addressing chronic pain.

The information technology infrastructure must allow all organizational phases (reservations, contacts, links for connection, and data collection) and guarantee data security. Operation includes a first visit usually performed in person. This face-to-face assessment is followed by the preparation for telemedicine (legal and regulatory issues, patient information, technical issues). Later, telemedicine is performed, and scheduled controls are programmed. New in-person visits can be required (e.g., to carry out minimally invasive procedures). IT, information technology.

4.1. IT Infrastructure Functioning

The COVID-19 pandemic and the need to provide alternative ways for the in-person visit has led to the creation of a wide range of information technology (IT) infrastructures. On the market, there are systems which have different complexities (and costs). In general terms, the platforms consist of an operating system for the management of the whole service, devices (e.g., laptops), and an integrated software system (software modules) for sending documents, reports, and imaging. In brief, the IT infrastructure must allow all organizational phases (reservations, contacts, links for connection, and data collection) and guarantee data security and privacy.

4.2. Operational Phase: First Face-to-Face Visit

Since programs of telemedicine should be based on good interpersonal relationships, a first visit should usually be performed in-person to:

- Establish a relationship with the patient, which is necessary for long-term reliance.
- Obtain a diagnosis of chronic pain.
- Assess and measure pain and HRQoL.
- Enable the physician to perform assessment maneuvers.
- Prepare the telemedicine visit.

An in-person contact provides the necessary and reciprocal confidence for further remote relationships. Indeed, patients must be evaluated for their ability to use telemedicine before a program is established. Most of the information infrastructures available for telemedicine require that patients and/or caregivers simply need to have an email address and a smartphone, iPad/tablet, or a personal computer with a camera and speakers. Based on the in-person visit, a remote follow-up and monitoring schedule can be prepared.

4.3. Operational Phase: Remote Follow-Up

For remote follow-up it is important to collect the right parameters that can be obtained remotely. All the different aspects of chronic pain must be monitored; pain intensity, therapeutic adherence, sleep quality, movement functionality, emotionality, and working abilities. Patients and caregivers need training in order enable them to focus on the relevant topics during the telemedicine visit; the physician will guide the visit and choose the relevant area to be investigated in the situation.

4.3.1. Clinical Assessment

The remote evaluation involves the study of the patient's medical record (imaging, laboratory tests, other documents) and the clinical–diagnostic phase. A comprehensive pain assessment is a crucial step in the management of a patient with pain. As already stated, chronic pain is a multidimensional experience resulting in impaired functioning in daily life and reduced quality of life and well-being of the patient, as it can be observed for chronic low back pain (cLBP) [19]. We suggest that physicians could assess pain severity mainly using the parameters of pain intensity, pain-related distress, and functioning.

Generally, assessment tools for telemedicine should be validated, and suitable; moreover, the same tools should always be used. In addition, other instruments can be used, for the objective evaluation of distress and functionality through web-based use. Based on the authors' experience, a combination of unidimensional and multidimensional tools can be adopted. The numerical rating scale (NRS) may easily be used by patients to assess pain intensity. The Brief Pain Inventory (BPI) is a validated, simple, and self-completed questionnaire (visual administration) that evaluates not only pain intensity, but also functionality, and provides long-term monitoring in patients with progressive conditions. The tool is reliable and valid for many clinical situations (e.g., cancer pain, and non-cancer pain conditions) and across cultures and languages. Functionality can be assessed by an ecological matrix scale [20–22], which considers the outer environment and the personality structure of the patient, the motives, the personal expectations, and needs, and helps the patient and the physician to identify treatment objectives that may be satisfactory for the patient.

In the setting of chronic low back pain, the Oswestry Disability Index and the Roland-Morris disability questionnaire may be useful for the functional evaluation of low back pain [22,23].

Other indices are available, such as the Low Back Pain Rating Scale (LBPRS), as an example, the Progressive Isoinertial Lifting Evaluation (PILE), and the Quebec Back Pain Disability Scale (QBPDS) [24]. In this respect, clinicians should also evaluate the capability of their patients to perform daily activities, the patient's emotional status, and their strength. Up to now, different proper instruments are available to evaluate HR-QoL, such as the EQ-5D from the EuroQol Research Foundation, and the short form (SF-12 scale) of the 36-item Health Survey instrument are available [25,26].

In addition to the evaluation through tools, all the anamnestic elements must be collected. Even if at a distance, the clinician will have to investigate the clinical elements of the painful symptomatology: location, intensity, triggering factors, therapies carried out, and comorbidities. Peculiar aspects, such as breakthrough cancer pain, drug effects, and clinical conditions, which may affect the use of particular categories of drugs (e.g., organ damage), must be evaluated.

Finally, the recommendations issued by the Initiative on Methods, Measurement, and Pain Assessment in Clinical Trials (IMMPACT) can also be used. A common tool would be used to evaluate the validity of telemedicine models. Notably, six core outcome domains were recommended by IMMPACT including pain, physical functioning, emotional functioning, participant ratings of global improvement, symptoms and adverse events, and patient disposition (adherence to the treatment regimen, reasons for withdrawal from treatment) [27]. A systematic review showed that eHealth and mHealth interventions had significant effects on multiple short- and intermediate-term outcome measures recommended in the IMMPACT guidelines [28].

4.3.2. Outcomes, Therapy, and Re-Evaluation

The televisit can give rise to different outcomes; it is possible to achieve clinical stability within the already known diagnostic framework. It can bring out the need for urgent access to diagnostic or therapeutic services, which require the patient to have a face-to-face consultation with the pain specialist. The third scenario includes the need for further examination to have a diagnosis, which the specialist will manage with the prescription of the necessary services.

During the televisit, the previous therapy can be either confirmed or changed; in this case, the specialist prescribes the drugs and sends the prescription to the patient as agreed with him and his caregiver. Although the primary objective of treatment should be control of pain intensity, functional recovery and general wellbeing are the overall aims of the patient's management. This means that pharmacological and non-pharmacological treatments must be considered, and that tailored objectives are to be pursued according to the expectations and needs of each patient [20–22].

If it is impossible to reach a diagnostic or therapeutic conclusion, the doctor will propose the execution of a further check-up in the times and the ways appropriate to the clinical situation, and the follow-up pathway will be planned. In some cases, possible adverse events may be mentioned by the physician, asking the patients whether any of them occurred in the last period; this may prove a simple and effective method to detect tolerability issues. Under certain circumstances, such as drug side effects, need for a physical examination or interventional procedures, an in-person visit may be necessary.

The remote visits should be tightly planned to ensure that monitoring can be performed adequately, and therapy can be adjusted. The data collected and outcome of the televisit will need to be recorded as a routine medical visit and filed according to local customs.

5. Open Issues and Suggestions

When developing a telemedicine plan for the care of patients with chronic pain, a number of challenges need to be considered. These include the risk of adverse effects of drugs prescribed in a remote visit, the correct management of patients with advanced age, cognitive impairment, emotional frailty, and the need to alternate remote and in-person visits. To address these challenges, some suggestions may be useful. Table 1 presents the recommended steps for the implementation of a telemedicine service for patients with chronic pain.

Table 1. Some recommended actions for the management of a telemedicine system.

Steps	Recommended Actions	What Is Needed
Preparation of a telemedicine system	Legal and regulatory issues	Acquire information about: • Local rules for requirements • Data protection • Remote informed consent • Clinical report and prescription release
	Technical equipment	An efficient connection to the internet A digital device with a webcam Web-based platform
	Medical skills	Identify the phases of the visit Identify the parameters suitable for remote assessments of pain intensity, personal requirements, functionality, sleep quality, adverse events, treatment adherence Identify diagnostic maneuvers that can be performed by caregivers in your place
Initiation of a telemedicine program	Prepare the patient	An in-person visit is usually necessary before any telemedicine program, to: • Diagnose the condition • Assess pain • Explain the program • Establish a relationship
Monitoring	Schedule the visits beforehand	Both the clinician and the patient need to know how much time is dedicated to the visits, to have the opportunity to continue the program long-term
	Provide the patient with tools and instructions Schedule the assessment frequency Provide instructions for emergencies	Dedicated digital platforms with assessment scales
Visits	Have a schedule and confirm each date Be aware that it may be necessary to alternate remote and in-person visits	• Be punctual • Let the patient speak first • Control times • Release a report

Note: BPI, brief pain inventory.

Prerequisites for remote care systems include correct and exhaustive information about legal issues and regulation, availability of suitable technical equipment, and specific medical skills for the remote management of patients with chronic pain so that a correct assessment is performed despite the lack of a physical inspection. Specifically, all local requirements for healthcare must be fulfilled by remote systems, as well as by traditional in-presence organizations, and some technical solutions must be found to this aim. Therefore, first of all, knowledge of regulation is needed so that medico-legal problems are detected and addressed, and then rules must be adapted to remote systems [29]. The main legal issues to be faced are related to data protection, privacy, and delivery of reports and prescriptions. Several online platforms are available, which respect such requirements.

A basic suitable technical equipment for the physician and the patient includes an efficient connection to the internet, a digital device (usually a PC) with a webcam, and a customized web-based platform. In addition, to set up a telemedicine system, a large proportion of the target population needs to have sufficient skills to use the proposed web-based platform.

This means that very simple digital tools are to be preferred. Common social media are often used, to facilitate patients and caregivers, but they would not fulfill privacy and data protection requirements. The current reference guidelines for telemedicine are obsolete [30]; some of those, for example, date back to the 1980s and were not followed by later recommendations [31–34]. Privacy regulations, technological opportunities, and problems related to the pandemic are open issues that must be urgently addressed. The use of validated and protected platforms available through public or private healthcare providers could help respond to these issues.

As an example, the Regional Health Service in Tuscany, Italy, made a digital platform for televisits available to all specialists, in June 2020. The online visit requires a PC (with Windows 7 or IOS 11, or later versions), or smartphone or tablet (with Android 5.0 or later version), or iPhone or iPad (with IOS 11 or later version). The platform was mainly used by diabetologists, rheumatologists, and cardiologists, and registered up to 5000 visits in the first month of activity. This platform is extremely user friendly, is linked to the online clinical records, and with the regional health booking service; digital prescriptions are delivered by a preexisting system, while a final report of each visit is provided on the platform, according to specific rules set by the regional healthcare system.

The last suggestion on this point is that patients and/or caregivers must be exhaustively trained to be able to use the telemedicine system.

In summary, before the initiation of the telemedicine program, the medical activities must be carefully planned and scheduled. Online performances will be very fast, and waiting times are very limited online; so, everything must be ready beforehand. In addition, it is important to identify the parameters deemed suitable for remote visits on one end and for telemonitoring on the other one. The patient and/or the caregiver must be selected and, if accepted, prepared. Selection will be based both on medical (as an example, chronic, stable conditions are more suitable than lately diagnosed, progressing patients) and cognitive qualities, and personal features. The patient/caregiver must accept the program and feel it as an opportunity for improved management, continuing assistance, and access to care. If a caregiver is necessary, the same person should be present at all remote visits, and the same person should oversee assessments for monitoring. If the patient is a child, special precautions should be used for an effective protection of her/his rights. A relationship between the patient and/or caregiver and the pain specialist should already exist, with a good therapeutic alliance. A first in-person visit will usually be performed to obtain a diagnosis, to prepare the patient, and state a reciprocal reliance. This phase is followed by telemonitoring and scheduled remote control visits. A strict and punctually respected schedule of the program will facilitate its long-term continuation.

Special attention is necessary for the organization of the clinical assessment. Firstly, suitable tools must be identified. As previously mentioned, simple tools are to be offered; NRS, BPI, self-evaluation numerical scales, and ecologic matrix scales may be used. The assessment frequency must be stated in advance and explained to the patient. A telemedicine system for chronic pain will improve continued assistance but does not usually provide an emergency service. Patients must be informed of the aim of the system and should know what to do in case of adverse events or serious pain episodes. If around-the-clock assistance is necessary according to the patient's conditions, a phone triage should be available to deliver primary information and refer the patient to the correct health operator. Finally, a telemonitoring program will be efficient if the patient and/or caregiver are empowered; so, great care is due to instruction, and information about the disease and the treatment.

During the online evaluation, the physician will not be able to perform physical maneuvers which are commonly used for in-presence assessment, but some simple ones may be proposed to the caregiver. Each visit will be mainly based on the evaluation of the report by the patient and the revision of data collected during the telemonitoring. The physician should develop his/her listening ability as much as possible, and empathy.

Punctuality, respect of visit duration, and the ability to listen to the patient and/or caregiver are very important. After each visit a medical report must be delivered; predisposed forms can be designed, and an identical delivery system, such as email, should be routinely used. As an example, patients with musculoskeletal chronic pain are very likely to benefit from an assiduous follow-up performed with a multidimensional assessment made possible and affordable by the introduction of telemedicine [35–37].

6. Limitations

We acknowledge that several problems limit the possibility of achieving our aim of providing useful hints for the implementation of telemedicine for chronic pain. No evidence is available on the efficacy of this model. Regulation is widely different worldwide, and this makes it impossible to address local issues. As an example, medical responsibility is criminal and not only civil in Italy. In addition, certified online platforms are very different from each other. All these discrepancies make it difficult to design the many care pathways that are needed to address a complex condition such as chronic pain.

7. Conclusions

Telemedicine seems to be promising for the efficient management of patients with chronic pain. This approach can deliver tailored pain management, providing improved access to health services and creating and maintaining a therapeutic alliance in the long term. Nevertheless, the effects of therapies provided via telemedicine on pain and pain-related conditions, such as disability, depression, and anxiety, are promising but not well documented yet. Furthermore, when approaching remote assistance for chronic pain, several issues are to be faced such as accurate diagnosis, assessment, monitoring and need to change treatment. Consequently, the implementation of new web-based systems for the management of chronic pain needs further evaluation and well-structured pathways must be necessarily built. Finally, as privacy regulation is incomplete worldwide, clinicians are bound to be extremely cautious about respecting the patient's rights.

Author Contributions: Conceptualization, M.C., F.M. and A.V.; methodology, M.C., F.M., A.V., C.S. and A.B.; writing—original draft preparation, M.C., F.M. and A.V.; writing—review and editing, M.C., F.M., A.V., L.M., C.S., M.P., A.B. and R.V.; visualization, M.C.; supervision, M.C., F.M., A.V. and R.V.; project administration, F.M., M.C. and A.V. All authors have read and agreed to the published version of the manuscript.

Funding: This research received no external funding.

Institutional Review Board Statement: Not applicable.

Informed Consent Statement: Not applicable.

Acknowledgments: Editorial assistance was provided by Laura Brogelli, PhD and Aashni Shah (Polistudium SRL, Milan, Italy). This assistance was supported by Grunenthal.

Conflicts of Interest: R.V., A.V., F.M., C.S., M.C., L.M. and M.P. report no conflict of interest in this work. A.B. is employed in Grunenthal Italia srl, Italy.

References

1. Nikolian, V.C.; Williams, A.M.; Jacobs, B.N.; Kemp, M.T.; Wilson, J.K.; Mulholland, M.W.; Alam, H.B. Pilot study to evaluate the safety, feasibility, and financial implications of a postoperative telemedicine program. *Ann. Surg.* **2018**, *268*, 700–707. [CrossRef]
2. Emerick, T.; Alter, B.; Jarquin, S.; Brancolini, S.; Bernstein, C.; Luong, K.; Morrisseyand, S.; Wasan, A. Telemedicine for Chronic Pain in the COVID-19 Era and Beyond. *Pain Med.* **2020**, *21*, 1743–1748. [CrossRef]
3. Mann, D.M.; Chen, J.; Chunara, R.; Testa, P.A.; Nov, O. COVID-19 transforms health care through telemedicine: Evidence from the field. *J. Am. Med. Inform. Assoc.* **2020**, *27*, 1132–1135. [CrossRef] [PubMed]
4. Puntillo, F.; Giglio, M.; Brienza, N.; Viswanath, O.; Urits, I.; Kaye, A.D.; Pergolizzi, J.; Paladini, A.; Varrassi, G. Impact of COVID-19 pandemic on chronic pain management: Looking for the best way to deliver care. *Best Pract. Res. Clin. Anaesthesiol.* **2020**, *34*, 529–537. [CrossRef] [PubMed]

5. Mao, B.; Liu, Y.; Chai, Y.H.; Jin, X.Y.; Lu, H.W.; Yang, J.W.; Gao, X.W.; Song, X.L.; Bao, H.; Wang, A.; et al. Assessing risk factors for SARS-CoV-2 infection in patients presenting with symptoms in Shanghai, China: A multicentre, observational cohort study. *Lancet Digit. Health* **2020**, *2*, e323–e330. [CrossRef]

6. Zajtchuk, R.; Gilbert, G.R. Telemedicine: A new dimension in the practice of medicine. *Dis. Mon.* **1999**, *45*, 197–262. [CrossRef]

7. Treede, R.D.; Rief, W.; Barke, A.; Aziz, Q.; Bennett, M.I.; Benoliel, R.; Cohen, M.; Evers, S.; Finnerup, N.B.; First, M.B.; et al. Chronic pain as a symptom or a disease: The IASP Classification of Chronic Pain for the International Classification of Diseases (ICD-11). *Pain* **2019**, *160*, 19–27. [CrossRef] [PubMed]

8. Basbaum, A.I.; Bautista, D.M.; Scherrer, G.; Julius, D. Cellular and molecular mechanisms of pain. *Cell* **2009**, *139*, 267–284. [CrossRef]

9. Morlion, B.; Coluzzi, F.; Aldington, D.; Kocot-Kepska, M.; Pergolizzi, J.; Mangas, A.C.; Ahlbeck, K.; Kalso, E. Pain chronification: What should a non-pain medicine specialist know? *Curr. Med. Res. Opin.* **2018**, *34*, 1169–1178. [CrossRef] [PubMed]

10. Goesling, J.; Clauw, D.J.; Hassett, A.L. Pain and Depression: An Integrative Review of Neurobiological and Psychological Factors. *Curr. Psychiatry Rep.* **2013**, *15*, 1–8. [CrossRef]

11. Hylands-White, N.; Duarte, R.V.; Raphael, J.H. An overview of treatment approaches for chronic pain management. *Rheumatol Int.* **2017**, *37*, 29–42. [CrossRef] [PubMed]

12. Zis, P.; Daskalaki, A.; Bountouni, I.; Sykioti, P.; Varrassi, G.; Paladini, A. Depression and chronic pain in the elderly: Links and management challenges. *Clin. Interv. Aging* **2017**, *12*, 709–720. [CrossRef]

13. Van Hecke, O.; Hocking, L.J.; Torrancem, N.; Campbell, A.; Padmanabhan, S.; Porteous, D.J.; McIntosh, A.M.; Burri, A.V.; Tanaka, H.; Williams, F.M.; et al. Chronic pain, depression and cardiovascular disease linked through a shared genetic predisposition: Analysis of a family-based cohort and twin study. *PLoS ONE* **2017**, *12*, e0170653. [CrossRef] [PubMed]

14. Thaler, A. Pain Comorbidities: Understanding and Treating the Complex Patient. *Pain Med.* **2013**, *14*, 763–764. [CrossRef]

15. Smith, D.; Wilkie, R.; Uthman, O.; Jordan, J.L.; McBeth, J. Chronic pain and mortality: A systematic review. *PLoS ONE* **2014**, *9*, e99048. [CrossRef]

16. American Medical Association. AMA Telehealth Quick Guide. Updated 10 August 2021. Available online: https://www.ama-assn.org/practice-management/digital/ama-telehealth-quick-guide (accessed on 2 November 2021).

17. Italian Health Ministry. Telemedicine. National Guidelines. Available online: https://www.salute.gov.it/imgs/C_17_pubblicazioni_2129_allegato.pdf (accessed on 30 September 2021).

18. Health Resources & Services Administration. Telehealth Programs. Available online: https://www.hrsa.gov/rural-health/telehealth (accessed on 22 July 2021).

19. Morlion, B. Chronic low back pain: Pharmacological, interventional and surgical strategies. *Nat. Rev. Neurol.* **2013**, *9*, 462–473. [CrossRef]

20. Kenrick, D.T.; Griskeviciusm, V.; Neuberg, S.L.; Schaller, M. Renovating the pyramid of needs: Contemporary extensions built upon ancient foundations. *Perspect. Psychol. Sci.* **2010**, *5*, 292–314. [CrossRef] [PubMed]

21. Thompson, J.R.; Tassé, M.J.; McLaughlin, C.A. Interrater reliability of the Supports Intensity Scale (SIS). *Am. J. Ment Retard* **2008**, *113*, 231–237. [CrossRef]

22. Fairbank, J.C.; Pynsent, P.B. The Oswestry Disability Index. *Spine (Phila PA 1976)* **2000**, *25*, 2940–2952, discussion 2952. [CrossRef]

23. Law, K.K.P.; Lee, P.L.; Kwan, W.W.; Mak, K.C.; Luk, K.D.K. Cross-cultural adaptation of Cantonese (Hong Kong) Oswestry Disability Index version 2.1b. *Eur. Spine J.* **2021**, *30*, 2670–2679. [CrossRef]

24. Smeets, R.; Köke, A.; Lin, C.W.; Ferreira, M.; Demoulin, C. Measures of function in low back pain/disorders: Low Back Pain Rating Scale (LBPRS), Oswestry Disability Index (ODI), Progressive Isoinertial Lifting Evaluation (PILE), Quebec Back Pain Disability Scale (QBPDS), and Roland-Morris Disability Questionnaire (RDQ). *Arthritis Care Res.* **2011**, *63*, S158–S173.

25. Rabin, R.; de Charro, F. EQ-5D: A measure of health status from the EuroQol Group. *Ann. Med.* **2001**, *33*, 337–343. [CrossRef]

26. Ware, J., Jr.; Kosinski, M.; Keller, S.D. A 12-Item Short-Form Health Survey: Construction of scales and preliminary tests of reliability and validity. *Med. Care* **1996**, *34*, 220–233. [CrossRef]

27. Dworkin, R.H.; Turk, D.C.; Wyrwich, K.W.; Beaton, D.; Cleeland, C.S.; Farrar, J.T.; Zavisic, S. Interpreting the Clinical Importance of Treatment Outcomes in Chronic Pain Clinical Trials: IMMPACT Recommendations. *J. Pain* **2008**, *9*, 105–121. [CrossRef] [PubMed]

28. Moman, R.N.; Dvorkin, J.; Pollard, E.M.; Wanderman, R.; Murad, M.H.; Warner, D.O.; Hooten, W.M. A Systematic Review and Meta-analysis of Unguided Electronic and Mobile Health Technologies for Chronic Pain–Is It Time to Start Prescribing Electronic Health Applications? *Pain Med.* **2019**, *20*, 2238–2255. [CrossRef]

29. Petrucci, E.; Vittori, A.; Cascella, M.; Vergallo, A.; Fiore, G.; Luciani, A.; Pizzi, B.; Degan, G.; Fineschi, V.; Marinangeli, F. Litigation in Anesthesia and Intensive Care Units: An Italian Retrospective Study. *Healthcare* **2021**, *9*, 1012. [CrossRef]

30. Kaplan, B. Revisiting health information technology ethical, legal, and social issues and evaluation: Telehealth/telemedicine and COVID-19. *Int. J. Med. Inform.* **2020**, *143*, 104239. [CrossRef]

31. Langarizadeh, M.; Moghbeli, F.; Aliabadi, A. Application of ethics for providing telemedicine services and information technology. *Med. Arch.* **2017**, *71*, 351–355. [PubMed]

32. Furtado, R. Telemedicine: The next-best thing to being there. *Dimens. Health Serv.* **1982**, *59*, 10–12.

33. Bashshur, R.L.; Armstrong, P.A. Telemedicine: A new mode for the delivery of health care. *Inquiry* **1976**, *13*, 233–244. [PubMed]

34. Michaels, E. Telemedicine: The best is yet to come, experts say. *CMAJ* **1989**, *141*, 612. [PubMed]

35. Vittori, A.; Petrucci, E.; Cascella, M.; Innamorato, M.; Cuomo, A.; Giarratano, A.; Petrini, F.; Marinangeli, F. Pursuing the Recovery of Severe Chronic Musculoskeletal Pain in Italy: Clinical and Organizational Perspectives from a SIAARTI Survey. *J. Pain Res.* **2021**, *14*, 3401–3410. [CrossRef] [PubMed]
36. Cuomo, A.; Cascella, M.; Forte, C.A.; Bimonte, S.; Esposito, G.; De Santis, S.; Crispo, A. Careful Breakthrough Cancer Pain Treatment through Rapid-Onset Transmucosal Fentanyl Improves the Quality of Life in Cancer Patients: Results from the BEST Multicenter Study. *J. Clin. Med.* **2020**, *9*, 1003. [CrossRef] [PubMed]
37. Cascella, M.; Crispo, A.; Esposito, G.; Forte, C.A.; Coluccia, S.; Porciello, G.; Cuomo, A. Multidimensional Statistical Technique for Interpreting the Spontaneous Breakthrough Cancer Pain Phenomenon. A Secondary Analysis from the IOPS-MS Study. *Cancers* **2021**, *13*, 4018. [CrossRef]

International Journal of
Environmental Research and Public Health

MDPI

Review

Efficacy of Web-Based Supportive Interventions in Quality of Life in COPD Patients, a Systematic Review and Meta-Analysis

Andrés Calvache-Mateo, Laura López-López, Alejandro Heredia-Ciuró, Javier Martín-Núñez, Janet Rodríguez-Torres, Araceli Ortiz-Rubio and Marie Carmen Valenza *

Department of Physical Therapy, Faculty of Health Sciences, University of Granada, 18007 Granada, Spain; andrescalvache@ugr.es (A.C.-M.); lauralopez@ugr.es (L.L.-L.); ahc@ugr.es (A.H.-C.); javimartinn29@gmail.com (J.M.-N.); jeanette92@ugr.es (J.R.-T.); aortiz@ugr.es (A.O.-R.)
* Correspondence: cvalenza@ugr.es; Tel.: +34-958-248-035

Abstract: Background: Adults living with Chronic Obstructive Pulmonary Disease (COPD) often have difficulties when trying to access health care services. Interactive communication technologies are a valuable tool to enable patients to access supportive interventions to cope with their disease. The aim of this revision and meta-analysis is to analyze the content and efficacy of web-based supportive interventions in quality of life in COPD. Methods: Medline (via PubMed), Web of Science, and Scopus were the databases used to select the studies for this systematic review. A screening, analysis, and assessment of the methodological quality was carried out by two independent researchers. A meta-analysis of the extracted data was performed. Results: A total of 9 of the 3089 studies reviewed met the inclusion criteria. Most repeated web content elements were educational and involved communication with healthcare professional content. Finally, seven of the nine studies were included in a quantitative analysis. Web-based supportive interventions significantly improved quality of life when added to usual care (SMD = −1.26, 95% CI = −1.65, −0.86; $p < 0.001$) but no significant differences were found when compared with an autonomous pedometer walking intervention ($p = 0.64$) or a face-to-face treatment ($p = 0.82$). Conclusion: This systematic review and meta-analysis suggests that web-based supportive interventions may complement or accompany treatments in COPD patients due to the advantages of online interventions. The results obtained should be treated with caution due to the limited number of studies in this area and methodological weaknesses.

Keywords: communication; COPD patients; educational content; supportive interventions; web-based

Citation: Calvache-Mateo, A.; López-López, L.; Heredia-Ciuró, A.; Martín-Núñez, J.; Rodríguez-Torres, J.; Ortiz-Rubio, A.; Valenza, M.C. Efficacy of Web-Based Supportive Interventions in Quality of Life in COPD Patients, a Systematic Review and Meta-Analysis. *Int. J. Environ. Res. Public Health* **2021**, *18*, 12692. https://doi.org/10.3390/ijerph182312692

Academic Editors: Shyamali Dharmage and Paul B. Tchounwou

Received: 27 September 2021
Accepted: 29 November 2021
Published: 2 December 2021

Publisher's Note: MDPI stays neutral with regard to jurisdictional claims in published maps and institutional affiliations.

1. Introduction

Chronic obstructive pulmonary disease (COPD) is a non-reversible inflammatory disease that causes progressive obstruction of the airways. According to the Global Initiative for Chronic Obstructive Lung Disease (GOLD) 2020 report, COPD is the leading lung disease in terms of mortality and morbidity worldwide [1,2]. Due to the increase in smoking and the progressive ageing of the population, the prevalence of COPD will increase in the coming years [3].

As the disease progresses, the symptoms become increasingly severe and complex. Often, the combination of psychological, emotional, and social factors with physical symptoms makes it difficult for patients and professionals to deal with the disease [4]. As a result, COPD patients experience significant impairment of disease-related quality of life as well as social isolation [5] that generates a significant burden of disability [6] and demands continuous health care [7].

Unfortunately, COPD patients face significant barriers when seeking access to appropriate health services to manage the disease, including living in medically underserved regions [8], language barriers [9], reduced mobility due to the disease itself, or other comorbidities, such as ageing and limited time [10]. In addition, due to the respiratory status of these patients and the potentially serious medical consequences for them, the risk of

351

COVID-19 infection should be minimized [11,12]. Despite all these obstacles, there are not many interventions to support COPD patients in dealing with their disease [8].

Technological development is a great opportunity to generate new tools to support COPD patients [8,13]. Those technologies have enabled existing therapies to be delivered online and allow for the development of new interventions tailored to patients' needs [14]. New technologies are increasingly being investigated with the aim of developing interventions that can adequately complement or replace interventions already provided in health services [15–17].

Rapid advances towards a more digitalized society as well as the rapid development of today's electronic devices have caused a significant rise in the availability of communication technologies applied to health services [18,19]. The different online health communication tools allow patients to access personalized content, disease self-management tools, and communication with healthcare professionals from the comfort and security of their own home [20–22].

The most recent systematic reviews and meta-analyses [23–33] on telehealth care analyze teleassessment, telephone assistance, mobile app development, and website assistance in depth, but they need to be analyzed separately [8].

Previous studies show chronic disease patients' need for personalized web-based interventions [34,35]. COPD patients demand access to information about their health status, related to the disease itself and to the improvement of quality of life [8]. Different mechanisms related to a perception of health-related needs, such as health education, self-management [36], and family and social support, have a significant influence on the quality of life of patients using web-based interventions [12,21,27,37–39].

Web-based interventions can encompass several distinct, often overlapping interventions, including: (1) tele-education content; (2) symptom and mood telemonitoring; (3) physical activity monitoring and personalized feedback to the patient; (4) tele-education in self-management skills; (5) tele-consultation with healthcare professionals; (6) tele-communication with other patients; (7) remote decision support systems; (8) tele-diagnosis; and (9) tele-rehabilitation [27,40,41].

The advantages offered by web-based interventions such as easy and on-demand access to health information content, interactive support with other patients, and tools for symptom self-management may have the potential to influence the different variables and symptoms of a COPD patient. There is a need to investigate whether these web-based interventions have an impact on the quality of life of COPD patients and determine which are the most appropriate contents. The aim of this revision and meta-analysis is to analyze the content and efficacy of web-based supportive interventions in quality of life in COPD.

2. Materials and Methods

2.1. Search Strategy and Eligibility Criteria

This systematic review was conducted according to the Preferred Reporting Items for Systematic Review and Meta-Analysis (PRISMA) statement guidelines [42] and its registration number in the International Prospective Register of Systematic Review (PROSPERO) is CRD42020211978. The Cochrane Collaboration guidelines for reviewing interventions were also closely followed [43]. Three databases were used for the electronic search: Medline (via Pubmed), Web of Science, and Scopus. The screening and analysis of the studies was conducted between November 2020 and March 2021. Relevant publications from inception to 1 March 2021 were included. A search strategy was created for Medline and then modified to be specific to each of the databases. The following Medical Subject Headings (MeSH) terms were used (Appendix A).

To adequately define the research question, the impact of patient, intervention, comparison, outcome (PICOS) strategy [44] was applied.

(P) Population: COPD patients over 18 years of age.

(I) Interventions: Studies that used web-based supportive interventions.

(C) Comparison: Non-web-based interventions.

(O) Outcome: Any outcome reporting quality of life (e.g., St. George Respiratory Questionnaire, Chronic Respiratory Disease Questionnaire).

(T) Timing: At any time.

(S) Setting: No restriction of setting.

Only full-text randomized controlled trials written in English, Spanish, and French were included in the systematic review. Systematic reviews and meta-analyses, observational studies, clinical practice guidelines, letters, abstracts, editorials, conference papers, theses, and dissertations were excluded. Studies in other languages were also considered for inclusion when translation was possible.

2.2. Study Selection and Data Extraction

After all studies had been retrieved from the three databases, duplicates were removed. To determine if the articles met the inclusion criteria for this systematic review, two independent investigators performed a first assessment of the title and abstract of all studies. If the article met the inclusion criteria, it was selected for a second phase in which the full text was analyzed and reviewed.

The Cochrane guidelines for systematic reviews were followed for data extraction [43]. A third reviewer was responsible for resolving any disagreement between the two main reviewers. The information extracted from the articles was: year of publication, main author, sample size, sample age, treatment status, severity of COPD, specific intervention for the control and experimental groups, web content elements, intervention duration, outcome measures, and main results. If the reviewers did not find any data during the analysis and review of the articles, they contacted the authors of the studies.

2.3. Assessment of Methodological Quality and Risk of Bias

The Downs and Black quality checklist was used to assess the methodological quality of the studies included in the review [45]. This assessment was carried out independently by the two principal investigators. This method contains 27 items divided into 5 subscales (study quality, external validity, study bias, confounding and selection bias, and study power). Due to its high reliability and validity, this scale is considered one of the six most appropriate scales to measure the quality of the studies included in a systematic review [46]. Studies are classified into four categories according to the score obtained: it will be classified as poor if its score is less than or equal to 14, fair if the score is between 15 and 19, good if the score is between 20 and 25, and excellent if the score is between 26 and 28 [46,47].

In addition to the methodological quality of the articles, the risk of bias was assessed using the Cochrane Risk of Bias Tool for Randomized Controlled Trials [43]. This measurement tool is divided into seven items that are subdivided into six subscales. The first subscale corresponds to the selection bias and is the only one with two items. The remaining subscales are called performance bias, detection bias, attrition bias, reporting bias, and other bias, and have only one item. When the reviewer determines that there is a low risk of bias for each of the items, the study is classified as high quality. When the reviewer determines that one of the items is not met because there is a high risk of bias or two of the items cannot be answered clearly, the study is classified as fair quality. When the reviewer determines that one of the items is not met because there is a high risk of bias or two of the items cannot be answered clearly and there are important limitations that may invalidate the results, the study is classified as poor quality. The study is also classified as poor quality when two or more items are not met.

2.4. Data Synthesis and Analysis

A meta-analysis was undertaken using Review Manager (RevMan v5.3; Cochrane Collaboration, Oxford, UK). All variables included were continuous data. Study authors were contacted by e-mail whenever data were insufficient for the purposes of meta-analysis (e.g., neither means nor standard deviation were provided). Authors were given 2 weeks to respond. If they had not responded within a week, they were written to again as a reminder. The embedded Review Manager calculator was used to calculate standard deviations whenever *p*-values or 95% confidence intervals were given [48].

The main outcome considered for this meta-analysis was quality of life. Standardized mean differences were used because all scales were assumed to measure the same underlying symptom or condition, but some studies measured outcomes on different scales and 95% confidence intervals (CI) were calculated for all outcomes [49]. Subgroup analysis was also used in this study to help clarify the different uses of web-based interventions.

When the studies presented different scales to measure quality of life, we selected the data provided by the Saint George Respiratory Questionnaire (SGRQ), since it is the most frequently used, disease-specific quality of life measure in this population group [50]. When studies did not use the SGRQ, scores from other disease-specific quality of life scales, such as the Chronic Respiratory Disease Questionnaire (CRQ), were used [50–52]. The scoring of the different scales was converted so that a lower score always indicated a better outcome.

The Q and I^2 statistics were calculated to examine statistical heterogeneity, and a visual inspection of the forest plots was also performed to identify outlier studies. The I^2 is a statistical value that is interpreted as the percentage of the total variation observed between studies that is due to the difference between them and not to sampling error (chance). An I^2 of $\geq$50%; I^2 >25% and < 50%; I^2 of $\leq$25% were considered to indicate high, moderate, and low heterogeneity, respectively. When the I^2 value is greater than 50%, the meta-analysis is considered heterogeneous and, therefore, a random effects analysis had to be used. Statistical significance was established as $p < 0.05$, which means that the effects differ significantly between the control and intervention groups. We also explored sources of heterogeneity and performed a sensitivity analysis excluding trials with high risk of attrition or detection bias [48].

3. Results

3.1. Study Selection

An initial search of the databases found 3089 records. After eliminating duplicates, a total of 1319 studies were selected. In the end, an overall total of 9 studies that analyzed a total of 1168 participants were included in this systematic review. The PRISMA flow diagram for the study selection process is shown in Figure 1.

3.2. Study Characteristics

Table 1 shows the main characteristics of the studies included in the systematic review. The included studies were published between 2013 and 2020, and assessed participants were aged between 66.1 [53] and 71.9 years [54]. All the studies except the study of Wang et al. (47.5%) [54] had a higher proportion of males than females in the study sample. Regarding COPD severity, five studies [55–59] included mild to very severe patients and four studies [53,54,60,61] included moderate to very severe patients. All studies included clinically stable patients, with the exception of Wang et al. [54] and Jiménez-Reguera et al. [61], which included patients after discharge.

The web-based supportive interventions of each study were covered in Table 2 by the content of the comparison group approach, the content of the experimental interventions, the intervention duration, the outcome measures, and main results. Table 2 also includes nine web content elements that were identified as important to the technical characteristics of internet-supported therapeutic interventions [27,62] as well as for evidence-based web interventions: 1, tele-education content; 2, symptom and mood telemonitoring; 3, physical activity monitoring and personalized feedback to the patient; 4, tele-education in self-management skills; 5, tele-consultation with healthcare professionals; 6, tele-communication with other patients; 7, remote decision support systems; 8, tele-diagnosis and 9, tele-rehabilitation [63,64].

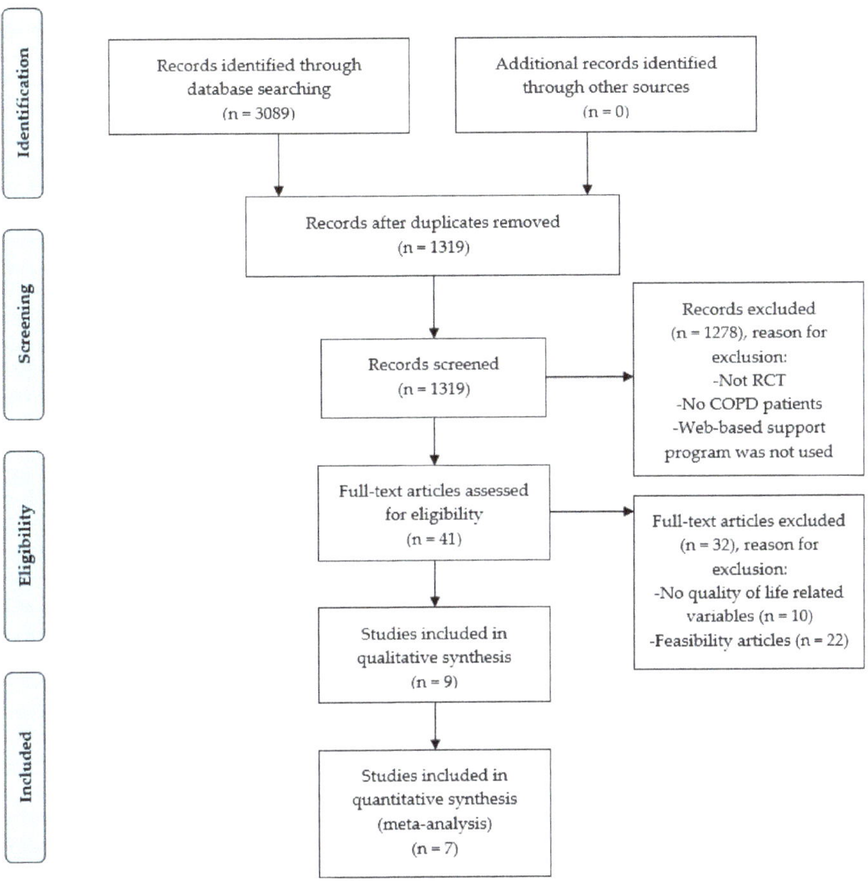

Figure 1. PRISMA flow chart of the literature screening process and results.

One study compared the usual care with a comparator group who received the usual care in addition to the web-based supportive program based on tele-education and tele-consultation with healthcare professionals [54]. Four studies compared a web-based supportive pedometer walking intervention based on physical activity monitoring, personalized feedback to the patient, and tele-education, with a pedometer walking intervention without web support [56–59].

Four studies attempted to demonstrate the non-inferiority of the web-based intervention when compared to a face-to-face program. For this purpose, the same intervention was carried out in both face-to-face and online modalities. Two studies were based on a telerehabilitation program [53,60], another in a self-management program [55], and the last one was based on tele-education and symptom and mood telemonitoring [61].

Most repeated web content elements were tele-education content, self-management skills training, and tele-consultation with healthcare professionals. Only one study [57] excluded educational content. Education in self-management skills and tele-communication with healthcare professionals were excluded by Jiménez-Reguera et al. [61]. in three of the studies [53,55,60].

Table 1. Characteristics of the included studies.

Study (Year)	Sample Size, Distribution and Sample Age n (% Men): (Mean $\pm$ SD)	Treatment Status	Severity	Downs and Black (Risk of Bias)
Nguyen et al. (2013) [55]	125 (54%) allocated randomly into: EG: 68.5 $\pm$ 11.0 CG1: 68.2 $\pm$ 9.9 CG2: 69.3 $\pm$ 8.0	Clinically stable	Mild to very severe	22 (Poor quality)
Moy et al. (2015) [56]	238 (93.7%) allocated randomly into: EG: 67.0 $\pm$ 8.6 CG: 66.4 $\pm$ 9.2	Clinically stable	Mild to very severe	22 (Poor quality)
Moy et al. (2016) [57]	238 (93.7%) allocated randomly into: EG: 67.0 $\pm$ 8.6 CG: 66.4 $\pm$ 9.2	Clinically stable	Mild to very severe	23 (Poor quality)
Wang et al. (2017) [54]	120 (47.5%) allocated randomly into: EG: 69.3 $\pm$ 7.8 CG: 71.9 $\pm$ 8.1	After discharge	Moderate to very severe	20 (Fair quality)
Wan et al. (2017) [58]	109 (98,2%) allocated randomly into: EG: 68.4 $\pm$ 8.7 CG: 68.8 $\pm$ 7.9	Clinically stable	Mild to very severe	23 (Fair quality)
Bourne et al. (2017) [60]	90 (65.56%) allocated randomly into: EG: 69.1 $\pm$ 7.9 CG: 71.4 $\pm$ 8.6	Clinically stable	Moderate to very severe	22 (Fair quality)
Chaplin et al. (2017) [53]	103 (68.93%) allocated randomly into: EG: 66.4 $\pm$ 10.1 CG: 66.1 $\pm$ 8.1	Clinically stable	Moderate to very severe	22 (Fair quality)
Wan et al. (2020) [59]	109 (98.17%) allocated randomly into: EG: 68.4 $\pm$ 8.7 CG: 68.7 $\pm$ 7.9	Clinically stable	Mild to very severe	23 (Fair quality)
Jiménez-Reguera et al. (2020) [61]	36 (61.11%) allocated randomly into: EG: 68.1 $\pm$ 6.6 CG: 68.1 $\pm$ 7.0	After discharge	Moderate to very severe	18 (Poor quality)

Notes: EG: experimental group; CG: control group; SD: standard deviation.

In each study, the mean duration of intervention was 7.9 months (ranging from 6 weeks to 15 months). Most of the studies conducted an intervention over one year [54,55,57,59]. One study conducted an intervention of 10 months [61] and 4 studies conducted an intervention of less than 4 months [53,56,58,60].

The included studies evaluated quality of life using different tools. Disease-specific tools, e.g., the St. George's Respiratory Questionnaire (SGRQ), Chronic Respiratory Questionnaire (CRQ), Chronic Obstructive Pulmonary Disease Assessment Test (CAT) and general tools, e.g., the Short Form 36-Item Health Survey (SF-36) and EuroQol 5-Dimension Questionnaire (EQ- 5D) were used. The most commonly reported outcome was SGRQ, which was followed by CRQ and CAT.

Other variables used in several studies were: self-efficacy, functional capacity, dyspnea, physical activity, lung function, anxiety, and depression. Self-efficacy was measured in four studies, with the most used tool being the Exercise Self-Regulatory Efficacy Scale (Ex-SRES). Functional capacity was the second most frequently measured variable after quality of life. Five studies measured functional capacity with the 6MWT being the most used tool [54,55,58,60,61]. Four studies measured dyspnea and physical activity [54,55,58,60], three studies measured anxiety and depression [53,58,60] and two studies measured lung function [54,61].

The results obtained in the majority of included RCTs show no significant differences between groups in quality of life. Only one study reaches significant results in quality of life when compared to control intervention [54]. This result can be due to the duration of the program (12 months) and the content related to coaching. Furthermore, the majority of included studies showed significant improvements among the group in quality of life outcomes after intervention [53,55–57,61]. In addition, some studies aimed to demonstrate that web-based intervention was not inferior to face-to-face intervention and found similar results in quality of life for the intervention and control groups [53,55,60,61].

Regarding the results of other outcomes, most of the included studies in this systematic review have significant results in a functional capacity. Four studies [54,56–58] were significant between group results in favor of the web-based intervention group and three studies were significant among group improvements in a functional capacity after intervention for the web-based group [53,55,61]. Studies intended to demonstrate the non-inferiority of web-based support intervention found similar functional capacity results for the intervention and control groups.

Nguyen et al. [55] showed a significant improvement in dyspnea compared with the baseline in the experimental group and Wang et al. [54] showed a significance between the group's difference in dyspnea and lung function in favor of the experimental group.

3.3. Risk of Bias

The Downs and Blacks scale scores are presented in Table 1. The average score of the included studies in this systematic review was 21.6 points. In accordance with the suggested cut-off points to grade studies according to methodological quality, one article was rated as "fair" (15–19 points) [61] and eight were categorized as "good" (20–25 points) [53–60]. Figure 2 shows, in detail, the scoring of the studies on the different items of the Cochrane Risk of Bias Tool for randomized trials.

3.4. Results of Meta-Analysis

Data from seven RCTs reporting results obtained in quality of life were included in the meta-analysis [54–58,60,61]. All the included studies use the SGRQ to measure quality of life, except for the study conducted by Nguyen et al. [55] which used the CRQ.

All studies that did not provide sufficient data on quality of life (means and standard deviations at baseline or after the intervention) and for which no response was received from the authors were excluded. Ultimately, the analysis has been performed on a total of 873 patients (359 for control and 514 for intervention).

Figure 3 depicts the forest plot. Due to the statistical heterogeneity of the results ($I^2 = 83\%$, $p < 0.001$), a statistical random effects model was applied. Patient quality of life was not significantly improved in the intervention groups in comparison with controls (SMD = -0.21, 95% CI = -0.56, 0.14).

When compared to usual care, the mean difference showed a significant overall effect with the addition of the web-based supportive program to usual care (SMD = -1.26, 95% CI = -1.65, -0.86; $p < 0.001$, one study [54]). When compared to a pedometer walking intervention without web-support with a web-based supportive pedometer walking intervention (SMD = -0.05, 95% CI = -0.28, 0.17; $p = 0.64$, three studies [56–58]) or a web-based supportive intervention with a face-to-face intervention (SMD = -0.03; 95% CI= -0.33, 0.26; $p = 0.82$, three studies [55,60,61]), the pooled SMD showed no significant overall effect.

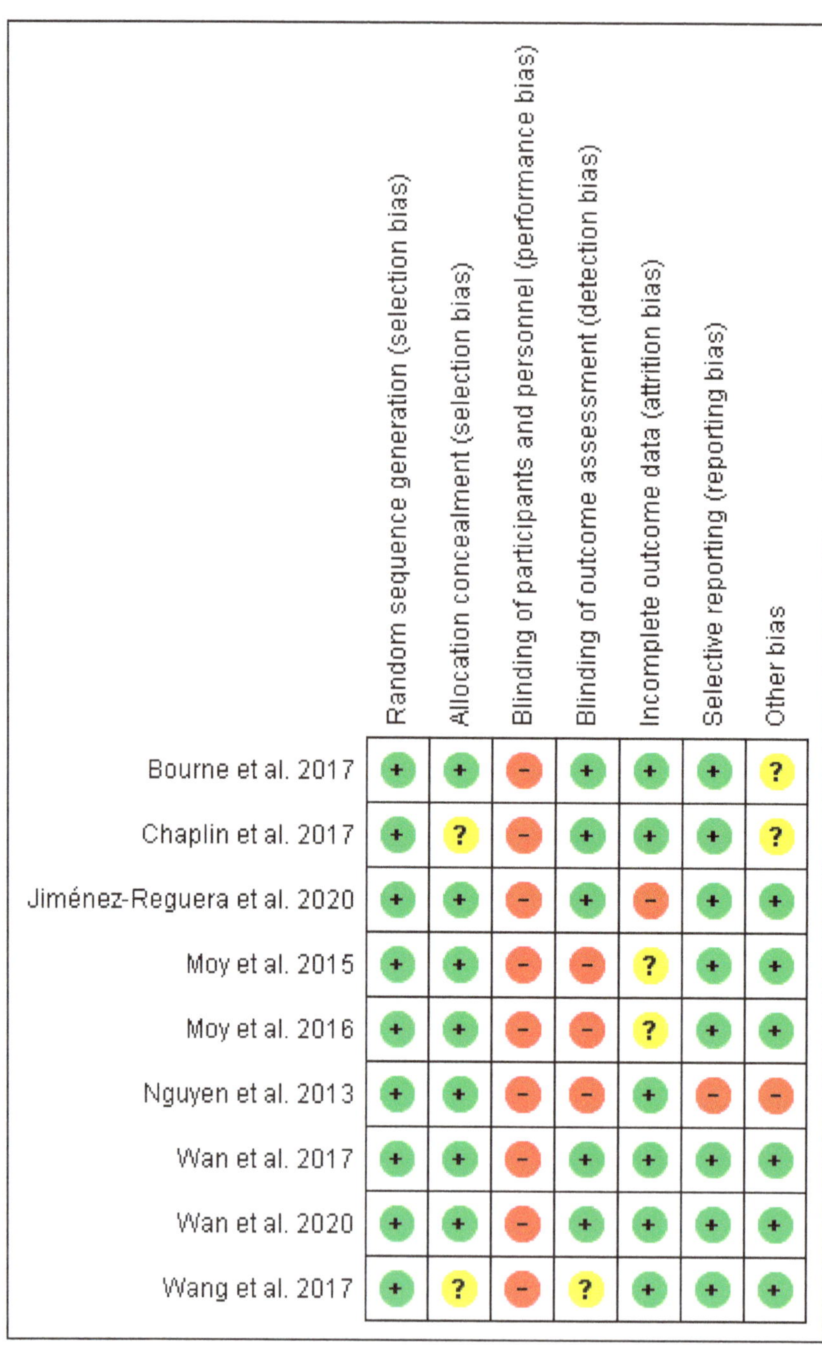

Figure 2. Risk of bias assessment of included studies. Notes: red, high risk of bias; yellow, moderate risk of bias; green, low risk of bias.

Table 2. Characteristics of the included studies in this systematic review.

Study	Interventions	Web Content Elements									Experimental Intervention Content	Intervention Duration	Outcomes Measures	Main Results
Nguyen et al. (2013) [55]	EG: internet-based dyspnea self-management program. CG1: face-to-face dyspnea self-management program. CG2: general health education.	1	2	3	4	5	6	7		9	Intervention included a personalized education program, dyspnea self-management training, exacerbation guidelines, personalized exercise with biweekly feedback and support, personal symptom and exercise log, real-time follow up, convenient access to information and support materials.	12 months	Quality of life measure(s): CRQ, SF-36. Other outcomes (measure(s)): self-efficacy (validated question); functional capacity (6MWT, ITT); dyspnea with activities (CRQ-D); arm endurance; adherence; satisfaction.	Quality of life results: No significant differences were found between groups in quality of life. EG participants had significant improvement in quality of life compared with baseline. Other outcomes results: Self-efficacy for managing dyspnea improved for the EG and CG1 compared with CG2. No significant differences were found in dyspnea and functional capacity between groups. EG participants had significant improvement in dyspnea and functional capacity compared with the baseline.
Moy et al. (2015) [56]	EG: web-based pedometer walking intervention. CG: pedometer walking intervention.	1	2	3	4	5	6				Step counting allowed for patient self-monitoring, new personalized weekly objectives were established, educational and motivational content to improve patient self-management, social support through an online forum.	4 months	Quality of life measure(s): SGRQ. Other outcomes (measure(s)): physical activity (pedometer); adherence; safety.	Quality of life results: No significant differences were found between groups in SGRQ total score. EG had significant improvement on symptoms and impact subscales compared to the CG. EG participants had significant improvement in SGRQ total score, symptoms, and impact compared with the baseline. Other outcomes results: EG had significant improvement on physical activity compared to the CG.
Moy et al. (2016) [57]	EG: web-based pedometer walking intervention. CG: pedometer walking intervention.		2	3	4	5	6				Step counting allowed for patient self-monitoring, new personalized weekly objectives were established, motivational content to improve patient self-management, social support through an online forum.	12 months	Quality of life measure(s): SGRQ. Other outcomes (measure(s)): physical activity (pedometer); adherence; safety.	Quality of life results: No significant differences were found between groups in quality of life. EG participants had significant improvement in SGRQ total score, symptoms, and impact compared with the baseline. Other outcomes results: Significant differences were found in physical activity between groups at month 4, but not in months 8 and 12.
Wang et al. (2017) [54]	EG: web based coaching program + routine care CG: routine care	1			4	5					These were used to manage patients' clinical and demographic variables and enabled communication between health care providers and patients. The patient was able to access disease information, pulmonary rehabilitation instructions, and particular management of the participant was determined according to the evolution of the disease.	12 months	Quality of life measure(s): SGRQ. Other outcomes (measure(s)): functional capacity (6MWT); dyspnea (MRC); lung function (spirometry).	Quality of life results: EG had significant improvement in the SGRQ total score, SGRQ symptoms, SGRQ activity and SGRQ impact compared to the CG. Other outcomes results: EG had significant improvement of lung function, functional capacity, and degree of dyspnea compared to CG.

Table 2. *Cont.*

Study	Interventions	Web Content Elements									Experimental Intervention Content	Intervention Duration	Outcomes Measures	Main Results
Wan et al. (2017) [58]	EG: web-based pedometer walking intervention. CG: pedometer walking intervention.	1	2	3	4	5	6				Step counting allowed for patient self-monitoring, new personalized weekly objectives were established, educational and motivational content to improve patient self-management, social support through an online forum.	3 months	Quality of life measure(s): SGRQ. Other outcomes (measure(s)): self-efficacy (Ex-SRES); functional capacity (6MWT); physical activity (pedometer); dyspnea (MRC); depression (BDI-II); COPD knowledge (BCKQ); social support (MOS-SSS); motivation and confidence to exercise; adherence.	Quality of life results: No significant differences were found between groups in quality of life. Other outcomes results: EG had significant improvement of daily step count compared to CG. No significant differences were found between groups in functional capacity, self-efficacy, dyspnea, depression, COPD knowledge, social support motivation, and confidence to exercise.
Bourne et al. (2017) [60]	EG: online supportive pulmonary rehabilitation. CG: face-to-face-supportive pulmonary rehabilitation.	1		3	4	5				9	Intervention included pulmonary online rehabilitation and educational videos to promote self-management.	6 weeks	Quality of life measure(s): SGRQ, CAT. Other outcomes (measure(s)): functional capacity (6MWT); dyspnea (MRC); anxiety and depression (HADS); adherence; safety.	Quality of life results: No significant differences were found between groups in quality of life. Other outcomes results: No significant differences were found between groups in exercise capacity, anxiety, and depression.
Chaplin et al. (2017) [53]	EG: web based pulmonary rehabilitation program. CG: face-to-face pulmonary rehabilitation program.	1	2		4	5	6	7		9	Intervention included education content, exacerbation guidelines, a home exercise program and goal setting, record of the progress, motivational interviewing techniques, and convenient access to information and support.	6–8 weeks	Quality of life measure(s): CRQ, CAT, EQ-5D. Other outcomes (measure(s)): self-efficacy (PRAISE); exercise capacity (ISWT, ESWT); anxiety and depression (HADS); COPD Knowledge (BCKQ).	Quality of life results: No significant differences were found between groups in quality of life. EG and CG participants had significant improvement in quality of life compared with the baseline. Other outcomes results: No significant differences were found between groups in any other outcome. EG and CG participants had significant improvement in functional capacity compared with the baseline.

Table 2. *Cont.*

Study	Interventions	Web Content Elements						Experimental Intervention Content	Intervention Duration	Outcomes Measures	Main Results
Wan et al. (2020) [59]	EG: web-based pedometer walking intervention. CG: pedometer walking intervention.	1	2	3	4	5	6	Step counting allowed for patient self-monitoring, new personalized weekly objectives were established, educational and motivational content provided to improve patient self-management, social support through an online forum.	15 months	Quality of life measure(s): SGRQ. Other outcomes (measure(s)): self-efficacy (Ex-SRES); physical activity (pedometer); acute exacerbations.	Quality of life results: No significant differences were found between groups in quality of life. CG participants had a significant worsening of quality of life compared with the baseline. There was no significant change in EG group, indicating no significant decline. Other outcomes results: No significant differences were found between groups in daily step count and self-efficacy. EG participants had significant improvement of acute exacerbations compared with baseline. EG participants had a minor decline that CG participants in daily step count compared with baseline.
Jiménez-Reguera et al. (2022) [61]	EG: web-based follow-up program. CG: face-to-face follow-up program.	1	2					Intervention included an educational program and data collection related to disease and physical activity, daily reminders of daily exercise, record of medication intake, daily mood, and level of tiredness.	10 months	Quality of life measure(s): SGRQ, CAT, EQ-5D. Other outcomes (measure(s)): functional capacity (6MWT); lung function (spirometry); adherence (CAP FISIO); adherence to physical activity (Morisky–Green Test).	Quality of life results: No significant differences were found between groups in quality of life. EG participants had a significant improvement of quality of life in compared with the baseline. Other outcomes results: No significant differences between the two groups were observed in functional capacity and lung function. EG participants had significant improvement of functional capacity in compared with baseline. EG participants had a significant improvement of adherence to the program and adherence to physical activity in compared with CG.

Notes: 1, tele-education content; 2, symptom and mood telemonitoring; 3, physical activity monitoring and personalized feedback to the patient; 4, tele-education in self-management skills; 5, tele-consultation with healthcare professionals; 6, tele-communication with other patients; 7, remote decision support systems; 9, tele-rehabilitation; EG, Experimental Group; CG, Control Group; CRQ, Chronic Respiratory Questionnaire; SF-36, Short Form 36 survey tool version 1; 6MWT, 6-Minute Walk Test; ITT, Incremental Treadmill Test; CRQ-D, Chronic Respiratory Questionnaire Dyspnea subscale; SGRQ, St. George's Respiratory Questionnaire; MRC, Medical Research Council scale; Ex-SRES, Exercise Self-Regulatory Efficacy Scale; BDI-II, Beck Depression Inventory-II; BCKQ, Bristol COPD Knowledge Questionnaire; MOS-SSS, Medical Outcomes Study Social Support Survey; CAT, COPD Assessment Test; HADS, Hospital Anxiety and Depression Scale; EQ-5D, EuroQol 5-Dimension questionnaire; PRAISE, PR Adapted Index of Self-Efficacy; ISWT, Incremental Shuttle Walk Test; ESWT, Endurance Shuttle Walk Test; CAP FISIO, Respiratory Physiotherapy Adherence self-report questionnaire.

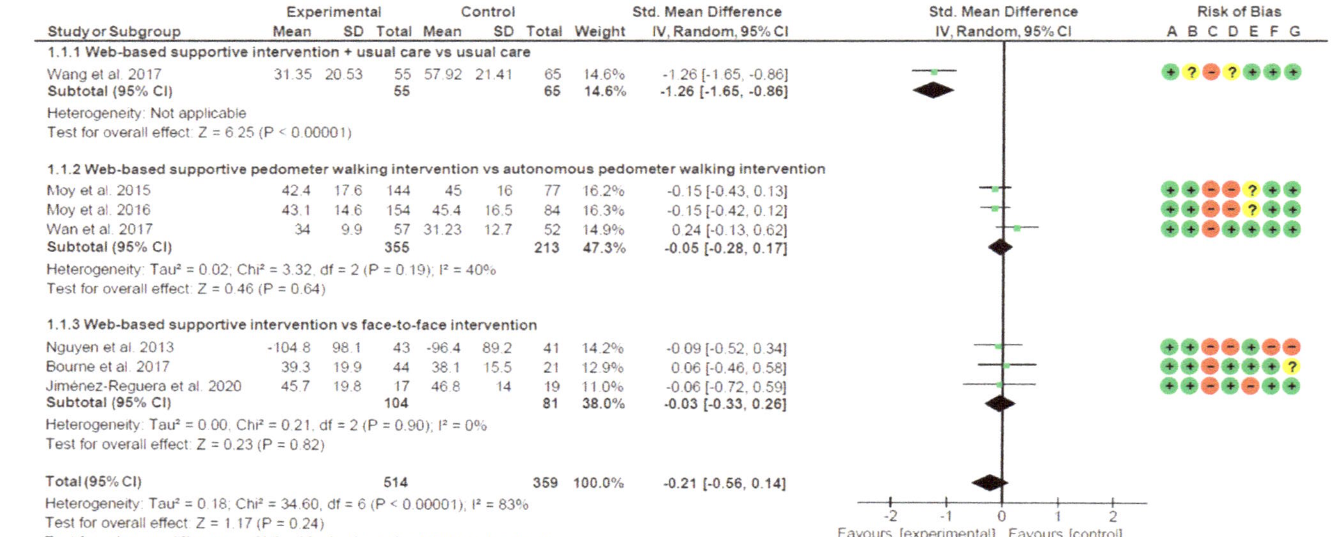

Figure 3. Forest plot of the effect of web-based supportive interventions on quality of life in COPD patients compared with the control group. Notes: Risk of bias color: red, high risk of bias; yellow, moderate risk of bias; green, low risk of bias.

4. Discussion

The continuous technological growth of today's society, the increasing use of online services, and patients' need for new supportive solutions have facilitated the creation of new web-based interventions that have not been properly tested yet. To the authors' knowledge, this is the first systematic review and meta-analysis evaluating the effects of web-based supportive interventions on quality of life in COPD patients.

Our results support the idea that web-based supportive interventions can improve the quality of life in COPD patients. Nevertheless, it is important to note that the systematic review of the literature related to the design of web-based supportive interventions must be correctly interpreted, considering the different sample sizes of the studies, the differences in length of therapy and follow up, and the differences in effect size of the included studies.

Our systematic review is the first one specifically exploring the effects of web-based supportive interventions in quality of life in COPD patients, with nine RCTs [53–61] included in the qualitative analysis. Our results are consistent with those of previous systematic reviews performed in COPD patients and other telehealth systems [23,34,65–68].

Internet-based interventions can, however, present a rather confusing picture as the only common ground is the delivery medium. The interventions may range from posting pamphlets online to dynamic combinations of text-based information and communicative features, such as forums, "ask an expert", or multimedia tools, to individually computer tailored content [69].

Regarding web components, Sobnath et al. [70] described the possible features that a potential supporting tool for COPD patients should have in their systematic review. The tools must be easily accessible both for patients and health professionals. In addition, they should be adapted to elderly patients with limited experience in the use of technology and have a user-friendly interface. According to previous literature, the tool should include a customized education section for each patient, with disease-specific information and self-management material, phycological motivation to encourage good adherence, electronic coaching, comment sections, and social networks to share information with health professionals [70,71].

Among the web-based supportive interventions analyzed, the educational content was the most used alone or in combination with other contents, and the most frequent comparison treatment was the same in a face-to-face format. When compared, web-based supportive interventions showed similar results in all measured variables.

The web-based support interventions analyzed in this systematic review used a variety of components of COPD patient support tools that were described by Sobnath et al. [70], such as personalized education sections and social networks to share information with medical professionals. Our results are in line with the previous systematic reviews conducted in patients with cancer in which the most common and promising interventions include a combination of effective communication with healthcare providers, customized educational strategies based on the patient's disease and condition, ongoing symptom monitoring, disease self-management tools, and automated feedback [72,73].

It is difficult to determine exactly which web elements are most important in designing an effective disease management tool, and to determine whether the effects are due to one or some of the elements, or to all of them together. Effective communication with healthcare providers is highly recommended content for web-based support intervention since patients have different characteristics, preferences, and needs [62,74] as seen in the Norwegian WebChoice study [75].

A Cochrane review identified that in improving the quality of life of COPD patients, the effects of technology-based interventions attenuated over time. Support interventions based on new technologies were found to be more effective in improving the quality of life of COPD patients than interventions based on face-to-face education and support materials even at six months, but not at one year. This is probably due to the fact that educational and motivational content were not updated during the maintenance phase [67,76], highlighting the importance of these elements.

Our systematic revision of web-based interventions in COPD, have shown additional improvements in dyspnea and physical activity in programs which include self-management components [54–56,58,59]. Different reviews [13,17,19,77] have reported the opportunities for telehealth interventions in increasing physical activity and symptoms when behavioral components are included.

Given the great heterogeneity and diversity of the studies included in this systematic review, it might not be recommended to perform a meta-analysis. However, a random-effects model was chosen to allow the pooling of more clinically heterogeneous studies [78]. Furthermore, to adequately answer the question discussed in this review, i.e., whether web-based support interventions are effective in improving the quality of life of COPD patients, and due to the great diversity of the studies published to date, it was necessary to use a wide range of studies in which these types of interventions were used. It is therefore required to adequately justify our findings.

The findings of our meta-analysis of pooled data do not identify statistically significant differences in the quality of life of COPD patients. Even though the results of this meta-analysis suggested that there is no evidence that web-based support interventions are effective in increasing the quality of life of COPD patients, the results should be analyzed by subgroups.

This meta-analysis supports the promising role and the feasibility of web-based supportive interventions in COPD patients to improve quality of life when added to the usual care, reaching the currently minimum significant established difference for SGRQ results in a mean COPD sample population of -4 points [79], but not when compared to an autonomous pedometer walking intervention or face-to-face treatment. These results are in line with the increasing evidence in literature on the success of telehealth interventions [64–67].

Four included studies used wearable systems like the pedometer in the web-based supportive programs [56–59]. Those programs showed similar results in quality of life to those using autonomous interventions. Those results can be due to the theory of self-regulation [80], in which the use of a pedometer (either web-based or autonomous) guides the patient to their own feelings, thoughts, and behaviors to achieve specific goals. In addition, blinding patients from the web-based supportive pedometer walking interventions would require giving a pedometer to the control group; this may cause the results of the control group to be altered, since the simple fact of having control of their daily steps may promote an increase in the physical activity of the patients.

Other studies have used web-based pulmonary rehabilitation programs compared to the same program developed face-to-face. The results obtained by Bourne et al. [60] show no significant differences between groups in quality of life. In the study by Jiménez-Reguera et al. and Nguyen et al. [55,61], the results show statistically significant improvement on the quality of life of the web-based group, but no differences between groups after intervention.

These studies support the argument that comparable results between web-based and face-to-face interventions, or the absence of impairment can be considered a success as seen in previous reviews [23,81], due to the opportunities for new technologies for at risk COPD patients [23,82]. In this line of thinking, web-based supportive interventions may complement routine care as no significant differences were found between the face-to-face and online modalities [70]. Some further advantages should be derived from the use of telehealth interventions for this argument to be valid and the extensive literature on this topic leaves no doubt. Telehealth intervention groups show better results than the control group in risk of exacerbation [83], costs of health care [84], hospitalization days [83], risk of hospitalizations, and risks of the emergency department visit, without the need for travel [85].

Our results are consistent with the increasing evidence in the literature on the efficacy of telehealth supportive interventions [23,34,65–67]. The use of web-based supportive interventions for COPD patients is not recommended if based solely on quality of life data, but there is also no argument against the use of these interventions.

Regarding the methodological quality, the seven RCTs included in the meta-analysis [54–58,60,61] were classified as "poor quality" according to the Downs and Black quality checklist. The main reason for the low quality of the studies included in this systematic review lies in methodological issues. For example, it has been shown in previous studies that selection bias in interventions based on technological tools is evident. The reason for this is that some patients are already used to the use of new technologies and the Internet, leading to the automatic preference of these over other tools [86].

In addition, web-based interventions appear to be unsuitable for all patients because the level of follow-up and adherence to treatments is often low [87]. Other factors that also increase the risk of bias involve the lack of patient blinding and not adequately describing the randomization method.

Strengths and Limitations

To start with, we need to assess the strengths of the present study. First, only RCTs were included to increase the quality of evidence, and second, we were able to pool data from seven studies in a meta-analysis.

Thirdly, in previous studies on the effects of e-health's intervention, web-based supportive intervention was not separately analyzed. In this study, web-based supportive intervention was first taken as a primary intervention.

The major weakness of this systematic review is the limited number of RCTs focused on web-based supportive interventions. However, the inclusion criteria enabled us to include articles with this type of intervention even if quality of life was not the main variable. There are no obvious reasons for the lack of research on COPD web-based supportive interventions but the issue of possible facilitators, such as a decreased burden of web-based interventions and the personalized nature and possible barriers including security and technical issues, should be addressed when performing these types of health interventions [88].

Other limitations need to be reported. First, one subgroup in our meta-analysis only had one study. Second, it should be noted that the diversity of the targeted interventions makes it difficult to distinguish whether the web-based supportive intervention was solely responsible for the observed effects. Third, since the authors were only fluent in French, English, and Spanish, they were only able to review research published or translated into these languages and not studies in other languages.

5. Conclusions

This systematic review and meta-analysis show the promising potential of web-based supportive interventions for improving quality of life in COPD patients. Due to the methodological limitations, the heterogeneity, and the limited number of studies in this field, the results should be treated with caution. Further randomized controlled studies are needed to evaluate the effect of web-based supportive interventions, with larger COPD populations and using appropriate interventions to blind the control group, thus increasing the evidence in this field of research.

Practical Implications

Our findings suggest that the most common and promising web-based supportive intervention content are the educational content as well as communication with healthcare professionals. This systematic review and meta-analysis suggest that web-based supportive interventions may complement or accompany treatments in COPD patients due to the advantages of online interventions.

Author Contributions: Conceptualization: A.C.-M., L.L.-L., J.M.-N., J.R.-T. and M.C.V.; methodology: A.C.-M., A.H.-C., J.M.-N., J.R.-T. and M.C.V.; software: A.C.-M., A.H.-C. and A.O.-R.; validation: A.C.-M., L.L.-L. and A.H.-C.; formal analysis: A.C.-M. and A.O.-R.; investigation: A.C.-M., L.L.-L., J.M.-N.; resources: A.C.-M.; data curation: A.C.-M. and A.O.-R.; writing—original draft preparation: A.C.-M., M.C.V. and L.L.-L.; writing—review and editing: A.C.-M., J.R.-T. and M.C.V.; visualization: L.L.-L., A.H.-C. and J.M.-N.; supervision: M.C.V.; project administration: M.C.V.; funding acquisition: M.C.V. All authors have read and agreed to the published version of the manuscript.

Funding: This research was funded by Ministerio de Ciencia, Innovación y Universidades, grant number FPU: 19/02609, FPU: 17/00408 and FPU: 16/01531 and by University of Granada, grant number FPU: PP20/05.

Institutional Review Board Statement: Not applicable.

Informed Consent Statement: Not applicable.

Conflicts of Interest: The authors declare no conflict of interest.

Appendix A

("Chronic Obstructive Lung Disease" OR Airflow Obstruction, Chronic" OR "Chronic Airflow Obstructions" OR "Chronic Airflow Obstruction" OR "COPD" OR "Chronic Obstructive Pulmonary Disease" OR "COAD" OR "Chronic Obstructive Airway Disease" OR "Airflow Obstructions, Chronic") AND ("eHealth" OR "ehealth" OR "e-Health" OR "e-health" OR "telemedicine" OR "tele-medicine" OR "Mobile Health" OR "Health, Mobile" OR "mHealth" OR "m-Health" OR "m-health" OR "telehealth" OR "tele-health" OR "telecare" OR "tele-care" OR "telemonitoring" OR "tele-monitoring" OR "teleconsultation" OR "tele-consultation" OR "health informatics" OR "internet" OR "mobile") AND ("Life Quality" OR "Health-Related Quality Of Life" OR "Health Related Quality Of Life" OR "HRQOL" OR "quality of life" OR "management" OR "adherence" OR "healthy lifestyle" OR "well-being").

References

1. Global Initiative for Chronic Obstructive Lung Disease. GOLD Report 2020. Available online: https://goldcopd.org/wp-content/uploads/2019/12/GOLD-2020-FINAL-ver1.2-03Dec19_WMV.pdf (accessed on 30 November 2021).
2. Gunasekaran, K.; Ahmad, M.; Rehman, S.; Thilagar, B.; Gopalratnam, K.; Ramalingam, S.; Paramasivam, V.; Arora, A.; Chandran, A. Impact of a Positive Viral Polymerase Chain Reaction on Outcomes of Chronic Obstructive Pulmonary Disease (COPD) Exacerbations. *Int. J. Environ. Res. Public Health* **2020**, *17*, 8072. [CrossRef]
3. Lopez, A.D.; Shibuya, K.; Rao, C.; Mathers, C.D.; Hansell, A.; Held, L.S.; Schmid, V.; Buist, S. Chronic obstructive pulmonary disease: Current burden and future projections. *Eur. Respir. J.* **2006**, *27*, 397–412. [CrossRef]
4. Disler, R.T.; Gallagher, R.D.; Davidson, P.M. Factors influencing self-management in chronic obstructive pulmonary disease: An integrative review. *Int. J. Nurs. Stud.* **2012**, *49*. [CrossRef] [PubMed]
5. Ghobadi, H.; Ahari, S.S.; Kameli, A.; Lari, S.M. The Relationship between COPD Assessment Test (CAT) Scores and Severity of Airflow Obstruction in Stable COPD Patients. *Tanaffos* **2012**, *11*, 22–26.
6. Izquierdo, J.L.; Morena, D.; Gonzalez-Moro, J.M.R.; Paredero, J.M.; Pérez, B.; Graziani, D.; Gutiérrez, M.; Rodríguez, J.M. Manejo clínico de la EPOC en situación de vida real. Análisis a partir de big data. *Arch. Bronconeumol.* **2021**, *57*, 94–100. [CrossRef]
7. Kosteli, M.-C.; Heneghan, N.R.; Roskell, C.; Williams, S.; Adab, P.; Dickens, A.P.; Enocson, A.; Fitzmaurice, D.A.; Jolly, K.; Jordan, R.; et al. Barriers and enablers of physical activity engagement for patients with COPD in primary care. *Int. J. Chronic Obstr. Pulm. Dis.* **2017**, *12*, 1019–1031. [CrossRef] [PubMed]
8. Stellefson, M.L.; Shuster, J.J.; Chaney, B.H.; Paige, S.R.; Alber, J.M.; Chaney, J.D.; Sriram, P.S. Web-based Health Information Seeking and eHealth Literacy among Patients Living with Chronic Obstructive Pulmonary Disease (COPD). *Health Commun.* **2017**, *33*, 1410–1424. [CrossRef] [PubMed]
9. Sadeghi, S.; Brooks, D.; Stagg-Peterson, S.; Goldstein, R. Growing Awareness of the Importance of Health Literacy in Individuals with COPD. *COPD: J. Chronic Obstr. Pulm. Dis.* **2013**, *10*, 72–78. [CrossRef]
10. Hernandez, P.; Balter, M.; Bourbeau, J.; Hodder, R. Living with chronic obstructive pulmonary disease: A survey of patients' knowledge and attitudes. *Respir. Med.* **2009**, *103*, 1004–1012. [CrossRef] [PubMed]
11. Leung, J.M.; Niikura, M.; Yang, C.W.T.; Sin, D.D. COVID-19 and COPD. *Eur. Respir. J.* **2020**, *56*, 2002108. [CrossRef] [PubMed]
12. Rutkowski, S. Management Challenges in Chronic Obstructive Pulmonary Disease in the COVID-19 Pandemic: Telehealth and Virtual Reality. *J. Clin. Med.* **2021**, *10*, 1261. [CrossRef]

13. Guerra-Paiva, S.; Dias, F.; Costaa, D.; Santos, V.; Santos, C. DPO2 Project: Telehealth to enhance the social role of physical activity in people living with COPD. *Proc. Comput. Sci.* **2021**, *181*, 869–875. [CrossRef]

14. Jansen, F.; Van Uden-Kraan, C.F.; Van Zwieten, V.; Witte, B.I.; Leeuw, I.M.V.-D. Cancer survivors' perceived need for supportive care and their attitude towards self-management and eHealth. *Support. Care Cancer* **2014**, *23*, 1679–1688. [CrossRef] [PubMed]

15. Murray, E. Web-Based Interventions for Behavior Change and Self-Management: Potential, Pitfalls, and Progress. *Medicine 2.0* **2012**, *1*, e3. [CrossRef]

16. Slev, V.N.; Mistiaen, P.; Pasman, H.R.W.; Leeuw, I.M.V.-D.; van Uden-Kraan, C.F.; Francke, A.L. Effects of eHealth for patients and informal caregivers confronted with cancer: A meta-review. *Int. J. Med. Inform.* **2016**, *87*, 54–67. [CrossRef]

17. Gaveikaite, V.; Grundstrom, C.; Winter, S.; Chouvarda, I.; Maglaveras, N.; Priori, R. A systematic map and in-depth review of European telehealth interventions efficacy for chronic obstructive pulmonary disease. *Respir. Med.* **2019**, *158*, 78–88. [CrossRef]

18. El-Gayar, O.; Timsina, P.; Nawar, N.; Eid, W. A systematic review of IT for diabetes self-management: Are we there yet? *Int. J. Med. Inform.* **2013**, *82*, 637–652. [CrossRef]

19. Bonnevie, T.; Smondack, P.; Elkins, M.; Gouel, B.; Medrinal, C.; Combret, Y.; Muir, J.-F.; Cuvelier, A.; Prieur, G.; Gravier, F.-E. Advanced telehealth technology improves home-based exercise therapy for people with stable chronic obstructive pulmonary disease: A systematic review. *J. Physiother.* **2020**, *67*, 27–40. [CrossRef] [PubMed]

20. Hall, A.K.; Stellefson, M.; Bernhardt, J. Healthy Aging 2.0: The Potential of New Media and Technology. *Prev. Chronic Dis.* **2012**, *9*. [CrossRef]

21. Stellefson, M.; Chaney, B.; Barry, A.; Chavarria, E.; Tennant, B.; Walsh-Childers, K.; Sriram, P.; Zagora, J. Web 2.0 Chronic Disease Self-Management for Older Adults: A Systematic Review. *J. Med. Internet Res.* **2013**, *15*, e35. [CrossRef] [PubMed]

22. Donner, C.F.; Raskin, J.; ZuWallack, R.; Nici, L.; Ambrosino, N.; Balbi, B.; Blackstock, F.; Casaburi, R.; Dreher, M.; Effing, T.; et al. Incorporating telemedicine into the integrated care of the COPD patient a summary of an inter-disciplinary workshop held in Stresa, Italy, 7–8 September 2017. *Respir. Med.* **2018**, *143*, 91–102. [CrossRef]

23. Polisena, J.; Tran, K.; Cimon, K.; Hutton, B.; McGill, S.; Palmer, K.; Scott, R.E. Home telehealth for chronic obstructive pulmonary disease: A systematic review and meta-analysis. *J. Telemed. Telecare* **2010**, *16*, 120–127. [CrossRef] [PubMed]

24. Udsen, F.W.; Hejlesen, O.; Ehlers, L.H. A systematic review of the cost and cost-effectiveness of telehealth for patients suffering from chronic obstructive pulmonary disease. *J. Telemed. Telecare* **2014**, *20*, 212–220. [CrossRef] [PubMed]

25. Liu, F.; Jiang, Y.; Xu, G.; Ding, Z. Effectiveness of Telemedicine Intervention for Chronic Obstructive Pulmonary Disease in China: A Systematic Review and Meta-Analysis. *Telemed. e-Health* **2020**, *26*, 1075–1092. [CrossRef]

26. Alwashmi, M.; Hawboldt, J.; Davis, E.; Marra, C.; Gamble, J.-M.; Abu Ashour, W. The Effect of Smartphone Interventions on Patients with Chronic Obstructive Pulmonary Disease Exacerbations: A Systematic Review and Meta-Analysis. *JMIR mHealth uHealth* **2016**, *4*, e105. [CrossRef]

27. Donner, C.; ZuWallack, R.; Nici, L. The Role of Telemedicine in Extending and Enhancing Medical Management of the Patient with Chronic Obstructive Pulmonary Disease. *Medicina* **2021**, *57*, 726. [CrossRef] [PubMed]

28. Jang, S.; Kim, Y.; Cho, W.-K. A Systematic Review and Meta-Analysis of Telemonitoring Interventions on Severe COPD Exacerbations. *Int. J. Environ. Res. Public Health* **2021**, *18*, 6757. [CrossRef]

29. Snoswell, C.L.; Stringer, H.; Taylor, M.L.; Caffery, L.J.; Smith, A.C. An overview of the effect of telehealth on mortality: A systematic review of meta-analyses. *J. Telemed. Telecare* **2021**. [CrossRef]

30. Metting, E.; Dassen, L.; Aardoom, J.; Versluis, A.; Chavannes, N. Effectiveness of Telemonitoring for Respiratory and Systemic Symptoms of Asthma and COPD: A Narrative Review. *Life* **2021**, *11*, 1215. [CrossRef]

31. Janjua, S.; Carter, D.; Threapleton, C.J.; Prigmore, S.; Disler, R.T. Telehealth interventions: Remote monitoring and consultations for people with chronic obstructive pulmonary disease (COPD). *Cochrane Database Syst. Rev.* **2021**, *2021*. [CrossRef]

32. Shaw, G.; Whelan, M.E.; Armitage, L.C.; Roberts, N.; Farmer, A.J. Are COPD self-management mobile applica-tions effective? A systematic review and meta-analysis. *NPJ Prim. Care Respir. Med.* **2020**, *30*, 11. [CrossRef]

33. Peters, G.M.; Kooij, L.; Lenferink, A.; van Harten, W.H.; Doggen, C.J.M. The Effect of Telehealth on Hospital Services Use: Systematic Review and Meta-analysis. *J. Med. Internet Res.* **2021**, *23*, e25195. [CrossRef]

34. Deng, N.; Gu, T.; Zhao, Q.; Zhang, X.; Zhao, F.; He, H. Effects of telephone support on exercise capacity and quality of life in patients with chronic obstructive pulmonary disease: A meta-analysis. *Psychol. Health Med.* **2018**, *23*, 917–933. [CrossRef] [PubMed]

35. Voncken-Brewster, V.; Tange, H.; de Vries, H.; Nagykaldi, Z.; Winkens, B.; van der Weijden, T. A randomized controlled trial evaluating the effectiveness of a web-based, computer-tailored self-management intervention for people with or at risk for COPD. *Int. J. Chronic Obstr. Pulm. Dis.* **2015**, *10*, 1061–1073. [CrossRef]

36. Samoocha, D.; Bruinvels, D.J.; Elbers, N.A.; Anema, J.R.; van der Beek, A.J. Effectiveness of web-based inter-ventions on patient empowerment: A systematic review and meta-analysis. *J. Med. Internet Res.* **2010**, *12*, e23. [CrossRef]

37. Pulman, A.; Lepping, P.; Paravastu, S.C.V.; Turner, J.; Billings, P.; Minchom, P. A patient centred framework for improving LTC quality of life through Web 2.0 technology. *Health Inform. J.* **2010**, *16*, 15–23. [CrossRef] [PubMed]

38. Murray, E.; Burns, J.; Tai, S.S.; Lai, R.; Nazareth, I. Interactive Health Communication Applications for people with chronic disease. *Cochrane Database Syst. Rev.* **2005**, CD004274. [CrossRef]

39. Bennett, G.G.; Glasgow, R.E. The delivery of public health interventions via the internet: Actualizing their po-tential. *Annu. Rev. Public Health* **2009**, *30*, 273–292. [CrossRef] [PubMed]

40. Selzler, A.M.; Wald, J.; Sedeno, M.; Jourdain, T.; Janaudis-Ferreira, T.; Goldstein, R.; Bourbeau, J.; Stickland, M.K. Telehealth pulmonary rehabilitation: A review of the literature and an example of a nationwide initiative to im-prove the accessibility of pulmonary rehabilitation. *Chron. Respir. Dis.* **2018**, *15*, 41–47. [CrossRef] [PubMed]

41. Ambrosino, N.; Vagheggini, G.; Mazzoleni, S.; Vitacca, M. Telemedicine in chronic obstructive pulmonary disease. *Breathe* **2016**, *12*, 350–356. [CrossRef] [PubMed]

42. Page, M.J.; McKenzie, J.E.; Bossuyt, P.M.; Boutron, I.; Hoffmann, T.C.; Mulrow, C.D.; Shamseer, L.; Tetzlaff, J.M.; Akl, E.A.; Brennan, S.E.; et al. The PRISMA 2020 statement: An updated guideline for reporting systematic reviews. *BMJ* **2021**, *372*, n71. [CrossRef] [PubMed]

43. Higgins, J.P.T.; Altman, D.G.; Gøtzsche, P.C.; Jüni, P.; Moher, D.; Oxman, A.D.; Savović, J.; Schulz, K.F.; Weeks, L.; Sterne, J.A.C.; et al. The Cochrane Collaboration's tool for assessing risk of bias in randomised trials. *BMJ* **2011**, *343*, d5928. [CrossRef]

44. Center for Reviews and Dissemination. *Systematic Reviews—CRD's Guidelines for Undertaking Reviews in Healthcare*; Center for Reviews and Dissemination: York, UK, 2009.

45. Downs, S.H.; Black, N. The feasibility of creating a checklist for the assessment of the methodological quality both of randomised and non-randomised studies of health care interventions. *J. Epidemiol. Community Health* **1998**, *52*, 377–384. [CrossRef] [PubMed]

46. Deeks, J.; Dinnes, J.; D'Amico, R.; Sowden, A.J.; Sakarovitch, C.; Song, F.; Petticrew, M.; Altman, D.G. Evaluating non-randomized intervention studies. *Health Technol. Assess.* **2003**, *7*, iii-173. [CrossRef]

47. Saunders, L.D.; Soomro, G.M.; Buckingham, J.; Jamtvedt, G.; Raina, P. Assessing the Methodological Quality of Nonrandomized Intervention Studies. *West. J. Nurs. Res.* **2003**, *25*, 223–237. [CrossRef]

48. Higgins, J.; Thomas, J.; Chandler, J.; Cumpston, M.; Li, T.; Page, M.; Welch, V. (Eds.) Cochrane Handbook for Systematic Reviews of Interventions Version 6.0 (Updated July 2019). Cochrane. 2019. Available online: www.training.cochrane.org/handbook (accessed on 30 November 2021).

49. Holroyd-Leduc, J.M.; Helewa, A.; Walker, J.M. Critical Evaluation of Research in Physical Rehabilitation: Towards Evidence-Based Practice. Philadelphia: WB Saunders Company, 2000. *Evid. Based Med.* **2002**, *7*, 135. [CrossRef]

50. Newham, J.J.; Presseau, J.; Heslop-Marshall, K.; Russell, S.; Ogunbayo, O.J.; Netts, P.; Hanratty, B.; Kaner, E. Features of self-management interventions for people with COPD associated with improved health-related quality of life and reduced emergency department visits: A systematic review and meta-analysis. *Int. J. Chronic Obstr. Pulm. Dis.* **2017**, *12*, 1705–1720. [CrossRef]

51. Blakemore, A.; Dickens, C.; Guthrie, E.; Bower, P.; Kontopantelis, E.; Afzal, C.; Coventry, P. Depression and anxiety predict health-related quality of life in chronic obstructive pulmonary disease: Systematic review and meta-analysis. *Int. J. Chronic Obstr. Pulm. Dis.* **2014**, *9*, 501–512. [CrossRef]

52. Cannon, D.; Buys, N.; Sriram, K.B.; Sharma, S.; Morris, N.; Sun, J. The effects of chronic obstructive pulmonary disease self-management interventions on improvement of quality of life in COPD patients: A meta-analysis. *Respir. Med.* **2016**, *121*, 81–90. [CrossRef]

53. Chaplin, E.; Hewitt, S.; Apps, L.; Bankart, J.; Pulikottil-Jacob, R.; Boyce, S.; Morgan, M.; Williams, J.; Singh, S. Interactive web-based pulmonary rehabilitation programme: A randomised controlled feasibility trial. *BMJ Open* **2017**, *7*, e013682. [CrossRef]

54. Wang, L.; He, L.; Tao, Y.; Sun, L.; Zheng, H.; Zheng, Y.; Shen, Y.; Liu, S.; Zhao, Y.; Wang, Y.; et al. Evaluating a Web-Based Coaching Program Using Electronic Health Records for Patients with Chronic Obstructive Pulmonary Disease in China: Randomized Controlled Trial. *J. Med. Internet Res.* **2017**, *19*, e264. [CrossRef]

55. Nguyen, H.Q.; Donesky, D.; Reinke, L.F.; Wolpin, S.; Chyall, L.; Benditt, J.O.; Paul, S.M.; Carrieri-Kohlman, V. Internet-Based Dyspnea Self-Management Support for Patients with Chronic Obstructive Pulmonary Disease. *J. Pain Symptom Manag.* **2012**, *46*, 43–55. [CrossRef]

56. Moy, M.L.; Collins, R.; Martinez, C.H.; Kadri, R.; Roman, P.; Holleman, R.G.; Kim, H.M.; Nguyen, H.Q.; Cohen, M.D.; Goodrich, D.; et al. An Internet-Mediated Pedometer-Based Program Improves Health-Related Quality-of-Life Domains and Daily Step Counts in COPD. *Chest* **2015**, *148*, 128–137. [CrossRef]

57. Moy, M.L.; Martinez, C.H.; Kadri, R.; Roman, P.; Holleman, R.G.; Kim, H.M.; Nguyen, H.Q.; Cohen, M.D.; Goodrich, D.E.; Giardino, N.D.; et al. Long-Term Effects of an Internet-Mediated Pedometer-Based Walking Program for Chronic Obstructive Pulmonary Disease: Randomized Controlled Trial. *J. Med. Internet Res.* **2016**, *18*, e215. [CrossRef]

58. Wan, E.S.; Kantorowski, A.; Homsy, D.; Teylan, M.; Kadri, R.; Richardson, C.R.; Gagnon, D.R.; Garshick, E.; Moy, M.L. Promoting physical activity in COPD: Insights from a randomized trial of a web-based intervention and pedometer use. *Respir. Med.* **2017**, *130*, 102–110. [CrossRef] [PubMed]

59. Wan, E.S.; Kantorowski, A.; Polak, M.; Kadri, R.; Richardson, C.R.; Gagnon, D.R.; Garshick, E.; Moy, M.L. Long-term effects of web-based pedometer-mediated intervention on COPD exacerbations. *Respir. Med.* **2020**, *162*, 105878. [CrossRef] [PubMed]

60. Bourne, S.; Devos, R.; North, M.; Chauhan, A.; Green, B.; Brown, T.; Cornelius, V.; Wilkinson, T. Online versus face-to-face pulmonary rehabilitation for patients with chronic obstructive pulmonary disease: Randomised controlled trial. *BMJ Open* **2017**, *7*, e014580. [CrossRef] [PubMed]

61. Jiménez-Reguera, B.; López, E.M.; Fitch, S.; Juarros-Monteagudo, L.; Sánchez-Cortés, M.; Rodríguez-Hermosa, J.L.; Calle-Rubio, M.; Hernández-Criado, M.T.; López-Martín, M.; Angulo, S.; et al. Development and Preliminary Evaluation of the Effects of an mHealth Web-Based Platform (HappyAir) on Adherence to a Maintenance Program After Pulmonary Rehabilitation in Patients with Chronic Obstructive Pulmonary Disease: Randomized Controlled Trial (Preprint). *JMIR mHealth uHealth* **2020**, *8*, e18465. [CrossRef]

62. Barak, A.; Klein, B.; Proudfoot, J.G. Defining Internet-Supported Therapeutic Interventions. *Ann. Behav. Med.* **2009**, *38*, 4–17. [CrossRef] [PubMed]
63. Kuijpers, W.; Groen, W.G.; Aaronson, N.K.; Van Harten, W.H. A Systematic Review of Web-Based Interventions for Patient Empowerment and Physical Activity in Chronic Diseases: Relevance for Cancer Survivors. *J. Med. Internet Res.* **2013**, *15*, e37. [CrossRef]
64. Gregersen, T.L.; Green, A.; Frausing, E.; Ringbæk, T.; Brøndum, E.; Ulrik, C.S. Do telemedical interventions im-prove quality of life in patients with COPD? A systematic review. *Int. J. COPD* **2016**, *11*, 809–822.
65. Janjua, S.; Banchoff, E.; Threapleton, C.J.D.; Prigmore, S.; Fletcher, J.; Disler, R.T. Digital interventions for the management of chronic obstructive pulmonary disease. *Cochrane Database Syst. Rev.* **2021**, *4*, CD013246. [CrossRef] [PubMed]
66. Gorst, S.L.; Armitage, C.J.; Brownsell, S.; Hawley, M.S. Home Telehealth Uptake and Continued Use Among Heart Failure and Chronic Obstructive Pulmonary Disease Patients: A Systematic Review. *Ann. Behav. Med.* **2014**, *48*, 323–336. [CrossRef] [PubMed]
67. Sul, A.R.; Lyu, D.-H.; Park, D.-A. Effectiveness of telemonitoring versus usual care for chronic obstructive pulmonary disease: A systematic review and meta-analysis. *J. Telemed. Telecare* **2018**, *26*, 189–199. [CrossRef] [PubMed]
68. Yohannes, A.M. Telehealthcare management for patients with chronic obstructive pulmonary disease. *Expert Rev. Respir. Med.* **2012**, *6*, 239–242. [CrossRef] [PubMed]
69. Kelders, S.M.; Kok, R.; Ossebaard, H.C.; Van Gemert-Pijnen, J.E. Persuasive System Design Does Matter: A Systematic Review of Adherence to Web-based Interventions. *J. Med. Internet Res.* **2012**, *14*, e152. [CrossRef]
70. Sobnath, D.D.; Philip, N.; Kayyali, R.; Nabhani-Gebara, S.; Pierscionek, B.; Vaes, A.W.; A Spruit, M.; Kaimakamis, E.; Mehring, M.; Ryan, D.; et al. Features of a Mobile Support App for Patients with Chronic Obstructive Pulmonary Disease: Literature Review and Current Applications. *JMIR mHealth uHealth* **2017**, *5*, e17. [CrossRef] [PubMed]
71. Mcdowell, J.E.; McClean, S.; FitzGibbon, F.; Tate, S. A randomised clinical trial of the effectiveness of home-based health care with telemonitoring in patients with COPD. *J. Telemed. Telecare* **2015**, *21*, 80–87. [CrossRef]
72. Fridriksdottir, N.; Gunnarsdottir, S.; Zoëga, S.; Ingadottir, B.; Hafsteinsdottir, E.J.G. Effects of web-based inter-ventions on cancer patients' symptoms: Review of randomized trials. *Support. Care Cancer* **2018**, *26*, 337–351. [CrossRef]
73. Triberti, S.; Savioni, L.; Sebri, V.; Pravettoni, G. eHealth for improving quality of life in breast cancer patients: A systematic review. *Cancer Treat. Rev.* **2019**, *74*, 1–14. [CrossRef]
74. Brunton, L.; Bower, P.; Sanders, C. The Contradictions of Telehealth User Experience in Chronic Obstructive Pulmonary Disease (COPD): A Qualitative Meta-Synthesis. *PLoS ONE* **2015**, *10*, e0139561. [CrossRef]
75. Ruland, C.M.; Maffei, R.M.; Børøsund, E.; Krahn, A.; Andersen, T.; Grimsbø, G.H. Evaluation of different features of an eHealth application for personalized illness management support: Cancer patients' use and appraisal of usefulness. *Int. J. Med. Inform.* **2013**, *82*, 593–603. [CrossRef] [PubMed]
76. McCabe, C.; McCann, M.; Brady, A.M. Computer and mobile technology interventions for self-management in chronic obstructive pulmonary disease. *Cochrane Database Syst. Rev.* **2017**, *5*, CD011425. [CrossRef]
77. Lundell, S.; Holmner, Å.; Rehn, B.; Nyberg, A.; Wadell, K. Telehealthcare in COPD: A systematic review and meta-analysis on physical outcomes and dyspnea. *Respir. Med.* **2015**, *109*, 11–26. [CrossRef] [PubMed]
78. Eysenbach, G.; Kummervold, P.E.; Ritterband, L. "Is Cybermedicine Killing You?"—The Story of a Cochrane Disaster. *J. Med. Internet Res.* **2005**, *7*, e21. [CrossRef]
79. Welling, J.B.A.; Hartman, J.; Hacken, N.H.T.; Klooster, K.; Slebos, D.-J. The minimal important difference for the St George's Respiratory Questionnaire in patients with severe COPD. *Eur. Respir. J.* **2015**, *46*, 1598–1604. [CrossRef] [PubMed]
80. Baumeister, R.F.; Vohs, K.D.; Nathan DeWall, C.; Zhang, L. How Emotion Shapes Behavior: Feedback, Anticipation, and Reflection, Rather Than Direct Causation. *Pers. Soc. Psychol. Rev.* **2007**, *11*, 167–203. [CrossRef]
81. De La Torre-Díez, I.; López-Coronado, M.; Vaca, C.; Aguado, J.S.; de Castro, C. Cost-utility and cost-effectiveness studies of telemedicine, electronic, and mobile health systems in the literature: A systematic review. *Telemed. e-Health* **2015**, *21*, 81–85. [CrossRef]
82. Barbosa, M.T.; Sousa, C.S.; Morais-Almeida, M.; Simões, M.J.; Mendes, P. Telemedicine in COPD: An Overview by Topics. *COPD J. Chronic Obstr. Pulm. Dis.* **2020**, *17*, 601–617. [CrossRef]
83. Kamei, T.; Yamamoto, Y.; Kajii, F.; Nakayama, Y.; Kawakami, C. Systematic review and meta-analysis of studies involving telehome monitoring-based telenursing for patients with chronic obstructive pulmonary disease. *Jpn. J. Nurs. Sci.* **2012**, *10*, 180–192. [CrossRef] [PubMed]
84. Cruz, J.; Brooks, D.; Marques, A. Home telemonitoring effectiveness in COPD: A systematic review. *Int. J. Clin. Pract.* **2014**, *68*, 369–378. [CrossRef] [PubMed]
85. Hong, Y.; Lee, S.H. Effectiveness of tele-monitoring by patient severity and intervention type in chronic obstruc-tive pulmonary disease patients: A systematic review and meta-analysis. *Int. J. Nurs. Stud.* **2019**, *92*, 1–15. [CrossRef] [PubMed]
86. Wantland, D.J.; Portillo, C.J.; Holzemer, W.L.; Slaughter, R.; McGhee, E.M. The Effectiveness of Web-Based vs. Non-Web-Based Interventions: A Meta-Analysis of Behavioral Change Outcomes. *J. Med. Internet Res.* **2004**, *6*, e40. [CrossRef] [PubMed]
87. Eysenbach, G. The law of attrition. *J. Med. Internet Res.* **2005**, *7*, e11. [CrossRef]
88. Bennett, A.V.; Jensen, R.E.; Basch, E. Electronic patient-reported outcome systems in oncology clinical practice. *CA A Cancer J. Clin.* **2012**, *62*, 336–347. [CrossRef] [PubMed]

International Journal of
Environmental Research and Public Health

Article

Eliciting Requirements for a Diabetes Self-Management Application for Underserved Populations: A Multi-Stakeholder Analysis

Samuel Bonet Olivencia [1], Arjun H. Rao [1], Alec Smith [1] and Farzan Sasangohar [1,2,*]

[1]   Department of Industrial & Systems Engineering, Texas A&M University, College Station, TX 77843, USA; Samuel089@tamu.edu (S.B.O.); Arjun.rao@collins.com (A.H.R.); Alec.smith@tamu.edu (A.S.)

[2]   Center for Outcomes Research, Houston Methodist, Houston, TX 77030, USA

*   Correspondence: sasangohar@tamu.edu

Abstract: Medically underserved communities have limited access to effective disease management resources in the U.S. Mobile health applications (mHealth apps) offer patients a cost-effective way to monitor and self-manage their condition and to communicate with providers; however, current diabetes self-management apps have rarely included end-users from underserved communities in the design process. This research documents key stakeholder-driven design requirements for a diabetes self-management app for medically underserved patients. Semi-structured survey interviews were carried out on 97 patients with diabetes and 11 healthcare providers from medically underserved counties in South Texas, to elicit perspectives and preferences regarding a diabetes self-management app, and their beliefs regarding such an app's usage and utility. Patients emphasized the need for accessible educational content and for quick access to guidance on regulating blood sugar, diet, and exercise and physical activity using multimedia rather than textual forms. Healthcare providers indicated that glucose monitoring, educational content, and the graphical visualization of diabetes data were among the top-rated app features. These findings suggest that specific design requirements for the underserved can improve the adoption, usability, and sustainability of such interventions. Designers should consider health literacy and numeracy, linguistic barriers, data visualization, data entry complexity, and information exchange capabilities.

Keywords: diabetes mellitus; self-management; blood glucose self-monitoring; mobile applications; medically underserved area; health literacy; telemedicine; disease management

Citation: Bonet Olivencia, S.; Rao, A.H.; Smith, A.; Sasangohar, F. Eliciting Requirements for a Diabetes Self-Management Application for Underserved Populations: A Multi-Stakeholder Analysis. *Int. J. Environ. Res. Public Health* **2022**, *19*, 127. https://doi.org/10.3390/ijerph19010127

Academic Editors: Irene Torres-Sanchez and Marie Carmen Valenza

Received: 1 November 2021
Accepted: 17 December 2021
Published: 23 December 2021

Publisher's Note: MDPI stays neutral with regard to jurisdictional claims in published maps and institutional affiliations.

1. Introduction

In 2020, over 30 million individuals in the United States suffered from diabetes, most (about 90%) with type 2 diabetes [1]. Rural/medically underserved areas—defined as populations with low access to primary care providers, high infant mortality, high poverty, and/or high elderly population [2]—have shown relatively poor diabetes outcomes compared to the urban/well-served areas [3]. Additionally, type 2 diabetes disproportionately affects people of certain racial and ethnic groups, many of whom may live in areas identified as rural/medically underserved [4], such as Hispanics/Latinx Americans. Recent estimates from the Centers for Disease Control showed that individuals from Hispanic/Latinx American heritage were more likely (17%) to suffer from diabetes than the non-Hispanic White population (8%) [5]. Additionally, data from the U.S. Department of Health and Human Services' Office of Minority Health revealed that diabetes was among the leading causes for mortality among the non-White population [6].

Effective self-management of diabetes can have a significant impact on health outcomes. Studies have shown that patients who received training in self-management were successful in regulating their blood glucose levels, dietary habits, and glycemic control [7–11]. However, several barriers restrict the ability of underserved patients to execute self-management effectively [12]. These include limited access to timely healthcare

371

services [13], limited financial resources [14], low literacy [15–17], and geographic barriers to seeking care from providers outside their community [18]. Therefore, there is a need to investigate methods or interventions that enable self-management by identifying and addressing such barriers systematically.

Telehealth, a type of health information technology, has received special attention in recent years for improving access to health care [19], and for supporting integrated care for chronic diseases by providing patient education and information transfer between patients and providers [20]. Recent advances in mobile health (mHealth) technologies, a modality of telehealth interventions, have shown promise in mitigating barriers related to accessibility. These technologies facilitate the self-management of diabetes, including discreet, cost-efficient, and non-invasive tools for monitoring health conditions, and a reliable platform for interactions between healthcare providers and patients [21,22]. A recent review of 11 mHealth apps for diabetes [23] revealed common features, such as setting reminders, tracking blood glucose and hemoglobin A1c (HbA1c), medication use, physical activity, and weight which support self-management. mHealth technologies, integrated with monitoring technologies, such as glucometers and continuous glucose monitors, have also shown promise in improving healthcare delivery [24]. Additionally, the recent integration of machine learning algorithms and artificial intelligence with mHealth played a vital role in the use of the data collected by these technologies for clinical decision-making [24].

While such characteristics make mHealth a promising method to address barriers to self-management in underserved populations, there is limited research documenting guidelines or mHealth design requirements for these populations. Previous research highlights the importance of supporting different languages and cultures for the improved adoption and sustainability in underserved populations. For example, Burner et al [25] and Williams et al. [26] discussed the need for providing basic features, such as educational content [25,26], reminders [26], and user interfaces in Spanish for Hispanic/Latinx users. In addition, glucometer connectivity functionalities [26], and the personalization of messages and content [25] are discussed as features that are important for users in underserved populations; however, such features are typically lacking in those apps available in the market [26]. Low health literacy and eHealth literacy have been identified as potential challenges for the sustained engagement of vulnerable populations with electronic, mobile, and telehealth tools; a systematic review shows that these factors have been underassessed in the published literature about the design of mobile interventions [27]. Additionally, research revealed that paid mobile apps are more likely to integrate strategies to engage low health literate populations, in comparison with free mobile apps [28]; however, cost has been identified as a major concern for people to download and adopt mHealth apps [29], and financial barriers can restrict underserved populations' self-management of their chronic conditions [14,30]. Age has been another factor highlighted in the literature affecting mHealth usage and adoption. While younger individuals have been identified as more likely to engage with mHealth apps, it is vital to assess design consideration for the elderly population [29]. For example, the use of simple, actionable, and information rich visualizations can help to address some of the design limitations of the low health literate and elderly population [31]. The patients' intrinsic level of motivation has also been linked to vulnerable the populations' level of engagement with mobile interventions [27]. Research has shown the need to apply design techniques, such as sequential multiple assignment randomized trials (SMART), to tailor self-monitoring mobile interventions to the patients' individual level of internal motivation [32].

Despite the evidence suggesting users' preferences for personalized nutritional and health behavior content [32], research [33] highlighted the lack of personalized feedback and significant usability issues, including ease of data entry and integration with patients and electronic health records, suggesting a potential gap in user-centered design (UCD) approaches. Indeed, usability tests on eight mHealth apps for diabetes revealed that more than two-thirds (6/8) were scored by patients as "marginal" or "not acceptable" [23]. This is supported by another study, in which about half of the participants reported stopping their use of mHealth apps due to a high data entry burden and loss of interest, among other factors [29]. UCD has shown promise in fostering user engagement and improving the perception of app effectiveness, with positive impact on sustained behavioral change [34,35]. To our knowledge, only a few attempts have been documented to utilize UCD to inform requirements for diabetes self-management apps (e.g., [36]) and no research has focused on the needs and expectations of the underserved. To address this gap, in this paper, we document the stakeholders' needs and expectations from a diabetes self-management app, by eliciting feedback from patients with diabetes and providers in several medically underserved areas in the United States.

2. Methods

Semi-structured interviews were conducted with a convenience sample of 97 patients and 11 healthcare providers from several medically underserved counties in South Texas. The interviews with patients were conducted by four nurse educators with graduate degrees in nursing or education during diabetes self-management education sessions, held between 8 April 2019 and 3 May 2019, as part of the Healthy Texas initiative. These education sessions aimed at educating patients with diabetes on practical strategies and tips for incorporating healthy behaviors in their daily activities, including effective nutrition, general health and wellness, the role of physical activity, and ways to mitigate the financial and physical burden of diabetes. Participants were informed about the study at the end of educational sessions and were selected if they met the participation criteria (aged 18+ and had diabetes). The authors FS and AR, who held doctoral degrees in Engineering and had extensive experience in qualitative research, provided these interviewers with training on conducting interviews. Providers were recruited and interviewed by the authors AR, SB, and FS during a diabetes education conference in South Texas. The research group used a booth at the conference exhibition room to recruit the participants. The study team did not establish a relationship with the participants prior to the study and no one other than the interviewers were present at the interview sessions. No potential participant refused or withdrew mid-study, and no repeated interviews were carried out.

Two interview protocols were developed for patients and providers, respectively, to reduce individual biases and assumptions and to standardize the interviews. Interviews with patients focused on understanding their expectations from a diabetes self-management app. The questions in the interview protocol for patients included topics, such as perceived barriers and limitations for diabetes self-management, the use of technology to manage diabetes, important characteristics in a technology for diabetes self-management, and preferences on features for an app for diabetes self-management. Similarly, interviews with providers focused on their expectations from a self-management app for diabetes both from the patients' perspective and the type of information or interactions providers expected from such a tool. The questions in the interview protocol for providers included topics, such as perceived barriers and limitations for patients to adopt and app for diabetes self-management, perceived barriers and limitations for providers to monitor patient who have adopted such technology, perceived importance on feature for an app for diabetes self-management, and preferences about data representation and data communication. The interviews took approximately 45 min for both the patients and providers. The patients and providers received a USD 25 or USD 50 gift card, respectively, for participation. The Texas A&M University Review Board reviewed and approved this study (IRB Protocol #IRB2018-1503D) and all participants provided informed consent.

The interviews were audio recorded and no field notes were made by the interviewers during or after the interviews. A transcription service, Temi, was used to transcribe the audio recorded interviews preceding analysis [37]. The thematic analysis of the interviews was conducted by two coders (AS and AR) [38,39]. The two coders completed the following steps, separately and sequentially, and then met to discuss any discrepancies: code creation, initial coding, and focused coding. The thematic coding process entailed a deeper discussion of the themes and constructs that emerged from the analysis. After coming to a consensus, the themes were discussed with the other authors (SB and FS) and changes were made, as necessary. MAXQDA 12 was used to complete the analysis [40]. AS and SB were doctoral students and had extensive experience in qualitative data analysis.

3. Results

3.1. Demographics of Participants

3.1.1. Patient Demographics

Table 1 presents the key demographics of the patients. A total of 100 patients participated in the interviews. After cleaning the data, removing incomplete entries, a total of 97 interviews were analyzed. The average age of the participants was 56.07 (SD = 13.10). A vast majority of the participants were Hispanic or Latinx (90%, 87/97). Most of the patients did not have a postsecondary degree, with 73.20% (71/97) of the respondents having either some college (no degree), a high school diploma, or less. Approximately half the respondents (50.51%; 49/97) had a household income of less than USD 30,000. About a fourth of participants reported not having medical insurance (24%; 23/97). A majority of the participants had type 2 diabetes (81%, 79/97). About a fourth of participants were diagnosed with diabetes within a year of the date of their participation in the study (25%, 24/97), and about 39% (38/97) of respondents reported having diabetes for more than 10 years.

Table 1. Demographic information of patients.

Characteristic	Number of Respondents	Percentage
Gender (*n* = 97)		
Female	71	73.20
Male	26	26.80
Income (*n* = 97)		
Less than USD 20,000	30	30.93
USD 20,000–USD 30,000	19	19.59
USD 30,000–USD 40,000	11	11.34
USD 40,000–USD 50,000	10	10.31
USD 50,000–USD 60,000	4	4.12
Above USD 60,000	8	8.25
Prefer not to answer	15	15.46
Race (*n* = 97)		
White (non-Hispanic or Latinx)	7	7.22
Hispanic or Latinx (White)	61	62.89
Hispanic or Latinx (non-White)	26	26.80
American Indian or Native	2	2.06
Two or more races	1	1.03
Education (*n* = 97)		
Less than high school diploma	16	16.50
High school diploma or GED	29	29.90
Some college, no degree	26	26.80
Associate degree	14	14.43
Bachelor's degree	9	9.28
Graduate or professional degree	3	3.09
Type of Diabetes (*n* = 97)		
Pre-diabetes	4	4.12
Type 1	9	9.28

Table 1. *Cont.*

Characteristic	Number of Respondents	Percentage
Type 2	79	81.44
Do not know	5	5.16
First Diagnosed with Diabetes (*n* = 97)		
Less than 6 months	12	12.37
6 months to 1 year	12	12.37
Greater than 1 year to 10 years	35	36.09
Greater than 10 years to 20 years	32	32.99
Greater than 20 years	6	6.18

3.1.2. Healthcare Provider Demographics

Eleven healthcare providers serving medically underserved communities in South Texas participated in the interviews. Table 2 presents the key demographics for the healthcare providers interviewed. On average, the physicians sampled had nearly 3 decades (mean = 28.86; SD = 7.75; and range: 10 to 38) of experience in their current roles. Most participants (9/11) practiced family medicine, one practiced general medicine, and one was a pediatric nurse practitioner. Two participants held leadership roles (president/CEO) in their respective organizations.

Table 2. Demographic information of healthcare providers.

Characteristic	Number of Respondents	Percentage
Gender (*n* = 11)		
Female	2	18.19
Male	9	81.81
Age (*n* = 11)		
45–54 years	1	9.09
55–64 years	5	45.45
65–74 years	5	45.45
Race (*n* = 11)		
White (Non-Hispanic or Latinx)	9	81.81
Hispanic or Latinx (non-White)	2	18.19
Nature of Experience (*n* = 11)		
Family medicine/practice	9	81.81
General medicine	1	9.09
Pediatric nurse practitioner	1	9.09

3.2. Participant Interview Themes

Patients were asked to specify features they desired in a diabetes self-management mobile app. A total of 97 participants responded to this question. The analysis of these responses resulted in five superordinate themes: (1) logging and tracking of blood sugar readings; (2) assistance with adopting a healthy lifestyle; (3) integration with the healthcare system; (4) reminders and alerts; and (5) usability and non-invasiveness. Almost 20% of the respondents (19.58%; 19/97) indicated that they did not know what features they would expect in a diabetes self-management app.

The healthcare providers were asked a series of questions about features they believed would benefit their patients and would improve their practice. The analysis of the responses from 11 physicians resulted in 5 superordinate themes: (1) dietary logs; (2) patient diabetes education; (3) reminders and alerts; (4) information communication and presentation; and (5) patient-related challenges and barriers. These themes and associated subthemes are discussed below. The proportion of participants whose response is captured by a theme or subtheme is presented with percentage (%) and counts (xx/XX). Some subtheme counts do not total 100% because some participants had responses in multiple subthemes.

3.2.1. Functional Requirements Suggested by the Patients

Logging and Tracking Blood Sugar Readings

Almost a quarter (24.74%; 24/97) of the patients in our sample expressed the need to be able to track and log their blood sugar readings. Two prominent subthemes were identified from the interviews: (a) logging readings and (b) assistance and insights from the readings.

Logging Readings: this subtheme captures the patients' desire for the app to help them log and recall their blood sugar readings. More than half of the patients (54.17%; 13/24) who expected this feature also highlighted the need to trace back to previous readings to check their well-being.

"Just to be able to keep track of myself . . . or tracking my glucose . . . without having to write it down"—P12

Assistance & Insights from the Readings: some patients (16.67%; 4/24) pointed out the need to understand what the entries mean. Specifically, they indicated the need for graphical interfaces to visualize the trends of their readings. For example, *"keeping track of history . . . so I can monitor for trends" (P19)*. The participants mentioned familiarity with similar visual trends, such as activity and expected similar visualization for sugar levels. In addition, some of the patients mentioned that descriptive statistics about their parameters would be useful in managing their condition, such *as "the daily average, and the weekly average" (P09)*.

Assistance with Adopting a Healthier Lifestyle

About a third (34.02%; 33/97) of the respondents indicated the need for assistance with managing their condition and adopting healthier choices, and demonstrated a willingness to learn about tips and techniques to manage their diabetes. Specifically, their responses were categorized into three subthemes: (a) diet regulation; (b) health tips; and (c) fitness and physical activity.

Diet Regulation: this theme captures the patients' desire to be provided with information on regulating their eating habits and food intake. Two thirds (66.67%; 22/33) of these participants wanted diet regulation features, including access to a list of the types of foods they can consume to maintain glycemic control. Furthermore, patients also wanted the app to help them to construct and adhere to a diet plan. Finally, some patients (based on the diabetes education they had received) indicated that they could benefit from having a carbohydrate "tracker".

"Like maybe like a diet plan, things to do or not to do you know that can lower your sugars if they're high."—P21

"[. . .] and a list of dos' and do not food, you know, like a list, an actual list."—P07

"How many carbs, I can [eat], you know, in, um, like in the mornings [. . . or] at lunchtime I'll have a sandwich [. . .] I think that's one of the reasons my diabetes goes up. It scares me, you know, to eat a lot of carbs."—P16

Health Tips: patients frequently (39.39%; 13/33) mentioned the need to easily access health-related resources. Although there was an interest in health resources in general, patients were particularly keen on accessing specific tips about nutrition. The participants also mentioned expecting prescriptive tips when presented with abnormal blood sugar values.

"There's a lot of things like for your heart and stuff [. . .] there's a lot of stuff out here that we eat and we're not supposed to because it's really damaging ourselves. So, you know some advice [. . .] give us something like that."—P11

"[…] to see, to measure if your sugar is high or low and to explain what things you can do to lower our sugar […]"—P08

Fitness and Physical Activity: several of these patients (15.15%; 5/33) indicated that, while there are several commercial apps for fitness and activity tracking, a fitness module integrated into the diabetes self-management app would be ideal, suggesting the perceived importance of the connection between physical activity and diabetes.

Reminders and Alerts

Some patients (11.34%; 11/97) suggested timely alerts or reminders would help them adhere to their medication regimen. Two subthemes emerged from patient responses: (a) reminders and scheduling, and (b) predictive capability.

Reminders and Scheduling: several of these patients (72.72%; 8/11) highlighted their busy lifestyle as a reason for forgetting to monitor their blood sugar levels. In addition to being reminded to monitor their health, patients also suggested that a scheduling tool would help them keep track of their appointments.

Predictive Capability: some of these patients (36.36%; 4/11) responded that they would like predictive features, such as the early detection of warning signs and monitoring trends, so they can mitigate any problems before they occur.

"[…] maybe signs to look for, like when you're going to have maybe an [hypoglycemia] episode, so like warning signs."—P39

Integration with Healthcare System

A few patients in our sample (3.09%; 3/97) highlighted the need for their diabetes self-management program to be integrated into their overall care system. The patients desired easy communication of diabetic parameters and progress reports with their healthcare provider.

"[…] being able to send it to the doctor, or bring a recording of the reading. That way they could keep track of it."—P06

Usability and Non-Invasiveness

Several patients (13.4%; 13/97) desired a system that would be easy to use and non-invasive. This superordinate theme can be categorized into two subthemes: (a) usability and (b) non-invasiveness.

Usability: Several patients raised concerns about their experiences with app usability (46.15%; 6/13) and expected an app that was reliable, accurate, and easy to use.

Non-invasiveness: this theme captures the patients' desire for a method of reading sugar levels without having to prick themselves, as commonly required by most glucometers. Most of these patients (53.84%; 7/13) were fatigued by the constant pricking for the blood sugar measurement and desired an app that would display blood sugar readings (potentially from an implantable continuous glucose monitor).

"Like I said, a feature that would allow you to check your glucose level without [pricking], … , I mean I don't know if they can make something like that without drawing your blood."—P14

3.2.2. Functional Requirements Suggested by Healthcare Providers
Dietary Logs

Healthcare providers (72.72%; 8/11) highlighted the importance of a diet/food log to keep track of what patients are consuming and to have patients engage in their treatment. Healthcare providers also suggested that the app should provide immediate feedback to the patients about the calorie density and quality of the food they are ingesting. However, providers cautioned against using food logs in clinical assessments as patients tended to be dishonest in their logs.

"So, I mean, if they want to write it down, that's fine [   ] if you're assuming perfect compliance and honesty. But my experience is that most patients aren't completely honest

with what they do. So, I guess in the ideal setting, a food log would be great. So, you can go, I see when you have that bowl of ice cream, you know, that wasn't broccoli, you know, then food log could be really important. So, I guess we could change it."—S01

"Food log with […] input about calorie and everything else. So, it'd be two-way […] Immediate feedback. Get pretty much immediate feedback. If they're going to go to the trouble of entering in all that food, they need to get, I don't want it just to be written down, you know, and just stored somewhere and they look it up. They're going to enter what they're going to eat in a food log. They need to get immediate feedback about the calories or, and this is on or off their diet or something like that."—S08

One physician alluded to the use of image processing and machine learning techniques to analyze a photograph of a plate of food. The results of the analysis provide a breakdown of calorie content and nutritional value.

"I thought it would be fantastic if a person sets their meal, their plate down, they take a photo of it. And artificial intelligence calculates the, based on the size of the plate, I mean, […], how much potatoes take up, how much the meat takes up. And it calculates […]. We load the fat amount, the protein amount [but] I don't think they have that yet."—S04

Patient Diabetes Education

Some providers (27.27%; 3/11) highlighted the low health literacy of patients they treated and encouraged the creation of an educational component in the diabetes self-management app for facilitating communication with educators.

"We have to give them the information … It's like a coach. This is the game plan … this is how you throw the ball and all that. [You have] repetition and they get better at it."—S08

Reminders and Alerts

Most providers (72.72%; 8/11) suggested that providing periodic reminders or alerts about multiple topics, including ingesting medication, diet adherence, activity reminders, and appointments, would benefit patients and help them in self-management.

"Self-management. So yeah, you get reminders. You got to do that for them. Probably about every two hours … you remind them about if your glucose is too high or too low … They could do a reading … to help them for self-management."—S02

Providers also cautioned designers about the tendency for patients to develop alarm fatigue leading to ignored reminders, thus highlighting the need to remind or alert only when appropriate.

"[…] there are patients who may feel like this is getting a little [annoying], and you're going to have to see everybody [feels] a little intruded."—S08

"[…] when you start getting emails that are 12 different things on the same subject, you just start going through them and not reading them. And that's what we're seeing. They will gloss over them."—S05

When discussing reminders, the healthcare providers' responses could be categorized into two subthemes: (a) medication intake recall and (b) activity reminder

Medication Intake Recall: several providers (50%; 4/8) indicated that some of their patients had trouble recalling the nature and amount of medication taken in a specific period. Therefore, the healthcare providers believed that including an easy to use and intuitive medication reminder feature in the app would remind patients about previously ingested or impending medication.

"Having that in the app, so they're documenting it […] I think from a provider standpoint it would be great, but from the patient standpoint, we can't get them to write it down in a book. It would have to be very simple. Like they go in and click a button or

two, you know, have their medications, already populated and they could just go in and go click, click, click."—S07

Activity Reminder: a few of the healthcare providers (37.5%; 3/8) suggested designing a feature that would help patients log their activities and remind them to exercise/stay active while giving the provider access to that information.

"[…] we need something to help them exercise on here and way of recording it. [Even though] those are already with Fitbit's and stuff, but that needs to be sent to the physician."—S04

Information Communication and Presentation

Although the healthcare providers encouraged open lines of communication with their patients, they highlighted key features relating to data communication and presentation, such as synchronous and asynchronous communication and information presentation.

Synchronous vs. Asynchronous Communication: two providers (18.2%; 2/11) stressed the need for both providers and patients to be able to communicate and exchange information, even for a self-management tool.

"Is there one-way or two-way communication with this app? It could be two-way. It has to be two. If it's two-way, I'd feel comfortable. If it's only one-way, it's not worth it."

An example of a such one-way or asynchronous communication method is the use of voice notes. The providers had mixed feelings about the use of voice notes as a means of communicating with patients. While some healthcare providers (27.27%; 3/11) believed that voice notes can be beneficial to patients who were visually challenged, the majority (72.72%; 6/11) were against the use of voice notes, citing issues with understanding patient accents and dialects.

"[…] for those with really poor eyesight, it's gonna have to be a voice [recording], in their language."—S04

"But you know as well as I do, there's so many dialects […] word accents. Sometimes you can't understand."—S03

The majority of providers (63.63%; 7/11) believed that sending text messages can be a useful medium to communicate specific, personalized, and urgent messages or instructions to patients, while a few preferred a chat feature.

"[…] texts would be for urgent things like too high or too low [blood sugar]."

"[…] what I like and what I think a lot of the younger crowd would like, would be, that "chat." […] You know, if you have questions, you're gonna chat"

Information Presentation: most providers were highly supportive (72.72%; 8/11) of having graphs in the mobile app. However, they remarked that some patients in their communities had challenges in comprehending the information provided in graphs and would often require the healthcare providers to describe it to them. The providers emphasized the need for the graphs to be simple, easy to read, with clearly displayed limits, and intuitive ways to warn patients about abnormal sugar levels.

"[…] People respond visibly very easily using warning colors. Green, good, red, bad. The line where yours is. Pictures and graphs are great and probably better than texts."—S08

Some healthcare providers indicated that adding appropriate pictures can help patients understand, interpret, and potentially maintain glycemic control.

"[…] they see somebody happy; they know their blood sugars in a happy range. Uh, see some blood sugars, they, they maybe they can follow it on a chart day to day. Happy face here means they're in control. A sad face here means they're out of control."—S04

"[…] every picture tells a story. I think they would like pictures. See where they were and where they're going."—S09

Of the six healthcare providers who responded to this question, 50% (3/6) were cautiously optimistic about the use of tables and charts to communicate clinical data to patients, while the other half felt that the patients might be overwhelmed.

Patient-Related Challenges or Barriers

When asked about the potential to implement a diabetes self-management app in a clinical setting, healthcare providers highlighted key barriers that could impede care. Their responses were classified into the following subthemes: (a) patient literacy levels; (b) privacy concerns; and (c) lack of motivation.

Patient Literacy Levels: several providers (45.45%; 5/11) emphasized the diversity in education levels of the patients and questioned the patients' ability to read and interpret graphs or other information on an application. Furthermore, providers raised concerns about the patients' general readiness to use technology-based interventions, age-related usability barriers, and the language barriers.

"Well, like I said, the people I'm going to use it on are usually older people and those people didn't grow up with technology."—S10

"Most of my people speak Spanish or Spanglish."—S02

Some providers (36.36%; 4/11) indicated that many of their patients can struggle with self-management, which can require them to visit the physician (in-person) to interpret their readings. This can in turn exacerbate issues related to access and geographical barriers. They suggested integrating the app with existing technologies, such as telemonitoring, or providing means of communicating relevant information to address this barrier.

"If they have to come to the office […] to present the data, that's a barrier. If it can be like the telehealth telemonitoring it's transmitted and that's not a barrier for them."—S04

"[…] Transportation is a big barrier to adopt something like this. [Because] they have to get to the office. They also have looking for rides and I'm in a neighborhood, lot of people walk to my place, well [those] people have to take a bus."—S02

Privacy Concerns: a few providers (18.18%; 2/11) cited privacy concerns, highlighting that patients can be unwilling to be monitored or reluctant to share data.

"I don't know if they'd be open to doing something like that. […] most of them don't […]. They don't want something intruding on their […] autonomy I guess."—S10

Lack of Motivation: a few providers (18.18%; 2/11) also indicated that some of the patients in underserved areas can lack motivation to adhere to their treatment. This concern in turn relates to sustained app use for effective adoption of a healthier lifestyle.

4. Discussion

Our qualitative investigation into the requirements for a diabetes self-management app provided rich data on the key features and functionalities for patient adoption and engagement. These data lend insights into the facilitators and barriers that can encourage or impede the patients' sustained use of a diabetes self-management app. Although there has been research on the preferences of medically underserved patients [25,26], our study adds the multi-stakeholder perspective of providers in medically underserved communities.

Our findings are consistent with the previous literature on essential features for a diabetes self-management app [41–43]. Evidence-based guidelines suggest that logging and tracking blood glucose levels are essential elements in diabetes management [44,45]. Complementing these guidelines, Chavez and colleagues stated that physical activity, nutrition, blood glucose testing, medication or insulin dosage, health feedback, and education were key diabetes management tasks [46]. Consistent with this literature, patients in our study highlighted the need for logging and monitoring blood glucose levels, including visual trends for such values, tips about health lifestyle choices (e.g., exercise and nutrition), and reminders—all key basic features in a diabetes self-management app [42,43]. In addition, patients also requested the creation of a schedule feature, which would likely help them track their medication intake. Moreover, patients emphasized their preference to have a list of the types of foods they can consume to maintain glycemic control and to have a feature that can serve as a carbohydrate "tracker." Research has shown that people with diabetes in low income and minority neighborhoods have limited access to healthy foods and limited discussions with healthcare providers about healthy eating [47]. In addition, it has been suggested that cultural factors, such as the preference for carbohydrate-rich foods, should be considered when understanding the prevalence of diabetes in Latinx/Hispanic communities in the U.S. [48]. Moreover, eating disorders are more prevalent in individuals with type 1 diabetes, in comparison to the general population [49]. Finally, while our findings did not include the desire for mental health support, previous work suggests that patients with diabetes support the inclusion of features for assistance with the psychological and emotional aspects of diabetes self-management, such as stress management and mechanisms to cope with negative emotions [43].

Patients were particularly interested in accessing educational resources to help them better self-manage their condition. Although patients interviewed in the study were part of a program that provided diabetes education, our results indicate their preference to have access to dietary tips and educational content in a mHealth app. Such a tool can remove the barriers related to access to educational resources and can serve as a central repository of educational information that patients can access on-demand with tailored content based on their preferences. The request for educational content was also supported by the healthcare providers in our sample, which highlights a potential gap and unmet need in existing apps, with education among the most underrepresented features [26,41]. Furthermore, it is important to note that content should be provided at levels consistent with the educational background of the population (primarily high school level or less in our sample). It is imperative that the content provided is aligned to recommendations in the literature to overcome the challenges and limitations related to the patients' literacy levels. Design implications, such as minimizing technical jargon [50–55], presenting simplified language into tangible units [50–53], explaining uncommon terms [50–53], aligning content to the patients' cultural background [50–52], and implementing visual and audible features over the use of text [50–56], must be considered when creating the educational content to be integrated in the mobile app. Williams and Schroeder [26] go on to state that the use of video-based educational material can complement text-based content since Hispanics are among the major consumers of online multimedia content [57]. Additionally, the availability of educational content in multiple languages seems to be essential to overcome language barriers, especially in those underserved communities in which patients with diabetes are predominantly non-English speakers. Finally, while participants mainly emphasized the need for educational content about healthy eating, further research is warranted to understand the perspectives about additional interventions to encourage behavioral change regarding eating habits. Participants also supported the integration of a fitness and physical activity module. However, given the availability and prevalence of apps for physical activity, further work is needed to understand the utilization of such apps among underserved populations and investigate unmet needs.

Our findings showed that patients were partial to understanding and interpreting their diabetic parameters through graphs and visualization. Although healthcare providers were generally supportive of these media, they cautioned against complex displays. Healthcare providers cautioned about the patients' literacy levels and technical literacy levels, questioning their ability to read graphs and information in the app. Additionally, research suggests that the limited health literacy influence patients sustained motivation for engaging in monitoring their condition through self-management [58]. This can suggest that the design of health graphs and visualizations should account for the users' literacy levels. This is in line with evidence suggesting the benefits to the patients' health outcomes, when interventions sensitive to low health literacy limitations are used [30]. One of the healthcare providers suggested the use of colors and imagery to convey meaning and urgency, a view supported by the findings of Desai and colleagues, who suggested the use of a traffic light representation and facial emotions among possible visualizations of blood glucose forecasts [31]. Furthermore, while some patients indicated that they would benefit from the blood sugar forecasting capabilities in the app, research has shown that individuals with low numeracy can find it difficult to interpret the uncertainties associated with a forecast, consequently leading to disengagement [31]. The research suggests that visualizations that are simple, information-rich, convey authority, and promote actionable and learning behaviors from users are more effective in assisting users with data interpretation [31]. Additionally, providing the user with step-by-step guidance (e.g., a welcome wizard) about the different screen or features in the app can help address this problem [59]. Additionally, research has shown that difficulties using a smartphone can also impact the use of mobile apps [60]. Therefore, emphasis should be given towards orienting users to the app when they seem to be newer adopters of smartphone technology.

Patients expressed an interest in logging data and notes relating to their diabetes self-management. Healthcare providers also encouraged patients to log information, in the belief that it would keep patients engaged in their treatment. However, research has shown that a common reason for abandoning the use of apps is due the time required to enter data [61], which our providers also considered as an issue. Self-management practices can be encouraged by reducing the burden of data entry, for instance through simplified interfaces with adjustable text icon sizes to cater to individuals with visual impairments [26]. Additionally, data entry tasks could be simplified by the integration of databases and auto-fill features that help to minimize the amount of information the patients must recall and enter. Furthermore, providers highlighted the importance of synchronous and asynchronous communication capabilities; however, healthcare providers highlighted difficulties that can arise due to language barriers. Therefore, the availability of communication functions should integrate translators. Additionally, ways to effectively manage synchronous or patient-initiated communications must be explored to avoid unnecessary burden on providers, since concerns about potential changes in workload and overwhelming number of notifications have been highlighted in the literature regarding provider support of mHealth interventions for diabetes self-management [62].

In our study, patients with diabetes expressed their preference for simple instructions for healthy diets (i.e., a list of do's and don'ts for healthy eating). This finding is consistent with the results of Turchioe et al. [32] that showed low-income diabetic patients' mixed attitudes towards goal setting for dietary intake, and the general lack of positive attitudes towards personalized decision support and self-discovery. This is in contrast to findings in studies with more advantaged populations (e.g., [63]), where more detailed information on the impact of dietary consumption on glucose levels were preferred. These differences warrant further research to investigate adaptive or personalized interventions that take the user's characteristic in mind and/or provide personalization capabilities for various user types.

The synergistic interplay between remote patient monitoring (RPM) systems and smartphone apps can play a key role in assisted diabetes self-management. Patients and healthcare providers agreed that seamless integration of the app with their healthcare infrastructure can improve monitoring and managing diabetes. Studies have shown that RPM systems facilitate monitoring vital signs and allow for early detection of potentially hazardous health conditions, allowing time for provider intervention and preventing expensive hospital admissions [64]. For example, a clinician can track an insulin-dependent patient's blood sugar profile, identify hypoglycemic tendencies, and make the necessary changes to the patient's insulin dosages. In addition, a recent study demonstrated the potential of an RPM-facilitated diabetes management program, which incorporated evidence-based lifestyle interventions [65]. However, enthusiasm to adopt RPM in medically underserved communities should be tempered by the patients' access to technology, user proficiency, and training requirements [66–68].

This work has certain limitations. First, our sample did not contain medically under-served patients beyond those residing in several counties in South Texas. Therefore, the results may not be generalizable to other medically-underserved communities. A similar limitation applies to the clinicians who were interviewed in our study. To overcome this limitation, we recommend carrying out the same study in multiple medically underserved communities across Texas and the U.S. Next, our sample consisted of only those patients who were seeking diabetes education and care through TAM-HST. It is conceivable that there are other patients with diabetes in that region who can have additional design requirements or have faced additional barriers not represented here. In addition, while in this study we identified patients by their diabetes type, no information was collected about usage of insulin, especially among patients with type 2 diabetes. Future work should investigate the differences between the requirements for users who do and do not use insulin. Next, our sample size for providers is small due to overall shortage of providers in the medically underserved counties and limited access. While we reached saturation with our current sample, future work should collect the perspectives of more providers in other underserved communities across Texas and the U.S. to validate and expand our findings. Finally, a significant proportion of the patients indicated that they were not knowledgeable about smartphone or self-management apps, thereby failing to provide design features. Although this is an unfortunate scenario, it presents an opportunity to educate patients about the capabilities of mHealth in assisting with diabetes self-management. It is imperative that these patients be included in formative usability tests to gather their perspective on the app design.

5. Conclusions

This paper highlights key features and functional requirements for the design of a diabetes self-management app tailored to the underserved community. Some of our findings are consistent with previous literature on essential features for a diabetes self-management app for the general population, such as including features to log and track blood glucose levels, physical activity, nutrition, and medication and insulin dosage, in addition to including reminders and educational content about healthy lifestyle choices [41,42,46]. In addition to these commonalities, with the published literature about the general guidelines for the design of self-management apps for diabetes, our review of relevant literature and interviews with patients and physicians in some representative underserved areas suggest that specific design requirements for the underserved can improve the adoption, usability, and sustainability of such interventions. Despite the prevalence of several self-management apps, the emergence of patient education as a desired feature suggests the need for designers to pay closer attention to the patients' linguistic abilities and health literacy levels. Both the patients and providers also strongly desired the use of appropriate visualizations of diabetes data. In this regard, we recommend further investigations into the types of visualizations that would facilitate the easy interpretation of diabetes data. The use of simplified interfaces, adjustable icons, databases, and auto-fill features were identified to simplify information visualization, information recall, and data entry tasks. The use of formative training and technology exposure were identified to address issues of low experience with technology and low knowledge about mHealth capabilities, which can affect adoption and sustained engagement. While our data suggest that patients in underserved communities desire educational content about healthy lifestyle choices (e.g., nutrition and exercise), it is important that this content is presented in a way that is sensitive to their social and economic limitations, cultural background, and promotes a healthy attitudes towards eating. The results from our study also provided insights into perceived patient adoption barriers, including health literacy levels, motivation, and privacy concerns. To mitigate these barriers, we recommend adopting a community-based participatory research approach to facilitate a grassroots-level education about the capabilities of the app being designed.

Author Contributions: Conceptualization, F.S.; methodology, F.S., A.H.R. and S.B.O.; data collection, F.S., A.H.R. and S.B.O.; formal analysis, A.H.R. and A.S.; writing—original draft preparation, A.H.R. and A.S.; writing—review and editing, F.S. and S.B.O. All authors have read and agreed to the published version of the manuscript.

Funding: This research was funded partly by the Robert J. Kleberg and Helen C. Kleberg Foundation and the National Science Foundation under Grant No. 1648451.

Institutional Review Board Statement: The study was conducted according to the guidelines of the Declaration of Helsinki, and approved by the Texas A&M University Review Board (IRB Protocol #IRB2018-1503D).

Informed Consent Statement: Informed consent was obtained from all subjects involved in the study.

Data Availability Statement: The data that support the findings of this study are available from the corresponding author, FS, upon reasonable request.

Acknowledgments: We thank Jacob M. Kolman, MA, research associate at Texas A&M University and scientific writer at Houston Methodist, for critical review and editing.

Conflicts of Interest: The authors declare no conflict of interest.

References

1. Centers for Disease Control and Prevention. National Diabetes Statistics Report. 2020. Available online: https://www.cdc.gov/diabetes/pdfs/data/statistics/national-diabetes-statistics-report.pdf (accessed on 18 October 2021).
2. Health Resources and Services Administration. U.S. Department of Health and Human Services. MUA Find. 2021. Available online: https://data.hrsa.gov/tools/shortage-area/mua-find (accessed on 21 October 2021).
3. Nichols, G.A.; McBurnie, M.; Paul, L.; Potter, J.E.; McCann, S.; Mayer, K.; Melgar, G.; D'Amato, S.; DeVoe, J.E. Peer reviewed: The high prevalence of diabetes in a large cohort of patients drawn from safety net clinics. *Prev. Chronic Dis.* **2016**, *13*, 160056. [CrossRef] [PubMed]
4. AuYoung, M.; Moin, T.; Richardson, C.R.; Damschroder, L.J. The diabetes prevention program for underserved populations: A brief review of strategies in the real world. *Diabetes Spectr.* **2019**, *32*, 312–317. [CrossRef] [PubMed]
5. Centers for Disease Control and Prevention. Hispanic/Latino Americans and Type 2 Diabetes. 2019. Available online: https://www.cdc.gov/diabetes/library/features/hispanic-diabetes.html (accessed on 18 October 2021).
6. Kochanek, K.D.; Murphy, S.L.; Xu, J.; Arias, E. Deaths: Final data for 2017. *Natl. Vital. Stat. Rep.* **2019**, *68*, 1–77.
7. Lirussi, F. The global challenge of type 2 diabetes and the strategies for response in ethnic minority groups. *Diabetes Metab. Res. Rev.* **2010**, *26*, 421–432. [CrossRef] [PubMed]
8. Lorig, K.R.; Holman, H. Self-management education: History, definition, outcomes, and mechanisms. *Ann. Behav. Med.* **2003**, *26*, 1–7. [CrossRef]
9. Lorig, K.R.; Ritter, P.L.; González, V.M. Hispanic chronic disease self-management: A randomized community-based outcome trial. *Nurs. Res.* **2003**, *52*, 361–369. [CrossRef]
10. Norris, S.L.; Engelgau, M.M.; Narayan, K.M. Effectiveness of self-management training in type 2 diabetes: A systematic review of randomized controlled trials. *Diabetes Care* **2001**, *24*, 561–587. [CrossRef]
11. Reyes, J.; Tripp-Reimer, T.; Parker, E.; Muller, B.; Laroche, H. Factors influencing diabetes self-management among medically underserved patients with type II diabetes. *Glob. Qual. Nurs. Res.* **2017**, *4*, 1–13. [CrossRef]
12. Heisler, M.; Faul, J.D.; Hayward, R.A.; Langa, K.M.; Blaum, C.; Weir, D. Mechanisms for racial and ethnic disparities in glycemic control in middle-aged and older Americans in the health and retirement study. *Arch. Intern. Med.* **2007**, *167*, 1853–1860. [CrossRef]
13. Rural Health Information Hub. Chronic Disease in Rural America. 2019. Available online: https://www.ruralhealthinfo.org/topics/chronic-disease (accessed on 18 October 2021).
14. Piette, J.D.; Wagner, T.H.; Potter, M.B.; Schillinger, D. Health insurance status, cost-related medication underuse, and outcomes among diabetes patients in three systems of care. *Med. Care* **2004**, *42*, 102–109. [CrossRef]
15. Rothman, R.; Malone, R.; Bryant, B.; Horlen, C.; DeWalt, D.; Pignone, M. The relationship between literacy and glycemic control in a diabetes disease-management program. *Diabetes Educ.* **2004**, *30*, 263–273. [CrossRef] [PubMed]
16. Schillinger, D.; Barton, L.R.; Karter, A.J.; Wang, F.; Adler, N. Does literacy mediate the relationship between education and health outcomes? A study of a low-income population with diabetes. *Public Health Rep.* **2006**, *121*, 245–254. [CrossRef] [PubMed]
17. Schillinger, D.; Grumbach, K.; Piette, J.; Wang, F.; Osmond, D.; Daher, C.; Palacios, J.; Diaz-Sullivan, G.; Bindman, A.B. Association of health literacy with diabetes outcomes. *J. Am. Med. Assoc.* **2002**, *288*, 475–482. [CrossRef] [PubMed]
18. Douthit, N.; Kiv, S.; Dwolatzky, T.; Biswas, S. Exposing some important barriers to health care access in the rural USA. *Public Health* **2015**, *129*, 611–620. [CrossRef]
19. Hjelm, N.M. Benefits and drawbacks of telemedicine. *J. Telemed. Telecare* **2005**, *11*, 60–70. [CrossRef]

20. Wootton, R. Twenty years of telemedicine in chronic disease management—An evidence synthesis. *J. Telemed. Telecare* **2012**, *18*, 211–220. [CrossRef]
21. Klonoff, D.C. The current status of mHealth for diabetes: Will it be the next big thing? *J. Diabetes Sci. Technol.* **2013**, *7*, 749–758. [CrossRef]
22. Kitsiou, S.; Paré, G.; Jaana, M.; Gerber, B. Effectiveness of mHealth interventions for patients with diabetes: An overview of systematic reviews. *PLoS ONE* **2017**, *12*, e0173160. [CrossRef]
23. Veazie, S.; Winchell, K.; Gilbert, J.; Paynter, R.; Ivlev, I.; Eden, K.B.; Nussbaum, K.; Weiskopf, N.; Guise, J.M.; Helfand, M. Rapid evidence review of mobile applications for self-management of diabetes. *J. Gen. Intern. Med.* **2018**, *33*, 1167–1176. [CrossRef]
24. El-Rashidy, N.; El-Sappagh, S.; Islam, S.M.; El-Bakry, H.; Abdelrazek, S. Mobile health in remote patient monitoring for chronic diseases: Principles, trends, and challenges. *Diagnostics* **2021**, *11*, 607. [CrossRef]
25. Burner, E.R.; Menchine, M.D.; Kubicek, K.; Robles, M.; Arora, S. Perceptions of successful cues to action and opportunities to augment behavioral triggers in diabetes self-management: Qualitative analysis of a mobile intervention for low-income Latinos with diabetes. *J. Med. Internet Res.* **2014**, *16*, e25. [CrossRef] [PubMed]
26. Williams, J.P.; Schroeder, D. Popular glucose tracking apps and use of mHealth by Latinos with diabetes: Review. *JMIR MHealth Uhealth* **2015**, *3*, e84. [CrossRef]
27. Parker, S.; Prince, A.; Thomas, L.; Song, H.; Milosevic, D.; Harris, M.F.; Group, I.S. Electronic, mobile and telehealth tools for vulnerable patients with chronic disease: A systematic review and realist synthesis. *BMJ Open* **2018**, *8*, e019192. [CrossRef]
28. Caburnay, C.A.; Graff, K.; Harris, J.K.; McQueen, A.; Smith, M.; Fairchild, M.; Kreuter, M.W. Evaluating diabetes mobile applications for health literate designs and functionality. *Prev. Chronic Dis.* **2015**, *12*, e61. [CrossRef]
29. Krebs, P.; Duncan, D.T. Health app use among US mobile phone owners: A national survey. *JMIR Mhealth Uhealth* **2015**, *3*, e101. [CrossRef] [PubMed]
30. Kim, S.H.; Lee, A. Health-literacy-sensitive diabetes self management interventions: A systematic review and metaanalysis. *Worldviews Evid. Based Nurs.* **2016**, *13*, 324–333. [CrossRef]
31. Desai, P.M.; Levine, M.E.; Albers, D.J.; Mamykina, L. Pictures worth a thousand words: Reflections on visualizing personal blood glucose forecasts for individuals with type 2 diabetes. In Proceedings of the 2018 CHI Conference on Human Factors in Computing Systems (CHI '18), Montreal, QC, Canada, 21–26 April 2018; Paper No. 538. Association for Computing Machine: New York, NY, USA, 2018; pp. 1–13.
32. Turchioe, M.R.; Heitkemper, E.M.; Lor, M.; Burgermaster, M.; Mamykina, L. Designing for engagement with self-monitoring: A user-centered approach with low-income, Latino adults with Type 2 Diabetes. *Int. J. Med. Inform.* **2019**, *130*. [CrossRef] [PubMed]
33. El-Gayar, O.; Timsina, P.; Nawar, N.; Eid, W. Mobile applications for diabetes self-management: Status and potential. *J. Diabetes Sci. Technol.* **2013**, *7*, 247–262. [CrossRef]
34. Cafazzo, J.A.; Leonard, K.; Easty, A.C.; Rossos, P.G.; Chan, C.T. The user-centered approach in the development of a complex hospital-at-home intervention. *Stud. Health Technol. Inform.* **2009**, *143*, 328–333.
35. McCurdie, T.; Taneva, S.; Casselman, M.; Yeung, M.; McDaniel, C.; Ho, W.; Cafazzo, J. mHealth consumer apps: The case for user-centered design. *Biomed. Instrum. Technol.* **2012**, *46*, 49–56. [CrossRef] [PubMed]
36. Goyal, S.; Morita, P.; Lewis, G.F.; Yu, C.; Seto, E.; Cafazzo, J.A. The systematic design of a behavioural mobile health application for the self-management of type 2 diabetes. *Can. J. Diabetes* **2016**, *40*, 95–104. [CrossRef]
37. Temi. Audio to Text Automatic Transcription Service & App [Internet]. 2020. Available online: https://www.temi.com/ (accessed on 21 October 2019).
38. Guest, G.; MacQueen, K.M.; Namey, E.E. *Applied Thematic Analysis*; Sage Publications: Thousand Oaks, CA, USA, 2012; ISBN 9781483384436.
39. Braun, V.; Clarke, V. Using thematic analysis in psychology. *Qual Res. Psychol.* **2006**, *3*, 77–101. [CrossRef]
40. VERBI Software. *MAXQDA [Software]*; VERBI Software GmbH: Berlin, Germany, 2020.
41. Chomutare, T.; Fernandez-Luque, L.; Arsand, E.; Hartvigsen, G. Features of mobile diabetes applications: Review of the literature and analysis of current applications compared against evidence-based guidelines. *J. Med. Internet Res.* **2011**, *13*, e65. [CrossRef] [PubMed]
42. Goyal, S.; Cafazzo, J.A. Mobile phone health apps for diabetes management: Current evidence and future developments. *QJM Int. J. Med.* **2013**, *106*, 1067–1069. [CrossRef] [PubMed]
43. Baptista, S.; Trawley, S.; Pouwer, F.; Oldenburg, B.; Wadley, G.; Speight, J. What do adults with type 2 diabetes want from the "perfect" app? Results from the second diabetes MILES: Australia (MILES-2) study. *Diabetes Technol. Ther.* **2019**, *21*, 393–399. [CrossRef]
44. American Diabetes Association. Introduction: Standards of medical care in diabetes—2018. *Diabetes Care* **2018**, *41* (Suppl. S1), S1–S2. [CrossRef]
45. National Institute for Health and Care Excellence. Type 2 Diabetes in Adults: Management. 2020. Available online: https://www.nice.org.uk/guidance/ng28 (accessed on 18 October 2021).
46. Chavez, S.; Fedele, D.; Guo, Y.; Bernier, A.; Smith, M.; Warnick, J.; Modave, F. Mobile apps for the management of diabetes. *Diabetes Care* **2017**, *40*, 145–146. [CrossRef] [PubMed]
47. Breland, J.Y.; McAndrew, L.M.; Gross, R.L.; Leventhal, H.; Horowitz, C.R. Challenges to healthy eating for people with diabetes in a low-income, minority neighborhood. *Diabetes Care* **2013**, *36*, 2895–2901. [CrossRef]

48. Aguayo-Mazzucato, C.; Diaque, P.; Hernandez, S.; Rosas, S.; Kostic, A.; Caballero, A.E. Understanding the growing epidemic of type 2 diabetes in the Hispanic population living in the United States. *Diabetes Metab. Res. Rev.* **2019**, *35*, e3097. [CrossRef]
49. Colton, P.; Rodin, G.; Bergenstal, R.; Parkin, C. Eating disorders and diabetes: Introduction and overview. *Diabetes Spectr.* **2009**, *22*, 138–142. [CrossRef]
50. Hill-Briggs, F.; Smith, A.S. Evaluation of diabetes and cardiovascular disease print patient education materials for use with low–health literate populations. *Diabetes Care* **2008**, *31*, 667–671. [CrossRef]
51. Centers for Disease Control and Prevention. *Scientific and Technical Information Simply Put*, 2nd ed.; Centers for Disease Control and Prevention: Atlanta, GA, USA, 1999.
52. Doak, C.C.; Doak, L.G.; Root, J.H. Teaching patients with low literacy skills. *Am. J. Nurs.* **1996**, *96*, 16M. [CrossRef]
53. National Cancer Institute. *Clear and Simple: Developing Effective Print Materials for Low-Literate Readers*; U.S. Department of Health and Human Services: Bethesda, MD, USA, 1994.
54. Huang, Y.M.; Shiyanbola, O.O.; Chan, H.Y. A path model linking health literacy, medication self-efficacy, medication adherence, and glycemic control. *Patient Educ. Couns.* **2018**, *101*, 1906–1913. [CrossRef]
55. Osborn, C.Y.; Cavanaugh, K.; Kripalani, S. Strategies to address low health literacy and numeracy in diabetes. *Clin. Diabetes* **2010**, *28*, 171–175. [CrossRef]
56. Carstens, A. Tailoring print materials to match literacy levels: A challenge for document designers and practitioners in adult literacy. *Lang. Matters* **2004**, *35*, 459–484. [CrossRef]
57. Google/OTX. Four Truths about US Hispanic Consumers. 2010. Available online: http://www.gstatic.com/ads/research/en/2010_FourTruthsAboutUSHispanics.pdf (accessed on 18 October 2021).
58. Ayre, J.; Bonner, C.; Muscat, D.M.; Bramwell, S.; McClelland, S.; Jayaballa, R.; Maberly, G.; McCaffery, K. Type 2 diabetes self-management schemas across diverse health literacy levels: A qualitative investigation. *Psychol. Health* **2021**, *27*, 1–21. [CrossRef]
59. Isaković, M.; Sedlar, U.; Volk, M.; Bešter, J. Usability pitfalls of diabetes mHealth apps for the elderly. *J. Diabetes Res.* **2016**, 1–9. [CrossRef] [PubMed]
60. Skrøvseth, S.O.; Årsand, E.; Godtliebsen, F.; Hartvigsen, G. Mobile phone-based pattern recognition and data analysis for patients with type 1 diabetes. *Diabetes Technol. Ther.* **2012**, *14*, 1098–1104. [CrossRef] [PubMed]
61. Brzan, P.P.; Rotman, E.; Pajnkihar, M.; Klanjsek, P. Mobile applications for control and self management of diabetes: A systematic review. *J. Med. Syst.* **2016**, *40*, 210. [CrossRef] [PubMed]
62. Ayre, J.; Bonner, C.; Bramwell, S.; McClelland, S.; Jayaballa, R.; Maberly, G.; McCaffery, K. Factors for supporting primary care physician engagement with patient apps for type 2 diabetes self-management that link to primary care: Interview study. *JMIR mHealth uHealth* **2019**, *7*, e11885. [CrossRef] [PubMed]
63. Tanenbaum, M.L.; Leventhal, H.; Breland, J.Y.; Yu, J.; Walker, E.A.; Gonzalez, J.S. Successful self-management among non-insulin-treated adults with Type 2 diabetes: A self-regulation perspective. *Diabet. Med.* **2015**, *32*, 1504–1512. [CrossRef] [PubMed]
64. Simons, D.; Egami, T.; Perry, J. Remote patient monitoring solutions. In *Advances in Health Care Technology Care Shaping the Future of Medical Care*; Spekowius, G., Wendler, T., Eds.; Springer: Dordrecht, The Netherlands, 2006; pp. 505–516. ISBN 978-1-4020-4384-0.
65. Michaud, T.L.; Siahpush, M.; Estabrooks, P.; Schwab, R.J.; LeVan, T.D.; Grimm, B.; Ramos, A.K.; Johansson, P.; Scoggins, D.; Su, D. Association between weight loss and glycemic outcomes: A post hoc analysis of a remote patient monitoring program for diabetes management. *Telemed. E-Health* **2019**, *26*, 621–628. [CrossRef] [PubMed]
66. Alvarado, M.M.; Kum, H.C.; Coronado, K.G.; Foster, M.J.; Ortega, P.; Lawley, M.A. Barriers to remote health interventions for type 2 diabetes: A systematic review and proposed classification scheme. *J. Med. Internet Res.* **2017**, *19*, e28. [CrossRef] [PubMed]
67. Christodoulakis, C.; Asgarian, A.; Easterbrook, S. Barriers to Adoption of Information Technology in Healthcare. In Proceedings of the 27th Annual International Conference on Computer Science and Software Engineering, Markham, ON, Canada, 6–8 November 2017; Available online: http://www.cs.toronto.edu/~{}christina/documents/ACM_CASCON2017.pdf (accessed on 18 October 2021).
68. Moiduddin, A.; Moore, J. *The Underserved and Health Information Technology—Issues and Opportunities*; Office of the Assistant Secretary for Planning and Evaluation (ASPE), U.S. Department of Health and Human Services: Washington, DC, USA, 2008. Available online: https://aspe.hhs.gov/system/files/pdf/134261/report.pdf (accessed on 18 October 2021).

MDPI

St. Alban-Anlage 66

4052 Basel

Switzerland

Tel. +41 61 683 77 34

Fax +41 61 302 89 18

www.mdpi.com

International Journal of Environmental Research and Public Health Editorial Office

E-mail: ijerph@mdpi.com

www.mdpi.com/journal/ijerph